ELECTRIC CIRCUITS
AC/DC An Integrated Approach

McGRAW-HILL SERIES IN ELECTRICAL ENGINEERING

Consulting Editor

STEPHEN W. DIRECTOR, *Carnegie-Mellon University*

Networks and Systems

Communications and Information Theory

Control Theory

Electronics and Electronic Circuits

Power and Energy

Electromagnetics

Computer Engineering and Switching Theory

Introductory and Survey

Radio, Television, Radar, and Antennas

PREVIOUS CONSULTING EDITORS

Ronald M. Bracewell, Colin Cherry, James F. Gibbons, Willis W. Harman, Hubert Heffner, Edward W. Herold, John G. Linvill, Simon Ramo, Ronald A. Rohrer, Anthony E. Siegman, Charles Susskind, Frederick E. Terman, John G. Truxal, Ernst Weber, and John R. Whinnery

ELECTRONICS AND ELECTRONIC CIRCUITS

Consulting Editor

STEPHEN W. DIRECTOR, *Carnegie-Mellon University*

ELECTRIC CIRCUITS

AC/DC An Integrated Approach

Charles I. Hubert

Professor of Electrical Engineering
United States Merchant Marine Academy

McGraw-Hill Book Company

New York   St. Louis   San Francisco   Auckland   Bogotá   Hamburg
Johannesburg   London   Madrid   Mexico   Montreal   New Delhi
Panama   Paris   São Paulo   Singapore   Sydney   Tokyo   Toronto

This book was set in Times New Roman.
The editors were Frank J. Cerra and Susan Hazlett;
the designer was Joan E. O'Connor;
the production supervisor was Leroy A. Young.
The drawings were done by J & R Services, Inc.
R. R. Donnelley & Sons Company was printer and binder.

ELECTRIC CIRCUITS AC/DC:
An Integrated Approach

1 2 3 4 5 6 7 8 9 0   DODO   8 9 8 7 6 5 4 3 2 1

ISBN 0-07-030845-4

Library of Congress Cataloging in Publication Data

Hubert, Charles I.
    Electric circuits AC/DC.

    (McGraw-Hill series in electrical engineering.
Electronics and electronic circuits)
        Includes index.
        1. Electric circuits.   I. Title.   II. Series.
TK3001.H78   621.319′2   81-5994
ISBN 0-07-030845-4   AACR2

CONTENTS

PREFACE

PHILOSOPHICAL APPROACH AND GENERAL COMMENTS

In these days of expanding technology, where more material needs to be covered in the available time, there is a need to present introductory electric circuit theory more concisely without losing effectiveness. This is accomplished by presenting a more direct approach to the steady-state analysis of electric circuits, in that the sinusoidal system is introduced earlier than usual, and DC is then treated as a sine wave whose period approaches infinity and whose frequency approaches zero. To obtain the steady-state DC response, it is only necessary to substitute zero for the frequency in all circuit calculations. With this integrated presentation, there is no need for separate (DC followed by AC) instruction in network analysis or network theorems. Total separation of AC circuits often leaves the student with the erroneous impression that separate rules and principles apply to AC circuits as compared to DC circuits. The AC/DC integration, with its many examples and problems, is done throughout the text immediately after the introduction of the sinusoidal system. It is not a token approach. Its beauty and simplicity manifests itself whenever the driving voltage is changed from sinusoidal to DC. Furthermore, it has been my experience, over the past 20 years of using the integrated technique, that students can absorb the same course material more efficiently and with a greater depth of understanding when the integrated approach is used. The class hours saved may be used for additional topics of choice, and/or treating the basic concepts in greater depth.

The text provides a built-in capability for several levels of instruction. Complex derivations that are not essential to the basic understanding and application of circuit theory are *blocked off*, but available for the student with the more advanced mathematical background. Thus, the teaching level may be selected by the instructor for a particular class of students. Omitting the blocked-off sections will not alter the smooth flow or effectiveness of the text. However, the presence of the blocked-off sections provides an upper level of instruction for those classes that can handle the math. Chapter 25 (circuits with nonsinusoidal drivers), Chap. 26 (transients in source-free circuits), and Chap. 27 (transients in driven systems) are included for the upper level student.

The applications orientation of the text gives the student a "feel" for the characteristic behavior of different types of circuits encountered in practice. The problem sets for each chapter typify circuit problems in the professional engineering world. As the student progresses, problems in later chapters require that the student apply the methods of that chapter as well as techniques studied in previous chapters.

The content, arrangement, and presentation of the material permit flexibility in topic selection, making the text adaptable to different courses.

CHAPTER COMMENTS

The first five chapters provide the background for analysis of series, parallel, and series-parallel resistive circuits. Included are: SI units; conductors, insulators, and semiconductors; current and voltage dividers; heat power and energy loss; maximum power transfer; voltage regulation of sources; a simple technique for determining the potential difference between any two points in a network; and such common circuit faults as shorts, opens, flashovers, grounds, and corona.

Chapters 6, 7, and 8 present the defining characteristics of capacitance and inductance, and develop the associated voltage, current, and energy relationships. Methods for the safe discharging of energy storage elements are introduced. Magnetic phenomena, the magnetic circuit, and the interaction of magnetic fields are included.

Chapters 9 and 10 introduce the sinusoidal system with DC as a special case of AC at 0 Hz. The effect of frequency on ideal R, L, and C is developed. Phasor and time-domain representation of current and voltage are stressed. Chapter 11 introduces the concept of the j-operator and develops the mathematics of complex algebra used in succeeding chapters. The steady-state analysis of series, parallel, and series- parallel circuits with sinusoidal drivers of all frequencies, including 0 Hz, is presented in Chap. 12.

Chapter 13 discusses power in the two-wire system for all frequencies and all circuit configurations. Included is the development of the relationships for average power, reactive power, apparent power, power factor, power-factor correction, phasor power, maximum power transfer, and the techniques for measuring active and reactive power. The simultaneous side-by-side development of active, reactive, and apparent power relationships for series and parallel circuits dramatizes the common relationships involved in the two-wire system, regardless of circuit configuration.

Chapter 14 develops loop and node analysis for general-type networks of any frequency including 0 Hz. A format for the rapid writing of loop and node equations eliminates confusion by providing a straightforward method for network analysis; the equations, written by inspection, are automatically in proper form for solution by determinants or computer. The simplicity of this format method helps avoid errors in sign and makes it easy to recheck equations before proceeding with the solution. Chapter 15 is devoted to network theorems and includes controlled sources.

Chapters 16 and 17 discuss resonance effects in electric circuits, and provide an introduction to filters.

Chapters 18 and 19 are devoted to coupled circuits and transformers. A simple procedure for determining polarity marks in multicoupled coils is presented, and the format developed in Chap. 14 for the rapid writing of loop equations is expanded to include coupling parameters. Chapter 20 derives the **z**, **y**, and **h** parameters for two-port networks and includes practical examples of its use.

Chapters 21 and 22 are devoted to the three-phase system. Included is a simple procedure for calculating feeder currents to different combinations of parallel-connected three-phase loads (balanced and unbalanced), the effects of phase reversal on the current in unbalanced systems, a simple accuracy check for feeding-current calculations, and three-phase to two-phase transformation. Methods for calculating and measuring active and reactive power in balanced and unbalanced three-phase circuits, and power-factor correction are included.

Chapter 23 introduces electromechanical forces and meter movements, and discusses the selection and application of commercial meters.

Chapter 24 develops methods for calculating current, voltage, average power, apparent power, and power factor in circuits with multifrequency drivers, and Chap. 25 applies the Fourier series to nonsinusoidal drivers to obtain multifrequency components, discusses symmetry, and introduces the frequency spectrum.

Chapters 26 and 27 provide a practical approach to the study of transients at the introductory level, enabling the student to avoid mathematical quagmires and conceptual errors while proceeding with the setup and solution of transient circuit problems. The method used draws from students' knowledge gained from previous chapters and appeals to their common sense. As the solutions to the many examples unfold, the significance of the solution, including such elusive points as why a resultant sign is negative, is explained. The examples and problems are real-world situations and are representative of professional engineering problems. A simple method for rapid and accurate sketching of complex waveforms is explained and demonstrated.

ACKNOWLEDGMENTS

The author takes this opportunity to acknowledge with gratitude the many helpful suggestions made by his colleagues, friends, and students during the preparation of this text. I am especially indebted to Professor Wallace H. McDonald, friend and colleague, who class-tested the manuscript over a 4-year period. His confidence in this method of presentation, his close support, his unwavering encouragement during the development and testing of the manuscript, his very detailed and incisive reviews and many helpful and significant suggestions are deeply appreciated.

An affectionate thanks to my wonderful wife Josephine for her encouragement, faith, counsel, patience, and companionship during the many years of preparing this and other manuscripts. Her early reviews of the manuscript, while in its formative stages, assisted in clarity of expression and avoidance of ambiguity. Her apparently endless years of pounding the typewriter for this and other texts were truly a work of love.

CHARLES I. HUBERT

ELECTRIC CIRCUITS
AC/DC An Integrated Approach

CHAPTER 1

UNITS, NOTATIONS, AND SIGNIFICANT FIGURES IN CIRCUIT ANALYSIS

Preliminary to the study of electric circuits is an understanding of the type of units used, the scientific and engineering notations generally encountered in practice, and the importance of significant figures in the mathematical solutions of problems.

1-1   SI UNITS

The international system of units, officially designated SI,† is the preferred system for published scientific and technical work. It is generally superior to other systems, particularly so in the fields of electrical science and technology, and will be used throughout the text. The most familiar SI units used in the electrical industry are ampere, volt, ohm, watt, meter, and second.

In those few sections of the text where it is necessary to relate to commonly used non-SI units such as horsepower, pound, and circular mil, both SI units and non-SI units will be used, with emphasis placed on the SI unit. A table of conversion ratios for commonly used non-SI units is on the inside back cover of the text.

Each conversion ratio and its reciprocal has a value of unity. Hence, it does not alter the value of the converted quantity; it merely changes the

---

† Système International d'Unités, abbreviated SI.

units. The conversion ratio must include the units involved and must be used in a manner that will cancel the undesired unit.

Example 1-1   Convert 56.8 inches to meters.

Solution
Selecting the conversion ratio (or its reciprocal, as appropriate) and multiplying,

$$56.8 \text{ in} \left(\frac{1 \text{ m}}{\text{in} \times 39.37} \right) = 1.44 \text{ m}$$

Example 1-2   Convert 0.0025 square meter to square millimeters.

Solution

$$0.0025 \text{ m}^2 \left(\frac{\text{mm}^2 \times 10^6}{1 \text{ m}^2} \right) = 2500 \text{ mm}^2$$

Note: In order to cancel the undesired unit, it was necessary to use the reciprocal of the conversion ratio listed in the table.

1-2   SCIENTIFIC NOTATION

Scientific notation provides a convenient means for comparing numbers as well as indicating significant figures. A number in scientific notation is written with the decimal point to the right of the first digit, and a multiplier 10^N is used to indicate the value of the number.

Example 1-3   The following numbers are expressed in both nonscientific and scientific notation:

		Scientific notation
0.016	=	1.6×10^{-2}
594	=	5.94×10^2
52.6	=	5.26×10
0.00045	=	4.5×10^{-4}
6,500,000	=	6.5×10^6

1-3   ENGINEERING NOTATION

Engineering notation is the application of decimal prefixes and their abbreviations to simplify language when dealing with very large or very small units.

The most common decimal prefixes used with SI units in engineering notation are given in Table 1-1. The prefixes represent powers of 10, with all exponents multiples of 3. Ten thousand volts is expressed as 10 kilovolts or 10 kV, one-millionth of an ampere is expressed as 1 microampere or 1 μA, etc.

Table 1-1

	PREFIX	
Power of 10	Name	Abbreviation
10^{12}	tera	T
10^9	giga	G
10^6	mega	M
10^3	kilo	k
10^{-3}	milli	m
10^{-6}	micro	μ
10^{-9}	nano	n
10^{-12}	pico	p

Example 1-4 The following voltages are expressed in both nonengineering and engineering notation:

Engineering notation

$$4000 \text{ V} = 4 \times 10^3 \text{ V} \qquad = 4 \text{ kV}$$
$$58{,}000 \text{ V} = 58 \times 10^3 \text{ V} \qquad = 58 \text{ kV}$$
$$735{,}000 \text{ V} = 735 \times 10^3 \text{ V} \qquad = 735 \text{ kV}$$
$$6{,}875{,}000 \text{ V} = 6.875 \times 10^6 \text{ V} \qquad = 6.875 \text{ MV}$$
$$0.0000000562 \text{ V} = 56.2 \times 10^{-9} \text{ V} \qquad = 56.2 \text{ nV}$$
$$0.0000017 \text{ V} = 1.7 \times 10^{-6} \text{ V} \qquad = 1.7 \text{ } \mu\text{V}$$
$$0.084 \text{ V} = 84 \times 10^{-3} \text{ V} \qquad = 84 \text{ mV}$$

The nameplate data of electrical apparatus and the solutions to engineering problems are generally expressed in engineering notation. However, *before engineering data are substituted into mathematical equations, all units must be converted to nonprefix form.* Failure to do so will result in serious errors.

Example 1-5 Determine the current I in a 400 MΩ resistor when 24 kV is applied. Use $I = V/R$ (Ohm's law).

Solution

$$R = 400 \text{ M}\Omega = 400 \times 10^6 \text{ }\Omega$$

$$V = 24 \text{ kV} = 24 \times 10^3 \text{ V}$$

$$I = \frac{V}{R} = \frac{24 \times 10^3}{400 \times 10^6} = 60 \times 10^{-6} = 60 \text{ }\mu\text{A}$$

1-4   **SIGNIFICANT FIGURES**

The numerical values of all *measurements* such as length, mass, voltage, current, and resistance are approximations. The accuracy of a measurement is determined by the particular measuring instrument, and no *physical measurement* can ever be absolutely precise.

For example, the voltage of a certain battery was measured with three different voltmeters. The respective measurements were 2.1 V, 2.08 V, and 2.075 V. The first voltmeter measured to the nearest tenth of a volt, the second to the nearest hundredth of a volt, and the third to the nearest thousandth of a volt. The number of digits in the measurements are called significant figures. Thus, in the above example, the significant figures in the *measured* data are 2, 3, and 4, respectively.

The last significant figure in any measured data is estimated and is therefore of doubtful accuracy. For this reason, *the number of significant figures retained in the numerical solution of a problem involving measured quantities should be rounded to match the number of significant figures in the least precise data.* Failure to do so will indicate greater or less accuracy in the solution than the measured data warrant.

Zeros as Significant Figures

Zeros in a number may or may not be significant; it all depends on how they are used.

Zeros appearing as the first figure of a number are not significant, since they serve only to locate the decimal point. For example,

Number	Significant figures
0.01	1
0.004	1
0.0035	2

Zeros between digits are always significant. For example,

Number	Significant figures
6008	4
9075	4
502	3

Zeros appearing as the last figures in a decimal are significant. For example,

Number	Significant figures
0.00010	2
100.040	6
50.00	4
6.0	2

Zeros appearing as the last figures in a *whole* number may or may not be significant. In such cases scientific or engineering notation should be used to provide the proper indicator. For example, the number 56,000 may have two, three, four, or five significant figures. If four significant figures is the correct precision, it should be written in scientific notation or engineering notation as

5.600×10^4 or   56.00 k

Addition and Subtraction

In adding or subtracting, the resultant should be rounded off to match the significant figures in the least precise number. This is illustrated in the following summations:

4.5⎪6⎪ ←——Least precise number	15⎪0⎪.8	
6.1⎪7⎪6	→17⎪0⎪.	
0.8⎪2⎪01	15⎪9⎪.98	
4.3⎪0⎪64	1⎪7⎪.542	
15.8⎪6⎪25	49⎪8⎪.322	

The answers rounded to match the significant figures of the least precise number are

15.86   and   498

In the following subtractions:

17⎪9⎪.04 ⟋Least precise number⟍	1768.⎪1⎪0
− 3⎪5⎪. ↖	− 34.⎪8⎪
14⎪4⎪.04	1733.⎪3⎪0

The answers rounded to match the significant figures of the least precise numbers are

144.   and   1733.3

Multiplication and Division

In multiplying or dividing, the resultant should be rounded off to match the number of significant figures in the least precise number. This is illustrated in the following examples, with all calculations made on an electronic calculator. The rounded figure is underlined.

Problem	Calculation	Least number of significant figures	Answer
65×0.384	24.96	2	2<u>5</u>
756×0.004	3.024	1	<u>3</u> or <u>3</u>.0†
0.173×0.066	0.0114180	2	0.01<u>1</u>
$\dfrac{1225}{0.026}$	47,115.38462	2	4<u>7</u> $\times 10^3$
$\dfrac{152}{987}$	0.154002026	3	0.15<u>4</u>

† Occasionally it may be desirable to include *one* additional figure to aid in evaluating the accuracy of the last significant figure; in such cases, it must be recognized that the *last two figures are in doubt.*

Defined Ratios and Multiplying Factors

Defined ratios and multiplying factors are infinitely accurate and are not subject to significant-figure scrutiny. The following examples should help to clarify this point.

Example 1-6   If the nameplate on an instrument indicates that the scale reading must be multiplied by 5, the multiplying factor should be viewed as

$$5.000000 \cdots 0$$

an infinitely accurate number. Assuming the instrument reads 14.3 on the scale, the actual value of the measured quantity is

$$5.000000 \cdots 0 \times 14.3 = 71.5$$

The number of significant figures was determined by the instrument reading.

Example 1-7   The primary to secondary voltage ratio V_P/V_S of a transformer is 6/1. If the primary-voltage measurement is 2346 V, determine the secondary voltage.

Solution

$$\frac{V_P}{V_S} = \frac{6}{1}$$

Substituting the primary voltage, and recognizing that the ratio is infinitely accurate,

$$\frac{2346}{V_S} = \frac{6.000000 \cdots 0}{1.000000 \cdots 0}$$

$$V_S = \frac{2346}{6} = 391 \tag{1-1}$$

Since the least number of significant figures is in the voltage measurement (four significant figures), the answer must be adjusted to four significant figures. Thus,

$$V_S = 391.0 \text{ V}$$

Example 1-8   The ohmic value of a resistor may be determined from the *measured* values of voltage and current. If the voltage measurement is 2346 V, the current measurement is 6.0 A, and the equation relating the variables is

$$\text{Resistance} = \frac{\text{voltage}}{\text{amperes}}$$

then, substituting the measured values,

$$\text{Resistance} = \frac{2346}{6.0} = 391 \tag{1-2}$$

Since the least number of significant figures is in the current measurement (two significant figures) the resistance calculation adjusted to two significant figures is

$$390 \ \Omega$$

Comparing Eq. (1-1) in Example (1-7) with Eq. (1-2) in Example (1-8) indicates identical answers. However, the numeral 6 in Eq. (1-1) is infinitely accurate, but 6.0 in Eq. (1-2) is accurate to only two significant figures.

If the current measurement in Example 1-8 was made to the nearest hundredth of an ampere (6.00), the resistance calculation, adjusted to three significant figures, would be 391 Ω. On the other hand, if the current measurement was made to the nearest ampere (6), the resistance calculation, adjusted to one significant figure, would be 400 Ω. Example 1-8 indicates the need for precision in measurement and recording of data.

-5   ROUNDING

When rounding answers to bring them in line with the required number of significant figures, the following procedures should be followed:

1.   If the first digit to be discarded is greater than 5, or if it is a 5 followed by at least one digit other than zero, the last digit to be retained should be increased by 1.

2. If the first digit to be discarded is less than 5, the last digit to be retained should not be changed.
3. If the first digit to be discarded is exactly 5 (or if it is a 5 followed by only zeros) the last digit should not be changed if it is even but should be increased by 1 if it is odd.

Example 1-9   The following numbers have been rounded to three significant figures. The digit to be rounded is circled, and the digits to be discarded are underlined.

0.0158324	0.0158
594287	594 × 10³
69.6512	69.7
7.13500	7.14
52.600	52.6
57863	579 × 10²
6.76500	6.76

In solving numerical problems that have several parts, when the solution to one part must be used to solve another part, avoid rounding until the problem is completed.

PROBLEMS

1-1   Express the following numbers in scientific notation:

659.4	0.00045
17.3	0.0008
100.750	0.106

1-2   Express the following voltages in engineering notation:

10,000 V	0.000006 V
500 V	0.0001258 V
10,000,000 V	0.000000001 V

1-3   Express the following numbers in two significant figures:

459,803	0.015
2500	1.632
6650	0.00042
3750	0.0081

1-4   Add the following numbers and round to the correct significant figures:

(a)	125.862	(b)	15.0085
	17.54		3.001
	3.4		16.0104
	1500		0.06

1-5 Perform the indicated operations and round to the correct number of significant figures. The circled factors are multipliers or ratios; all other factors are measurements.

(a) $\dfrac{14.3 \times 17.04 \times ③}{0.0032 \times 1.2}$

(b) $\dfrac{1500 \times 36.52}{40.158 \times ⑩⓪}$

(c) $\dfrac{(17.8)^2 1.84}{5400 \times ①/③}$

(d) $\dfrac{(1700 + 18.54)16}{41.2}$

CHAPTER 2

CURRENT, VOLTAGE, INSULATORS, AND CONDUCTORS

2-1 CURRENT AND VOLTAGE

An electric current is a time rate of movement of electric charges. The current may be a movement of positive charges, a movement of negative charges, or a combination of positive and negative charges moving in opposite directions. In metallic conductors the current is a movement of small negatively charged particles called electrons. In gases the current is a movement of negatively charged electrons in one direction and a drift of positively charged ions in the opposite direction (an ion is a charged particle of matter). In salt solutions the current is a movement of positive ions and negative ions in opposite directions. In semiconductors the current is a movement of electrons in one direction and a movement of positively charged holes in the opposite direction.

Any material that allows the essentially free passage of current when connected to a battery or other source of electric energy is called a conductor. In metallic conductors, *free electrons* move randomly about the crystal structure of the material until a driving voltage, also called a voltage source or electromotive force (emf), is applied. Some common sources of voltage are batteries, electromechanical generators, solar cells, and fuel cells. Figure 2-1a shows the random motion of the free electrons when no voltage is applied. Figure 2-1b shows the effect of a driving voltage; *the free electrons are forced in the direction of the driving voltage.* The direction of this electron

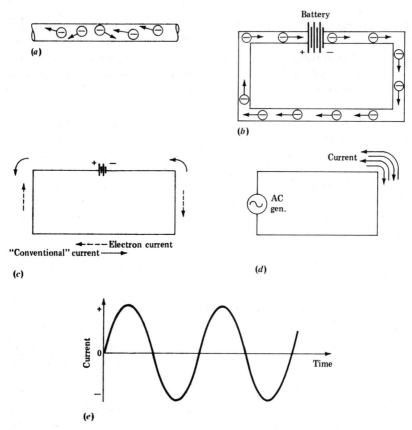

Figure 2-1 (*a*) Random motion of free electrons; (*b*) effect of a driving voltage on free electrons; (*c*) conventional and electron currents; (*d*) alternations of current caused by an alternating driving voltage; (*e*) sinusoidally varying current.

movement is from the negative (−) terminal of the driver, through the conductor, to the positive (+) terminal of the driver. The closed loop formed by the battery and the conductor is called an *electric circuit*.

The driver does not supply the electrons. The electrons are always present in the conductors as free electrons and are caused to move by the application of a driving voltage. The negative terminal of the driving source repels the free electrons, and the positive terminal attracts them. As long as the circuit is closed, the electrons continue to circle around the loop; each time an electron enters the positive terminal of the driver, another electron leaves the negative terminal for its journey around the loop. Thus the total number of free electrons in the conductor is always the same. This unidirectional movement of electrons, caused by the application of a unidirectional generator or battery, is called direct current.

Although the actual direction of current is from the negative terminal of the driver to the positive terminal, to conform with conventional practice and to avoid confusion with most other literature, *the direction of current in all circuits throughout this text has been standardized as going from positive to negative.* Using this conventional direction of current, instead of the actual direction of electron flow, will in no way affect the solution of electric-circuit problems. The conventional direction was established before the electron was discovered, and to attempt to change it now is to beat one's head needlessly against a stone wall. Figure 2-1c contrasts the conventional direction of current (+ to −) with the actual direction of electron movement.

The application of an alternating voltage to a conductor causes the electrons to flow first in one direction and then in the opposite direction, oscillating continuously about their central position. This repetitive back-and-forth movement of electric charges constitutes an *alternating current.* Thus, any apparatus that generates a repetitive alternating driving voltage is called an *alternating-current (AC) generator.* Figure 2-1d illustrates the alternations of current supplied by an AC generator, and Fig. 2-1e is a graph showing how the magnitude and direction of a sinusoidally varying alternating current change with time.

Current is a flow rate, and is defined as the time rate of flow of electric charges past a specified point in a given direction. The unit of electric current is the *ampere* (A) and the unit of electric charge is the *coulomb* (C). One coulomb of electric charge is equivalent to the charge produced by 6.25×10^{18} electrons. Hence, a current of one ampere is equivalent to 6.25×10^{18} electrons passing a specified point in one second. Expressed mathematically,

$$i = \frac{dq}{dt} \tag{2-1}$$

where i = current, amperes (A)

$\dfrac{dq}{dt}$ = rate of flow of electric charge, coulombs per second (C/s)

dt = infinitesimally small period of time

dq = amount of charge in coulombs that passes a specified point during time dt

2-2 ELECTRICAL INSULATION

Materials that possess extremely low electrical conductivity are classified as nonconductors and are better known as insulators or *dielectrics.* A vacuum is the only known perfect dielectric.† All other insulating materials, such as

---

† In vacuum tubes, such as diodes, triodes, and other multielement vacuum tubes, a filament heated to incandescence emits electrons that provide a conducting path.

paper, cloth, wood, bakelite, rubber, plastic, glass, mica, ceramic, shellac, varnish, air, and special-purpose oils, are imperfect dielectrics. The function of electrical insulation is to prevent the passage of current between two or more conductors. To do this, the conductors may be covered with cotton, plastic or rubber tape, mica tubing, ceramic tubing, etc. The specific selection depends on the electrical, thermal, and mechanical stresses involved, and whether in wet or dry locations. The twin conductors in ordinary 120-V lamp cord are separated by a rubber or plastic coating, but the conductors on long-distance high-voltage transmission lines are insulated from one another by air. The desired spacing of the transmission-line conductors is obtained with porcelain insulators.

The electrons in an insulating material are tightly bound in the molecular structure and are not free to wander within the material, as they do in a conductor. Hence, at normal room temperatures, the application of rated to moderately high driving voltages to the insulating material causes only a very few electrons to break away. However, if a driver of sufficiently high voltage is applied, e.g., lightning or other high-voltage surges, the electrons will literally be torn away from their molecular bonds, destroying the insulating

Table 2-1   Insulation classification

Trade name	Type letter	Max. operating temp.	Application
Thermoplastic	T	60°C 140°F	Dry locations
Heat-resistant rubber	RH	75°C 167°F	Dry locations
Varnished cambric	V	85°C 185°F	Dry locations
Thermoplastic and asbestos	TA	90°C 194°F	Switchboard wiring
Asbestos and varnished cambric	AVL	110°C 230°F	Dry and wet locations
Asbestos	AIA	125°C 257°F	Dry locations within apparatus
Asbestos	A	200°C 392°F	Dry locations within apparatus
Extruded polytetra fluoroethylene	TFE	250°C 482°F	Dry locations within apparatus

Extracted by permission from the 1978 National Electrical Code, Table 310–2a, copyright 1977, National Fire Protection Association, Boston, Mass.

properties of the material. In the case of organic materials such as wood or rubber, this electron avalanche through the material generates sufficient heat to cause carbonization. The voltage at which breakdown occurs is called the dielectric strength of the insulation.

A listing of trade names, type letter, maximum operating temperature, and corresponding applications of some insulated conductors is given in Table 2-1. A word of warning about one type of thermoplastic insulation called polyvinyl chloride (PVC): When heated to about 230°C, PVC insulation gives off hazardous hydrogen chloride gas (HCl). The gas appears as a white mist that has an irritating and corrosive effect on the respiratory system. At much higher temperatures, dark sooty smoke is released.

2-3   CONDUCTORS FOR ENERGY TRANSMISSION

Conductors selected for the efficient transmission of electrical energy must have good electrical conductivity; that is, they must be able to pass the current with little opposition. The relatively high electrical conductivity and good heat-conduction capability of copper compared with other metals of similar cost result in its extensive use in electronic circuits, motors, generators, controls, switches, cables, buses, etc. Buses are solid or hollow conductors, generally of large cross-sectional area, from which taps are made to supply three or more circuits. Aluminum, because of its light weight, good casting properties, and high conductivity (although not so good as copper), is used in squirrel-cage rotors, cables, bus bars, and long-distance transmission lines.

The ampacity (allowable current-carrying capacity) of a wire or cable depends on the thermal stability of the insulation covering the conductor, the heat-dissipation capability to the surrounding media, and the particular combination of work and rest periods. Ampacity tables for different conditions of insulation and surrounding media are given in Appendixes 1 and 2. More extensive tables for copper and aluminum conductors are available in book form from the Insulated Power Cable Engineers Association.

Cadmium and silver are used as contacts in control equipment, and gold plating is sometimes used to provide the excellent mating surfaces of bayonet types of contacts in electronic equipment. Although silver has a higher conductivity than copper, the relatively high cost makes its general use prohibitive. Carbon, in very short lengths and large cross-sectional areas, has applications as sliding contacts for rheostats, commutators, and slip rings.

For safety reasons all water must be treated as having good conductivity. Although distilled water is not a conductor, any water around electrical equipment such as tap water and rainwater is contaminated with salts and conducts electricity.

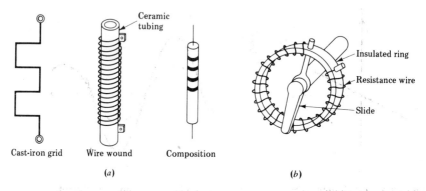

Figure 2-2 (a) Resistors; (b) rheostat.

2-4 **CONDUCTORS FOR LIMITING CURRENT AND FOR CONVERTING ELECTRICAL ENERGY TO HEAT ENERGY**

Conductors of low *conductivity* are used in those applications where the magnitude of the current must be limited or where it is desired to convert electrical energy to heat energy (toaster, electric range, etc.). Examples of relatively low conductivity materials are iron, tungsten, nickel alloys, and compositions of carbon and inert materials. Figure 2-2a illustrates some different arrangements of current-limiting conductors, *called resistors*, and Fig. 2-2b shows an adjustable resistor called a *rheostat*.

2-5 **RESISTANCE**

When electrons are driven through a conductor, they collide with one another and with other parts of the atoms that make up the material. Such collisions interfere with the free movement of the electrons and generate heat. This property of a material that limits the magnitude of a current and converts electrical energy to heat energy is called *resistance*.

For a given length and cross-sectional area, materials of low conductivity have a higher resistance to current than materials of high conductivity; thus, to cause the same current in both, the material of low conductivity (higher resistance) requires a higher driving voltage.

Resistance may be measured with an ohmmeter and is generally expressed in ohms (Ω), kilohms (kΩ), or megohms (MΩ). The resistance of all conductors is temperature-dependent to some extent. Hence, valid measurements call for the determination of conductor temperature at the time of resistance measurement. Most conducting materials, such as copper, aluminum, iron, nickel, and tungsten, increase in resistance with increasing temperature. Carbon, on the other hand, decreases in resistance with increasing temperature.

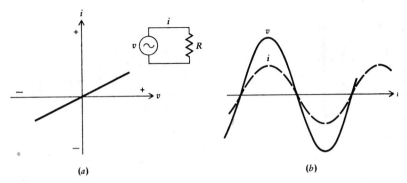

Figure 2-3 (a) Current versus voltage for a linear bilateral resistor; (b) current and voltage as functions of time.

Resistors that maintain an essentially constant resistance over a wide range of temperature, and are not affected by the direction or magnitude of the applied voltage or current, are called *linear bilateral resistors*; linear because a graph of current versus voltage is a straight line, and bilateral because it is unaffected by the direction of the applied voltage. Figure 2-3*a* shows a graph of the current versus voltage for a linear bilateral resistor when an AC voltage is applied, and Fig. 2-3*b* shows the current and voltage as functions of time.

2-6   TEMPERATURE COEFFICIENT OF RESISTANCE

If the resistance at one temperature is known, the resistance at some other temperature may be determined by substituting into the following formula:

$$R_H = R_L[1 + \alpha(T_H - T_L)] \qquad\qquad (2\text{-}2)$$

where R_H = resistance at higher temperature, ohms (Ω)
R_L = resistance at lower temperature, ohms (Ω)
T_H = higher temperature, °C
T_L = lower temperature, °C
α = temperature coefficient of resistance

The temperature coefficients of some of the more commonly used conductors are given in Table 2-2.

Although the temperature coefficients in Table 2-2 are for 20°C, they can be used in calculations over the normal range of operating temperatures

Table 2-2 Temperature coefficients of resistance of conductor materials

Material	Temperature coefficient α, Ω per °C per Ω at 20°C
Aluminum	0.0039
Antimony	0.0036
Bismuth	0.004
Brass	0.002
Constantan (60% Cu, 40% Ni)	0.000008
Copper, annealed	0.00393
Copper, hard-drawn	0.00382
German silver	0.0004
Iron	0.005
Lead	0.0041
Magnesium	0.004
Manganin (84% Cu, 12% Mn, 4% Ni)	0.000006
Mercury	0.00089
Molybdenum	0.0034
Monel metal	0.002
Nichrome	0.0004
Nickel	0.006
Platinum	0.003
Silver (99.98% pure)	0.0038
Steel, soft	0.0042
Tin	0.0042
Tungsten	0.0045
Zinc	0.0037

Source: Values taken from Smithsonian Physical Tables.

associated with electric circuits and machinery without introducing significant errors. For example, the temperature coefficient for annealed copper is 0.00393 at 20°C and 0.0038 at 100°C, a difference of only 0.00013. Annealed copper (heat-treated) is used in most electrical equipment, such as electronic circuits, motors, generators, and transformers. Hard-drawn copper is used in some telephone lines to reduce the sag between poles.

To convert degrees Fahrenheit to degrees Celsius, and vice versa, use the following formulas:

$$°C = (°F + 40) \times \tfrac{5}{9} - 40$$

$$°F = (°C + 40) \times \tfrac{9}{5} - 40$$

(2-3)

Example 2-1 The resistance of a certain annealed-copper cable is 10.0 mΩ at 20°C. During normal current-carrying operations the cable temperature rises to 80°C. Determine its resistance at the operating temperature.

Solution
From Table 2-2, the temperature coefficient for annealed copper is 0.00393. Substituting into Eq. (2-2),

$$R_H = R_L[1 + \alpha(T_H - T_L)]$$

$$R_H = 0.0100[1 + 0.00393(80 - 20)]$$

$$R_H = 12.4 \times 10^{-3}\ \Omega = 12.4\ \text{m}\Omega$$

Example 2-2 Determine the resistance of the cable in Example 2-1 if it is cooled to -50°C.

Solution

$$R_H = R_L[1 + \alpha(T_H - T_L)]$$

$$0.0100 = R_L\{1 + 0.00393[20 - (-50)]\}$$

$$R_L = 7.84 \times 10^{-3} = 7.84\ \text{m}\Omega$$

2-7 **PREFERRED RESISTANCE VALUES**
When designing electronic circuits, primary consideration should be given to the selection of commercially standardized components. This reduces cost and makes replacement parts readily available. The *preferred resistance values* given in Table 2-3 were standardized by the electronics industry, and are commercially available in ± 5 percent tolerance. A tolerance of 5 percent means that the resistor selected may be 5 percent less or 5 percent greater than the *nominal value* given in the table. Resistance values printed in heavy type (boldface) are available also in ± 10 percent tolerance.

The standard color code for small tubular composition resistors is given in Table 2-4. Bands A and B represent the first and second figures, respectively, band C indicates the number of zeros, and band D indicates the tolerance. For example, if bands A, B, C, and D are yellow, green, orange, and gold, respectively, the resistance value is

45,000 Ω $\pm$ 5 percent

Table 2-3   Preferred resistance values

Ω			kΩ			MΩ	
1.0	**10**	**100**	**1.0**	**10**	**100**	**1.0**	**10.0**
1.1	11	110	1.1	11	110	1.1	11.0
1.2	**12**	**120**	**1.2**	**12**	**120**	**1.2**	**12.0**
1.3	13	130	1.3	13	130	1.3	13.0
1.5	**15**	**150**	**1.5**	**15**	**150**	**1.5**	**15.0**
1.6	16	160	1.6	16	160	1.6	16.0
1.8	**18**	**180**	**1.8**	**18**	**180**	**1.8**	**18.0**
2.0	20	200	2.0	20	200	2.0	20.0
2.2	**22**	**220**	**2.2**	**22**	**220**	**2.2**	**22.0**
2.4	24	240	2.4	24	240	2.4	
2.7	**27**	**270**	**2.7**	**27**	**270**	**2.7**	
3.0	30	300	3.0	30	300	3.0	
3.3	**33**	**330**	**3.3**	**33**	**330**	**3.3**	
3.6	36	360	3.6	36	360	3.6	
3.9	**39**	**390**	**3.9**	**39**	**390**	**3.9**	
4.3	43	430	4.3	43	430	4.3	
4.7	**47**	**470**	**4.7**	**47**	**470**	**4.7**	
5.1	51	510	5.1	51	510	5.1	
5.6	**56**	**560**	**5.6**	**56**	**560**	**5.6**	
6.2	62	620	6.2	62	620	6.2	
6.8	**68**	**680**	**6.8**	**68**	**680**	**6.8**	
7.5	75	750	7.5	75	750	7.5	
8.2	**82**	**820**	**8.2**	**82**	**820**	**8.2**	
9.1	91	910	9.1	91	910	9.1	

Table 2-4   Standard color code for small tubular resistors

Color	Band A, first figure	Band B, second figure	Band C, remaining figures	Band D, tolerance
Black	0	0		
Brown	1	1	0	
Red	2	2	00	
Orange	3	3	000	
Yellow	4	4	0,000	
Green	5	5	00,000	
Blue	6	6	000,000	
Violet	7	7	0,000,000	
Gray	8	8	00,000,000	
White	9	9	000,000,000	
Gold	...	...	....	± 5%
Silver	...	...	....	±10%
No color	...	...	....	±20%

2-8   **RESISTIVITY**

The resistivity of a material, also called its specific resistance, is the resistance of a specified unit length and unit cross section of that material.

   If the resistivity of a material is known, the resistance of a conductor of any length and any cross section may be determined by substituting in the following formula:

$$R = \frac{\rho\ell}{A}$$ (2-4)

where R = resistance of conductor, ohms (Ω)
$\quad\quad\quad \ell$ = length of conductor
$\quad\quad\quad A$ = cross-sectional area of conductor
$\quad\quad\quad \rho$ = resistivity whose units are RA/ℓ

The length, cross-sectional area, and resistivity in Eq. (2-4) *must have matching units.* A listing of resistivities for conductors frequently used in electrical systems is given in Table 2-5.

   In the SI system, ℓ is in meters, A is in square meters, and the derived units for ρ is

$$\rho = \frac{RA}{\ell} = \text{ohm} \cdot \text{meter}^2/\text{meter} = \text{ohm} \cdot \text{meter} \ (\Omega \cdot \text{m})$$

If the circular mil (cmil) system is used, ℓ is in feet, A is in circular mils, and the derived unit for ρ is

$$\rho = \frac{RA}{\ell} = \Omega \cdot \text{cmil/ft}$$

Example 2-3   Determine the resistance of a 2.0-m length of 10.0-mm-diameter round aluminum conductor.

Solution
From Table 2-5, the resistivity of aluminum is $2.826 \times 10^{-8}\ \Omega \cdot \text{m}$. The cross-sectional area is

$$A = \frac{\pi D^2}{4} = \frac{\pi(0.01)^2}{4} = 78.54 \times 10^{-6}\ \text{m}^2$$

$$R = \frac{\rho\ell}{A} = \frac{2.826 \times 10^{-8}(2)}{78.54 \times 10^{-6}} = 719 \times 10^{-6}$$

$$R = 719\ \mu\Omega$$

Table 2-5 Resistivity of common metallic conductors 20°C (68°F)

Material	$\Omega \cdot m \times 10^{-8}$	$\Omega \cdot cmil/ft$
Aluminum	2.826	17.0
Brass	7.0	42.1
Constantan (40% Ni, 60% Cu)	49.0	294.7
Copper (annealed)	1.724	10.371
Copper (hard-drawn)	1.77	10.65
Gold	2.44	14.68
Iron (99.98% pure)	10.0	60.2
Lead	22.0	132.3
Mercury	95.77	576.1
Nichrome	100.0	601.5
Nickel	7.80	46.9
Platinum	10.0	60.2
Silver	1.60	9.6
Tin	11.50	69.2
Tungsten (drawn)	5.60	33.7
Zinc	5.80	34.9

2-9 CIRCULAR MIL AREA

A non-SI unit of cross-sectional area commonly used for specifying conductor sizes is the circular mil (cmil). One mil is 0.001 inch, and one circular mil is defined as the area of a circle whose diameter is one mil. The equivalent area of one circular mil in square inches is

$$1 \text{ cmil} = \frac{\pi D^2}{4} = \frac{\pi (0.001)^2}{4}$$

$$\boxed{1 \text{ cmil} = 7.854 \times 10^{-7} \text{ in}^2} \tag{2-5}$$

To determine the circular mil area of any conductor, regardless of its shape, calculate its cross-sectional area in *square inches*, and then divide it by the square inches in one circular mil as obtained from Eq. (2-5). Thus, for any cross-sectional area,

$$\text{cmil} = \frac{A}{7.854 \times 10^{-7}}$$

$$\boxed{\text{cmil} = 1.273 A (10)^6} \tag{2-6}$$

where A = cross-sectional area of any shape conductor, in^2.

Circular Cross Sections

For circular cross sections, A in Eq. (2-6) becomes

$$A = \frac{\pi D^2}{4}$$

and Eq. (2-6) reduces to

$$\text{cmil} = 1.273 \frac{\pi D^2}{4} \times 10^6$$

$$\boxed{\text{cmil} = D^2(10)^6} \tag{2-7}$$

where D = diameter of round conductor, in

Example 2-4 Determine the resistance of a 10-ft length of $\frac{1}{2}$-in by 6-in annealed-copper bus.

Solution
From Eq. (2-6),

$$\text{cmil} = 1.273 A(10^6)$$

$$\text{cmil} = 1.273(\tfrac{1}{2} \times 6)10^6$$

$$\text{cmil} = 3.819(10^6)$$

$$R = \frac{\rho \ell}{A}$$

$\rho = 10.371 \ \Omega \cdot \text{cmil/ft}$ (from Table 2-5)

$$R = \frac{10.371(10)}{3.819(10^6)} = 0.000027 \ \Omega = 27 \ \mu\Omega$$

2-10   **WIRE TABLES**

Copper wire is the most commonly used conductor in electric circuits, transformers, machinery, controls, etc. Table 2-6† provides very useful data concerning solid annealed-copper wire in metric gage and American wire gage (AWG). The resistance of each size is given in Ω/km and Ω/kft; this information is very helpful when determining the resistance of a coil of copper wire, a long extension cord, a length of cable, etc.

† This is a partial listing. For complete tables of properties of copper, aluminum, and other conductors, see Fink, *Standard Handbook for Electrical Engineers*, McGraw-Hill Book Company, New York, 1980.

Table 2-6 Properties of annealed solid copper wire conductors 20°C (68°F)

WIRE GAGE			AREA		RESISTANCE	
AWG	Metric	Diam., mm	cmil	mm²	Ω/kft	Ω/km
0000		11.68	211,600	107.2	0.04901	0.1608
000		10.40	167,800	85.01	0.06181	0.2028
	100	10.0	155,000	78.54	0.06691	0.2195
00		9.266	133,100	67.43	0.07793	0.2557
	90	9.0	125,500	63.62	0.08264	0.2711
0		8.252	105,600	53.49	0.09821	0.3222
	80	8.0	99,200	50.27	0.10455	0.3430
1		7.348	83,690	42.41	0.12392	0.4066
	70	7.000	75,950	38.49	0.13655	0.4480
2		6.543	66,360	33.62	0.15628	0.5128
	60	6.0	55,800	28.27	0.18586	0.6098
3		5.827	52,620	26.67	0.19709	0.6467
4		5.189	41,740	21.15	0.24847	0.8152
	50	5.0	38,750	19.63	0.26764	0.8781
5		4.620	33,090	16.77	0.31342	1.028
	45	4.5	31,390	15.90	0.33039	1.084
6		4.115	26,240	13.30	0.39524	1.297
	40	4.0	24,800	12.57	0.41819	1.372
7		3.665	20,820	10.55	0.49813	1.634
	35	3.5	18,990	9.622	0.54613	1.792
8		3.264	16,510	8.367	0.62816	2.061
	30	3.0	13,950	7.068	0.74344	2.439
9		2.906	13,090	6.631	0.79228	2.599
10		2.588	10,380	5.261	0.99913	3.278
	25	2.5	9,687	4.908	1.0706	3.513
11		2.304	8,230	4.170	1.2602	4.135
12		2.05	6,530	3.310	1.5882	5.211
	20	2.0	6,200	3.142	1.6727	5.488
13		1.83	5,180	2.630	2.0021	6.569
	18	1.8	5,022	2.545	2.0651	6.776
14		1.63	4,110	2.08	2.5234	8.279
	16	1.6	3,968	2.011	2.6137	8.576
15		1.45	3,260	1.650	3.1813	0.438
	14	1.4	3,038	1.539	3.4138	11.201
16		1.29	2,580	1.310	4.0198	13.189
	12	1.2	2,232	1.131	4.6465	15.245
17		1.150	2,050	1.040	5.0590	16.599
18		1.02	1,620	0.823	6.4019	21.005
	10	1.00	1,550	0.7854	6.6910	21.953

Example 2-5 A certain field coil for a DC motor is wound with exactly 0.50 km of size 10 (metric) solid annealed-copper wire. Determine (a) the resistance of the coil at 20°C; (b) the resistance of the coil at 40°C.

Solution
(a) From Table 2-6 the resistance of size 10 (metric) is 21.953 Ω/km at 20°C. Hence, the resistance of 0.50 km at 20°C is

$$R_{20°C} = 0.50 \times 21.953 = 10.976 = 11.0 \ \Omega$$

(b)

$$R_H = R_L[1 + \alpha(T_H - T_L)]$$

From Table 2-2, $\alpha = 0.00393$,

$$R_H = 11.0[1 + 0.00393(40 - 20)]$$

$$R_H = 11.8 \ \Omega$$

2-11 CONDUCTANCE

Conductance is a measure of the ease with which a material conducts an electric current. It is defined as the reciprocal of resistance, is assigned the letter G, and is expressed in *siemens*.†

$$G = \frac{1}{R} \tag{2-8}$$

where R = resistance, ohms (Ω)
   G = conductance, siemens (S)

Example 2-6 Determine the conductance of each of the following resistors: 25 Ω, 60 kΩ, 1.2 MΩ.

Solution

$$G = \frac{1}{25} = 0.040 \ S$$

$$G = \frac{1}{60,000} = 16.667 \times 10^{-6} \ S = 16.7 \ \mu S$$

$$G = \frac{1}{1.20 \times 10^6} = 833 \times 10^{-9} \ S = 833 \ nS$$

Conductance is used extensively in parallel-circuit calculations.

† Previously called the mho.

2-12 SEMICONDUCTORS AND DIODES

A semiconductor material, such as silicon or germanium, has a conductivity about midway between good conductors and good insulators. Useful semiconductor devices are grown as crystals from a melt of semiconductor material to which extremely small amounts of impurities have been added. Some of the semiconductor devices grown in this manner are diodes, silicon controlled rectifiers (called thyristors or SCRs), and transistors. A diode has the characteristics of a conductor when the applied voltage is in the *forward direction* and the characteristics of an insulator when the applied voltage is in the *reverse direction*. This is illustrated in Fig. 2-4 for forward and reverse directions of applied voltage (also called forward and reverse bias). In the

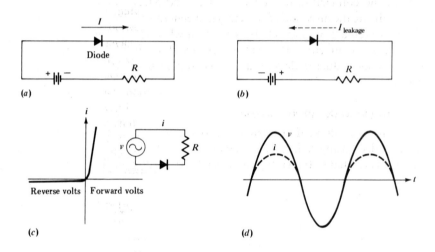

Figure 2-4 (*a*) Diode biased in forward direction; (*b*) diode biased in the reverse direction; (*c*) current versus voltage; (*d*) current and voltage as functions of time.

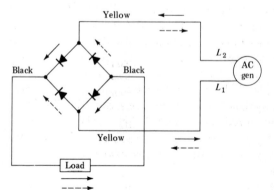

Figure 2-5 Full-wave rectifier circuit.

forward direction (Fig. 2-4a), the diode offers very little resistance to the current. In the reverse direction (Fig. 2-4b), the diode offers considerable resistance to the current, allowing only a very small *leakage* current. Hence, if an alternating driving voltage is applied, the diode will act as a rectifier, blocking the current when the voltage is in the reverse direction.

Figure 2-4c shows a graph of current versus voltage for a circuit containing a diode and an AC driving voltage. Figure 2-4d shows the same current and voltage as functions of time. Note that the diode blocks the current when the driving voltage is reversed.

The use of four diodes, in what is called a *bridge type* of full-wave rectifier circuit, provides unidirectional current to the load, even though the generator current is alternating. This is shown in Fig. 2-5. The identifying colors of the connecting wires represent the NEMA† standard. The solid arrows indicate the direction of current when generator terminal L_2 is positive, and the broken arrows indicate the direction when L_1 is positive. Note that the direction of current in the load is the same in both cases. Other applications of semiconductor devices are discussed in connection with specific apparatus.

2-13   **VARISTOR (NONLINEAR RESISTOR)**

The varistor is a voltage-dependent metal-oxide material that has the property of sharply decreasing resistance with increasing voltage. The current through a varistor is related to the voltage across it by the following formula:

$$i = ke^{\eta} \qquad (2\text{-}9)$$

where i = instantaneous current, A
       e = instantaneous voltage, V
       k = constant

The exponent η is dependent on the metal oxides used. Zinc-oxide-based varistors can be designed to have a value of η between 25 and 50 or more. Silicon-carbide-based varistors can be designed to have a value of η between 2 and 6. The zinc-oxide-based varistors are used primarily for protection of solid-state power supplies against low and medium surge voltages developed within residential and industrial power circuits. Silicon-carbide varistors provide protection against high-voltage surges caused by lightning and by the discharge of energy stored in the magnetic fields of large coils.

Figure 2-6a illustrates the current-voltage characteristic for a silicon-carbide-based varistor, and Fig. 2-6b illustrates the shape of the current wave

† National Electrical Manufacturers Association.

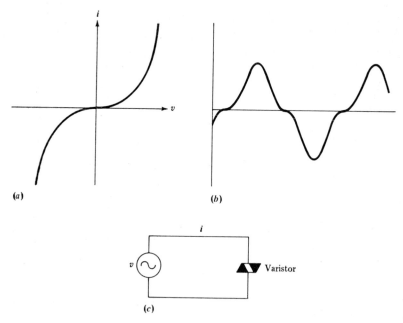

Figure 2-6 Silicon-carbide varistor: (a) current-voltage characteristic; (b) current versus time; (c) elementary diagram.

through the varistor when driven by a sinusoidal voltage wave. The standard graphical symbol for a varistor is shown in Fig. 2-6c.

2-14 **THERMISTOR**

A thermisotor is a resistor that has a negative temperature coefficient of resistance. That is, its resistance decreases with increasing temperatures. The resistance of these thermally sensitive resistors changes markedly with changes in temperature as shown in Fig. 2-7a.

Thermistors are manufactured from manganese, nickel, and cobalt oxides, along with suitable binding materials. The thermistor elements are sintered under controlled temperature and atmospheric conditions to obtain the desired characteristics.

Thermistors are used for temperature measurements and control, high-temperature alarms for machinery, and many other applications. Figure 2-7b shows a simple alarm circuit to detect overheating of a motor bearing. With the thermistor fastened to the bearing, a rise in bearing temperature will cause the thermistor resistance to decrease. When the resistance decreases to a value that causes the relay to operate, the alarm will sound.

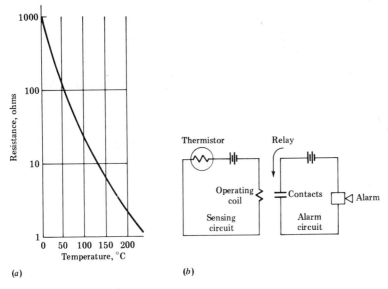

(a)

(b)

Figure 2-7   Thermistor: (a) resistance-temperature characteristic; (b) alarm circuit.

2-15   PHOTOCONDUCTORS

Photoconductors, also called photoresistors, are semiconductors whose resistance values decrease with increasing light intensity. Photoconductors are used in circuits for automatic control of street lighting, brightness control for TV sets that adjust automatically with changes in room light, photographic exposure meters, etc.

Figure 2-8 shows a simple photoconductor circuit for controlling a lighting circuit. The photoconductor used in this application has a resistance of less than 500 Ω in daylight and over 20 MΩ when in the dark. During daylight hours, the relatively low resistance of the photoconductor allows sufficient current in the relay coil to cause the normally closed contacts to be

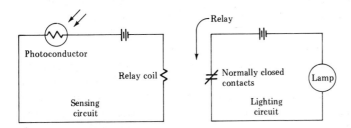

Figure 2-8   Photoconductor circuit for controlled lighting.

pulled apart, and the light is deenergized. During the dark period, the extremely high resistance of the photoconductor does not permit sufficient current in the relay coil; the contacts fall to the closed position and the lamp is energized.

SUMMARY OF FORMULAS

$$i = \frac{dq}{dt} \qquad i = ke^n$$

$$R_H = R_L[1 + \alpha(T_H - T_L)]$$

$$°C = (°F + 40)\tfrac{5}{9} - 40$$

$$°F = (°C + 40)\tfrac{9}{5} - 40$$

$$R = \frac{\rho\ell}{A}$$

$$\text{cmil} = 1.273A(10^6) = D^2(10)^6$$

$$G = \frac{1}{R}$$

PROBLEMS

2-1   The resistance of a certain length of nichrome wire is 50 Ω at 20°C. What is its resistance at 60°C?

2-2   The resistance of a length of aluminum conductor is 80 Ω at 40°C. What is its resistance at 20°C?

2-3   The copper field windings of a generator have a resistance of 162 Ω at 60°C. What is its resistance at 20°C?

2-4   Determine the resistance of a 564-m length of aluminum conductor whose rectangular cross section is 40 mm by 20 mm.

2-5   Determine (a) the circular mil area of an annealed-copper bus whose cross section is 0.50 in by 6.0 in; (b) the resistance of a 200-ft length of bus.

2-6   Determine the resistance of a 30-m length of aluminum bus whose dimensions are 20 mm by 200 mm.

2-7   Determine (a) the circular mil area of a round aluminum conductor whose diameter is 4.0 in; (b) the resistance of 1000 ft of the conductor.

2-8   Determine the resistance of a 10-km length of size 30 (metric) annealed-copper conductor.

2-9   Determine the resistance of a 6542-ft length of size 00 AWG annealed-copper cable.

2-10   Determine the conductance of each of the following resistors: 50 kΩ, 465 Ω, 0.018 Ω, 0.00040 Ω, 52 MΩ.

2-11   Determine the resistance of each of the the following conductances: 0.0041 S, 3.25 S, 180.55 S, 1000 S, 400 mS.

CHAPTER 3
ELECTRIC CIRCUITS

An electric circuit is any arrangement of conducting parts that form a closed loop. A few examples of simple electric circuits are shown in Fig. 3-1. A closed loop constitutes a circuit whether or not it includes a useful load or a driver. In the case of Fig. 3-1a there is no current in the loop unless it is induced by a changing magnetic field, such as that caused by moving magnets, a lightning discharge, or a changing current in a nearby circuit. The current (I) in the electric circuits shown in Fig. 3.1b and c is the result of the application of a DC voltage source.

3-1 DC VOLTAGE SOURCES

The most common sources of DC voltage are electrochemical cells, DC generators, and AC generator-rectifier sets. Other DC voltage sources used in special applications are fuel cells and solar cells. A battery is a set of cells (generally identical) connected in series, parallel, or series-parallel combinations.

The electrochemical cell stores chemical energy for conversion later to electrical energy. The two basic types of electrochemical cells are the primary cell and the secondary cell.

A *primary* cell is chemically irreversible. Such cells cannot be recharged and must be discarded when no longer usable; their active material is consumed during discharge.

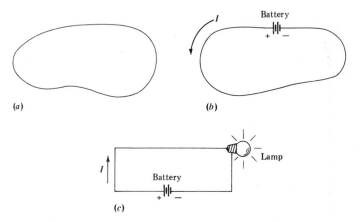

Figure 3-1   Examples of simple electric circuits: (*a*) closed loop without a driver or useful load; (*b*) closed loop with a battery; (*c*) closed loop with a battery and a lamp.

The *secondary cell*, known as a *storage cell*, is chemically reversible; its active materials are not consumed. Such cells may be recharged by passing current through the battery in the reverse direction.

The capacity of a battery, expressed in *ampere-hours* (A · h), is a measure of its ability to deliver energy. It is the product of the current in amperes times the hours of discharge. Unfortunately, the A · h capacity of a battery varies with discharge rate, specific gravity, temperature, and final voltage. The discharge rates commonly used as a basis for battery comparison are the 8-h rate for stationary batteries, 20-h rate for automotive batteries, and 6-h rate for electric-vehicle batteries. For example, a 100-A · h battery at the 8-h rate will deliver 12.5 A for 8 h before the cell voltage reduces to its minimum useful value (1.75 V for a 2-V lead-acid cell).

Dry Cell

The dry cell, shown in Fig. 3-2, is the most common type of primary cell. It has a carbon rod for the anode (positive terminal) and a zinc container for the cathode (negative terminal). The zinc container has an inner lining of absorbent material, and the space between the electrodes is filled with a mixture of crushed coke, manganese dioxide, and graphite, saturated with a solution of sal ammoniac and zinc chloride. The zinc-carbon cell has an open-circuit voltage of about 1.6 V and a capacity of approximately 30 A · h when supplying current at the rate of 0.2 A, 4 h/day. The life of a dry cell is limited. It gradually discharges by internal chemical action, called local action, even though no load is connected to its terminals. For this reason, some manufacturers stamp a service date on the outer covering of each cell. Dry cells should be stored in a cool, dry place.

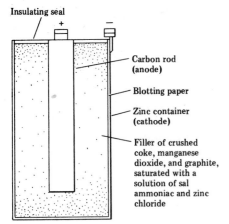

Insulating seal

Carbon rod
(anode)

Blotting paper

Zinc container
(cathode)

Filler of crushed
coke, manganese
dioxide, and graphite,
saturated with a
solution of sal
ammoniac and zinc
chloride

Figure 3-2 Dry cell.

Lead-Acid Cell

The lead-acid cell, illustrated in Fig. 3-3a, is the most common type of secondary cell. It has positive plates of lead peroxide and negative plates of sponge lead. The plates are kept apart by separators made of microporous rubber, fiberglass, perforated plastic or hard rubber, or resin-impregnated cellulose. The electrolyte is a solution of sulfuric acid and water. During discharge, the positive and negative plates are chemically changed to lead sulfate. The charging process restores the plates to their original chemical makeup of sponge lead and lead peroxide. The open-circuit voltage of a fully charged cell is about 2.05 V and varies slightly with the temperature and specific gravity of the electrolyte.

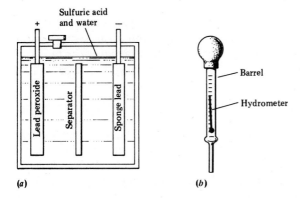

Sulfuric acid
and water

Lead peroxide

Separator

Sponge lead

Barrel

Hydrometer

(a) (b)

Figure 3-3 (a) Lead-acid cell; (b) hydrometer.

Lead-acid batteries should never be left in a discharged condition for any great length of time, because the lead sulfate may harden and become nonporous, preventing the battery from accepting a full charge. Because the specific gravity of the electrolyte changes with the state of charge or discharge, the battery condition may be determined with a hydrometer, shown in Fig. 3-3b. Enough electrolyte should be drawn into the glass barrel to make the hydrometer float. Hydrometer readings should always be taken before the addition of distilled water. If the level of the electrolyte is low, water should be added, and the battery should be charged before a hydrometer reading is taken.

The temperature of the electrolyte has considerable effect on its hydrometric values. High temperatures cause the volume of the electrolyte to increase, resulting in lower values of specific gravity, whereas low temperatures cause the volume of the electrolyte to decrease, resulting in higher values of specific gravity. Hence, if the specific-gravity readings are to be of any value in determining the condition of a battery, corrections to some reference temperature must be made. The specific gravity of a fully charged lead-acid battery is 1.285, and that of a discharged battery is 1.125, both measured at 27°C (80°F).

When lead-acid batteries are operated in extremely low temperatures, it is imperative that they be kept in a fully charged condition at all times. Low temperatures reduce the useful capacity of a battery, and at -18°C (0°F), its capacity is reduced to one-half its value at 27°C (80°F). Furthermore, the freezing point of the electrolyte is affected by the condition of charge. A fully charged battery has a freezing point of -68°C (-90°F), whereas a dead battery freezes at -6°C (22°F). Freezing of the electrolyte damages the plates and ruptures the case, thus rendering the battery useless.

The lowering of electrolyte level in normal usage is caused by evaporation of water, and the restoration to the correct level should be made only with distilled water. However, if electrolyte is lost because of spilling or bubbling over, it should be replaced with sulfuric acid of the correct specific gravity.†

DC Generator

The DC generator is an electromechanical machine that provides a DC output voltage when driven by a prime mover. Conventional prime movers are diesel engines, gasoline engines, steam turbines, water wheels, windmills, etc. The DC generator, with an associated voltage regulator, maintains an essentially constant voltage for all load conditions, from no load to full load;

---

† For detailed information on the maintenance and charging of different types of storage batteries, see C. I. Hubert, *Preventive Maintenance of Electrical Equipment*, McGraw-Hill Book Company, New York, 1969. This section reproduced with permission.

the load may be lamps, heaters, motors, electronic equipment, etc. The standard ratings for commercial DC generators are 120 V, 240 V, and 600 V.

3-2   **AC VOLTAGE SOURCES**

Alternating voltages may be produced by electromechanical machines called AC generators or by electronic oscillators. An AC generator operates on the same principle as the DC machine except that its output voltage is alternating (AC). The advantage of an AC voltage source is that the voltage may be raised or lowered through the medium of a transformer. A DC voltage may be obtained from an AC machine by means of a rectifier. The AC voltage is fed into a rectifier, as previously shown in Fig. 2-5, and the rectified output is connected directly to the load or is fed through a filter connected between the rectifier and the load. A properly designed filter will produce a DC voltage with essentially zero ripple.

3-3   **VOLTAGE SOURCES IN SERIES AND IN PARALLEL**

Depending on the voltage, ampere-hour, or other requirements, voltage sources may be connected in series, parallel, or series-parallel arrangements. For example, flashlight batteries are connected in series to obtain the required voltage; electromechanical generators are often connected in parallel to supply the many loads in a distribution system; and batteries used for propulsion systems may be connected in series, parallel, or series-parallel arrangements to meet the voltage and ampere-hour requirements of the load.

Series Connection

When voltage sources are connected in series, the net output voltage is the *algebraic sum* of the component voltages. Figure 3-4a shows five voltage sources connected in series; the arrows associated with each source indicate

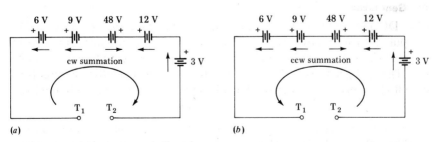

Figure 3-4   Series connection of voltage sources: (*a*) cw summation of voltages; (*b*) ccw summation of voltages.

the direction of the respective driving voltages. To make the summation, "walk" through the circuit from T_1 to T_2, or vice versa, adding voltages as you proceed. If travel through a source is in the direction of its arrow, the source voltage is considered positive; if travel through a source is opposite to its arrow, the source voltage is considered negative. Thus, referring to Fig. 3-4a and summing in the clockwise (cw) direction,

$$V_{\text{out}} = -6 - 9 + 48 - 12 - 3 = +18 \ V$$

The positive sign indicates that the net *direction* of the driving voltage is in the direction of travel. Hence T_2 is the positive terminal.

  If the summation is in the counterclockwise direction (ccw), as shown in Fig. 3-4b,

$$V_{\text{out}} = +3 + 12 - 48 + 9 + 6 = -18 \ V$$

The negative sign indicates that the net *direction* of the driving voltage is opposite to the direction of travel. Hence T_2 is the positive terminal.

Parallel Connection

The parallel connection of voltage sources requires that all sources have essentially identical voltages. Voltage sources of different voltage ratings may cause the higher-voltage sources to feed into the lower-voltage sources, and permanent damage to the sources may occur. When paralleling voltage sources, all positive terminals must be connected together, and all negative terminals connected together, as shown in Fig. 3-5.

Series-Parallel Connection

Figure 3-6a shows twelve 6-V batteries arranged to obtain an output voltage of 12 V. Figure 3-6b shows the same batteries arranged for an 18-V output, and Fig. 3-6c shows an arrangement for a 36-V output.

Example 3-1 Design a series-parallel arrangement of forty 6-V batteries that will provide an output voltage of 24 V.

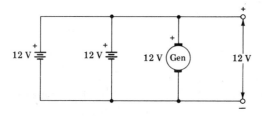

Figure 3-5 Parallel-connected voltage sources.

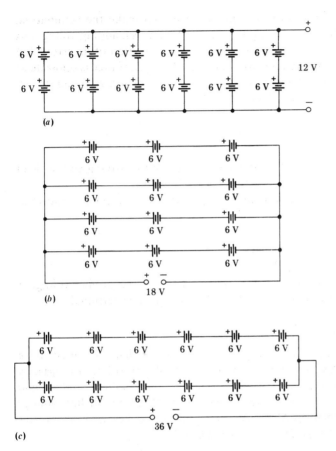

Figure 3-6 Series-parallel connections of voltage sources to obtain different output voltages.

Solution
Since each battery is rated at 6 V, the number of series-connected batteries per branch required to obtain 24 V is

$$\frac{24}{6} = 4 \text{ batteries per branch}$$

The number of parallel branches is

$$\frac{40}{4} = 10 \text{ branches}$$

Thus, the required arrangement is 10 parallel branches, each containing 4 batteries in series.

3-4 OHM'S LAW

The current in an electric circuit is directly proportional to the driving voltage and inversely proportional to the opposition that the circuit offers to the current. This relationship, called Ohm's law, expressed mathematically for a resistive circuit, is

$$I = \frac{E}{R} \qquad\qquad (3\text{-}1)$$

where E = driving voltage, volts (V)
  R = resistance, ohms (Ω)
  I = current, amperes (A)

Example 3-2 A 12-V source is connected to a circuit whose resistance is 48 Ω. Determine the current.

Solution

$$I = \frac{E}{R} = \frac{12}{48} = 0.25 \text{ A}$$

3-5 RESISTANCE MEASUREMENT

Resistance may be measured directly with an ohmmeter or indirectly by means of the voltmeter-ammeter method shown in Fig. 3-7. The ammeter must always be connected in series with the resistor, and the voltmeter must always be connected in parallel with the resistor.

A fuse or circuit breaker is used to prevent damage to the circuit in the event of an accidental short circuit.† A short circuit, commonly called a *short*, is an electrical connection that bypasses part or all of an electric circuit, thus providing an additional route for the current.

In the voltmeter-ammeter method, resistance is determined from Ohm's law by dividing the voltmeter indication by the ammeter indication:

$$R = \frac{E}{I} = \frac{\text{voltmeter indication}}{\text{ammeter indication}}$$

---

† For proper selection of circuit protective devices, see C. I. Hubert, *Preventive Maintenance of Electrical Equipment*, McGraw-Hill Book Company, New York, 1969.

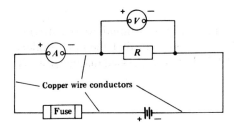

Figure 3-7 Voltmeter-ammeter method for measuring resistance.

The measurements should be made quickly to avoid changes in meter readings caused by changes in resistance due to heating.

To avoid a reverse deflection of the meter and the possible consequences of a bent pointer, the positive terminal of the voltmeter should always be connected to the positive terminal of the resistor. To make the correct connection, the *relative polarity* of the resistor terminals must be determined. The relative polarity of the two terminals of a resistor or other apparatus may be determined by an observer "standing" at each resistor terminal in turn, "facing" the driver terminal to which it is connected. The polarity of the driver terminal that the observer faces is the polarity of the resistor terminal in question. Figure 3-8 shows the relative polarity of the terminals for three resistors, and the correct connections to the voltmeter and ammeter. Note that the positive terminal of the ammeter must always face the positive terminal of the driver.

A voltmeter measures the difference in electrical potential (difference in voltage) between two points in an electric circuit; it cannot measure the voltage at a single point. In Fig. 3-8, each voltmeter *measures the potential difference* across a specific resistor.

The graphical symbol for a resistor is either a rectangle with the value of resistance or the letter *R* inserted as shown in Fig. 3-7, or the zigzag symbol shown in Fig. 3-8. Both are American National Standard graphic symbols, and both will be used in this text.

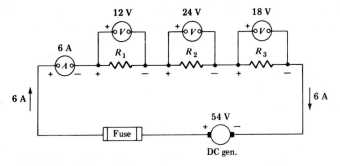

Figure 3-8 Circuit for Example 3-3.

Example 3-3 Using the voltmeter-ammeter method, determine the resistance of each resistor in Fig. 3-8.

Solution
As indicated in the diagram, a current of 6 A is passing through each of the resistors, and the voltages measured across R_1, R_2, and R_3 are 12 V, 24 V, and 18 V, respectively. From Ohm's law,

$$R_1 = \frac{V_1}{I} = \frac{12}{6} = 2 \, \Omega$$

$$R_2 = \frac{24}{6} = 4 \, \Omega \qquad R_3 = \frac{18}{6} = 3 \, \Omega$$

3-6 ENERGY AND POWER IN ELECTRIC CIRCUITS

In its most elementary sense, work is done when a force moves a body through some distance. *The amount of energy expended in moving the body is equal to the work done.* In the SI system of units, one joule of energy represents the amount of work done by a force of one newton acting through a distance of one meter. Expressed mathematically,

$$W = Fd$$

where W = energy, joule (J)
 F = force, newtons (N)
 d = distance, meters (m)

Similarly, work is done when a voltage source causes electric charges to move through some potential difference. The amount of energy expended in moving the electric charges is equal to the work done. In the SI system of units, *one joule of energy represents the amount of work done by a voltage source that moves one coulomb of electric charge across a potential difference of one volt.* Expressed mathematically,

$$W = Vq \qquad\qquad\qquad (3\text{-}2)$$

where W = energy, joules (J)
 q = electric charge, coulombs (C)
 V = voltage, volts (V)

The rate at which energy is used, stored, or transferred is called power, and is expressed in joules per second or watts. *One watt is equal to one joule per second.*

A mathematical expression for electrical power may be obtained by dividing both sides of Eq. (3-2) by time t. Thus,

$$\frac{W}{t} = \frac{Vq}{t} \tag{3-3}$$

where W/t = power, J/s
   q/t = current, C/s

Designating $P = W/t$, and recognizing that $I = q/t$, Eq. (3-3) becomes

$$P = VI \tag{3-4}$$

where P = power, watts (W)
   V = voltage, volts (V)
   I = current, amperes (A)

Thus electrical power may be determined from the known values of current and voltage.

3-7   POWER LOSS AND ENERGY LOSS IN RESISTORS

Power in an electric circuit is defined as the time rate of transfer of energy and is generally expressed in watts or joules per second. At any instant of time the electric power supplied to a circuit is equal to the product of the input current in amperes and the driving voltage in volts:

$$\boxed{P = VI} \tag{3-5}$$

where P = power, W
   V = volts impressed across the circuit, V
   I = current in the circuit, A

The power loss in a *resistor* is equal to the product of the voltage across the resistor times the current to the resistor:

$$\boxed{P_R = V_R I_R} \tag{3-6}$$

From Ohm's law,

$$V_R = I_R R \tag{3-7}$$

Substituting Eq. (3-7) into Eq. (3-6),

$$\boxed{P_R = I_R^2 R}$$

(3-8)

Equation (3-8) provides a very convenient means for determining the heat-power loss in resistors. The heat is generated within the resistor by electron collisions with atomic particles of the material.

As indicated in Eq. (3-8), *the rate of generation of heat energy is a function of the resistance and the square of the current through it.*

A third formula for the calculation of heat-power loss in a resistor may be obtained by solving Eq. (3-7) for the current, and then substituting it in Eq. (3-6). The resulting equation is

$$\boxed{P_R = \frac{V_R^2}{R}}$$

(3-9)

In calculations involving large amounts of power, a larger unit such as the *kilowatt* (kW) or *megawatt* (MW) is used:

$$1 \text{ kW} = 1000 \text{ watts} = 10^3 \text{W}$$

$$1 \text{ MW} = 1{,}000{,}000 \text{ watts} = 10^6 \text{W}$$

Example 3-4 A 240-V DC source is connected to an 8-Ω resistor. Determine the heat power expended in the resistor, using the three power formulas.

Solution

$$P = \frac{V^2}{R} = \frac{(240)^2}{8} = 7200 \text{ W}$$

$$I = \frac{V}{R} = \frac{240}{8} = 30 \text{ A}$$

$$P = I^2 R = (30)^2(8) = 7200 \text{ W}$$

$$P = VI = (240)(30) = 7200 \text{ W}$$

Although power equations (3-6), (3-8), and (3-9) are useful for calculating the *rate* at which electrical energy is converted to heat energy, the *actual energy expended* is equal to power × time. Thus, with the power expressed in watts and the time in seconds, the expended energy in watt-seconds or joules is

$$W_R = P_R t \qquad (3\text{-}10)$$

One watt-second of energy is called one joule (J).

Substituting Eqs. (3-6), (3-8), or (3-9) into Eq. (3-10) results in three other useful expressions for the heat energy expended in a resistor:

$$W_R = V_R I_R t \qquad (3\text{-}11)$$

$$W_R = I_R^2 R t \qquad (3\text{-}12)$$

$$W_R = \frac{V_R^2 t}{R} \qquad (3\text{-}13)$$

If the power is expressed in watts and the time in hours (h), the energy expended may be expressed in watt-hours (W · h). If the power is expressed in kilowatts and the time in hours, the energy may be expressed in kilowatt-hours (kW · h).

The factors used to convert from watt-seconds to watt-hours and kilowatt-hours are:

$$\frac{\text{watt-seconds}}{3600} = \text{W} \cdot \text{h}$$

$$\frac{\text{watt-seconds}}{1000(3600)} = \text{kW} \cdot \text{h}$$

Depending on the $I^2 R$ losses in a conductor, the total operating time, and the environmental conditions, the temperature of the conductor can range from warm to the melting point of the material. A well-known example of heat loss in conductors is the heating of extension cords and cables when overloaded. Almost everyone has observed the heating of a plug where it mates with a wall receptacle for such household appliances as toasters, grills, and coffee pots; if the contact resistance between the two mating surfaces is excessively high, the heat produced can start a fire. Mating surfaces for plug-in devices and contactors depend on spring pressure and clean surfaces for low contact resistance. Corroded or dirty surfaces, worn contacts, or weak spring pressure due to loss of temper result in a higher than normal contact resistance.

Example 3-5 Which of the following combinations of current and time will result in the greater expenditure of heat energy in a 0.50-Ω resistor: 150 A for 1 s, 60 A for 3 s, or 25 A for 40 s?

Solution

Combination	$W_R = I_R^2 Rt$
150 A, 1 s	$W_R = (150)^2(0.5)(1) = 11{,}250 \text{ J} = 3.13 \text{ W} \cdot \text{h}$
60 A, 3 s	$W_R = (60)^2(0.5)(3) = 5400 \text{ J} = 1.50 \text{ W} \cdot \text{h}$
25 A, 40 s	$W_R = (25)^2(0.5)(40) = 12{,}500 \text{ J} = 3.47 \text{ W} \cdot \text{h}$

The 25-A 40-s combination resulted in the greater expenditure of heat energy.

Example 3-6 Figure 3-9 shows a plot of the voltage impressed across a 10-Ω resistor. Determine the total heat energy expended by the resistor in 8 s.

Solution

$$W_R = Pt = \frac{V_R^2}{R} t$$

$$W_R = \frac{(30)^2(2)}{10} + \frac{(40)^2(4)}{10} + \frac{(10)^2(2)}{10}$$

$$W_R = 180 + 640 + 20 = 840 \text{ J}$$

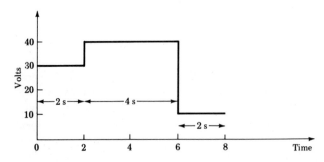

Figure 3-9 Circuit for Example 3-6.

SUMMARY OF FORMULAS

$$I = \frac{E}{R} \text{ or } \frac{V}{R}$$

$$P = V_R I_R = I_R^2 R = \frac{V_R^2}{R}$$

$$W = Pt = V_R I_R t = I_R^2 Rt = \frac{V_R^2}{R} t$$

PROBLEMS

3-1 Design a series-parallel arrangement of fifty 12-V batteries that will provide an output voltage of 120 V for an emergency lighting system. Sketch the circuit.

3-2 Design a series-parallel arrangement of thirty-six 6-V batteries that will provide an output voltage of 18 V. Sketch the circuit.

3-3 Determine the voltage required to send 15 A through a coil wound with 600 m of size 20 (metric) annealed-copper wire.

3-4 A 240-V DC source is connected to a 56-Ω resistor. Determine the current.

3-5 A 6.0-V battery, a 12-V battery, and a 360-Ω resistor are connected in series. Determine the current if (a) the battery voltages are additive; (b) the battery voltages are subtractive. Sketch the two circuits.

3-6 If the voltage measured across a resistor is 156 V and the measured current to the resistor is 4.00 mA, determine the ohmic value of the resistor.

3-7 Determine the voltage required to cause 150 μA in a 10-MΩ resistor.

3-8 Determine the current required to obtain 420 V across a 16-MΩ resistor.

3-9 An electric heater draws 2000 W from a 100-V DC system. Sketch the circuit and determine (a) the current drawn by the heater; (b) the resistance of the heater; (c) the total heat energy expended in 8 h.

3-10 An electric soldering iron uses 6 kW·h of energy in 12 h when connected to a 120-V source. Determine (a) the power rating of the soldering iron; (b) the current; (c) the resistance of the soldering iron.

3-11 Which of the following combinations of current and time will result in the greatest expenditure of heat energy in a 10-Ω resistor: 50 A for 1 s, 80 A for 0.5 s, 10 A for 10 s, 1 A for 1000 s?

CHAPTER 4

SERIES AND PARALLEL RESISTIVE CIRCUITS WITH DC DRIVERS

All electric circuits include series-connected sections, parallel-connected sections, or both. It is the fundamental laws that govern the behavior of these sections that provide the basic tools for the analysis of all circuits, no matter how complex.

4-1   SERIES CIRCUIT

A series circuit, such as that shown in Fig. 4-1, is a *closed-loop arrangement of drivers and circuit elements that has the same current passing through every component in the loop.* The battery, acting as an electron pump, causes the same rate of flow of electrons through all parts of the circuit.

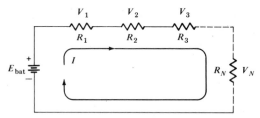

Figure 4-1   Series circuit.

The current-voltage relationships inherent in series-connected resistors are developed with the aid of the fundamental laws of Ohm and Kirchhoff.

KIRCHHOFF'S VOLTAGE LAW

Kirchhoff's voltage law, a very powerful tool for solving circuit problems, has its greatest applications in network analysis. However, the basic concept is so broad that it can be applied to any circuit, from the simple series circuit to the most complex network. Kirchhoff's voltage law states:

In any closed loop formed by circuit elements, voltage sources, or both, the algebraic sum of the driving voltages around the loop is equal to the algebraic sum of the voltage drops across the elements in the loop, with all voltages considered at the same instant of time. Stated briefly,

$$\sum \text{driving voltages} = \sum \text{voltage drops}$$

A driving voltage, also called a voltage rise or voltage source, is defined as the voltage produced by a battery or a generator, and a voltage drop is defined as the voltage across an element caused by current to or through the element. All generators and batteries are *sources of voltage*, and all circuit elements, such as resistance, inductance, and capacitance, *cause voltage drops.*

Unless otherwise specified, in this and other chapters, the resistance of batteries, generators, connecting wires, switches, fuses, and circuit breakers is assumed to have negligible values compared with the values of the concentrated blocks of resistance R_1, R_2, R_3, etc., generally found in an electric circuit.

Referring to Fig. 4-1,

$$\sum \text{voltage rises} = E_{\text{bat}}$$

$$\sum \text{voltage drops} = V_1 + V_2 + V_3 + \cdots + V_N$$

Applying Kirchhoff's voltage law to the circuit in Fig. 4-1.

$$E_{\text{bat}} = V_1 + V_2 + V_3 + \cdots + V_N \tag{4-1}$$

From Ohm's law, the voltage drop across a resistor is equal to the product of the resistance and the current through it. Thus,

$$V_1 = IR_1 \qquad V_2 = IR_2 \qquad V_3 = IR_3 \qquad V_N = IR_N$$

Substituting these IR drops into Eq. (4-1), and factoring out the common current,

$$E_{bat} = IR_1 + IR_2 + IR_3 + \cdots + IR_N$$

$$E_{bat} = I(R_1 + R_2 + R_3 + \cdots + R_N) \tag{4-2}$$

The sum of the series-connected resistors in Eq. (4-2) is called the *equivalent resistance of the series-connected resistors*, and is designated $R_{eq,S}$.

$$\boxed{R_{eq,S} = R_1 + R_2 + R_3 + \cdots + R_N} \tag{4-3}$$

Example 4-1 A motor field circuit, consisting of four coils of wire connected in series, is connected to a 240-V DC driver. Each coil has a resistance of 26 Ω. Sketch the circuit and determine (*a*) the resistance of the field circuit; (*b*) the circuit current; (*c*) the total heat-power loss; (*d*) the voltage drop across each coil; (*e*) the total heat energy expended in 8 h of operation.

Solution
The circuit is shown in Fig. 4-2.

(*a*) $R_{eq,S} = 26 \times 4 = 104 \ \Omega$

(*b*) $I = \dfrac{E}{R} = \dfrac{240}{104} = 2.308$

$\quad I = 2.31$ A

(*c*) $P = I^2 R = (2.308)^2(104) = 553.9$

$\quad P = 554$ W

(*d*) $V = IR = (2.308)(26) = 60$ V

(*e*) $W = Pt = 554(8) = 4432$

$\quad W = 4.43$ kW $\cdot$ h

Figure 4-3*a* shows a series circuit containing three voltage sources and four resistors; the arrows above the sources indicate the directions of the respective driving voltages. Assuming the sum of voltages E_1 and E_2 is greater than voltage E_3, the direction of the net driving voltage, and hence the direction of the current, will be clockwise.

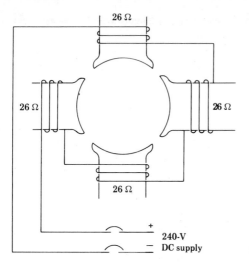

Figure 4-2 Circuit for Example 4-1.

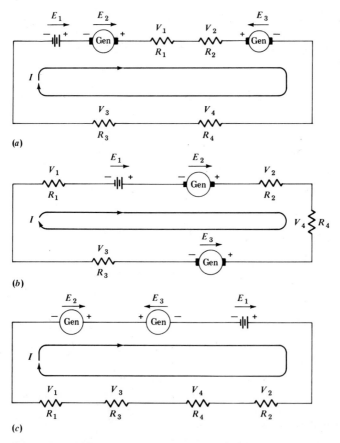

Figure 4-3 Different arrangements of the same components in a
series circuit.

Applying Kirchhoff's voltage law to Fig. 4-3a,

$$E_1 + E_2 - E_3 = V_1 + V_2 + V_3 + V_4 \qquad (4\text{-}4)$$

Expressing the voltage drops in terms of the common current and the respective resistances, and then solving for the current,

$$E_1 + E_2 - E_3 = IR_1 + IR_2 + IR_3 + IR_4$$

$$E_1 + E_2 - E_3 = I(R_1 + R_2 + R_3 + R_4) \qquad (4\text{-}5)$$

$$I = \frac{E_1 + E_2 - E_3}{R_1 + R_2 + R_3 + R_4}$$

Figure 4-3b and c shows different series circuit arrangements of the same components that are in Fig. 4-3a. In each case the magnitude and direction of the net driving voltage, the equivalent series resistance, and the magnitude and direction of the circuit current are identical to the corresponding values obtained for the circuit in Fig. 4-3a. *The equivalent resistance of a series circuit is independent of the location of the sources and the resistors, and the net driving voltage is independent of the arrangement of the generators provided that the directions of the respective driving voltages are not changed.*

Example 4-2 Assume the voltage sources in Fig. 4-3a are 100 V, 200 V, and 125 V for E_1, E_2, and E_3, respectively, and resistors R_1, R_2 R_3, and R_4 are, respectively, 5 Ω, 10 Ω, 20 Ω, and 15 Ω. Determine (a) the current; (b) the voltage drop across each resistor.

Solution
(a) Applying Kirchhoff's voltage law,

$$100 + 200 - 125 = 5I + 10I + 20I + 15I$$

$$175 = 50I$$

$$I = 3.5 \text{ A}$$

(b) $V_1 = IR_1 = 3.5(5) = 17.5$ V

$\quad V_2 = IR_2 = 3.5(10) = 35$ V

$\quad V_3 = IR_3 = 3.5(20) = 70$ V

$\quad V_4 = IR_4 = 3.5(15) = 52.5$ V

4-3 **VOLTAGE DIVIDERS**

A voltage divider, also called a potential divider, is a device used to provide an output voltage that is lower than the input or driving voltage. The common types of adjustable voltage dividers use the three-terminal *rotary-switch* rheostats shown in Fig. 4-4a or the three-terminal *slide-wire* rheostats shown in Fig. 4-4b and c. The individual turns of resistance wire in the slide-wire rheostat correspond to the individual resistors in the rotary-switch rheostat.

A fixed voltage divider uses two or more series-connected resistors whose input and output terminals are connected as shown in Fig. 4-4d.

Theory of the Voltage Divider

The applied driving voltage causes a current that is the same throughout the entire resistance of the voltage-divider circuit. Hence the voltage across any section of the rheostat is equal to the current multiplied by the resistance of that section.

$$V_{section} = IR_{section}$$

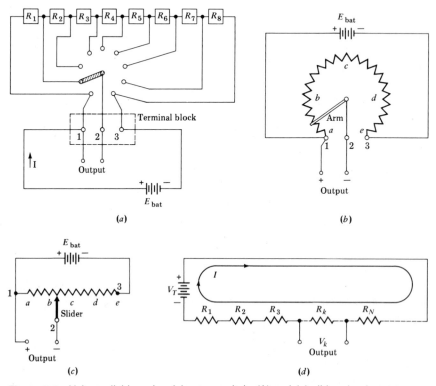

Figure 4-4 Voltage divider using (a) rotary switch; (b) and (c) slide-wire rheostats; (d) fixed resistors.

If it is desired to obtain an output voltage equal to 20 percent of the input voltage, the voltage across the embraced section must be 20 percent of the input voltage. Referring to Fig. 4-4a, the current in the series circuit formed by the rheostat and the battery is

$$I = \frac{E_{bat}}{R_T} \tag{4-6}$$

where R_T = total resistance of all resistors in series

From Eq. (4-6),

$$E_{bat} = IR_T$$

20 percent E_{bat} = 20 percent (IR_T)

$$0.20\ E_{bat} = 0.20\ IR_T$$

$$0.20\ E_{bat} = I\,(0.20\ R_T) \tag{4-7}$$

Hence, as indicated in Eq. (4-7), to obtain an output voltage equal to 20 percent of the input voltage, the resistance of the embraced section must be 20 percent of the total rheostat resistance.

Example 4-3   It is desired to obtain 48 V from a 240-V system using a 1000-Ω rheostat. Determine the required slide setting.

Solution

$$\frac{48}{240} = 0.20\text{ or } 20\text{ percent of the input voltage}$$

Thus, the resistance embraced by the output terminals must be

$$0.20\ (1000) = 200\ \Omega$$

A very useful voltage-divider equation is derived from the circuit shown in Fig. 4-4d. The circuit current, as determined from Ohm's law, is

$$I = \frac{V_T}{R_T} \tag{4-8}$$

where $R_T = R_1 + R_2 + R_3 + \cdots + R_k + \cdots + R_N$
V_T = input voltage

The voltage drop across resistor R_k is

$$V_K = IR_k \tag{4-9}$$

Substituting Eq. (4-8) into Eq. (4-9) results in the following voltage-divider equation:

$$\boxed{V_k = V_T \frac{R_k}{R_T}} \tag{4-10}$$

Example 4-4 (a) Using the voltage-divider equation, determine the voltage between terminals 3 and 4 in Fig. 4-5; (b) determine the voltage between terminals 5 and 7.

Solution

(a) $R_T = 50 + 20 + 70 + 80 + 60 = 280 \ \Omega$

$$V_{3,4} = V_T \frac{R_{3,4}}{R_T} = 9.0 \ \frac{20}{280} = 0.64 \ V$$

(b) $V_{5,7} = V_T \frac{R_{5,7}}{R_T} = 9.0 \ \frac{140}{280} = 4.5 \ V$

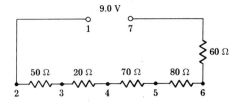

Figure 4-5 Circuit for Example 4-4.

4-4 **PARALLEL CIRCUIT**

If the connections of circuit elements and drivers are such that their respective terminals connect to the same two junctions, as shown in Fig. 4-6a, then the elements and the drivers are in parallel, and the voltage across every paralleled branch is identical. The line connecting the source to the junction is generally called the feeder, and a junction is often called a node. For two or more branches to be in parallel, the endpoints of each branch must be connected directly to the corresponding endpoints of the other branches. For example, in Fig. 4-6b, resistors R_3, R_4, and R_5 are in parallel, and resistors R_6 and R_7 are in parallel. The two groups of paralleled resistors are in series with R_1 and R_2.

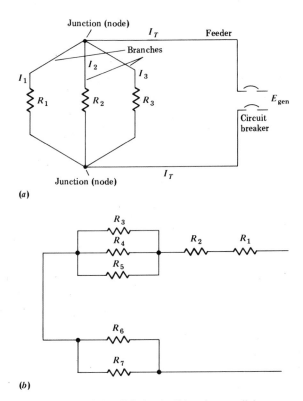

Junction (node)

I_T Feeder

Branches

I_2

I_1 I_3

R_1 R_2 R_3 E_{gen}

Circuit
breaker

I_T

Junction (node)

(a)

R_3

R_4 R_2 R_1

R_5

R_6

R_7

(b)

Figure 4-6   (a) Parallel circuit; (b) series-parallel
connections.

4-5   KIRCHHOFF'S CURRENT LAW

The currents entering and leaving a junction are related to each other by
Kirchhoff's current law, which states:

*The algebraic sum of all the currents going to a junction is equal to the algebraic
sum of all the currents leaving the junction, with all currents considered at the
same instant of time.*

Stated briefly,

$$\sum \text{currents to the junction} = \sum \text{currents leaving the junction}$$

thus, applying Kirchhoff's current law to Fig. 4-6a,

$$I_T = I_1 + I_2 + I_3$$

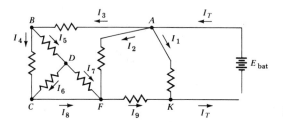

Figure 4-7 Complex circuit.

Kirchhoff's current law is a very simple but powerful tool that has considerable application in the analysis of complex circuits such as that shown in Fig. 4-7. This circuit is neither a series circuit nor a parallel circuit. Applying Kirchhoff's current law to the nodes in Fig. 4-7,

Node A: $I_T = I_1 + I_2 + I_3$

Node B: $I_3 = I_4 + I_5$

Node C: $I_4 + I_6 = I_8$

Node D: $I_5 = I_6 + I_7$

Node F: $I_2 + I_7 + I_8 = I_9$

Node K: $I_9 + I_1 = I_T$

4-6   EQUIVALENT RESISTANCE OF PARALLELED RESISTORS

The equivalent resistance of a group of paralleled resistors is a value of resistance that will draw the same total current from the source as do the paralleled resistors it replaces. For example, if $R_{eq,\,P}$ in Fig. 4-8b is the equivalent resistance of all the paralleled resistors in Fig. 4-8a, and the battery voltages are equal, then the two feeder currents are equal. An expression for the equivalent resistance, in terms of the component paralleled resistances, may be obtained by solving for the feeder current in each of the two circuits, and then equating the currents. Thus, applying Ohm's law to the circuit in Fig. 4-8b,

$$I_T = \frac{E_{bat}}{R_{eq,\,P}} \tag{4-11}$$

Applying Kirchhoff's current law to Fig. 4-8a,

$$I_T = I_1 + I_2 + I_3 + I_4 + \cdots + I_N$$

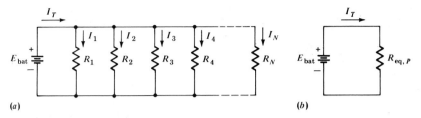

(a)   (b)

Figure 4-8   (a) Parallel circuit; (b) equivalent circuit.

Expressing the branch currents in terms of the common driving voltage and the respective branch resistances,

$$I_T = \frac{E_{bat}}{R_1} + \frac{E_{bat}}{R_2} + \frac{E_{bat}}{R_3} + \frac{E_{bat}}{R_4} + \cdots + \frac{E_{bat}}{R_N}$$

Factoring out E_{bat}

$$I_T = E_{bat}\left(\frac{1}{R_1} + \frac{1}{R_2} + \frac{1}{R_3} + \frac{1}{R_4} + \cdots + \frac{1}{R_N}\right) \qquad (4\text{-}12)$$

Substituting Eq. (4-11) into Eq. (4-12), and simplifying,

$$\frac{E_{bat}}{R_{eq,P}} = E_{bat}\left(\frac{1}{R_1} + \frac{1}{R_2} + \frac{1}{R_3} + \frac{1}{R_4} + \cdots + \frac{1}{R_N}\right)$$

$$\boxed{\frac{1}{R_{eq,P}} = \frac{1}{R_1} + \frac{1}{R_2} + \frac{1}{R_3} + \frac{1}{R_4} + \cdots + \frac{1}{R_N}} \qquad (4\text{-}13)$$

Equation (4-13) provides a means for determining the equivalent resistance of a group of paralleled resistors.

If all paralleled resistances are equal, Eq. (4-13) reduces to the following special-case equation:

$$\frac{1}{R_{eq,P}} = N\frac{1}{R}$$

$$\boxed{R_{eq,P} = \frac{R}{N}} \qquad (4\text{-}14)$$

where R = resistance of each branch
        N = total number of branches

Example 4-5   (*a*) Determine the equivalent resistance of a group of paralleled resistors whose values are 5 kΩ, 10 kΩ, 2 kΩ and 50 kΩ; (*b*) determine the equivalent resistance if the resistors are connected in series.

Solution

(a) $\dfrac{1}{R_{eq}} = \dfrac{1}{5000} + \dfrac{1}{10{,}000} + \dfrac{1}{2000} + \dfrac{1}{50{,}000}$

$\dfrac{1}{R_{eq}} = 200 \times 10^{-6} + 100 \times 10^{-6} + 500 \times 10^{-6} + 20 \times 10^{-6}$

$\dfrac{1}{R_{eq}} = 820 \times 10^{-6}$

$R_{eq} = 1219.5\ \Omega$

(b) $R_{eq} = 5\ \text{k}\Omega + 10\ \text{k}\Omega + 2\ \text{k}\Omega + 50\ \text{k}\Omega$

$R_{eq} = 67\ \text{k}\Omega$

4-7   CONDUCTANCE METHOD FOR PARALLEL-RESISTANCE CALCULATIONS

The conductance method for parallel-resistance calculations was developed to avoid awkward equations such as Eq. (4-13). The conductance method is the recommended method for solving parallel-resistance problems and is particularly advantageous when the circuit includes three or more paralleled branches.

Referring to Fig. 4-9, the total current supplied by the generator is

$$I_P = I_1 + I_2 + I_3 + \cdots + I_N$$

Expressing the branch currents in terms of the common driving voltage and respective branch resistances,

$$I_P = \frac{V_P}{R_1} + \frac{V_P}{R_2} + \frac{V_P}{R_3} + \cdots + \frac{V_P}{R_N}$$

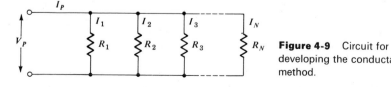

Figure 4-9   Circuit for developing the conductance method.

Factoring out the common driving voltage,

$$I_P = V_P\left(\frac{1}{R_1} + \frac{1}{R_2} + \frac{1}{R_3} + \cdots + \frac{1}{R_N}\right)$$

Substituting the conductance for the reciprocal of the resistance,

$$I_P = V_P(G_1 + G_2 + G_3 + \cdots + G_N) \qquad (4\text{-}15)$$

Defining,

$$\boxed{G_P = G_1 + G_2 + G_3 + \cdots + G_N} \qquad (4\text{-}16)$$

Equation (4-15) becomes

$$\boxed{I_P = V_P G_P} \qquad (4\text{-}17)$$

where I_P = feeder current, A
 V_p = voltage across the paralleled resistors, V
 G_P = total conductance of paralleled resistors, S
 $G_1, G_2, \ldots, G_N$ = conductance of individual branches, S

The conductance of a circuit is a measure of the ease with which the circuit can conduct current. Circuits with higher values of conductance will have higher values of current.

To obtain the equivalent resistance of a group of paralleled resistors, calculate the total conductance and then take the reciprocal to determine $R_{eq,P}$.

$$\boxed{R_{eq,P} = \frac{1}{G_P}} \qquad (4\text{-}18)$$

4-8   **SPECIAL-CASE FORMULA FOR TWO RESISTORS IN PARALLEL**

A simple special-case formula that *applies only to two resistors in parallel* is derived from Eq. (4-16) as follows:

$$G_{P2} = G_A + G_B$$

$$R_{P2} = \frac{1}{G_{P2}} = \frac{1}{G_A + G_B} = \frac{1}{\dfrac{1}{R_A} + \dfrac{1}{R_B}}$$

Simplifying,

$$R_{P2} = \frac{R_A R_B}{R_A + R_B}$$

(4-19)

where R_{P2} is the equivalent resistance of two paralleled resistors.

Example 4-6 (a) Using the special-case formula, determine the equivalent resistance of a 2400-Ω resistor in parallel with a 9100-Ω resistor; (b) determine the equivalent resistance using the admittance method as expressed in Eq. (4-16).

Solution

(a) $R_{p2} = \dfrac{R_A R_B}{R_A + R_B} = \dfrac{2400(9100)}{2400 + 9100} = 1899 \ \Omega$

(b) $G_P = G_1 + G_2 = \dfrac{1}{2400} + \dfrac{1}{9100}$

$\quad G_P = 416.67 \times 10^{-6} + 109.89 \times 10^{-6}$

$\quad G_P = 526.55 \times 10^{-6} \ \text{S}$

$\quad R_{eq,P} = \dfrac{1}{G_P} = \dfrac{1}{526.55 \times 10^{-6}} = 1899 \ \Omega$

4-9 **CURRENT-DIVIDER EQUATION FOR PARALLEL-CONNECTED RESISTORS**
A simple formula for determining the current drawn by any one of many paralleled resistors can be derived from Fig. 4-10. The total current supplied to the parallel circuit is

$$I_P = V_P G_P$$

(4-20)

where

$$G_P = G_1 + G_2 + \cdots + G_k + \cdots + G_N$$

(4-21)

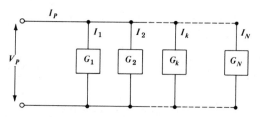

Figure 4-10 Circuit for deriving the current-divider equation.

The current to conductance G_k is

$$I_k = V_P G_k \qquad (4\text{-}22)$$

Solving Eq. (4-20) for V_P and substituting it into Eq. (4-22) results in the current-divider equation

$$\boxed{I_k = I_P \frac{G_k}{G_P}} \qquad (4\text{-}23)$$

Example 4-7 (a) Using the current-divider equation, determine the current in the 6200-Ω resistor in Fig. 4-11; (b) determine the voltage drop across the paralleled section; (c) the current drawn by the 564-Ω resistor; (d) the power drawn by the 6200-Ω resistor.

Solution
(a) The total conductance of the parallel section is

$$G_P = \frac{1}{3920} + \frac{1}{6200} + \frac{1}{564} + \frac{1}{7500}$$

$$G_P = 255.1 \times 10^{-6} + 161.3 \times 10^{-6} + 1773.0 \times 10^{-6} + 133.3 \times 10^{-6}$$

$$G_P = 2322.7 \times 10^{-6}\ \text{S}$$

$$I_k = I_P \frac{G_k}{G_P} = 0.560 \frac{161.3 \times 10^{-6}}{2322.7 \times 10^{-6}}$$

$$I_k = 38.9 \times 10^{-3}\ \text{A} = 38.9\ \text{mA}$$

(b) $I_P = V_P G_P$

$$0.560 = V_P(2322.7 \times 10^{-6})$$

$$V_P = 241\ \text{V}$$

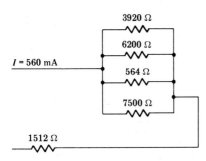

Figure 4-11 Circuit for Example 4-7.

(c) Since the voltage across the parallel section is known, the current through the 564-Ω resistor may be determined by Ohm's law:

$$I = \frac{V}{R} = \frac{241}{564} = 0.427 \text{ A} = 427 \text{ mA}$$

(d) $P = I^2 R = (0.0389)^2(6200)$

$$P = 9.38 \text{ W}$$

4-10   **CIRCUIT FAULTS†**

Circuit faults generally fall in one of the following categories: short circuits, flashovers, grounds, opens, and corona.

Short Circuit

A short circuit, commonly called a short, is an electrical connection that bypasses part or all of an electric circuit, thus providing an additional route for the current. Figure 4-12 illustrates some examples of undesirable shorts. Figure 4-12a shows a *solid short*, where a solid metallic connection across the lamp may result in a blown fuse or tripped circuit breaker. Figure 4-12b shows a *partial short* between two commutator bars, where an accumulated mixture of carbon dust and oil vapors formed a conducting path; it is called a partial short, or *high-resistance short*, because the resistance of the con- ducting path is higher than that for a metal-to-metal contact. Figure 4-12c shows a shorted turn of wire in a field coil; the insulation, worn away through vibration, wind erosion, or excess heat, caused adjacent wires to make contact through the medium of a dirt path. If electric conduction between adjacent wires of the coil is due to dirt, it is called a high-resistance or partial short; if due to metal-to-metal contact, it is called a *dead short* or *solid short*.

Example 4-8 A 240-V DC generator supplies power through a 300-m cable to a heating load whose resistance is 5.76 Ω. The two conductors that make up the cable are each size AWG 8 copper wire. Determine (a) the current in the cable; (b) the voltage drop in the cable; (c) the power loss in the cable; (d) the power drawn by the load; (e) the current in the cable if an accidental solid short occurs across the heater terminals.

Solution
(a) The circuit for part a is shown in Fig. 4-13a. From Table 2-6, the resistance of AWG 8 copper wire is 2.061 Ω/km.

$$R_{cable} = 2(0.300)(2.061) = 1.237 \ \Omega$$

---

† For a detailed analysis of specific faults, see C. I. Hubert, *Preventive Maintenance of Electrical Equipment*, McGraw-Hill Book Company, New York, 1969. This section reproduced with permission.

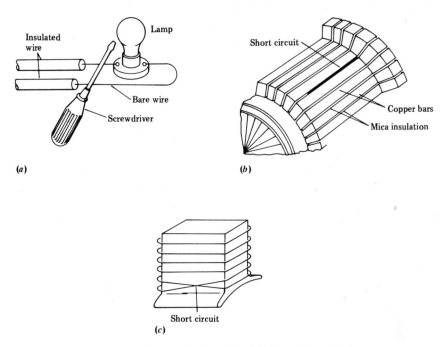

Figure 4-12 Examples of short circuits: (a) solid short; (b) partial short; (c) shorted turn of wire in a field coil.

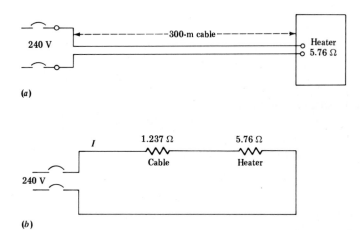

Figure 4-13 (a) Circuit for Example 4-8; (b) equivalent series circuit.

An equivalent series circuit model representing Fig. 4-13*a* is shown in Fig. 4-13*b*. The total resistance of the series circuit is

$$R_{eq,S} = 1.237 + 5.76 = 6.997 \; \Omega$$

Applying Ohm's law,

$$I = \frac{V}{R} = \frac{240}{6.997} = 34.3 \text{ A}$$

(*b*) $V_{cable} = IR_{cable} = 34.3(1.237) = 42.4$ V

(*c*) $P_{cable} = I^2 R_{cable} = (34.3)^2(1.237) = 1455$ W

(*d*) $P_{load} = I^2 R_{load} = (34.3)^2(5.76) = 6777$ W

(*e*) A solid short across the heater terminals bypasses the heater, leaving only the cable resistance to limit the current.

$$I = \frac{V}{R} = \frac{240}{1.237} = 194 \text{ A}$$

Flashover

A flashover is a violent disruptive discharge around or across the surface of a solid or liquid insulator. Flashovers occur suddenly, involve heavy currents, and generally cause considerable damage. A flashover is always preceded by ionization, which is a process by which the surrounding atmosphere is made into a conductor. This process may be caused by creepage paths developed across dirty and moist insulators; very high temperatures, such as those caused by a nearby arc; switching overvoltages; or lightning. Although the current through creepage paths may not be appreciable, the arcing and sparking produced can ionize the surrounding atmosphere and cause a flashover. Electric arcs are highly mobile and can travel away from the initial striking point, causing extensive damage to nearby equipment. Furthermore, when the arcing occurs in metal-enclosed switchgear, the arc can be expected to spread to ground.

The shortest distance between two bare conductors of opposite polarity or between a conductor and ground, measured along the surface of an insulating material, is called the *creepage distance*.

Ground Fault

A ground fault is an accidental electrical connection between the wiring of an apparatus and its metal framework or enclosure, the wiring and a water pipe, the wiring and the armored sheath of a cable, the wiring and the chassis

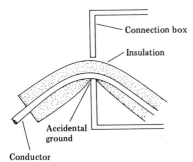

Connection box

Insulation

Accidental ground

Conductor

Figure 4-14 Accidental ground.

of a car, the wiring and the hull of a ship, etc. Figure 4-14 illustrates an accidental ground caused by the chafing of the insulation against the sharp edge in the opening of a connection box. Ground faults can also occur in electrical machinery when the effects of age, heat, and vibration damage the insulation, permitting the entry of conducting dust; if the dust forms a bridge between the exposed conductor and the frame, the wiring is grounded. Apparatus exposed to atmospheric conditions, even though totally enclosed and "weatherproofed," sometimes develops ground faults due to breathing. During the heat of the day the air inside the apparatus expands, forcing some internal air through the seals to the outside. In the evening, when the apparatus cools, a partial vacuum is formed inside the apparatus, and cool moist air is sucked in. The moisture condenses inside the apparatus, and after many heating and cooling cycles, a small pool of water begins to form. Eventually enough water may collect to make an electrical connection between the wiring and the frame, grounding the equipment.

Bonding

Bonding is the electrical interconnecting of metallic conduit, metal armor or metal sheath of adjacent cables, the framework of electrical apparatus, metallic connection boxes, metallic framework of switchboards, and metallic water pipes. Bonding and grounding to earth through metallic water pipes or other means reduce shock hazards to personnel and minimize electrostatic noise in electronic equipment.

Open

An open, or open circuit, is a break in the continuity of the circuit. This break can be deliberate or accidental. Deliberate opens are made by the manual opening of a switch or circuit breaker. Accidental opens may be caused by the blowing of a fuse, the melting of a soldered connection, the blowing out of a lamp, breaking of a wire or resistor due to vibration, etc.

Corona

Corona is a luminous, and sometimes audible, electric discharge caused by ionization of the atmosphere surrounding high-voltage conductors. Its most harmful effect is the production of ozone and nitrogen oxides, which cause chemical deterioration of the organic materials used in insulation. Ozone, a form of oxygen (O_3), is an extremely powerful oxidizing agent and, in addition to damaging insulation, readily oxidizes such metals as copper and iron.

Another adverse effect of corona is severe local heating of insulation resulting in carbonization (called *corona burning*) that produces leakage tracks (paths) in organic materials. The telltale marks of corona discharge on varnished insulation are usually in the form of white or gray dust spots. Each spot, which occurs at a point of high-voltage stress, has a matching spot on an opposite surface.

A particularly annoying problem associated with corona is radio and television interference.

4-11   **EFFECT OF INTERNAL RESISTANCE ON THE OUTPUT OF A VOLTAGE SOURCE**

All voltage sources have internal resistance, called source resistance, that causes a drop in output voltage when the source is supplying current to a circuit.

If the resistance of the voltage source is negligible compared with the resistance of the circuit, the resistance of the source may be omitted without introducing any significant error. However, in those applications where the resistance of the source is not negligible, it must be accounted for in the circuit analysis.

To account for source resistance, each generator or battery is replaced in the circuit diagram with an *equivalent-series-circuit model* that includes an

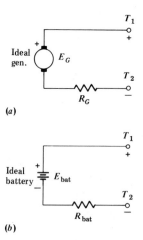

(*a*)

(*b*)

Figure 4-15   Equivalent-series-circuit model of (*a*) generator; (*b*) battery.

ideal constant-voltage generator or *ideal constant-voltage battery* in series with the respective source resistance. This is shown in Fig. 4-15, where T_1 and T_2 represent the output terminals of the equivalent-circuit model.

The ideal generator, or ideal battery, is an imaginary voltage source that maintains a constant voltage regardless of the amount of current that it supplies to the circuit. The source resistance accounts for the drop in output voltage when the ideal voltage source supplies current to the circuit.

The voltage assigned to the ideal generator or ideal battery is the *open-circuit voltage* of the actual generator or battery that it replaces; it is the voltage at the terminals of the actual source when the source is not supplying current.

Example 4-9 A 12-V DC source with an internal resistance of 2 Ω is connected through a switch to an 18-Ω load as shown in Fig. 4-16a. The voltmeter connected across terminals T_1, T_2 is a digital voltmeter (DVM) whose internal resistance of 2 MΩ draws negligible current from the source. (*a*) With the switch closed, as shown in Fig. 4-16a, determine the current to the load, the voltmeter reading, and the voltage drop within the source; (*b*) with the switch open, as shown in Fig. 4-16b, determine the voltmeter reading.

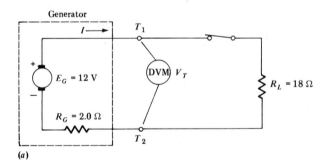

(*a*)

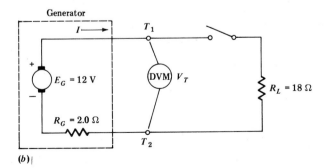

(*b*)

Figure 4-16 Circuits for Example 4-9.

Solution

(a) Since the DVM draws insignificant current, it can be neglected when making circuit calculations. Applying Ohm's law to the series circuit,

$$I = \frac{V}{R} = \frac{12}{2 + 18} = 0.60 \text{ A}$$

Or, using Kirchhoff's voltage law,

$$\sum \text{voltage rises} = \sum \text{voltage drops}$$

$$12 = 18I + 2I$$

$$I = 0.60 \text{ A}$$

The voltmeter reads the voltage drop across the T_1, T_2 terminals, which is equal to the IR drop across the 18-Ω resistor. Thus

$$V_T = (0.60)(18) = 10.8 \text{ V}$$

The voltage drop within the source is

$$V_{\text{drop, source}} = (0.60)(2) = 1.2 \text{ V}$$

(b) With the switch open, the circuit reduces to that of Fig. 4-16b. Applying Kirchhoff's voltage law to the loop,

$$\sum \text{voltage rises} = \sum \text{voltage drops}$$

The only voltage rise is that of the 12-V source; the voltage drops are the IR drop in the 2-Ω source resistance and the voltage drop across the 2 MΩ of the voltmeter.

$$12 = 2I + 2 \times 10^6 I$$

$$I = 6 \times 10^{-6} \text{ A} = 6 \, \mu\text{A}$$

The voltage drop across the DVM is

$$V = IR = 6 \times 10^{-6}(2 \times 10^6) = 12 \text{ V}$$

Note: The calculations for Fig. 4-16b can be simplified if we correctly assume that the current drawn by the DVM is negligible. Using this assumption, and applying Kirchhoff's voltage law,

$$12 = 2I + V_T$$

where V_T = voltage drop across the DVM

With

$I \approx 0$,

$12 = 2(0) + V_T$

$V_T = 12 \text{ V}$

As demonstrated in this example, the open-circuit voltage at terminals T_1, T_2 is 12 V, and the voltage with the switch closed is 10.8 V.

4-12   VOLTAGE-CURRENT CHARACTERISTIC OF A VOLTAGE SOURCE

The voltage-current characteristic of a source is a plot of voltage versus current at the terminals of the source. Referring to Fig. 4-17, the volt-ampere characteristic is a plot of V_T versus I_T as the rheostat resistance is adjusted from maximum to minimum resistance. The equation representing the V_T versus I_T characteristic may be derived by applying Kirchhoff's voltage law to the circuit and then solving for V_T. Thus,

$$\sum \text{voltage rises} = \sum \text{voltage drops}$$

$$E_G = V_T + I_T R_G$$

$$\boxed{V_T = E_G - I_T R_G} \tag{4-24}$$

where V_T = voltage at terminals of source, V
I_T = current from source, A
E_G = open-circuit voltage of source, V
R_G = source resistance, Ω

Example 4-10   Plot the voltage-current characteristic of a source that has an open-circuit voltage of 24 V, a source resistance of 0.20 Ω, and a current rating of 30 A.

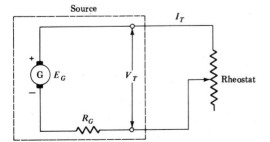

Figure 4-17   Circuit for determining the voltage-current characteristic of a voltage source.

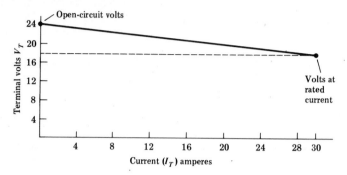

Figure 4-18 Characteristic curve for Example 4-10.

Solution
Substituting into Eq. (4-24),

$$V_T = 24 - 0.20 I_T$$

Plotting V_T versus I_T as I_T increases from zero to 30 A results in the characteristic shown in Fig. 4-18.

4-13   VOLTAGE REGULATION OF A VOLTAGE SOURCE

The voltage regulation of a voltage source is defined as the percent change in terminal voltage from no load to full load with respect to the full-load value. Expressed mathematically,

$$\text{Percent Reg.} = \frac{V_{NL} - V_{FL}}{V_{FL}} \times 100 \qquad (4\text{-}25)$$

where V_{NL} = terminal voltage at no load
V_{FL} = terminal voltage at full load

A knowledge of voltage regulation is a necessary consideration when selecting DC generators for operation in parallel. Generators with the same voltage and kilowatt ratings but different voltage regulation will not take equal shares of the load current.

Voltage regulation is a necessary consideration when designing or applying voltage sources to transistor circuits. Power supplies with too high a regulation may cause damage to a transistor when the transistor is not carrying current. The higher output voltage of the source caused by a drop in current may exceed the voltage rating of the transistor.

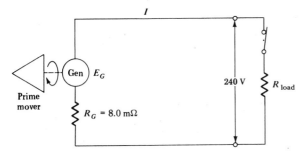

Figure 4-19   Circuit for Example 4-11.

Example 4-11   A certain DC generator, shown in Fig. 4-19, has a terminal voltage of 240 V when delivering its rated 200 kW to a load. The internal resistance of the generator is 8.00 mΩ. Determine (*a*) the rated current (full-load current); (*b*) the open-circuit voltage; (*c*) the voltage regulation of the generator.

Solution

(*a*) $P = VI$

   $200,000 = 240\ I$

   $I = 833.3\ \text{A}$

(*b*) The open-circuit voltage is E_G. Applying Kirchoff's voltage law,

$E_G = 240 + 0.008(833.3)$

$E_G = 246.7\ \text{V}$

Percent Reg. $= \dfrac{V_{NL} - V_{FL}}{V_{FL}} \times 100$

Percent Reg. $= \dfrac{246.7 - 240}{240} \times 100 = 2.79\ \text{percent}$

4-14   **MAXIMUM-POWER-TRANSFER THEOREM**

In low-energy circuits, such as in very low level generators or amplifiers, the amount of energy transferred to the load is more important than the overall efficiency of the operation. The maximum-power-transfer theorem states that *for maximum power to be transferred from a source to a load, the load resistance*

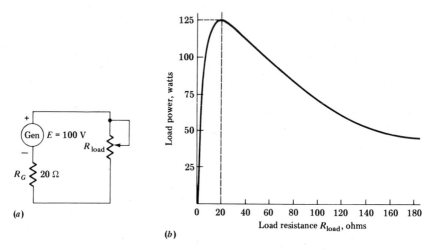

Figure 4-20 (a) Generator supplying power to an adjustable resistance load; (b) curve of load power versus load resistance.

must be equal to the source resistance. This phenomenon can be explained with the aid of Fig. 4-20a, which shows a generator supplying power to an adjustable resistance load. The current to the load is

$$I_L = \frac{E_G}{R_G + R_L} \tag{4-26}$$

The power drawn by the load is

$$P_L = I^2 R_L \tag{4-27}$$

Substituting Eq. (4-26) into Eq. (4-27) and simplifying,

$$P_L = \left(\frac{E_G}{R_G + R_L}\right)^2 R_L$$

$$P_L = \frac{E_G^2 R_L}{(R_G + R_L)^2} \tag{4-28}$$

A plot of Eq. (4-28) for values of R_L ranging from zero to 180 Ω is shown in Fig. 4-20b; the assumed values for E_G and R_G are 100 V and 20 Ω, respectively. The plot indicates that maximum power delivery occurs when the load resistance is equal to the source resistance.

Mathematical Proof for the General Case of the Maximum-Power-Transfer Theorem

The relationship that corresponds to the general case for any DC source and a resistance load is given in Eq. (4-28). The general shape of the curve corresponding to Eq. (4-28) is similar to that shown in Fig. 4-20b. The power curve rises to some peak value and then falls off with increasing values of load resistance. As indicated, the slope of the curve is zero when the power delivered to the load reaches its maximum value.

The value of resistance at which the slope is zero may be obtained by taking the derivative of Eq. (4-28) with respect to R_L, equating it to zero, and then solving for R_L. Thus, rearranging terms in Eq. (4-28) and differentiating,

$$P_L = E_G^2 R_L (R_G + R_L)^{-2}$$

$$\frac{dP_L}{dR_L} = E_G^2 [R_L(-2)(R_G + R_L)^{-3} + (R_G + R_L)^{-2}] = 0$$

$$-2R_L + R_G + R_L = 0$$

$$R_L = R_G$$

SUMMARY OF FORMULAS

$$R_{eq,S} = R_1 + R_2 + R_3 + \cdots + R_N$$

$$V_k = V_T \frac{R_k}{R_T}$$

$$\frac{1}{R_{eq,P}} = \frac{1}{R_1} + \frac{1}{R_2} + \frac{1}{R_3} + \cdots + \frac{1}{R_N}$$

$$G_P = G_1 + G_2 + G_3 + \cdots + G_N$$

$$R_{eq,P} = \frac{1}{G_P}$$

$$R_{eq,P2} = \frac{R_A R_B}{R_A + R_B}$$

$$I_k = I_P \frac{G_k}{G_P}$$

$$\text{Percent Reg.} = \frac{V_{NL} - V_{FL}}{V_{FL}} \times 100$$

PROBLEMS

4-1 Determine the equivalent resistance of the following series-connected resistors: 6 Ω, 10 Ω, 5 Ω, 30 Ω.

4-2 The equivalent resistance of six series-connected resistors is 10 kΩ. If the values of five of the resistors are 1000 Ω, 2400 Ω, 1800 Ω, 4000 Ω, and 500 Ω, respectively, determine the resistance of the sixth resistor.

4-3 A 120-V source of negligible resistance is connected in series with the following resistors: 20 Ω, 200 Ω, 2000 Ω, and 20 kΩ. Determine (*a*) the current; (*b*) the voltage drop across each resistor; (*c*) the power supplied by the source.

4-4 Referring to the circuit in Fig. 4-21, determine (*a*) the current; (*b*) the voltage drop across the 6-Ω resistor; (*c*) the power drawn by the 6-Ω resistor; (*d*) the energy expended in the 6-Ω resistor if the circuit was energized for 20 h; (*e*) the current if the 24-V generator is reversed.

4-5 Determine the slide setting required on a 4000-Ω rheostat in order to get a 9-V output from a 12-V battery. Sketch the circuit.

4-6 A 10,000-Ω slide-wire potential divider rheostat is adjusted so that the output terminals embrace 8000 Ω. Sketch the circuit and determine the input voltage required to obtain a 50-V output.

4-7 Using the voltage-divider equation, determine the voltage between points *c* and *d* in Fig. 4-22.

4-8 Determine the equivalent resistance of the following parallel-connected resistors: 6 Ω, 10 Ω, 5 Ω, 30 Ω.

4-9 The equivalent resistance of six paralleled resistors is 0.45 Ω. If five of the resistors are 3 Ω, 4 Ω, 2 Ω, 8 Ω, and 6 Ω, respectively, determine the resistance of the sixth resistor.

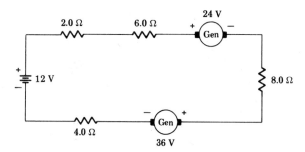

Figure 4-21 Circuit for Prob. 4-4.

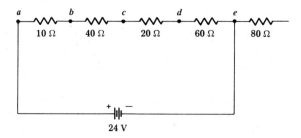

Figure 4-22 Circuit for Prob. 4-7.

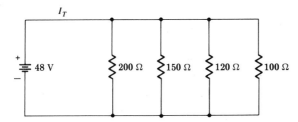

Figure 4-23   Circuit for Prob. 4-10.

4-10 (*a*) Using the conductance method, determine the battery current in Fig. 4.23; (*b*) determine the power drawn by the 150-Ω resistor.

4-11   Using the special-case formula, determine the equivalent resistance of each of the following sets of paralleled resistors: (*a*) 100 Ω, 80 Ω; (*b*) 5 Ω, 500 Ω; (*c*) 2 kΩ, 2 kΩ; (*d*) 0.01 Ω, 100 Ω.

4-12   Referring to Fig. 4-24, determine (*a*) the current in the 40-Ω resistor; (*b*) the heat power expended in the 40-Ω resistor; (*c*) the voltage drop across the 20-Ω resistor; (*d*) the voltage drop across the parallel section.

4-13   Twenty 12-Ω resistors are connected in parallel, and the combination is connected to a 40-V source. Determine (*a*) the equivalent parallel resistance; (*b*) the total current supplied by the driver.

4-14   A 120-V DC generator supplies power to an electromagnet through a 150-m cable. The two conductors that make up the cable are each size AWG 6 copper wire. The resistance of the electromagnet is 6.2 Ω. Sketch the circuit and determine (*a*) the current in the magnet; (*b*) the voltage drop in the cable; (*c*) the voltage across the magnet; (*d*) the power loss in the cable; (*e*) the current in the cable if an accidental solid short occurs across the magnet terminals.

4-15   A 50-V DC generator with an internal resistance of 0.26 Ω is connected to a 12-Ω resistor through a switch. Sketch the circuit and determine the voltage at the generator terminals with (*a*) the switch closed; (*b*) the switch open; (*c*) the percent voltage regulation of the source.

4-16   A certain DC generator is delivering rated 100 A at 200 V to a load. The resistance of the generator is 0.2 Ω. Sketch the circuit, showing a switch between the generator and the load, and determine (*a*) the voltage drop in the generator; (*b*) the open-circuit voltage; (*c*) the voltage regulation.

4-17   An audio amplifier acts as a source when it supplies power to a speaker. If the resistance of the amplifier is 8 Ω, what should be the resistance of the speaker in order that maximum power may be delivered?

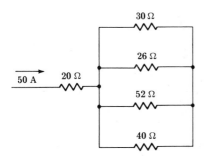

Figure 4-24   Circuit for Prob. 4-12.

CHAPTER 5

SERIES-PARALLEL RESISTIVE NETWORKS WITH DC DRIVERS

This chapter introduces circuits that combine series sections and parallel sections in various network configurations. Although the terms network and circuit are synonymous, circuits that are more complex than the simple series and simple parallel circuits discussed in Chap. 4 are generally called networks.

5-1   INPUT RESISTANCE OF SERIES-PARALLEL CIRCUITS

A series-parallel circuit such as that shown in Fig. 5-1a is a combination of series sections and parallel sections. The equivalent resistance of such configurations may be obtained by using techniques developed for the simple series and simple parallel circuits. The equivalent resistance of a circuit *containing one driver*, whether series, parallel, series-parallel, or some other configuration, is called the *input resistance* R_{in} or *driving-point resistance*. It is the resistance as viewed from the source (driver).

The solution of problems in series-parallel circuits generally requires the use of conductance-resistance conversions. To reduce the risk of error, such calculations should be carried out to at least four decimal places.

Example 5-1 Referring to the circuit diagram in Fig. 5-1a, determine (a) the input resistance; (b) the feeder current; (c) the current in the 15-Ω resistor; (d) the voltage drop across the 11-Ω resistor; (e) the power drawn by the 11-Ω resistor.

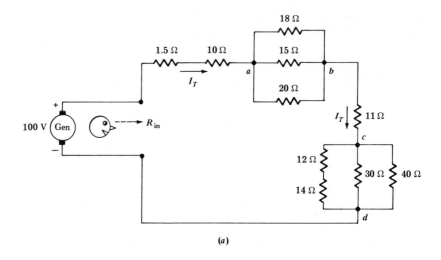

(a)

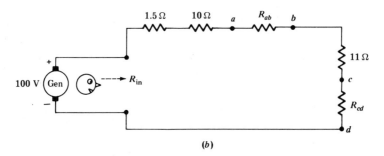

(b)

Figure 5-1 Circuits for Example 5-1.

Solution
(a) The equivalent resistance of the paralleled sections is determined and substituted into the circuit as shown in Fig. 5-1b; the equivalent resistance of the resultant series circuit is the input resistance of the series-parallel circuit. Thus, referring to Fig. 5-1a, for parallel section *ab*,

$$G_{ab} = \frac{1}{18} + \frac{1}{15} + \frac{1}{20}$$

$$G_{ab} = 0.0556 + 0.0667 + 0.0500 = 0.1723 \text{ S}$$

$$R_{ab} = \frac{1}{G_{ab}} = \frac{1}{0.1723} = 5.8 \ \Omega$$

For parallel section cd, the 12-Ω and 14-Ω resistors in series total $12 + 14 = 26\ \Omega$. The conductance of the parallel combination of 26 Ω, 30 Ω, and 40 Ω is

$$G_{cd} = \frac{1}{26} + \frac{1}{30} + \frac{1}{40}$$

$$G_{cd} = 0.0385 + 0.0333 + 0.0250 = 0.0968\ \text{S}$$

$$R_{cd} = \frac{1}{G_{cd}} = \frac{1}{0.0968} = 10.3\ \Omega$$

Substituting R_{ab} and R_{cd} for the respective parallel sections results in the equivalent series circuit shown in Fig. 5-1b. Thus the overall resistance of the circuit, called the input resistance or driving-point resistance, is

$$R_{in} = 1.5 + 10 + 5.8 + 11 + 10.3 = 38.6\ \Omega$$

(b) $I_T = \dfrac{V}{R_{in}} = \dfrac{100}{38.6} = 2.59\ \text{A}$

(c) Using the current-divider equation,

$$I_{15} = I_T \frac{G_{15}}{G_{ab}} = 2.59\,\frac{0.0667}{0.1723} = 1.00\ \text{A}$$

(d) $V_{11} = I_{11}R_{11} = 2.59 \times 11 = 28.5\ \text{V}$

(e) Using three different methods,

$$P_{11} = V_{11}I_{11} = 28.5(2.59) = 73.8\ \text{W}$$

$$P_{11} = I_{11}^2 R_{11} = (2.59)^2 11 = 73.8\ \text{W}$$

$$P_{11} = \frac{V_{11}^2}{R_{11}} = \frac{(28.5)^2}{11} = 73.8\ \text{W}$$

5-2   **LADDER NETWORKS**

Ladder networks such as that shown in Fig. 5-2a have alternate series and parallel elements. They are called ladder networks because of their geometry; they look like ladders.

To determine the input resistance of a ladder network, *start at the section of the network farthest from the driver, combining elements as you progress toward the input terminals.*

Example 5-2   Referring to Fig. 5-2a, determine (a) the driving-point impedance; (b) feeder current; (c) the voltage drop across the 60-Ω resistor.

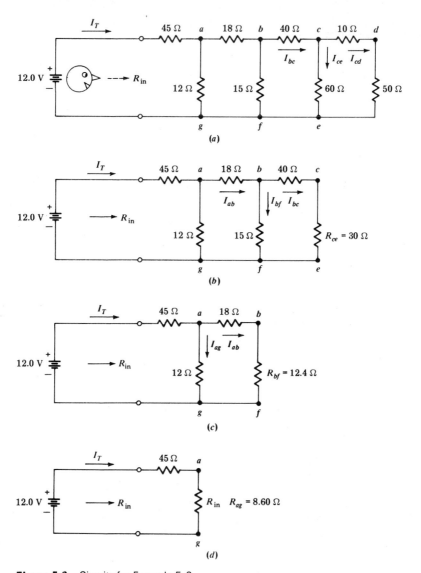

Figure 5-2 Circuits for Example 5-2.

Solution

(a) Starting at the section farthest from the driver, the series section composed of a 10-Ω resistor and a 50-Ω resistor is in parallel with the 60-Ω resistor. Using the special-case formula for two resistors in parallel,

$$R_{ce} = \frac{(10 + 50)(60)}{(10 + 50) + 60} = 30 \ \Omega$$

Substituting 30 Ω for the respective series-parallel section in Fig. 5-2*a* results in Fig. 5-2*b*.

Using the same procedure, the series section composed of the 40-Ω resistor and the 30-Ω resistor in Fig. 5-2*b* is in parallel with the 15-Ω resistor. Thus,

$$R_{bf} = \frac{(40 + 30)(15)}{(40 + 30) + 15} = 12.4\ \Omega$$

Substituting 12.4 Ω for the respective series-parallel section in Fig. 5-2*b* results in Fig. 5-2*c*.

Continuing, the series section composed of the 18-Ω resistor and the 12.4-Ω resistor in Fig. 5-2*c* is connected in parallel with the 12-Ω resistor. Thus,

$$R_{ag} = \frac{(18 + 12.4)(12)}{(18 + 12.4) + 12} = 8.60\ \Omega$$

Substituting 8.60 Ω for the respective series-parallel section in Fig. 5-2*c* results in Fig. 5-2*d*. Referring to Fig. 5-2*d*,

$$R_{in} = 45 + 8.6 = 53.6\ \Omega$$

(*b*) $I_T = \dfrac{V}{R} = \dfrac{12.0}{53.6} = 0.224\ \text{A} = 224\ \text{mA}$

(*c*) To determine the voltage drop (*IR* drop) across the 60-Ω resistor, it is necessary to determine the current through it. Although there are several methods for attacking the problem, a very straightforward method is to apply Kirchhoff's current law to each node in turn, starting from the node nearest the driver and working out to the node that is directly connected to the 60-Ω branch (node *c* in Fig. 5-2*a*). Applying Kirchhoff's current law to node *a* in Fig. 5-2*c*,

$$I_T = I_{ag} + I_{ab}$$

where I_{ag} = current in branch *ag*
I_{ab} = current in branch *abf*

Using the current-divider equation,

$$I_{ab} = I_T \frac{G_{abf}}{G_P}$$

$$G_P = G_{ag} + G_{abf} = \frac{1}{12} + \frac{1}{18 + 12.4}$$

$$G_P = 0.0833 + 0.0329 = 0.1162\ \text{S}$$

$$G_{abf} = 0.0329\ \text{S}$$

$$I_{ab} = 0.224\ \frac{0.0329}{0.1162} = 0.0634\ \text{A}$$

Applying Kirchhoff's current law to node b in Fig. (5-2b),

$$I_{ab} = I_{bf} + I_{bc}$$

where I_{bf} = current in branch bf
I_{bc} = current in branch bce

Using the current-divider equation,

$$I_{bc} = I_{ab} \frac{G_{bce}}{G_P}$$

$$G_P = G_{bf} + G_{bce} = \frac{1}{15} + \frac{1}{40 + 30}$$

$$G_P = 0.0667 + 0.0143 = 0.0810 \text{ S}$$

$$I_{bc} = 0.0634 \frac{0.0143}{0.0810} = 0.0112 \text{ A}$$

Applying Kirchhoff's current law to node c in Fig. 5-2a,

$$I_{bc} = I_{ce} + I_{cd}$$

where I_{ce} = current in branch ce
I_{cd} = current in branch cde

Using the current-divider equation,

$$I_{ce} = I_{bc} \frac{G_{ce}}{G_P}$$

$$G_P = G_{ce} + G_{cde} = \frac{1}{60} + \frac{1}{10 + 50}$$

$$G_p = 0.0167 + 0.0167 = 0.0334$$

$$G_{ce} = 0.0167 \text{ S}$$

$$I_{ce} = 0.0112 \frac{0.0167}{0.0334} = 0.00560 \text{ A}$$

Hence the voltage drop across the 60-Ω resistor is

$$V_{60\Omega} = I_{60\Omega} R_{60\Omega} = 0.00560 \times 60 = 0.336 \text{ V}$$

$$V_{60\Omega} = 336 \text{ mV}$$

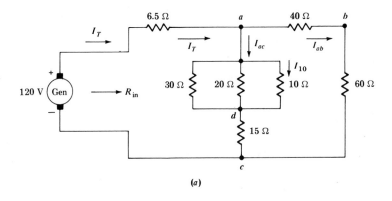

(a)

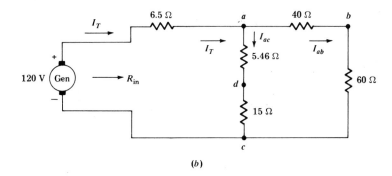

(b)

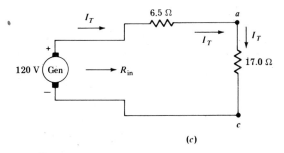

(c)

Figure 5-3 Circuits for Example 5-3.

Example 5-3 For the circuit shown in Fig. 5-3a, determine (a) the input resistance; (b) the feeder current; (c) the voltage drop across the 10-Ω resistor.

Solution
Although this circuit is not a ladder network, the principles used in the solution of ladder-network problems are applicable. (a) The circuit is first simplified by replacing

each of the three parallel resistors with its respective equivalent resistance. Using the conductance method,

$$G_P = G_{30} + G_{20} + G_{10} = \frac{1}{30} + \frac{1}{20} + \frac{1}{10}$$

$$G_P = 0.0333 + 0.0500 + 0.1000 = 0.1833 \text{ S}$$

$$R_{eq, P} = \frac{1}{G_P} = \frac{1}{0.1833} = 5.46 \ \Omega$$

Substituting 5.46 Ω for the section of paralleled resistors in Fig. 5-3a reduces the network to that shown in Fig. 5-3b.

Combining the two paralleled branches in Fig. 5-3b,

$$G_P = G_{adc} + G_{abc}$$

$$G_P = \frac{1}{5.46 + 15} + \frac{1}{40 + 60}$$

$$G_P = 0.0489 + 0.0100 = 0.0589 \text{ S}$$

$$R_{eq, P} = \frac{1}{G_P} = \frac{1}{0.0589} = 17.0 \ \Omega$$

Substituting 17.0 Ω for the paralleled branches in Fig. 5-3b results in the equivalent series circuit shown in Fig. 5-3c. The input resistance, as determined from Fig. 5-3c, is

$$R_{in} = 6.5 + 17.0 = 23.5 \ \Omega$$

(b) The feeder current is

$$I_T = \frac{V}{R_{in}} = \frac{120}{23.5} = 5.11 \text{ A}$$

(c) To obtain the voltage drop (IR drop) across the 10-Ω resistor in Fig. 5-3a, it is necessary to determine the current through it. Applying the current-divider equation to the three paralleled resistors,

$$I_{10} = I_{ac} \frac{G_{10}}{G_P} \tag{5-1}$$

Conductances G_P and G_{10} for the *three* paralleled branches were determined in part a to be 0.1833 S and 0.1000 S, respectively. Substituting these values into Eq. (5-1),

$$I_{10} = I_{ac} \frac{0.1000}{0.1833}$$

$$I_{10} = I_{ac}(0.5455) \tag{5-2}$$

Current I_{ac} may be calculated by applying the current-divider equation to the two paralleled branches in Fig. 5-3b.

$$I_{ac} = I_T \frac{G_{adc}}{G_P} \tag{5.3}$$

Conductances G_P and G_{adc} for the *two* paralleled branches were determined in part *a* to be 0.0589 S and 0.0489 S, respectively.

Substituting the known values into Eq. (5-3), and evaluating,

$$I_{ac} = 5.11 \frac{0.0489}{0.0589} = 4.24 \text{ A}$$

Substituting 4.24 for I_{ac} in Eq. (5-2),

$$I_{10} = 4.24(0.5455) = 2.31 \text{ A}$$

$$V_{10} = I_{10} R_{10} = 2.31(10) = 23.1 \text{ V}$$

5-3   **DETERMINATION OF THE POTENTIAL DIFFERENCE BETWEEN ANY TWO POINTS IN A CIRCUIT**

The potential difference, or voltage drop, between two points (A and B) in an electric circuit may be determined by "traveling" *any path* connecting the two points, *subtracting the voltage drops encountered along the path from the driving voltages along the same path.* That is,

$$\boxed{V_{AB} = (\textstyle\sum \text{ driving voltages}) - (\textstyle\sum \text{ voltage drops})} \tag{5-4}$$

Equation (5-4) is called the *potential-difference equation* or PD equation, and is merely a rewording of Kirchhoff's current law.

When driving voltages and voltage drops are entered in the potential-difference equation, a driving voltage is multiplied by (-1) if its respective directional arrow is opposite to the direction of travel, and a voltage drop (IR) is multiplied by (-1) if its respective branch-current arrow is opposite to the direction of travel.

Example 5-4 Figure 5-4 shows a simple series-parallel circuit with the directions and magnitudes of the currents indicated on the diagram. Using the potential-difference equation, determine the potential difference, or voltmeter reading, between points A and B. Assume the voltmeter draws negligible current.

Solution
The problem will be solved by taking a short path, and then solved again by taking a long path.

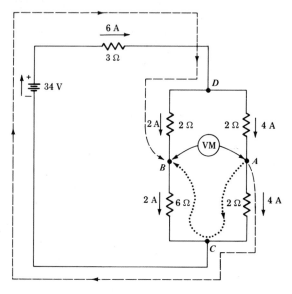

Figure 5-4 Circuit for Example 5-4.

Short path (dotted line): There is no driving voltage along the short path, and all voltage drops are *IR* drops. Thus,

$$V_{AB} = (\sum \text{driving voltages}) - (\sum \text{voltage drops})$$

$$V_{AB} = 0 - [4 \times 2 + (-2) \times 6] = 4 \text{ V}$$

Long path (dashed line): This path includes the driving voltage.

$$V_{AB} = (\sum \text{driving voltages}) - (\sum \text{voltage drops})$$

$$V_{AB} = 34 - (4 \times 2 + 6 \times 3 + 2 \times 2) = 4 \text{ V}$$

As indicated in Example 5-4, calculation of the potential difference between any two points in a circuit is independent of the chosen path of travel.

Example 5-5 For the network shown in Fig. 5-5, determine (*a*) the input voltage; (*b*) the potential difference between nodes *a* and *c*; (*c*) the potential difference between nodes *e* and *b*.

Solution
(*a*) Taking path *abde*,

$$V_{ae} = (\sum \text{driving voltages}) - (\sum \text{voltage drops})$$

$$V_{ae} = 0 - (10 \times 15 + 2.5 \times 11 + 10 \times 13) = -307.5 \text{ V}$$

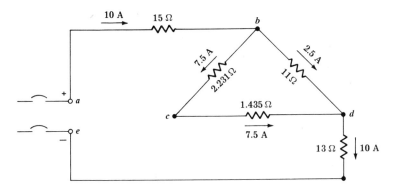

Figure 5-5 Circuit for Example 5-5.

The sign of the answer (+ or −) depends on the direction of travel. If the direction of travel was *from e to a* along path *edba*,

$V_{ea} = 0 - [(-10) \times 13 + (-2.5)(11) + (-10) \times 15] = 307.5$ V

(*b*) Taking path *abc*

$V_{ac} = (\sum \text{driving voltages}) - (\sum \text{voltage drops})$

$V_{ac} = 0 - [10 \times 15 + 7.5 \times 2.231] = -166.73$ V

If we take path *abdc*

$V_{ac} = 0 - [10 \times 15 + 2.5 \times 11 + (-7.5) \times 1.435] = -166.73$ V

(*c*) Taking path *edb*,

$V_{eb} = 0 - [(-10) \times 13 + (-2.5) \times 11] = 157.5$ V

5-4   **BRIDGE NETWORKS**

Bridge networks are series-parallel circuits that have many applications in instrumentation and control. The basic bridge circuit, called the Wheatstone bridge, is shown in Fig. 5-6. The Wheatstone bridge provides a very effective means for measuring the resistance of an unknown resistor.

The Wheatstone-bridge relationship is developed by applying the *potential-difference equation* to terminals *g* and *h*, using two different paths, *gbadh* and *gbcdh*, and then manipulating the resultant equations. Thus:

For path *gbadh*,

$V_{gh} = (\sum \text{driving voltages}) - (\sum \text{voltage drops})$

$V_{gh} = 0 - [(-I_a)R_1 + I_b R_3]$ (5-5)

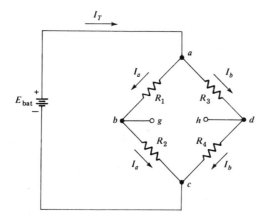

Figure 5-6   Wheatstone-bridge network.

For path $gbcdh$,

$$V_{gh} = 0 - [I_a R_2 + (-I_b)R_4] \tag{5-6}$$

Assuming the selection of resistors is such that $V_{gh} = 0$ (balanced bridge), Eq. (5-5) and Eq. (5-6) reduce to

$$I_a R_1 = I_b R_3 \tag{5-7}$$

$$I_a R_2 = I_b R_4 \tag{5-8}$$

Dividing Eq. (5-7) by Eq. (5-8), and simplifying,

$$\frac{I_a R_1}{I_a R_2} = \frac{I_b R_3}{I_b R_4}$$

$$\frac{R_1}{R_2} = \frac{R_3}{R_4} \tag{5-9}$$

Equation (5-9) is the Wheatstone-bridge relationship.

 Figure 5-7 illustrates an elementary Wheatstone-bridge network for resistance measurement. The unknown resistor R_X is connected between nodes c and d, and a zero-center galvanometer is connected between nodes b and d. The galvanometer is generally an uncalibrated microammeter with a zero center, which can deflect in either direction.

 To measure the resistance of an unknown resistor, the slide-wire rheostat is adjusted to cause the galvanometer to obtain a zero indication (null reading). When the null reading is obtained, the bridge is balanced, and the resistance of the unknown may be determined.

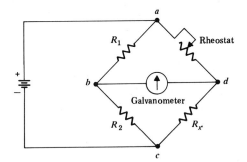

Figure 5-7 Wheatstone-bridge network for resistance measurement.

Applying Eq. (5-9) to the bridge circuit in Fig. 5-7 and solving for R_X,

$$\frac{R_1}{R_2} = \frac{R_{\text{rheo}}}{R_X}$$

$$\boxed{R_X = R_{\text{rheo}} \frac{R_2}{R_1}} \qquad (5\text{-}10)$$

where R_X = unknown resistor, Ω
    R_{rheo} = rheostat setting that results in a balanced bridge ($V_{bd} = 0$), Ω
    $\dfrac{R_2}{R_1}$ = bridge ratio

Example 5-6   Assuming R_2 and R_1 have resistance values of 1000 Ω and 10,000 Ω, respectively, and the rheostat setting that balances the bridge is 1242 Ω, calculate the resistance of the unknown.

Solution

$$R_X = R_{\text{rheo}} \frac{R_2}{R_1}$$

$$R_X = 1242 \, \frac{1000}{10,000} = 124.2 \, \Omega$$

SUMMARY OF FORMULAS

$$V_{AB} = \left(\sum \text{driving voltages}\right) - \left(\sum \text{voltage drops}\right)$$

$$R_X = R_{\text{rheo}} \frac{R_2}{R_1}$$

PROBLEMS

5-1 For the circuit in Fig. 5-8 determine (a) the input resistance; (b) the feeder current; (c) voltage drop across each resistor; (d) current through the 21-Ω resistor.

5-2 Determine the current through the 12-Ω resistor in Fig. 5-9.

5-3 Referring to Fig. 5-10, determine (a) the input resistance; (b) the feeder current; (c) the current in the 500-Ω resistor; (d) the power supplied by the generator; (e) the total kilowatt-hours of energy expended in 36 h of operation.

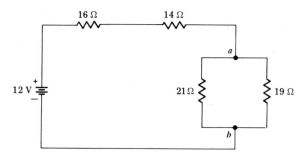

Figure 5-8 Circuit for Prob. 5-1.

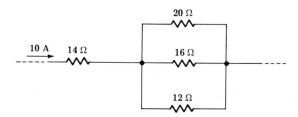

Figure 5-9 Circuit for Prob. 5-2.

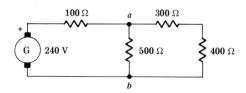

Figure 5-10 Circuit for Prob. 5-3.

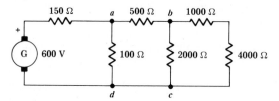

Figure 5-11 Circuit for Prob. 5-4.

5-4 For the circuit in Fig. 5-11, determine (*a*) the input resistance; (*b*) the feeder current; (*c*) the current in the 2000-Ω resistor; (*d*) the heat power expended in the 2000-Ω resistor; (*e*) the potential difference between nodes *a* and *c*.

5-5 Determine the input resistance of the network in Fig. 5-12.

5-6 For the network in Fig. 5-13 determine (*a*) the input resistance; (*b*) the feeder current; (*c*) the current through the 2-kΩ resistor; (*d*) the voltage drop across the 500-Ω resistor; (*e*) the potential difference between nodes *a* and *c*.

5-7 Assuming the bridge in Fig. 5-14 is balanced, determine R_X.

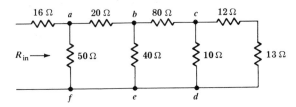

Figure 5-12   Circuit for Prob. 5-5.

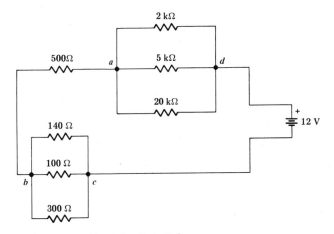

Figure 5-13   Circuit for Prob. 5-6.

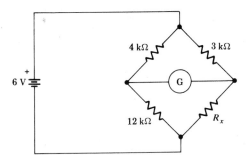

Figure 5-14   Circuit for Prob. 5-7.

CHAPTER 6

CAPACITORS

Capacitance is that property of an electric circuit, or circuit element, that delays a change in the voltage across it. The delay is caused by the absorption or release of energy, and is associated with a change in the electric charge.

CHARGING ACTION

If two conductors separated by a nonconducting material, such as air, paper, rubber, plastics, or glass, are connected to a DC generator or battery, as shown in Fig. 6-1a, the "free electrons" in the conducting material drift in the direction of the driving voltage. The battery, acting as an "electron pump," transfers some of the "free electrons" from conductor A to conductor B. The transfer of electrons causes conductor B to become increasingly negative, and conductor A increasingly positive. Thus, a voltage difference builds up between the conductors. The material that lost electrons is said to be *positively charged*, and that which gained electrons is said to be *negatively charged*. As the charging process continues, conductor B eventually becomes sufficiently negative to prevent any additional transfer of electrons. When this occurs, the voltage measured from conductor A to conductor B is equal and opposite to the driving voltage. This opposing voltage is called a *countervoltage* and designated e_C. A graph of voltage buildup versus time is shown in Fig. 6-1b.

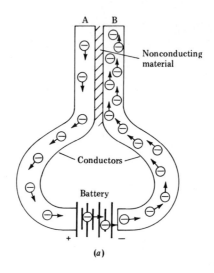

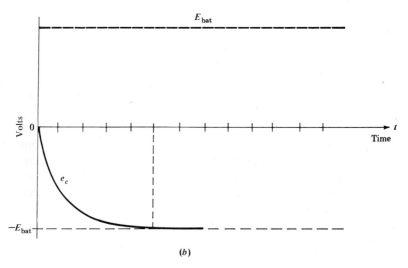

Figure 6-1   (*a*) Capacitor formed by two conductors separated by a nonconducting material; (*b*) voltage buildup versus time.

The rate of electron movement is limited by the resistance of the conducting materials. Hence *the charging process will take longer if higher-resistance materials are used.* The use of longer conductors of the same material and same cross section will increase the capacity for electron storage. This will permit more of the "free electrons" to be shifted from *A* to *B* before the accumulated charge is sufficiently concentrated to build up an equal and opposite voltage. In contrast, shorter conductors have less capacity for electron storage; hence the same electron density (concentration of electrons per unit area) and the same countervoltage will occur with less electron transfer and in less time than with longer conductors. The capacitance property of a circuit does not prevent the voltage from changing; it merely *delays the change*.

For a given geometry and spacing of the conductors in Fig. 6-1a, the *ratio of the accumulated electric charge in the conductors, to the voltage across the conductors, is constant and independent of the driving voltage*. This ratio is called capacitance, and is designated by the letter C. Thus

$$C = \frac{q}{V_c} \qquad\qquad (6\text{-}1)$$

where C = capacitance, farads (F)
       q = accumulated charge, coulombs (C)
      V_C = voltage measured between conductors of opposite polarity, volts (V)

The farad is a rather large unit, and in most circuit applications, the capacitance is on the order of microfarads, nanofarads, or picofarads (μF, nF, pF, respectively). However, when substituting into mathematical equations, all units must be converted to nonprefix form.

The capacitance between two or more conductors separated by a dielectric is always present, and must often be accounted for in the design of electrical systems. The capacitance effect occurs between turns of insulated wire in a coil, between the turns in a coil and the steel framework of a motor, between collector, base, and emitter of a transistor, between electrodes in electronic tubes, between two telephone wires, etc. These *stray* capacitances can have a profound effect when energized from high-frequency systems. The effect of frequency on capacitors is discussed in Chap. 10.

Example 6-1 Determine the capacitance of the parallel conductors in Fig. 6-1a, if a 240-V battery causes an accumulated charge of 25 nanocoulombs (25 nC).

Solution

$$C = \frac{q}{V} = \frac{25 \times 10^{-9}}{240} = 104 \times 10^{-12}$$

$$C = 104 \text{ pF}$$

·2 **LUMPED CAPACITANCE**

In those circuit applications where it is desirable to increase the time delay in the buildup of voltage, or where more energy-storage capability is desired,

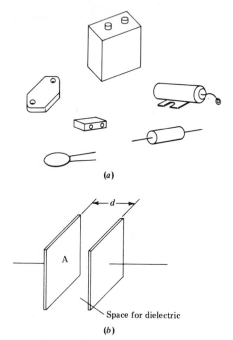

(a)

(b)

Figure 6-2 (a) Some of the varieties of commercially available capacitors; (b) construction details of a simple two-plate capacitor.

a *lumped capacitance, called a capacitor,* is added to the circuit. Some of the many varieties of capacitors are shown in Fig. 6-2a.

Commercial capacitors are composed of two parallel conductors called plates, generally metal foil, separated by an insulating material called a dielectric. The various types of capacitors differ primarily in the type of dielectric used. Air, paper, mica, ceramic, and electrochemical (electrolytic) dielectrics are some of the more common types used in the construction of capacitors. Figure 6-2b shows the construction details of a simple parallel-plate capacitor.

The capacitance of a parallel-plate capacitor is related to the surface area of the plates, the thickness of the dielectric, and the type of dielectric used, by the following equation:

$$C = \frac{KA(8.854 \times 10^{-12})}{d}$$

(6-2)

where C = capacitance, farads (F)
  A = surface area common to both plates or foils, square meters (m^2)
  d = thickness of dielectric, meters (m)
  K = dielectric constant of insulating material

Some representative dielectric constants are 1.0 for air, 5 for mica, 6 for glass, and 7500 for ceramic (barium-strontium titanate).

Example 6-2 Determine the capacitance of the parallel-plate capacitor shown in Fig. 6-2b. Assume the plates are each 20 mm by 20 mm, and the dielectric is 0.50-mm thick ceramic.

Solution

$A = (0.020 \times 0.020) = 4.0 \times 10^{-4} \text{ m}^2$

$d = 0.00050 \text{ m}$

$$C = \frac{7500(4.0 \times 10^{-4}) \times 8.854 \times 10^{-12}}{0.00050}$$

$C = 53.12 \times 10^{-9} = 53.1 \text{ nF}$

Capacitors are essential to the operation of a large variety of electronic and power equipment. In power applications, capacitors are used for motor starting, power-factor improvement, energy absorbers to reduce sparking, energy storage for laser power supplies, flash photography, hotshot wind tunnels, etc.

Electrolytic capacitors utilize metal foils such as magnesium, titanium, and aluminum for the plates and a thin metallic-oxide coating, formed during the application of voltage, for the dielectric. The oxide film, produced electrochemically, offers exceedingly high resistance to the current. Electrolytes such as boric acid and ethylene glycol are generally used to saturate a porous paper or gauze separator placed between the sections of foil. At an applied voltage of 100 V, the oxide coating on an aluminum plate is approximately 10^{-4} mm thick. The close spacing of the plates, brought about by the thinness of the dielectric, results in a high value of capacitance for a small volume of material.

Electrolytic capacitors offer high resistance to electron flow in one direction but act as good conductors when the applied voltage is reversed. Hence, when used in DC circuits, such capacitors must be connected in accordance with the indicated polarity markings. When designed for use in AC systems, two capacitors of the same rating are connected in series, back to back. In this manner one unit is charging while the other conducts.

With the exception of electrolytic capacitors, the leakage current through the dielectric is insignificant. Hence, when making circuit calculations involving capacitors, very little error will be introduced if the capacitors are considered to be ideal elements.

6-3 **VOLTAGE, CHARGE, AND CURRENT CHARACTERISTICS OF A CAPACITOR**

Figure 6-3a shows an *uncharged* capacitor connected in series with a battery, a resistor, and a switch. When the switch is closed, the driving voltage causes a transfer of electrons from foil A to foil B. The *current*, which is the rate of transfer of electrons, *depends on the driving voltage, the initial charge on the capacitor, and the resistance of the circuit.*

From Kirchhoff's voltage law the algebraic sum of the voltage drops must equal the net driving voltage. Thus, for the circuit in Fig. 6-3a, when the switch is closed,

$$E_{bat} = v_C + v_R \tag{6-3}$$

$$E_{bat} = v_C + i_C R$$

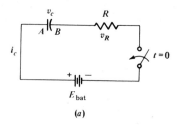

(a)

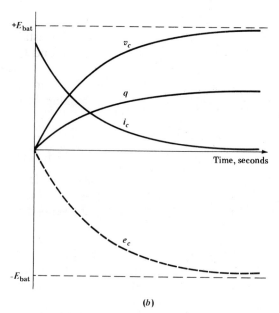

(b)

Figure 6-3 (a) RC circuit; (b) characteristic curves.

Solving for the current,

$$i_C = \frac{E_{bat} - v_C}{R} \tag{6-4}$$

where i_C = instantaneous current, amperes (A)
E_{bat} = battery voltage, volts (V)
v_C = instantaneous voltage drop across the capacitor, volts (V)
R = overall resistance of circuit, ohms (Ω)

If the capacitor has no residual charge from some previous application of a driving voltage, that is, the initial charge is zero, then the voltage across the capacitor will be zero before the switch is closed. Since a capacitor delays a change in the voltage across it, the voltage across the capacitor at the instant the switch is closed ($t = 0+$) must be the same value it had before the switch was closed ($t = 0-$). Thus, for the circuit under discussion, at the instant the switch is closed,

$$v_C\big|_{t=(0+)} = 0$$

and Eq. (6-4) becomes

$$i_C\big|_{t=(0+)} = \frac{E_{bat} - 0}{R} \tag{6-5}$$

As indicated in Eq. (6-5), the current to an uncharged capacitor, at the instant the switch is closed, is determined solely by the circuit resistance and the magnitude of the driving voltage at that instant of time.

Although the uncharged capacitor in Fig. 6-3a has no effect on the initial value of the current, as the charging progresses, the buildup of an opposing voltage across the plates reduces the current to zero. A capacitor is in fact a discontinuity (break or open) in an otherwise closed loop, and is further evidence that any current must be transitory, decaying to zero as the charging progresses.

When the capacitor is fully charged, there will be no further transfer of charge, the steady-state current will be zero, and the steady-state value of voltage across the capacitor terminals will be E_{bat}. This may be verified by substituting zero for the current in Eq. (6-4), and solving for v_C,

$$0 = \frac{E_{bat} - v_{C, ss}}{R}$$

$$v_{C, ss} = E_{bat}$$

The accumulated charge in a capacitor may be determined for any instant of time by using Eq. (6-1).

$$q = Cv_C \qquad (6\text{-}6)$$

Note that the accumulated charge at any instant is equal to the product of the capacitance and the instantaneous voltage drop across it.

Example 6-3 Assume the circuit components in Fig. 6-3a are a 13.2-V battery, a 10-kΩ resistor, and an uncharged 150-μF capacitor. Determine (a) the voltage across the capacitor at $t = (0+)$; (b) the current at $t = (0+)$; (c) the current and voltage at steady state; (d) the charge in the capacitor at steady state; (e) the voltage across the resistor at steady state.

Solution

(a) $v_C|_{t=(0+)} = v_C|_{t=(0-)} = 0$

(b) Applying Kirchhoff's voltage law,

$$E_{bat} = v_C + iR \qquad [\text{at} \quad t = (0+), v_C = 0]$$

Thus,

$$E_{bat} = iR$$

$$13.2 = i(10,000)$$

$$i|_{t=(0+)} = 1.32 \times 10^{-3} = 1.32 \text{ mA}$$

(c) At steady state, the capacitor is fully charged. Hence,

$$v_{C, ss} = E_{bat} = 13.2 \text{ V}$$

$$i_{ss} = 0$$

(d) $q_{ss} = CV_{ss}$

$$q_{ss} = 150 \times 10^{-6}(13.2) = 1.98 \times 10^{-3} \text{ C}$$

$$q_{ss} = 1.98 \text{ mC}$$

(e) $v_{R, ss} = i_{ss}R$

$$v_{R, ss} = 0 \times 10,000 = 0 \text{ V}$$

Figure 6-3b shows the characteristic curves of e_C, v_C, i_C, and q as functions of time for the circuit in Fig. 6-3a. Curve e_C shows the countervoltage generated in opposition to the driving voltage; voltage v_C is that part of the driving voltage that is "used up" in overcoming e_C. Voltage v_C is equal to e_C but opposite in direction.

Closing the switch will not result in an instantaneous buildup of voltage across a capacitor, and likewise, opening the switch will not result in an instantaneous drop in voltage. By virtue of its ability to store electric charge, a capacitor serves to *oppose, and hence delay*, any change in the voltage across its terminals.

If a discharged capacitor is connected across a battery or generator, its instantaneous action is equivalent to that of a short circuit. With no charge in the capacitor, its countervoltage at the instant of connection is zero, and a relatively large current appears in the circuit. Although a discharged capacitor offers little opposition to the current when initially connected to a DC source, the rapid buildup of charge with its attendant voltage drop serves to decrease the current quickly from its initial high value to zero.

The rate of accumulation of electric charge with respect to time is directly proportional to the rate of change of voltage across the capacitor. This stems from Eq. (6-6):

$$q = Cv_C$$

Expressed as a rate of change with respect to time,

$$\frac{dq}{dt} = C\frac{dv_C}{dt} \tag{6-7}$$

where $\dfrac{dq}{dt}$ = rate of change of q with respect to time, coulombs/second (C/s)

$\dfrac{dv_C}{dt}$ = rate of change of voltage across the capacitor with respect to time, volts/second (V/s)

However, the rate of change, or rate of transfer, of electric charge is defined as the electric current.

$$i = \frac{dq}{dt} \tag{6-8}$$

Substituting Eq. (6-8) into Eq. (6-7),

$$i = C\frac{dv_C}{dt} \tag{6-9}$$

Equation (6-9) expresses the current to a capacitor in terms of the rate of change of voltage across it.

The voltage across a capacitor in terms of the current to the capacitor may be obtained by solving Eq. (6-6) for v_C, and then expressing the charge q in terms of the current. Thus, from Eq. (6-6),

$$v_C = \frac{1}{C} q \tag{6-10}$$

where q = total accumulated charge at some time T. Referring to Fig. 6-4, the accumulated charge at the instant of time t_a is the sum of all the increments of charge accumulated from time $t = 0$ to time $t = t_a$. Thus,

$$q_a = \sum_{t=0}^{t=t_a} \Delta q_1 + \Delta q_2 + \Delta q_3 + \cdots + \Delta q_n \tag{6-11}$$

where Δq_1 = increment of charge accumulated in increment of time Δt_1
$\sum$ means "the sum of"

If the ΔT increments of time are made infinitesimally small (called dt), the corresponding increments of charge will be very small (dq). Substituting dq for Δq in Eq. (6-11),

$$q_a = \sum_{t=0}^{t=t_a} dq_1 + dq_2 + dq_3 + \cdots + dq_n \tag{6-12}$$

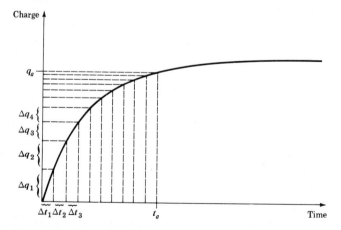

Figure 6-4   Curve showing charge accumulation in a capacitor.

Substituting Eq. (6-12) into Eq. (6-10), the voltage across the capacitor at $t = t_a$ is

$$v_C\bigg|_{t=t_a} = \frac{1}{C} \sum_{t=0}^{t=t_a} dq_1 + dq_2 + dq_3 + \cdots + dq_n \tag{6-13}$$

From Eq. (6-8),

$$dq = i\, dt \tag{6-14}$$

Substituting Eq. (6-14) into (6-13),

$$v_C\bigg|_{t=t_a} = \frac{1}{C} \sum_{t=0}^{t=t_a} (i\, dt)\text{'s} \tag{6-15}$$

As indicated in Eq. (6-15), the voltage across the capacitor is equal to $1/C$ multiplied by the sum of all the $(i\, dt)$'s from $t = 0$ to $t = t_a$.

Whenever the summation is made by taking an infinite number of infinitesimally small increments, as shown in Eq. (6-15), the $\sum$ sign is generally replaced by the $\int$ sign (called the integral sign), and written as

$$v_C\bigg|_{t=t_a} = \frac{1}{C} \int_{t=0}^{t=t_a} i\, dt \tag{6-16}$$

6-4 EQUIVALENT CAPACITANCE OF TWO OR MORE CAPACITORS IN SERIES

Applying Kirchhoff's voltage law to the circuit in Fig. 6-5a,

$$v_T = v_1 + v_2 + v_3 + \cdots + v_n \tag{6-17}$$

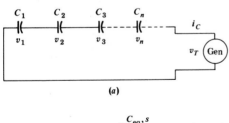

(a)

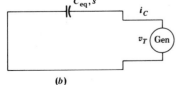

(b)

Figure 6-5 (a) Capacitors in series; (b) equivalent circuit.

From Eq. (6-10),

$$v_1 = \frac{q_1}{C_1} \qquad v_2 = \frac{q_2}{C_2} \qquad v_3 = \frac{q_3}{C_3} \qquad v_n = \frac{q_n}{C_n} \tag{6-18}$$

Substituting Eq. (6-18) into Eq. (6-17),

$$v_T = \frac{q_1}{C_1} + \frac{q_2}{C_2} + \frac{q_3}{C_3} + \cdots + \frac{q_n}{C_n} \tag{6-19}$$

Since the current to all elements in a series circuit is the same, and the current is the rate of transfer of charge ($i = dq/dt$), the accumulation of charge in every capacitor must be the same. Thus,

$$q_1 = q_2 = q_3 = q_n = q$$

and Eq. (6-19) becomes

$$v_T = \frac{q}{C_1} + \frac{q}{C_2} + \frac{q}{C_3} + \cdots + \frac{q}{C_n} \tag{6-20}$$

$$v_T = q\left(\frac{1}{C_1} + \frac{1}{C_2} + \frac{1}{C_3} + \cdots + \frac{1}{C_n}\right) \tag{6-21}$$

For the capacitor in Fig. 6-5b to be the equivalent of the series-connected group in Fig. 6-5a, it must acquire the same charge when connected to the same driving voltage. Thus, in terms of charge and capacitance,

$$v_T = q\left(\frac{1}{C_{eq,s}}\right) \tag{6-22}$$

The q in Eq. (6-22) is identical to the q in each of the series-connected capacitors; *it is not the sum of the individual q's.* That is,

$$\boxed{q_S = q_1 = q_2 = q_3 = \cdots = q_N}$$

$$q_S \neq q_1 + q_2 + q_3 + \cdots + q_n$$

Substituting Eq. (6-22) into Eq. (6-21) and simplifying results in a formula for determining the equivalent capacitance of capacitors in series,

$$\boxed{\frac{1}{C_{eq,s}} = \frac{1}{C_1} + \frac{1}{C_2} + \frac{1}{C_3} + \cdots + \frac{1}{C_n}} \tag{6-23}$$

As indicated in Eq. (6-23), increasing the number of series-connected capacitors decreases the value of the equivalent capacitance.

6-5   **EQUIVALENT CAPACITANCE OF TWO OR MORE CAPACITORS IN PARALLEL**

Applying Kirchhoff's current law to the circuit in Fig. 6-6a,

$$i_T = i_1 + i_2 + i_3 + \cdots + i_n \tag{6-24}$$

The current to each capacitor is a function of the rate of change of voltage across the capacitor. Thus, from Eq. (6-9),

$$i_1 = C_1 \frac{dv_1}{dt} \qquad i_2 = C_2 \frac{dv_2}{dt} \qquad i_3 = C_3 \frac{dv_3}{dt} \qquad i_n = C_n \frac{dv_n}{dt} \tag{6-25}$$

Substituting Eq. (6-25) into Eq. (6-24), and recognizing that the voltage across each element in a parallel circuit is the same and equal to the driving voltage,

$$i_T = C_1 \frac{dv_T}{dt} + C_2 \frac{dv_T}{dt} + C_3 \frac{dv_T}{dt} + \cdots + C_n \frac{dv_T}{dt}$$

Factoring,

$$i_T = (C_1 + C_2 + C_3 + \cdots + C_n) \frac{dv_T}{dt} \tag{6-26}$$

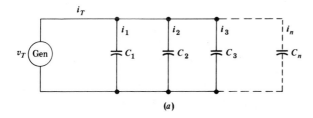

(a)

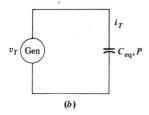

(b)

Figure 6-6   (a) Capacitors in parallel; (b) equivalent circuit.

For a single capacitor such as that shown in Fig. 6-6b to be the equivalent of the parallel-connected capacitors shown in Fig. 6-6a, its current must equal the total current drawn by the paralleled group of capacitors. Expressing the current to the equivalent capacitor in terms of its capacitance and the rate of change of voltage across it,

$$i_T = C_{eq,P} \frac{dv_T}{dt} \tag{6-27}$$

Substituting Eq. (6-27) into Eq. (6-26) and simplifying results in a formula for determining the equivalent capacitance of capacitors in parallel,

$$\boxed{C_{eq,P} = C_1 + C_2 + C_3 + \cdots + C_n} \tag{6-28}$$

As indicated in Eq. (6-28), increasing the number of parallel-connected capacitors increases the value of the equivalent capacitance.

Substituting $C = q/V$ in Eq. (6-28), and recognizing that the voltage across each capacitor is the same,

$$\frac{q_P}{V} = \frac{q_1}{V} + \frac{q_2}{V} + \frac{q_3}{V} + \cdots + \frac{q_n}{V}$$

$$\boxed{q_P = q_1 + q_2 + q_3 + \cdots + q_n} \tag{6-29}$$

Example 6-4   Given the following capacitors: 125 μF, 65.0 μF, and 425 μF. Determine (a) the equivalent capacitance if connected in series; (b) the equivalent capacitance if connected in parallel.

Solution

(a) $\dfrac{1}{C_{eq,S}} = \dfrac{1}{C_1} + \dfrac{1}{C_2} + \dfrac{1}{C_3}$

$\dfrac{1}{C_{eq,S}} = \dfrac{1}{125 \times 10^{-6}} + \dfrac{1}{65 \times 10^{-6}} + \dfrac{1}{425 \times 10^{-6}}$

$C_{eq,S} = 38.9 \ \mu F$

(b) $C_{eq,P} = C_1 + C_2 + C_3$

$C_{eq,P} = 125 \times 10^{-6} + 65.0 \times 10^{-6} + 425 \times 10^{-6}$

$C_{eq,P} = 615 \times 10^{-6} = 615 \ \mu F$

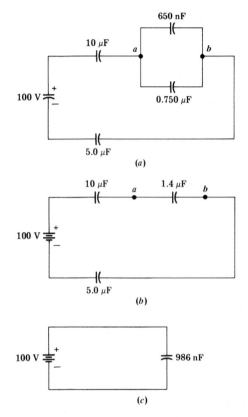

Figure 6-7   Circuits for Example 6-5.

Example 6-5   Referring to the circuit in Fig. 6-7a, determine (a) the equivalent capacitance of the series-parallel connection; (b) the voltage drop across each capacitor.

Solution
(a) Combining the two parallel branches,

$$C_{eq, P} = 650 \times 10^{-9} + 0.750 \times 10^{-6} = 1.40 \times 10^{-6} \text{ F}$$

Substituting the value of $C_{eq, P}$ for the paralleled capacitors results in Fig. 6-7b. Combining the three series-connected capacitors in Fig. 6-7b,

$$\frac{1}{C_{eq, S}} = \frac{1}{10 \times 10^{-6}} + \frac{1}{1.40 \times 10^{-6}} + \frac{1}{5.0 \times 10^{-6}}$$

$$\frac{1}{C_{eq, S}} = 1014 \times 10^{-3}$$

$$C_{eq, S} = 986 \times 10^{-9} = 986 \text{ nF}$$

The reduced circuit is shown in Fig. 6-7c.

(b) The accumulated charge is

$$q = CV = 986 \times 10^{-9}(100)$$

$$q = 98.6 \times 10^{-6} = 98.6 \ \mu C$$

Since the three series capacitors in Fig. 6-7b must have the same q, and $V = q/C$,

$$V_{10} = \frac{q_{10}}{C_{10}} = \frac{98.6 \times 10^{-6}}{10 \times 10^{-6}} = 9.86 \ V$$

$$V_5 = \frac{q_5}{C_5} = \frac{98.6 \times 10^{-6}}{5 \times 10^{-6}} = 19.7 \ V$$

$$V_{1.4} = \frac{q_{1.4}}{C_{1.4}} = \frac{98.6 \times 10^{-6}}{1.4 \times 10^{-6}} = 70.4 \ V$$

Note: $V_{1.4}$ is the voltage across the parallel group. Hence, referring to Fig. 6-7a,

$$V_{650} = 70.4 \ V$$

$$V_{0.750} = 70.4 \ V$$

Example 6-6   Assume the switch in Fig. 6-8a is closed and the circuit has reached its steady-state condition. Determine (a) the current in each circuit element; (b) the voltage drop across each element; (c) the voltage across the capacitor at the instant the switch is opened; (d) the discharge current at the instant the switch is opened.

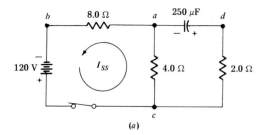

(a)

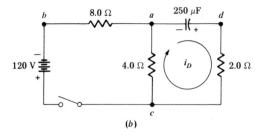

(b)

Figure 6-8   Circuits for Example 6-6.

Solution

(*a*) At steady state, the capacitor is fully charged, and its charging current is reduced to zero. Hence, there can be no current in the 2-Ω resistor; the charged capacitor acts as an open circuit. The only steady-state current in the circuit is through the loop formed by the battery, the 8-Ω resistor, and the 4-Ω resistor. Applying Ohm's law,

$$I_{ss} = \frac{E_{bat}}{R} = \frac{120}{8 + 4} = 10 \text{ A}$$

(*b*) The voltage drops across the resistors, as determined from Ohm's law, are

$$V_8 = I_8 R_8 = 10(8) = 80 \text{ V}$$

$$V_4 = I_4 R_4 = 10(4) = 40 \text{ V}$$

$$V_2 = I_2 R_2 = 0(2) = 0 \text{ V}$$

The voltage drop across the capacitor may be determined by applying the PD equation to nodes *a* and *d*. Thus,

$$V_{ad} = \left(\sum \text{driving voltages} \right) - \left(\sum \text{voltage drops} \right)$$

Traveling counterclockwise (ccw) along path *acd*,

$$V_{ad} = 0 - [(-10) \times 4 + (0) \times 2] = 40 \text{ V}$$

(*c*) The circuit with the switch open is shown in Fig. 6-8*b*. Since the capacitance property of a capacitor delays a change in the voltage across it,

$$v_C|_{t=(0+)} = v_C|_{t=(0-)} = 40 \text{ V}$$

(*d*) The current in the discharge loop, at the instant the switch is opened, may be determined by applying Kirchhoff's voltage law to that loop. Thus,

$$\sum \text{driving voltages} = \sum \text{voltage drops}$$

With the switch open, the driving voltage is zero. Hence,

$$0 = 2i_D + 4i_D + v_C$$

At $t = (0+)$,

$$0 = (2 + 4)i_D + 40$$

$$i_D|_{t=(0+)} = -6.7 \text{ A}$$

The minus sign indicates that the current has reversed.

6-6 TIME CONSTANT OF AN RC CIRCUIT†

The equation of the current and voltage curves in Fig. 6-3b and Fig. 6-9c is

$$i_C = \frac{E_{bat}}{R}(\varepsilon^{-t/RC}) \tag{6-30}$$

$$v_C = E_{bat}(1 - \varepsilon^{-t/RC}) \tag{6-31}$$

where i_C = instantaneous current, A
v_C = instantaneous voltage, V
E_{bat} = battery voltage, V
R = resistance of circuit, Ω
C = capacitance of circuit, F
t = time, s
ε = 2.718 (base for natural logarithms)

when $t = (0+)$

$$i_C\Big|_{t=0+} = \frac{E_{bat}}{R}(\varepsilon^0) \qquad\qquad v_C\Big|_{t=0+} = E_{bat}(1 - \varepsilon^0)$$

$$i_C\Big|_{t=0+} = \frac{E_{bat}}{R} \qquad\qquad v_C\Big|_{t=0+} = E_{bat}(1 - 1) = 0$$

As t increases, the exponential factor $\varepsilon^{-t/RC}$ gets smaller; current i_C gets smaller, and voltage v_C gets larger. As t approaches infinity, the exponential term approaches zero, causing v_C and i_C to approach their steady-state values. Thus, as $t \to \infty$,

$$i_C\Big|_{t\to\infty} = \frac{E_{bat}}{R}(\varepsilon^{-\infty/RC}) \qquad\qquad v_C\Big|_{t\to\infty} = E_{bat}(1 - \varepsilon^{-\infty/RC})$$

$$i_C\Big|_{t\to\infty} = \frac{E_{bat}}{R}(0) \qquad\qquad v_C\Big|_{t\to\infty} = E_{bat}(1 - 0)$$

$$i_C\Big|_{t\to\infty} = 0 \qquad\qquad v_C\Big|_{t\to\infty} = E_{bat}$$

Regardless of the values of R and C, the total time required to reach the absolute steady state will be the same, infinite.

However, a comparison of the voltage versus time curves for circuits with different products of R and C shows that the initial rise in voltage across

† A complete analysis of the response of an RC circuit to a driving voltage is given in Chap. 27.

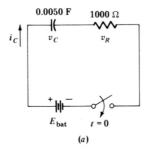

(a)

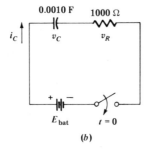

(b)

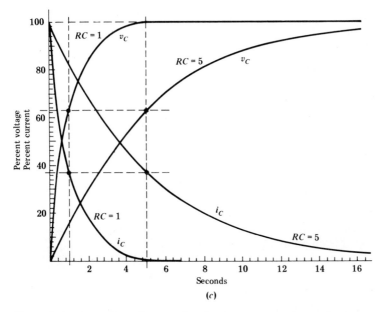

(c)

Figure 6-9 Comparison of curves showing current versus time and voltage versus time for *RC* circuits with different time constants.

the capacitor takes a longer time in those circuits that have higher RC values. This is shown in Fig. 6-9c. *The RC product is called the time constant of an RC circuit*; it is equal to the time in seconds that it takes for the voltage across the capacitor and the current to the capacitor to have completed 63.2 percent of their respective total changes.

$$\tau = RC \qquad\qquad (6\text{-}32)$$

where τ = one time constant, s
  C = circuit capacitance, F
  R = circuit resistance, Ω

Verification of the 63.2 percent change can be accomplished by substituting $t = RC$ in Eqs. (6-30) and (6-31) and evaluating. Thus,

$$i_C = \frac{E_{bat}}{R}(\varepsilon^{-t/RC}) \qquad\qquad v_C = E_{bat}(1 - \varepsilon^{-t/RC})$$

$$i_C = \frac{E_{bat}}{R}(\varepsilon^{-RC/RC}) \qquad\qquad v_C = E_{bat}(1 - \varepsilon^{-RC/RC})$$

$$i_C = \frac{E_{bat}}{R}(\varepsilon^{-1}) \qquad\qquad v_C = E_{bat}(1 - \varepsilon^{-1})$$

$$i_C = \frac{E_{bat}}{R}(0.368) \qquad\qquad v_C = E_{bat}(1 - 0.368)$$

$$i_C = 0.368\frac{E_{bat}}{R} \qquad\qquad v_C = 0.632\,E_{bat}$$

$$i_C = 36.8 \text{ percent}\frac{E_{bat}}{R} \qquad\qquad v_C = 63.2 \text{ percent } E_{bat}$$

Note that i_C changed from an initial current of E_{bat}/R at $t = (0+)$ to 36.8 percent E_{bat}/R at $t = 1$ time constant, a change of 63.2 percent. The capacitor voltage changed from an initial value of zero volts at $t = (0+)$ to 63.2 percent E_{bat} at $t = 1$ time constant.

In five time constants (5τ), the current will have dropped to

$$i_C = \frac{E_{bat}}{R}(\varepsilon^{-5RC/RC}) = \frac{E_{bat}}{R}(\varepsilon^{-5})$$

$$i_C = 0.0067\frac{E_{bat}}{R}$$

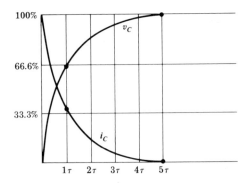

Figure 6-10 Graphical method for approximating the transient behavior of an *RC* circuit.

In the same five time constants, the voltage v_C will have risen to

$$v_C = E_{bat}(1 - \varepsilon^{-5RC/RC}) = E_{bat}(1 - \varepsilon^{-5})$$

$$v_C = E_{bat}(1 - 0.0067) = 0.993E_{bat}$$

Hence, for all practical purposes, after an elapsed time equal to five time constants, the capacitor may be considered fully charged, at which time v_C and i_C are essentially at steady state.

Figure 6-10 illustrates a method that can be used to approximate closely the graph of the transient behavior of an *RC* circuit. Draw four equally spaced horizontal lines to represent 0, 33.3 percent, 66.6 percent, and 100 percent of the initial current to the capacitor. Draw six equally spaced vertical lines to represent 0, $1\tau, 2\tau, 3\tau, 4\tau$, and 5τ. Plot the points that represent 1τ and 5τ (shown with dots), and then approximate the curve.

Example 6-7 A 5-μF capacitor is connected in series with a 2-MΩ resistor, a 120-V DC source, and a switch. The circuit is similar to that shown in Fig. 6-9a and b. Determine (*a*) the time constant; (*b*) the current at $t = 1\tau$; (*c*) the voltage drop across the capacitor at $t = 1\tau$; (*d*) the voltage drop across the resistor at $t = 1\tau$; (*e*) instantaneous power expended in the resistor at $t = 1\tau$; (*f*) the electric charge at $t = 1\tau$; (*g*) the current at $t = 1$ s; (*h*) the voltage drop across the capacitor at $t = 1$ s.

Solution

(*a*) $\tau = RC = 2 \times 10^6 \times 5 \times 10^{-6} = 10$ s

(*b*) $i_C|_{t=1\tau} = 0.368 \dfrac{E}{R} = 0.368 \dfrac{120}{2 \times 10^6} = 22.1 \times 10^{-6}$ A $= 22.1$ μA

(*c*) $v_C|_{t=1\tau} = 0.632E = 0.632(120) = 75.8$ V

(d) From Kirchhoff's voltage law,

$E_{bat} = v_C + v_R$

At $t = 1\tau$,

$120 = 75.8 + v_R$

$v_R = 44.2$ V

(e) $p_R = v_R i_R$

At $t = 1\tau$,

$p_R = 44.2 \times 22.1 \times 10^{-6} = 977 \ \mu W$

Using the V_R^2/R relationship	*Using the $I_R^2 R$ relationship*
$p_R = \dfrac{(44.2)^2}{2 \times 10^6} = 977 \ \mu W$	$p_R = (22.1 \times 10^{-6})^2 (2 \times 10^6)$
	$= 977 \ \mu W$

(f) $q = Cv$

At $t = 1\tau$,

$q = 5 \times 10^{-6}(75.8) = 379 \ \mu C$

(g) $i_C = \dfrac{E}{R}(\varepsilon^{-t/RC}) = \dfrac{120}{2 \times 10^6}\left(\exp \dfrac{-1}{2 \times 10^6 \times 5 \times 10^{-6}}\right)$

$i_C = 54.3 \ \mu A$

(h) $v_C = E(1 - \varepsilon^{-t/RC}) = 120\left(1 - \exp \dfrac{-1}{2 \times 10^6 \times 5 \times 10^{-6}}\right)$

$v_C = 11.4$ V

6-7   GRAPHICAL DETERMINATION OF TIME CONSTANTS

The time constant for an RC circuit whose parameters are not known can be obtained from an oscilloscope recording or xy plot of v_C versus time. This is accomplished by first drawing a tangent line to the voltage curve at $t = (0+)$, as shown in Fig. 6-11. Then, from the point where the tangent line intersects the E_{bat} line, drop a perpendicular to the time axis. The point of intersection with the time axis is the time constant in seconds. Thus, ΔT in Fig. 6-11 is the time constant.

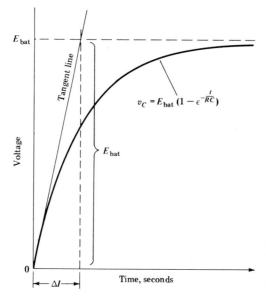

Figure 6-11 Determining the time constant of an RC circuit from an oscilloscope recording or an xy plot.

Justification for Graphical Determination of Time Constants

The slope of the voltage curve at $t = (0+)$ may be obtained algebraically by taking the derivative of Eq. (6-31) with respect to time, and then evaluating it at $t = (0+)$. Thus, from Eq. (6-31),

$$v_C = E_{bat}(1 - \varepsilon^{-t/RC})$$

$$\frac{dv_C}{dt} = E_{bat}\left[0 - \left(-\frac{1}{RC}\varepsilon^{-t/RC}\right)\right]$$

$$\frac{dv_C}{dt} = \frac{E_{bat}}{RC}\varepsilon^{-t/RC}$$

$$\left.\frac{dv_C}{dt}\right|_{t=(0+)} = \frac{E_{bat}}{RC}$$

The slope of the voltage obtained graphically from Fig. 6-11 at $t = (0+)$ is

$$\left.\text{Slope}\right|_{t=(0+)} = \frac{E_{bat}}{\Delta T}$$

Since the slope obtained graphically must be equal to the slope obtained algebraically,

$$\frac{E_{\text{bat}}}{RC} = \frac{E_{\text{bat}}}{\Delta T}$$

Solving for ΔT,

$$\Delta T = RC$$

6-8   ENERGY STORAGE IN A CAPACITOR

The process of transferring electric charge from one plate of a capacitor to the other plate results in the accumulation of energy. This energy, in the form of displaced electric charges (static electricity), remains stored for some time after the driving voltage is disconnected. The amount of energy stored in the capacitor depends on the capacitance times the square of the voltage developed across it. Thus

$$\boxed{W_C = \tfrac{1}{2}Cv_C^2} \tag{6-33}$$

where W_C = energy accumulated in capacitor, joules (J)
$\quad\quad\quad C$ = capacitance, farads (F)
$\quad\quad\quad v_C$ = voltage measured between plates of opposite polarity, volts (V)

The total energy accumulated in the capacitor at steady state is

$$W_{C,\text{ss}} = \tfrac{1}{2}Cv_{C,\text{ss}}^2 \tag{6-34}$$

The energy stored in a capacitor is not released at the instant the capacitor is disconnected from the generator. The duration of the charge, whether for minutes, hours, or days, depends on such factors as the resistance of the dielectric, dielectric constant, surface leakage, humidity, and radioactivity of the environment, all of which serve to gradually dissipate the stored energy. For safety in handling, large capacitors are equipped with parallel-connected "bleeder" resistors that serve to drain the stored charge in 1 to 5 min after being disconnected from the line.

Derivation of Equation (6-33)

Referring to Fig. 6-12a, the instantaneous power or rate of transfer of energy from the battery to the capacitor is equal to the voltage drop across the capacitor multiplied by the instantaneous current. Thus,

$$p = v_C i_C \quad\quad \text{watts or joules/second}$$

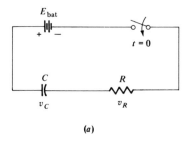

(a)

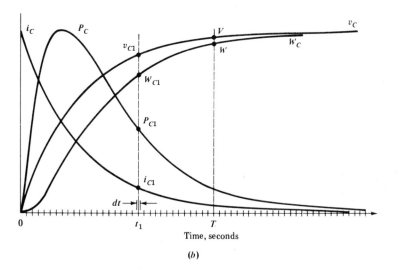

(b)

Figure 6-12   Energy and power relationships in an *RC* circuit.

The increment of energy accumulated in some infinitesimal increment of time *dt* is

$$dw = p \, dt$$

$$dw = v_C i_C \, dt \qquad \text{joules}$$

where v_C and i_C are the respective instantaneous values of voltage and current that occur at a specific instant of time. For example, in Fig. 6-12b, v_{C1} and i_{C1} correspond to the instant of time t_1.

The total energy accumulated in the capacitor in time *T* may be determined by integrating all the increments of energy from $t = 0$ to $t = T$. Thus,

$$W = \int_0^T v_C i_C \, dt \qquad\qquad (6\text{-}35)$$

From Eq. (6-9),

$$i_C = C\frac{dv_C}{dt} \tag{6-36}$$

Substituting Eq. (6-36) into Eq. (6.35),

$$W = \int_0^T v_C C\frac{dv_C}{dt}\, dt$$

Changing the limits and integrating,

$$W = C\int_0^V v_C\, dv_C$$

$$W = \tfrac{1}{2}C(v_C^2)_0^V$$

$$W = \tfrac{1}{2}CV^2$$

Example 6-8   A 6.0 Ω resistor is connected in series with a 2000-μF capacitor, a 48-V battery, and a switch. Assume the capacitor has zero initial charge. Determine (a) the voltage across the capacitor at the instant the switch is closed ($t = 0+$); (b) the current to the capacitor at the instant the switch is closed ($t = 0+$); (c) the voltage across the capacitor at $t = 1$ time constant; (d) the current at $t = 1$ time constant; (e) the rate of transfer of energy at $t = 1$ time constant; (f) the energy stored in the capacitor at steady state; (g) the voltage across the resistor at $t = 0.02$ s.

Solution
The circuit is identical to that shown in Fig. 6-12a.
(a) Since a capacitor delays a change in the voltage across it, and the capacitor has zero initial charge,

$$v_C\Big|_{t=(0+)} = v_C\Big|_{t=(0-)} = 0\text{V}$$

(b) Applying Kirchhoff's voltage law,

$$E_{bat} = v_C + v_R$$

$$48 = 0 + 6i$$

$$i\Big|_{t=(0+)} = 8\text{ A}$$

(c) $v_C\Big|_{t=1\tau} = 0.632E_{bat} = 0.632(48) = 30.3$ V

(d) $i_C\Big|_{t=1\tau} = 0.368\dfrac{E_{bat}}{R} = \dfrac{0.368(48)}{6.0} = 2.94$ A

(e) $P_C| = (v_C|)\,(i_C|)$
$\quad\;_{t=1\tau}\quad\;_{t=1\tau\,t=1\tau}$

$P_C| = (30.3)(2.94) = 89.1 \text{ W}$
$\;_{t=1\tau}$

(f) $W_C = \frac{1}{2}Cv_C^2$

$W_{C,ss} = \frac{1}{2}(2000 \times 10^{-6})(48)^2 = 2.30 \text{ J}$

(g) $E_{bat} = v_C + v_R$

$\qquad v_R = E_{bat} - v_C$

$\qquad v_R = E_{bat} - E_{bat}(1 - \varepsilon^{-t/RC})$

$\qquad v_R = 48 - 48\left(1 - \exp\dfrac{-0.02}{6(2000 \times 10^{-6})}\right)$

$\qquad v_R = 48 - 38.93$

$\qquad v_R = 9.07 \text{ V}$

Example 6-9 Assuming a certain driving voltage causes the current to an initially discharged 1000-μF capacitor to increase at a constant rate of 0.06 A/s ($i_C = 0.06t$), determine the voltage across the capacitor at $t = 10$ s.

Solution

$$v_C = \frac{1}{C}\int_0^T i\,dt$$

$$v_C = \frac{1}{1000(10^{-6})}\int_0^{10} 0.06t\,dt$$

$$v_C = 1000(0.06)\left(\frac{t^2}{2}\right)_0^{10}$$

$$v_C = 60\left(\frac{10^2}{2}\right)$$

$$V_C = 3000 \text{ V}$$

6-9 **DISCHARGING A CAPACITOR**

When a capacitor is discharged, the energy stored in the capacitor is dissipated as heat energy in the resistance of the connecting wires, capacitor plates, discharge resistor, and spark.

Large capacitors can store a considerable amount of energy a long time. Hence, before cleaning, touching, or disconnecting a large capacitor, open the switch or circuit breaker, and discharge the capacitor through a heavy-duty 50,000-Ω resistor; the discharge time should be $\geq$ five time constants. Discharging should be done between terminals and between terminals and ground. Although small capacitors may be discharged by short-circuiting their terminals with a copper strap or heavy cable, large capacitors may be damaged by this practice; the very large mechanical forces caused by the high discharge current may destroy them. Under no circumstances should a bank of large capacitors be discharged through short circuiting. The sudden release of a large amount of energy could vaporize or explode the shorting device, causing injury to personnel as well as damage to the capacitors.

Figure 6-13a shows the direction of charging current when the switch connects the battery to the RC circuit, and Fig. 6-13b shows the direction of current when the switch is thrown to the discharge position. If the direction of capacitor current during charge is considered positive, the direction of

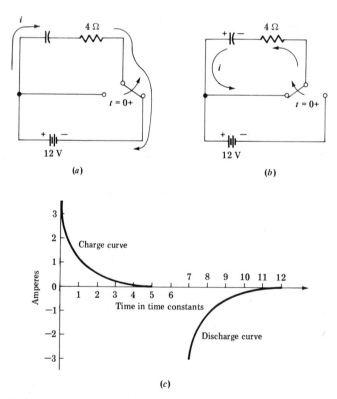

Figure 6-13 (a) Charging a capacitor; (b) discharging a capacitor; (c) current curves during charge and discharge.

capacitor current during discharge is negative. Figure 6-13c shows the charge and discharge curves for the RC circuit.

Example 6-10 Assuming the resistance and capacitance in Fig. 6-13a are $4\,\Omega$ and $1000\,\mu F$, respectively, determine (a) the current at $t = (0+)$; (b) Repeat part a for the discharge conditions in Fig. 6-13b.

Solution
(a) Applying Kirchhoff's voltage law,

$$E_{bat} = v_C + v_R$$

$$v_C\big|_{t=(0+)} = v_C\big|_{t=(0-)} = 0$$

$$12 = 0 + 4i$$

$$i\big|_{t=(0+)} = 3\,A$$

(b) Assuming the capacitor is fully charged before the switch is thrown to the discharge position,

$$v_C\big|_{t=(0+)} = v_C\big|_{t=(0-)} = 12\,V$$

Applying Kirchhoff's voltage law, and recognizing that there is no voltage source in the loop,

$$0 = v_C + v_R$$

$$0 = 12 + 4i$$

$$i = -3\,A$$

Example 6-11 A $400\text{-}\mu F$ capacitor charged to 2500 V is to be discharged through a $50\text{-}k\Omega$ resistor. Determine (a) the recommended minimum discharge time; (b) the capacitor voltage for the conditions in part a.

Solution
(a) $5\tau = 5RC = 5(50,000)(400 \times 10^{-6}) = 100\,s$
(b) The discharge loop is similar to that shown in Fig. 6-13b. Applying Kirchhoff's voltage law,

$$v_{C,dis} + v_R = 0$$

$$v_{C,dis} + (-iR) = 0$$

The negative sign indicates a discharge current. Thus,

$$v_{C,dis} = iR$$

Substituting,

$$i = I_0 \varepsilon^{-t/RC}$$

$$v_{C,\text{dis}} = I_0 R \varepsilon^{-t/RC}$$

Defining,

$$V_0 = I_0 R$$

$$v_{C,\text{dis}} = V_0 \varepsilon^{-t/RC}$$

where

$$I_0 = i_C \text{ at } t = (0+)$$

$$V_0 = v_C \text{ at } t = (0+)$$

In five time constants,

$$v_{C,\text{dis}} = 2500\varepsilon^{-5RC/RC} = 2500\varepsilon^{-5} = 2500(0.0067) = 16.8 \text{ V}$$

SUMMARY OF FORMULAS

$$v_C = \frac{1}{C} \int_0^T i_C \, dt \qquad i_C = C \frac{dv_C}{dt} \qquad C = \frac{KA(8.854 \times 10^{-12})}{d}$$

$$v_{C,\text{dis}} = V_0 \varepsilon^{-t/RC}$$

$$v_C = E_{\text{bat}}(1 - \varepsilon^{-t/RC}) \qquad i_C = \frac{E_{\text{bat}}}{R}(\varepsilon^{-t/RC}) \qquad C = \frac{q}{v_C}$$

$$\frac{1}{C_{\text{eq},S}} = \frac{1}{C_1} + \frac{1}{C_2} + \frac{1}{C_3} + \cdots + \frac{1}{C_n} \qquad q_S = q_1 = q_2 = q_3 = q_N$$

$$C_{\text{eq},P} = C_1 + C_2 + C_3 + \cdots + C_n \qquad q_P = q_1 + q_2 + q_3 + \cdots + q_n$$

$$W = \tfrac{1}{2}Cv_C^2 \qquad p = v_C i_C \qquad \tau = RC$$

PROBLEMS

6-1 Determine the capacitance of a parallel-plate capacitor that has plate dimensions of 80 mm × 100 mm, and a mica dielectric 0.01 mm thick.

6-2 Determine the capacitance of a parallel-plate capacitor that has circular plates 150 mm in diameter, and a glass dielectric 0.10 mm thick.

6-3 Determine the equivalent capacitance of the following series-connected capacitors: 2 μF, 18 μF, 60 μF, and 140 μF.

6-4 If the capacitors in Prob. 6-3 are connected in parallel, determine the equivalent capacitance of the parallel combination.

6-5 A set of parallel capacitors are connected in series with 300-μF and 100-μF capacitors. The paralleled capacitors are 200 μF, 50 μF, and 150 μF. Sketch the circuit and determine the equivalent capacitance of the series-parallel combination.

6-6 For the circuit shown in Fig. 6-14, determine (*a*) the equivalent capacitance of the circuit; (*b*) the voltage drop across each capacitor.

6-7 Assume the switch in Fig. 6-15 is closed and the circuit has reached its steady-state condition. Determine (*a*) the voltage drop across the capacitor; (*b*) the accumulated charge; (*c*) the voltage across the capacitor at the instant the switch is opened; (*d*) the discharge current at the instant the switch is opened.

6-8 The circuit shown in Fig. 6-16 is at steady state. Determine (*a*) the potential difference across the capacitor; (*b*) the accumulated charge; (*c*) the discharge current at the instant the switch is opened.

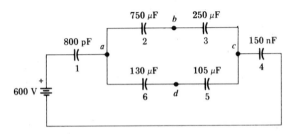

Figure 6-14   Circuit for Prob. 6-6.

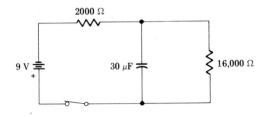

Figure 6-15   Circuit for Prob. 6-7.

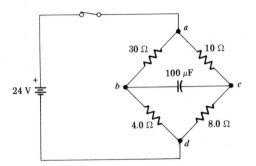

Figure 6-16   Circuit for Prob. 6-8.

6-9 A 3000-μF capacitor is connected in series with a 10-MΩ resistor, a 60-V battery, and a switch. Sketch the circuit and determine (a) the time constant; (b) i_C, v_C, and v_R at $t = (0+)$; (c) i_C, v_C, and v_R at $t = 1\tau$; (d) the accumulated charge at $t = 1\tau$; (e) the energy stored at steady state.

6-10 A 20-kΩ resistor is connected in series with a 1000-μF capacitor, a 20-V battery, and a switch. Sketch the circuit and determine (a) the time constant; (b) i_C, v_C, and v_R at $t = 1\tau$; (c) i_C, v_C, and v_R at steady state; (d) the rate of transfer of energy to the capacitor at steady state; (e) the energy stored in the capacitor at steady state.

6-11 A 2500-μF capacitor is connected in series with a 16-Ω resistor, a 240-V battery, and a switch. The switch is closed at $t = 0$. Sketch the circuit and determine (a) the current at $t = (0+)$; (b) the voltage across the capacitor at $t = (0+)$; (c) the voltage across the resistor at $t = (0+)$; (d) the time constant; (e) current at $t = 1\tau$; (f) voltage across the resistor at $t = 1\tau$; (g) voltage across the capacitor at $t = 1\tau$; (h) rate of transfer of energy to the capacitor at $t = 1\tau$; (i) energy stored in the capacitor at steady state; (j) the voltage across the capacitor 0.06 s after the switch is closed.

6-12 A 24-V battery is connected in series with a 150-μF capacitor, a 1-kΩ resistor, and a switch. Sketch the circuit and determine (a) i_C and v_C at $t = (0+)$; (b) the steady-state current; (c) the current at $t = 1\tau$; (d) the IR drop at $t = 1\tau$; (e) the voltage across the capacitor and the voltage across the resistor at steady state; (f) the accumulated energy at steady state; (g) the accumulated charge at steady state; (h) the rate of transfer of energy to the capacitor at $t = 1\tau$; (i) the heat power dissipated by the resistor at $t = 1\tau$.

6-13 A 100-V battery is connected in series with a 10-Ω resistor and a capacitor. If the total energy accumulated in the capacitor at steady state is 200 J, determine (a) the capacitance of the capacitor; (b) the accumulated charge in coulombs.

6-14 A 600-μF capacitor charged to 400 V is to be discharged through a 2-kΩ resistor and a switch. Determine (a) the current at $t = (0+)$; (b) the capacitor voltage at $t = 3s$.

CHAPTER 7

INTRODUCTION TO
MAGNETIC PHENOMENA

The presentation of magnetic phenomena in this chapter provides the student with the necessary background for an understanding of inductance and mutual coupling in electric circuits, and the magnetic relationships involved in electrical machinery and control.

7-1 MAGNETIC FIELD

A magnetic field can be best described as a *magnetic influence* resulting from electric charges in motion, and may be detected by a magnetic compass some distance away from the moving charges. For example, a flash of lightning between clouds, or between a cloud and earth, generates a magnetic field that can be detected miles away from the actual stroke. A similar but much lesser magnetic field will be produced if a body is given an electric charge by rubbing, and the body is then hurled across the room; the magnetic field will be present only while the electric charge is in motion. Figure 7-1 illustrates the magnetic field surrounding a lightning stroke, and Fig. 7-2 illustrates the magnetic field around a charged body moving through dry air or a vacuum. The concentric circles around the moving charges represent the magnetic field, and are called *magnetic flux lines.*

   In a similar manner, the movement of electric charges in a conductor, caused by the application of a driving voltage, will generate a magnetic field around the conductor. This is shown in Fig. 7-3. If no driving voltage is

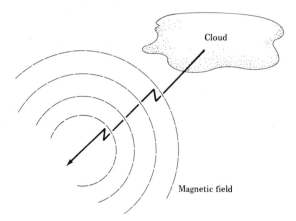

Figure 7-1 Magnetic field surrounding a lightning stroke.

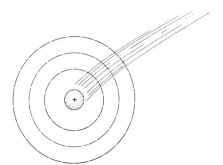

Figure 7-2 Magnetic field around a charged body moving through dry air or a vacuum.

applied, the net effect of the random movement of the electric charges in the conductor will result in no magnetic field.

Since the net movement of electric charges in a given direction constitutes an electric current, *it is the current in the conductor that establishes the magnetic field.* Referring to Fig. 7-3, the battery voltage is the driver or electromotive force (emf) that causes the current, and the current is the driver or magnetomotive force (mmf) that causes the magnetic field.

The direction of the magnetic field around a current is dependent on the direction of the current, and may be determined by the *right-hand rule*: grasp the conductor with the right hand with the thumb pointing in the direction of the current, as shown in Fig. 7-3, and the fingers will point in the direction of the magnetic field around the conductor.

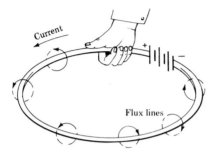

Figure 7-3 Right-hand rule for determining the direction of flux for a given direction of current, and vice versa.

MAGNETIC FLUX DENSITY

Magnetic flux density is a measure of the concentration of flux passing through a unit area, and is the fundamental magnetic field quantity. The magnitude of the flux density provides an indication of the effectiveness of the magnetic field in causing a torque on a test magnet placed within the field. Flux density can be measured with a flux-density meter, which consists of a tiny cylindrical magnetic probe with magnetic poles on opposite ends of its diameter, attached to a nonmagnetic shaft, a pointer, and a control spring. The measurement is made by inserting the probe in the magnetic field, and then rotating the entire instrument until the pointer indicates maximum deflection on the scale. The flux density is the position of maximum deflection, and can be read directly from the scale. The unit of magnetic flux density is the tesla (T).

The density of the earth's magnetic field varies throughout the world and is approximately 5.7×10^{-5} T in the Washington, D.C., region. Magnetic flux densities of up to 17 T are available in commercial magnets using superconductors.

The total magnetic flux in a region of uniform flux density may be determined by the following formula:

$$\Phi = BA \qquad\qquad (7\text{-}1)$$

where A = area, square meters (m²)
   B = density of magnetic flux in area A, teslas (T)
   Φ = total flux in area A, webers (Wb)

Solving Eq. (7-1) for B shows that a flux density of 1 T is equal to 1 Wb/m².

$$B = \frac{\Phi}{A} \quad \text{Wb/m}^2$$

For example, a 0.50-m² area with a uniform flux density of 6.0 T has a total magnetic flux of

$$\Phi = BA = 6(0.5) = 3 \text{ Wb}$$

7-3   **MAGNETIC FLUX DENSITY INSIDE AND OUTSIDE A CONDUCTOR**

A plot of the flux density inside and outside a current-carrying round conductor of radius R is shown in Fig. 7-4. As indicated in the plot, the flux density increases linearly from zero tesla at the center of the conductor to some maximum value at the surface, and then decreases inversely with increased distance from the conductor. The equations representing the two sections of the curve are

$$B = \frac{\mu_0 I_T}{2\pi d} \qquad d \geq R \tag{7-2}$$

$$B = \frac{\mu_0 d I_T}{2\pi R^2} \qquad d < R \tag{7-3}$$

where I_T = total current, (A)
   R = radius of conductor, meters (m)
   d = distance from center of conductor, meters (m)
   B = flux density, teslas (T)
   $\mu_0 = 4\pi(10^{-7})$ henry per meter (H/m)

The constant μ_0 (pronounced mu) is *called the permeability of a vacuum.*

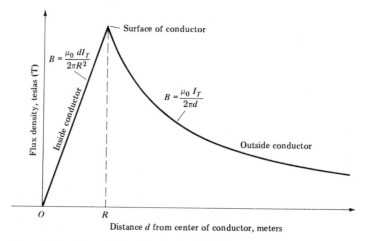

Figure 7-4   Magnetic flux density inside and outside a conductor.

Example 7-1 Given a conductor of circular cross section, 40.0 mm in diameter, carrying a current of 1715 A. Determine (a) the flux density at the surface; (b) the flux density at a point 1 m from the center of the conductor; (c) the flux density 12.7 mm from the center of the conductor.

Solution

$$R = \frac{0.040}{2} = 0.020 \text{ m}$$

(a) Flux density at the surface of the conductor:

$$B = \frac{\mu_0 I_T}{2\pi d} \qquad d \geq R$$

$$B_{\Big|} = \frac{4\pi(10^{-7})(17.15)}{2\pi(0.020)} = 0.0172 \text{ T}$$

$$d = 0.020 \text{ m}$$

(b) Flux density 1 m from the center of the conductor:

$$B = \frac{\mu_0 I_T}{2\pi d} \qquad d \geq R$$

$$B_{\Big|} = \frac{4\pi(10^{-7})(1715)}{2\pi(1)} = 0.000343 \text{ T}$$

$$d = 1 \text{ m}$$

(c) Flux density 0.0127 m from the center of the conductor:

$$B = \frac{\mu_0 d I_T}{2\pi R^2} \qquad d < R$$

$$B_{\Big|} = \frac{4\pi(10^{-7})(0.0127)(1715)}{2\pi(0.020)^2} = 0.0109 \text{ T}$$

$$d = 0.0127 \text{ m}$$

Derivation of Equations (7-2) and (7-3)

Figure 7-5 shows a cylindrical conductor of radius R carrying a current of I A uniformly distributed throughout its cross section. Contour C_i is a circular contour located *inside* the conductor d_i m from its center, and contour C_o is a circular contour located outside the conductor a distance d_o m from the center.

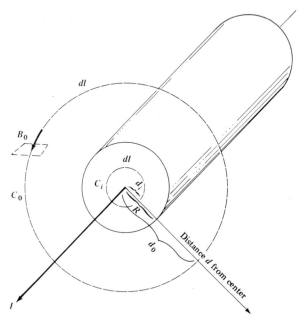

Figure 7-5 Application of Ampere's law to circular contours around a conductor of circular cross section.

From *Ampere's law, applied to circular contours around a circular conductor,* as shown in Fig. 7-5,

$$B \oint_C d\ell = \mu_0 I_{\text{enc}} \tag{7-4}$$

where $d\ell$ = differential increment of length on the contour, m

$\oint_C d\ell$ = length of contour C, also called the line integral around contour C, m

I_{enc} = current *enclosed by contour C*, A

The length of any circular contour located d m from the center of the conductor is the circumference $2\pi d$ m. Thus, for any circular contour

$$\oint_C d\ell = 2\pi d \tag{7-5}$$

Substituting Eq. (7-5) into Eq. (7-4),

$$B2\pi d = \mu_0 I_{\text{enc}}$$

$$B = \frac{\mu_0 I_{\text{enc}}}{2\pi d} \tag{7-6}$$

Equation (7-6) is an expression for the flux density at any point on a circular contour, d m from the center of a conductor, *that encloses a current* I_{enc}.

Contours *located on the surface of the conductor, or outside the conductor, enclose the total current*, and Eq. (7-6) becomes

$$B = \frac{\mu_0 I_T}{2\pi d} \qquad d \geq R$$

where I_T = total current in the conductor

However, *for contours located inside the conductor ($d < R$), the current enclosed by the contour is only a fraction of the total current*. The fraction of the total current enclosed by an inner contour is

$$\frac{I_{enc}}{I_T} = \frac{\pi d^2}{\pi R^2} \qquad d < R$$

$$I_{enc} = \frac{d^2 I_T}{R^2} \qquad d < R \tag{7-7}$$

Substituting Eq. (7-7) into Eq. (7-6),

$$B = \frac{\mu_0}{2\pi d} \frac{d^2}{R^2} I_T \qquad d < R$$

$$B = \frac{\mu_0 d I_T}{2\pi R^2} \qquad d < R \tag{7-8}$$

7-4 MAGNETIC FIELD ABOUT A COIL

Coiling a conductor into two or more turns has the effect of using the same current more than once. For example, a four-turn coil carrying a current of 10 A produces the same magnetic flux density as a one-turn coil carrying 40 A. Hence, the magnetomotive force produced by a current in a coil is equal to the current times the number of turns in the coil. Expressed mathematically,

$$\boxed{\mathscr{F} = NI} \tag{7-9}$$

where $\mathscr{F}$ = magnetomotive force, ampere-turns (A · t)
         N = number of series-connected turns in the coil
         I = current in the coil, amperes (A)

The direction of the magnetic field about a coil may be determined by using a variation of the right-hand rule described in Sec. 7-1. Grasp the coil with the *right hand* with the fingers curled in the direction the current takes around the loops, and the thumb will then point in the direction of the magnetic field inside the coil. This is illustrated in Fig. 7-6.

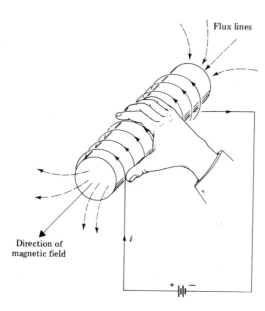

Flux lines

Direction of
magnetic field

i

$+ \; -$

Figure 7-6   Right-hand rule for determining the direction of the magnetic field through the window of a çoil for a given direction of coil current, and vice versa.

It should be noted that the flux lines representing the magnetic field around a wire or around a coil are continuous lines; there is no starting point and no finishing point.

7-5   FERROMAGNETISM

A magnetic field produced within a ferromagnetic material such as iron or steel is caused by the uncompensated spinning of electrons about their own axis within the atomic structure of the material, and by the parallel alignment of these electrons with similar uncompensated electron spins in adjacent atoms. Groups of adjacent atoms with parallel electron spins, shown in Fig. 7-7, are called *magnetic domains*; the arrows indicate the directions of the respective magnetic fields of the individual spinning electrons. Each domain is in itself a region of strong magnetization.

In unmagnetized material (Fig. 7-7a), the random orientation of the domains results in no net external field. However, if the magnetic material is placed in a region possessing a magnetic field, for example, the earth's magnetic field, or the magnetic field around a current, the domains will tend to align themselves in the direction of the field as shown in Fig. 7-7b. When removed from the field, the domains tend to go back to a random pattern. In materials used for permanent magnets, the magnetic domains are difficult to align; but once aligned, and then removed from the external magnetizing field, a good amount of alignment is retained. This is called residual magnetism.

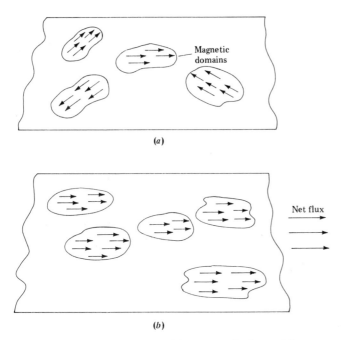

Figure 7-7 Magnetic domains: (*a*) unmagnetized material;
(*b*) magnetized material.

7-6   MAGNETIC CIRCUIT

A magnetic circuit as used in transformers, machines, and other magnetic apparatus is an arrangement of magnetic materials (called the core) which forms a path to contain the flux and guide it in a specific direction. Magnetic materials provide a means by which excellent control may be exercised over the magnitude, density, and direction of this flux. Some examples of magnetic circuits are shown in Fig. 7-8.

The flux within the core of a magnetic circuit is directly proportional to the applied magnetomotive force, and inversely proportional to the reluctance of the materials that makes up the magnetic circuit. *Reluctance is a measure of the opposition that the magnetic circuit offers to the flux.* Expressed mathematically,

$$\Phi = \frac{\mathscr{F}}{\mathscr{R}}$$

(7-10)

where $\mathscr{F}$ = mmf, ampere-turns (A · t)
         $\mathscr{R}$ = reluctance, ampere-turns per weber (A · t/Wb)
         Φ = flux, webers (Wb)

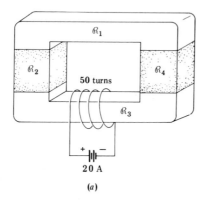

50 turns

20 A

(a)

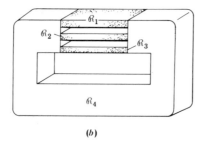

(b)

Figure 7-8 Examples of magnetic circuits: (*a*) series; (*b*) parallel-series.

Equation (7-10) may also be written as

$$\Phi = \frac{NI}{\mathscr{R}}$$

The relationship expressed in Eq. (7-10) is called the *magnetic-circuit law*, and is analogous to the Ohm's law equation for the electric circuit

$$I = \frac{E}{R}$$

Because of this similarity, Eq. (7-10) is sometimes referred to as *Ohm's law of the magnetic circuit.*

Example 7-2 Figure 7-9 shows a doughnut-shaped magnetic circuit whose reluctance is given as 5000 A · t/Wb. The five-turn coil has a resistance of 0.20 Ω. Calculate the flux in the core.

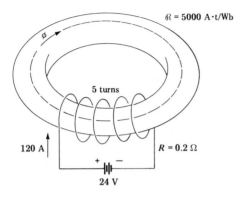

$\mathcal{R} = 5000 \text{ A·t/Wb}$

5 turns

120 A

$R = 0.2 \ \Omega$

24 V

Figure 7-9   Circuit for Example 7-2.

Solution

$$I = \frac{E}{R} = \frac{24}{0.20} = 120 \text{ A}$$

$$\mathcal{F} = NI = 5(120) = 600 \text{ A·t}$$

$$\Phi = \frac{\mathcal{F}}{\mathcal{R}} = \frac{600}{5000} = 0.120 \text{ Wb}$$

The reluctance of a ferromagnetic circuit or section of a ferromagnetic circuit is dependent on its length and cross-sectional area, and the permeability of the material used. Expressed mathematically,

$$\mathcal{R} = \frac{\ell}{\mu A} \qquad\qquad (7\text{-}11)$$

where $\mathcal{R}$ = reluctance, ampere-turns per weber (A·t/Wb)
   ℓ = length of section, meters (m)
   A = cross-sectional area of section, square meters (m²)
   μ = permeability of material, henrys per meter (H/m)

Permeability is a measure of the ease with which a material conducts magnetic flux. A material of high permeability is a good conductor of magnetic flux.

Reluctances in Series and Parallel

The equivalent value of reluctances in series or reluctances in parallel is calculated in a manner similar to that for resistances in series or resistances in parallel. Thus, *for a series magnetic circuit,*

$$\mathscr{R}_{eq,S} = \mathscr{R}_1 + \mathscr{R}_2 + \mathscr{R}_3 + \cdots + \mathscr{R}_n \qquad (7\text{-}12)$$

For a parallel magnetic circuit, or the parallel section of a series-parallel circuit,

$$\frac{1}{\mathscr{R}_{eq,P}} = \frac{1}{\mathscr{R}_1} + \frac{1}{\mathscr{R}_2} + \frac{1}{\mathscr{R}_3} + \cdots + \frac{1}{\mathscr{R}_n} \qquad (7\text{-}13)$$

Example 7-3 Calculate the magnetic flux in the series magnetic circuit shown in Fig. 7-8a. Assume

$\mathscr{R}_1 = 4000 \text{ A} \cdot \text{t/Wb}$

$\mathscr{R}_2 = 5000 \text{ A} \cdot \text{t/Wb}$

$\mathscr{R}_3 = 3000 \text{ A} \cdot \text{t/Wb}$

$\mathscr{R}_4 = 8000 \text{ A} \cdot \text{t/Wb}$

Solution

$\mathscr{R}_{eq,S} = 4000 + 5000 + 3000 + 8000$

$\mathscr{R}_{eq,S} = 20{,}000 \text{ A} \cdot \text{t/Wb}$

$$\Phi = \frac{\mathscr{F}}{\mathscr{R}} = \frac{50(20)}{20000} = 0.05 \text{ Wb}$$

Example 7-4 Calculate the reluctance of the magnetic circuit in Fig. 7-8b. Assume:

$\mathscr{R}_1 = 2000 \text{ A} \cdot \text{t/Wb} \qquad \mathscr{R}_2 = 3000 \text{ A} \cdot \text{t/Wb}$

$\mathscr{R}_3 = 7000 \text{ A} \cdot \text{t/Wb} \qquad \mathscr{R}_4 = 2500 \text{ A} \cdot \text{t/Wb}$

Solution
The equivalent reluctance of the three paralleled reluctances is

$$\frac{1}{\mathscr{R}_{eq,P}} = \frac{1}{2000} + \frac{1}{3000} + \frac{1}{7000}$$

$$\frac{1}{\mathscr{R}_{eq,P}} = 0.00050 + 0.000333 + 0.000143$$

$$\frac{1}{\mathscr{R}_{eq,P}} = 0.000976$$

$$\mathscr{R}_{eq,P} = 1024.6 \text{ A} \cdot \text{t/Wb}$$

The equivalent parallel reluctance is in series with $\mathscr{R}_4$. Hence,

$$\mathscr{R}_{eq} = 1024.6 + 2500$$

$$\mathscr{R}_{eq} = 3524.6 \text{ A} \cdot \text{t/Wb}$$

-7   MAGNETIC FIELD INTENSITY

Magnetic field intensity is a measure of the intensity of the magnetomotive force applied to a magnetic circuit; it has significant applications in magnetic-circuit design by providing a relatively simple method for solving magnetic-circuit problems. The relationship between magnetic field intensity and magnetic flux density is obtained by substituting Eqs. (7-9) and (7-11) into Eq. (7-10):

$$\Phi = \frac{\mathscr{F}}{\mathscr{R}} = \frac{NI}{\ell/\mu A} = \frac{\mu ANI}{\ell}$$

Rearranging the terms and substituting B for Φ/A,

$$\frac{\Phi}{A} = \frac{\mu NI}{\ell}$$

$$B = \frac{\mu NI}{\ell}$$

Defining,

$$\boxed{H = \frac{NI}{\ell}} \tag{7-14}$$

and substituting into the preceding equation,

$$\boxed{B = \mu H} \tag{7-15}$$

where B = flux density in a *section* of the ferromagnetic circuit, T
H = magnetic field intensity, ampere-turns per unit length of the *section*, A · t/m
μ = permeability of the substance that makes up the *section*, H/m.

Graphs of Eq. (7-15) for some commonly used ferromagnetic materials are shown in Fig. 7-10a. These graphs are called *BH* curves or magnetization curves. *The nonlinearity of the curves indicates that the permeability μ of the magnetic material is not constant.* The permeability of a magnetic material operating at some flux density B_1 may be determined by obtaining the corresponding H_1 from the *BH* curve, and then substituting the values into Eq. (7-15). Thus, if cast steel is to be operated at a flux density of 1.25 T, the corresponding magnetic field intensity obtained from the curve will be 1500 A · t/m. From Eq. (7-15),

$$\mu = \frac{B}{H} = \frac{1.25}{1500} = 0.000833 \text{ H/m}$$

Magnetic permeability is the ratio of the density of flux produced in a substance to the intensity of the magnetizing force that caused it.

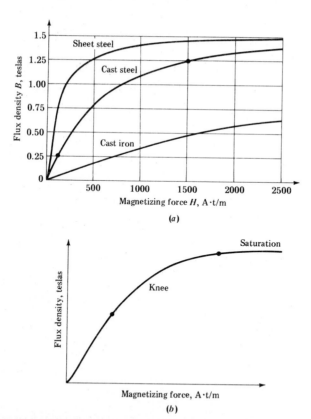

Figure 7-10 (*a*) Magnetization curves (*BH*) for different ferromagnetic materials; (*b*) general shape of *BH* curves for ferromagnetic materials.

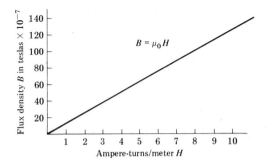

Figure 7-11 *BH* curve for free space.

Figure 7-10*b* illustrates the general shape of *BH* curves for ferromagnetic materials used in electrical machinery and transformers. The curves are slightly concave up for "low" flux densities, exhibit a straight-line characteristic for "medium" flux densities, and are then concave down for "high" flux densities. The part of the curve that is concave down is known as the *knee* of the curve. If further increases in magnetizing force produce no "useful" increase in flux density, the material is said to be *saturated*, and the *BH* curve will be almost flat. Magnetic saturation is the condition that exists when all the magnetic domains are oriented in the direction of the applied magnetomotive force. The *BH* curve for free space is a straight line, as shown in Fig. 7-11, because the permeability of free space is constant.

Example 7-5 Referring to Fig. 7-12, determine the current required in the 100-turn coil in order to obtain a flux density of 0.25 T in the air gap. The ferromagnetic core is cast steel.

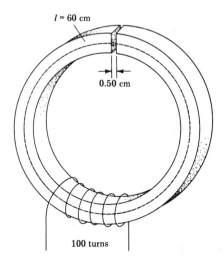

Figure 7-12 Circuit for Example 7-5.

Solution
The magnetic circuit is in two sections, a magnetic core and an air gap. Neglecting the small amount of fringing around the ends of the core, the flux density in the core is approximately the same as in the air gap, 0.25 T. The magnetizing force required to obtain 0.25 T in cast steel is obtained from Fig. 7-10a:

$$H_C = 125 \text{ A} \cdot \text{t/m}$$

The length of the core is given as 60 cm or 0.60 m. Hence, the ampere-turns required to produce a flux density of 0.25 T in the core alone are

$$H_C = \frac{\mathscr{F}_C}{\ell_C}$$

$$125 = \frac{\mathscr{F}_C}{0.60}$$

$$\mathscr{F}_C = 75 \text{ A} \cdot \text{t}$$

The magnetizing force required to obtain a flux density of 0.25 T in the air gap is determined from Eq. (7-15):

$$B_0 = \mu_0 H_0$$

$$0.25 = 4\pi 10^{-7} H_0$$

$$H_0 = 198{,}944 \text{ A} \cdot \text{t/m}$$

Since the length of the air gap is 0.5 cm or 0.005 m, the ampere-turns required to produce a flux density of 0.25 T in the air gap alone are

$$H_0 = \frac{\mathscr{F}_0}{\ell_0}$$

$$198{,}944 = \frac{\mathscr{F}_0}{0.005}$$

$$\mathscr{F}_0 = 994.72 \text{ A} \cdot \text{t}$$

Thus, the total ampere-turns required are

$$\mathscr{F}_C + \mathscr{F}_0 = 75 + 994.72 = 1069.72 \text{ A} \cdot \text{t}$$

$$\mathscr{F} = NI$$

$$1069.72 = 100 I$$

$$I = 10.70 \text{ A}$$

-8   RELATIVE PERMEABILITY

The relative permeability of a material is defined as the ratio of its permeability
to the permeability of free space:

$$\boxed{\mu_r = \frac{\mu}{\mu_0}} \qquad\qquad (7\text{-}16)$$

where μ_r = relative permeability, a dimensionless constant
μ = permeability of the material, H/m
μ_0 = permeability of free space ($4\pi 10^{-7}$ H/m)

Using the definition in Eq. (7-16), the relative permeability of free space is

$$\mu_{r(\text{free space})} = \frac{\mu_0}{\mu_0} = 1$$

The relative permeability of any material is a measure of the relative ease
with which that material conducts magnetic flux compared with the conduction of flux in free space. The relative permeability of air is so very nearly
equal to that of a vacuum that for all practical purposes it is considered
equal to 1.

Materials that have relative permeabilities less than that of air are
called diamagnetic. Examples of diamagnetic materials are carbon, copper,
silver, and bismuth. Diamagnetic materials are very weakly repelled by a
magnet. Materials that have a relative permeability only slightly greater than
that of air, platinum, for example, are called *paramagnetic*. Materials that
have a relative permeability much greater than that of air are called *ferromagnetic*. Examples of ferromagnetic materials are iron, nickel, and cobalt.
Ferromagnetic materials are used for the cores of all electromagnetic equipment such as machinery, transformers, and magnets. Whenever reference is
made to magnetic materials, a ferromagnetic substance is implied.

7-9   MAGNETIC HYSTERESIS

If an *alternating magnetomotive force* is applied to a magnetic material and the
flux density B is plotted against the magnetic field intensity H, the resultant
magnetization curve is called a *hysteresis loop*. The hysteresis loop and the
circuit used to obtain it are shown in Fig. 7-13. The double-pole double-
throw switch (DPDT) is used to reverse the direction of current through the
coil, and the rheostat is used to vary the magnitude of the current from zero

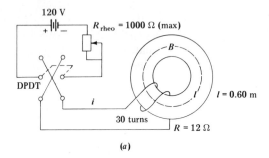

(a)

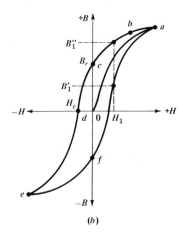

Figure 7-13 (a) circuit for obtaining a hysteresis loop; (b) hysteresis loop.

(b)

amperes with the circuit open to a maximum of 10 A with the circuit closed and the rheostat set for zero resistance.

$$I_{max} = \frac{E_{bat}}{R_{coil} + R_{rheo}} = \frac{120}{12 + 0} = 10 \text{ A}$$

The corresponding magnetic field intensity is

$$H_{max} = \frac{NI_{max}}{\ell} = \frac{30(10)}{0.60} = 500 \text{ A} \cdot \text{t/m}$$

Assuming an unmagnetized core ($H = 0$, $B = 0$), increasing the current from zero to its maximum of 10 A, decreasing it to zero, building it up to 10 A in the reverse direction, decreasing to zero, and then back up to 10 A in the forward direction generates the BH curve shown in Fig. 7-13b; the excursion of the BH curve follows the sequence of points 0, a, b, c, d, e, f, a. Note that the curve does not return to the initial starting values of zero H and zero B.

Point c on the curve designates the residual flux density B_r, called *residual magnetism*, that can be retained in a *closed magnetic circuit* after the current in the coil is reduced to zero. Point d designates negative magnetomotive force H_c, called *coercive force*, required to reduce the residual magnetism to zero. Further examination of the hysteresis loop shows that, for a given magnetizing force, the flux density will be different when the current in the coil is increasing than when the current is decreasing. For example, for the same magnetizing force of H_1 in Fig. 7-13b, the flux density will be B'_1 when the magnetizing force is increasing and will be B''_1 when the magnetizing force is decreasing. This phenomenon, called *hysteresis*, is exhibited by all ferromagnetic materials.

The hysteresis effect in a magnetic material is due to the opposition offered by the magnetic domains to the turning moment of a magnetizing force and is evidenced by the lagging of the flux behind the magnetizing force. Hard steels such as high-carbon steels resist magnetization to a considerable extent. Low-carbon steels with a high silicon content offer little opposition to change and are easily magnetized and demagnetized. Steels with high values of hysteresis are useful for permanent magnets, for when finally magnetized the domains do not easily disorient themselves. Magnetic alloys such as alnico retain a considerable proportion of their flux after removal of the magnetizing force.

Magnetic hysteresis affects the rate of response of magnetic flux to a magnetizing force. In electrical apparatus such as transformers, in which the desired characteristic necessitates quick response of flux to a change in mmf, with little residual magnetism, a high-grade silicon steel is used. On the other hand, machines such as self-excited generators require steel that retains sufficient residual magnetism to permit the buildup of voltage. Hence, the choice of magnetic materials is dictated by the application.

7-10   **FORMATION OF POLES IN A MAGNETIC CIRCUIT**

Magnetic poles that form in parts of a magnetic circuit are manifestations of changes in the reluctance of the circuit. If the reluctance of a magnetic circuit is uniform throughout, no poles will be evidenced. For example, consider the homogeneous iron doughnut shown in Fig. 7-14a. The application of a magnetomotive force causes flux within the iron core, but the iron does not exhibit poles until the ring is cut as shown in Fig. 7-14b. Cutting the doughnut introduces a path of higher reluctance that is manifested by the formation of poles at the open ends. Likewise, a conductor carrying an electric current is surrounded by a magnetic field, as shown in Fig. 7-14c. If the reluctance of the surrounding medium remains uniform, air, for example, no magnetic poles will be formed. The introduction of a block of iron in the vicinity of the wire sets up areas in which changes in reluctance occur. This manifests itself by the formation of poles, as shown in Fig. 7-14d.

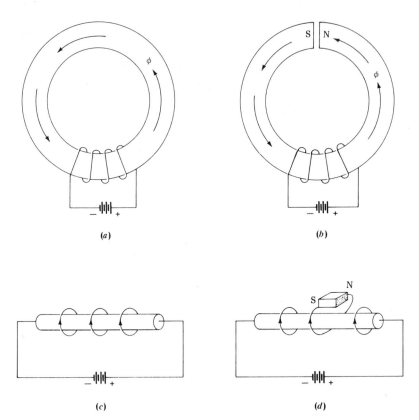

Figure 7-14 Formation of poles in a magnetic circuit: (*a*), (*b*) caused by a cut in the iron doughnut; (*c*), (*d*) caused by the introduction of a block of iron.

7-11 INTERACTION OF MAGNETIC FIELDS

When two or more *sources* of magnetic fields *are arranged so that their fluxes, or a component of their fluxes are parallel, within a common region, a mechanical force will be produced that tends to either push the sources of flux together or to push them apart.* The relationship between the direction of the two magnetic fields and the direction of the mechanical force produced by the interaction of the two fields can be established with the aid of the magnets shown in Fig. 7-15.

Through experiment, it is learned that like poles of magnets repel each other, whereas unlike poles attract. A comparison of Fig. 7-15*a* and Fig. 7-15*b* indicates that a *repelling or separating force occurs when the two magnets have components of flux in the same region that are parallel and in the same direction,* whereas *a force of attraction occurs when the two magnets have components of flux in the same region that are parallel but in opposite directions.*

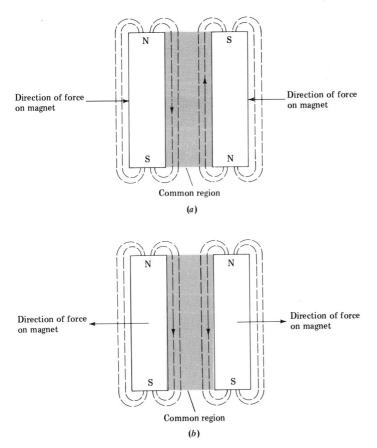

Figure 7-15 Mechanical force produced by parallel magnets: (*a*) force of attraction; (*b*) force of repulsion.

Thus, when determining the direction of the mechanical force produced by the interaction of the magnetic fields of two currents, or the mechanical force produced by the interaction of the magnetic field of a magnet and the magnetic field of a current, the only information necessary is the relative directions of the fluxes produced in a common region by the two sources.

Figure 7-16 shows the direction of the mechanical forces produced by the magnetic fields of adjacent current-carrying conductors. If the currents in adjacent conductors are in the same direction, as shown in Fig. 7-16*a*, the fluxes *in the common region will be parallel and in opposite directions, causing a force of attraction.* However, if the currents in adjacent conductors are in opposite directions, as shown in Fig. 7-16*b, the respective fluxes in the common region will be parallel and in the same direction, causing a separating force to be produced.*

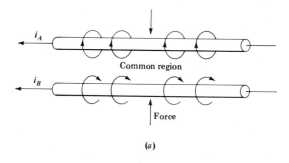

(a)

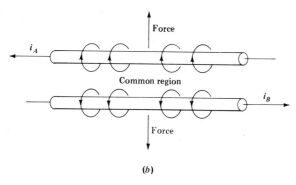

(b)

Figure 7-16 Mechanical force produced by the magnetic fields of adjacent current-carrying conductors: (a) force of attraction; (b) force of repulsion.

The forces between adjacent conductors, under conditions of very high currents such as may occur during fault conditions, can literally bend the conductors out of shape and even tear them from their supporting structures. Thus, in those applications where the *available short-circuit current* is of such a magnitude as to cause destruction of equipment if a fault occurs, special fuses (called *current limiters*) are installed. These fuses can open the circuit in less than 4 ms, preventing the current from attaining damaging values.†

Figure 7-17a illustrates the interaction of the magnetic field produced by the current in a conductor and the magnetic field produced by a magnet. The flux from the magnet is upward on both sides of the conductor, whereas the flux produced by the conductor is downward on the left side of the conductor and upward on the right side. In the common region to the left of the conductor the respective fluxes are in opposite directions, and in the common region to the right of the conductor the respective fluxes are in the same

† C. I. Hubert, *Preventative Maintenance of Electrical Equipment*, McGraw-Hill Book Company, New York, 1969.

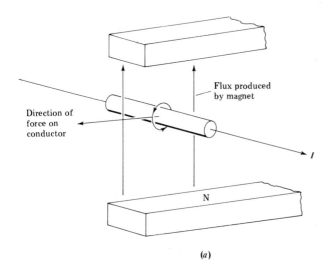

(a)

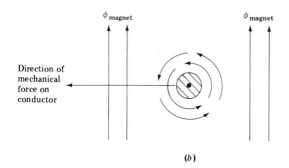

(b)

Figure 7-17 (*a*) Interaction of the magnetic field produced by
current in a conductor, situated in and perpendicular to the
flux lines of a magnet; (*b*) end view of conductor.

direction. Figure 7-17*b* is an end view of the conductor showing the flux due
to the magnet and the flux due to the current in the conductor. The direction
of the force on the conductor will be toward the left, and the direction of the
force on the magnet will be toward the right. If the magnet is locked in
position, and the mechanical force is great enough, the conductor will be
thrown out of the field.

If the conductor is arranged so that its axis lies parallel to the flux lines
of the magnet, as shown in Fig. 7-18, *all the flux lines of the magnet will be
perpendicular to all the flux lines of the current, and no mechanical force will be
produced.*

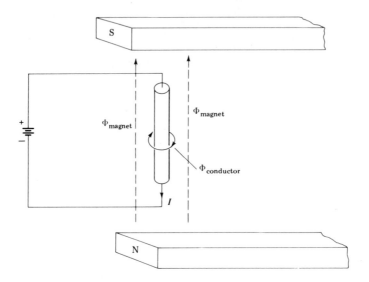

Figure 7-18 Conductor with current axis parallel to the flux of the magnet.

SUMMARY OF FORMULAS

$$\Phi = B \times A \qquad \Phi = \frac{\mathscr{F}}{\mathscr{R}}$$

$$\mathscr{R} = \frac{\ell}{\mu A} \qquad \mathscr{F} = NI$$

$$\mathscr{R}_{eq, S} = \mathscr{R}_1 + \mathscr{R}_2 + \mathscr{R}_3 + \cdots + \mathscr{R}_n$$

$$\frac{1}{\mathscr{R}_{eq, P}} = \frac{1}{\mathscr{R}_1} + \frac{1}{\mathscr{R}_2} + \frac{1}{\mathscr{R}_3} + \cdots + \frac{1}{\mathscr{R}_n}$$

$$B = \frac{\mu_0 I_T}{2\pi d} \qquad d \geq R$$

$$B = \frac{\mu_0 d I_T}{2\pi R^2} \qquad d < R$$

$$B = \mu H \qquad H = \frac{NI}{\ell} \qquad \mu_r = \frac{\mu}{\mu_0} \qquad \mu_0 = 4\pi(10^{-7})$$

PROBLEMS

7-1 A round copper conductor 2.0 cm in diameter is carrying a current of 1240 A. Determine the flux density (*a*) at the surface of the conductor; (*b*) 0.8 cm from the center of the conductor; (*c*) 0.60 m from the center of the conductor.

7-2 An aluminum conductor 7.5 cm in diameter is carrying a current of 2000 A. Determine the flux density (*a*) at the center of the conductor; (*b*) 3.0 cm from the center of the conductor; (*c*) 20 cm from the center of the conductor.

7-3 Two hundred turns of copper wire are wound around a doughnut-shaped ferromagnetic core. A 6-V battery connected to the coil causes a current of 3 A. Sketch the circuit and determine (*a*) the resistance of the coil; (*b*) the flux in the ferromagnetic core if the reluctance of the magnetic circuit is 1500 A · t/Wb.

7-4 Determine the permeability of a ferromagnetic circuit that has a mean length of 0.2 m, a cross section of 0.025 m^2, and a reluctance of 3200 A · t/Wb.

7-5 A ferromagnetic circuit has a mean length of 2.0 m and a cross-sectional area of 0.25 m^2. A 20-turn coil carrying a current of 30 A is wound around the core. Assuming the permeability is 1.131×10^{-3} H/m for the given current, determine (*a*) the reluctance of the magnetic circuit; (*b*) the core flux; (*c*) the flux density.

7-6 A cast-iron ring has an average length of 0.060 m and a cross-sectional area of 0.20 m^2. A 15-turn coil wound on the ring has a resistance of 0.80 Ω. Determine the flux in the core if a 6.0-V battery is connected to the coil.

7-7 A magnetic circuit has a mean length of 1.40 m and a cross-sectional area of 0.25 m^2. Current in a 100-turn 60-Ω coil wound around the core causes a core flux of 0.060 Wb. Assuming the reluctance of the magnetic circuit is 2000 A · t/Wb, determine (*a*) the current required to produce the flux; (*b*) the voltage required to produce the current; (*c*) the permeability of the core.

7-8 A magnetic circuit composed of two half rings is joined together at the ends to form a doughnut. The mean length of each half ring is 0.4 m and the cross-sectional area of each is 0.0010 m^2. A 400-turn coil is wound around the doughnut. If one half ring is cast steel and the other is sheet steel, determine the current required to establish a flux of 5×10^{-4} Wb in the core.

7-9 Determine the ampere requirements for the 200-turn coil in Fig. 7-19, in order to obtain a flux density of 0.50 T in the air gap. Assume the ferromagnetic material is cast iron.

7-10 Repeat Problem 7-9, assuming the ferromagnetic material is sheet steel.

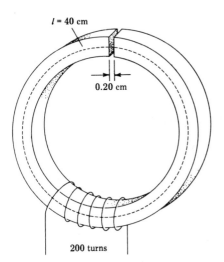

l = 40 cm

0.20 cm

200 turns

Figure 7-19 Figure for Prob. 7-9.

CHAPTER 8

INDUCTANCE

Inductance (also called self-inductance) *is that property of a circuit or circuit element that delays a change in the current through it. The delay is accompanied by the absorption or release of energy, and is associated with a change in the magnitude of the magnetic field surrounding the conductors.*

In any electric circuit, all the magnetic flux surrounding the current-carrying conductors passes in the same direction through the *window* formed by the circuit. This is shown in Fig. 8-1a for a single-loop circuit and in Fig. 8-1b for a three-loop coil.

When the switch to an electric circuit is closed, the *buildup of current in the circuit causes a buildup of flux through the window. The changing flux generates a voltage in the circuit that acts in opposition to the changing current. This opposing action is a manifestation of Lenz's law. In accordance with Lenz's law, any magnetically induced voltage will always be generated in a direction to oppose the action that caused it.*

8-1 VOLTAGE AND CURRENT CHARACTERISTICS OF AN INDUCTOR

The magnitude of the voltage induced in any coil, by a changing magnetic flux, is proportional to the number of turns of wire in the coil and the rate of change of

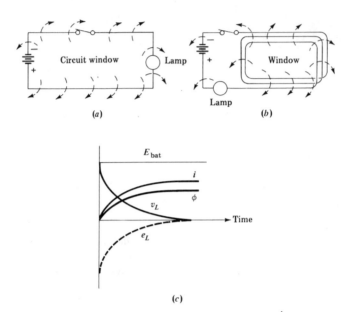

Figure 8-1 Inductance property of a circuit: (*a*) flux through window of a single loop; (*b*) flux through window of a multiloop coil; (*c*) representative curves for *i*, ϕ, v_L, and e_L.

flux through its window. This relationship is known as *Faraday's law.* Expressed mathematically,

$$e_L = -N\frac{d\phi}{dt}$$ (8-1)

where e_L = voltage induced in the coil, volts (V)

      N = number of series-connected turns in the coil

      $\dfrac{d\phi}{dt}$ = rate of change of flux through the window of the coil, webers per second (Wb/s)

The *minus* sign comes from Lenz's law, and indicates that the voltage is generated in a direction to oppose the change in flux that caused it. Because of its opposing action, a magnetically induced voltage is often called a counter-voltage or counter-emf (cemf).

    When an electric circuit is connected to a generator or battery, the driving voltage has to force the current against the opposing voltage induced by the changing flux. In accomplishing this, part of the driving voltage is "used up" in overcoming the induced voltage. This "loss of voltage," called a *voltage drop* (v_L), *is equal in magnitude but opposite in direction to the induced voltage*. That is,

$$v_L = -e_L$$

Thus *the voltage drop*, caused by the induced voltage, is

$$v_L = N \frac{d\phi}{dt} \tag{8-2}$$

The self-induced voltage, called the voltage of self-induction, *does not prevent the current from changing; it serves only to delay the change*. Thus, after the switch is closed, the current will rise with time, from zero amperes at time-zero to its final (steady-state) value equal to E_{bat} divided by the circuit resistance.

$$i_{ss} = \frac{E_{bat}}{R} \tag{8-3}$$

A change in the magnitude or direction of the current in any conductor or coil will always set up a voltage in a direction to oppose the change. Hence, *the direction of the induced voltage will be dependent on whether the current is increasing or decreasing.*

The inductive effect is somewhat analogous to the flywheel effect in a mechanical system. A flywheel has a fixed value of inertia, which is a function of its mass and physical dimensions. As long as its velocity does not change, the opposition to motion offered by the flywheel is zero. However, if any attempt were made to change the velocity, a force or reactance would be set up by the flywheel to oppose the change. *The opposing force serves merely to delay the change, but cannot prevent it.*

Likewise, any attempt to change the rate of flow of electrons in a conductor or coil would set up a voltage that would delay but not prevent the change.

The value of the current at any instant of time before steady state depends on the instantaneous magnitude of the opposing self-induced voltage, as well as on the battery voltage and the circuit resistance. From Kirchhoff's voltage law, the algebraic sum of the voltage drops must equal the net driving voltage. Hence, for Fig. 8-1a and b, when the switch is closed,

$$E_{bat} = v_L + iR \tag{8-4}$$

Solving for the current,

$$i = \frac{E_{bat} - v_L}{R} \qquad (8\text{-}5)$$

where $\quad i =$ instantaneous current, A
$\quad E_{bat} =$ battery voltage, V
$\quad v_L =$ instantaneous self-induced voltage drop, V
$\quad R =$ overall resistance of the circuit, Ω

Defining $t = (0-)$ to be an infinitesimal period of time before the switch is operated and $t = (0+)$ to be an infinitesimal period of time after the switch is operated,

$$i\big|_{t=(0+)} = i\big|_{t=(0-)} = 0$$

The time delay imposed by the inductance causes the value of the current at an infinitesimal time after the switch is closed to be equal to the value of the current at an infinitesimal time before the switch was closed. Hence, at $t = (0+)$, Eq. (8-5) reduces to

$$0 = \frac{E_{bat} - v_L}{R}$$

solving for v_L,

$$v_L\big|_{t=(0+)} = E_{bat} \qquad (8\text{-}6)$$

Equation (8-6) indicates that, at the instant the switch is closed, the voltage drop caused by the emf of self-induction will be equal to the battery voltage.

Representative curves for i, ϕ, v_L, and e_L are shown in Fig. 8-1c; note that when the current (and hence the flux) has the greatest rate of change, the induced voltage has its greatest value. When the current reaches its steady-state value, the flux through the window will have attained its maximum value and will no longer be changing. Thus the self-induced voltage, which depends on a changing flux, will be zero at steady state.

8-2   LUMPED INDUCTANCE

The self-inductance of a circuit may be increased by adding a lumped inductance, called an *inductor*, in series with the circuit as shown in Fig. 8-2a. The lumped inductance consists of a coil of wire with or without a ferromagnetic core. If the lumped inductance is very much greater than the self-inductance produced by the connecting wires, which is generally the case, the inductance of the connecting wires may be neglected.

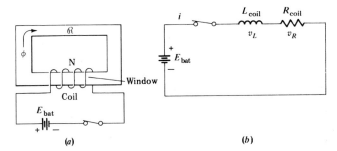

Figure 8-2 (*a*) Lumped inductance; (*b*) equivalent circuit of a lumped inductance.

The flux through the window of the coil in Fig. 8-2*a* is directly related to the current in the coil by the magnetic-circuit law developed in Sec. 7-6.

$$\phi = \frac{Ni}{\mathscr{R}} \tag{8-7}$$

where ϕ = flux through the coil window, webers (Wb)
i = current in the coil, amperes (A)
$\mathscr{R}$ = reluctance of the magnetic circuit, ampere-turns per weber $(A \cdot t/Wb)$
N = number of series-connected turns

Assuming the reluctance is constant, the rate of change of flux is proportional to the rate of change of current. Thus from Eq. (8-7),

$$\frac{d\phi}{dt} = \frac{N}{\mathscr{R}}\frac{di}{dt} \cdot \tag{8-8}$$

where $\dfrac{d\phi}{dt}$ = rate of change of flux

$\dfrac{di}{dt}$ = rate of change of current

Substituting Eq. (8-8) into Eq. (8-2),

$$v_L = N\frac{d\phi}{dt} = N\frac{N}{\mathscr{R}}\frac{di}{dt}$$

$$v_L = \frac{N^2}{\mathscr{R}}\frac{di}{dt} \tag{8-9}$$

Defining,

$$L = \frac{N^2}{\mathcal{R}} \qquad (8\text{-}10)$$

Equation (8-9) becomes

$$v_L = L\frac{di}{dt} \qquad (8\text{-}11)$$

Parameter L is called the self-inductance of the inductor, and is expressed in henrys (H).

A more meaningful expression for L is obtained by equating the two expressions for the generated voltages: Eqs. (8-11) and (8-2). Thus,

$$L\frac{di}{dt} = N\frac{d\phi}{dt}$$

Solving for L,

$$L = N\frac{d\phi}{di} \qquad (8\text{-}12)$$

where $\dfrac{d\phi}{di}$ represents the rate of change of flux with respect to current.

Equation (8-12) provides a very powerful tool for the analysis of nonlinear magnetic components.

When analyzing circuits containing lumped inductance, the resistance of the coil, and the voltage drop due to that resistance, must be considered. As an aid in the analysis, the inductance and resistance properties of the coil are generally represented as separate series-connected elements in an *equivalent circuit*, as shown in Fig. 8-2b. The circuit shown in Fig. 8-2b is the equivalent circuit of Fig. 8-2a. If the resistance of the coil is appreciably greater than the resistance of the connecting wires, the resistance of the connecting wires may be neglected.

Example 8-1 A 300-turn coil has a resistance of 6.0 Ω and an inductance of 0.50 H. Determine the new resistance and the new inductance if one-third of the turns are removed. Assume all turns have the same circumference.

Solution
Since the resistance of a coil is proportional to the length of wire in the coil,

$$\frac{R_1}{R_2} = \frac{N_1}{N_2}$$

$$\frac{6}{R_2} = \frac{300}{200}$$

$$R_2 = 4\ \Omega$$

The inductance may be determined from Eq. (8-10),

$$\frac{L_1}{L_2} = \frac{(N^2/\mathscr{R})_1}{(N^2/\mathscr{R})_2}$$

Assuming the reluctance is constant,

$$\frac{0.5}{L_2} = \frac{(300)^2}{(200)^2}$$

$$L_2 = 0.22\ \text{H}$$

8-3   NORMAL INDUCTANCE

The relationship between the inductance of a coil and the reluctance of its magnetic circuit, as expressed by Eq. (8-10), is

$$L = \frac{N^2}{\mathscr{R}}$$

If $\mathscr{R}$ is constant, which is the case for an air-core inductor, then the *magnetization curve* will be a straight line (linear) as shown in Fig. 8-3a. The magnetization curve is a plot of Eq. (8-7).

However, if the reluctance is not constant over the full range of current, the flux will not be proportional to the current, and the magnetization curve will be nonlinear as shown in Fig. 8-3b. The magnetization curve shown in Fig. 8-3b is characteristic of ferromagnetic cores used in electrical machinery and transformers.

The inductance of a coil is related to the slope, or steepness, of its magnetization curve by Eq. (8-12), where

$$L = N\frac{d\phi}{di}$$

The slope is represented by the ratio $d\phi/di$, which is the rate of change of flux with respect to current. The slope of the magnetization curve in Fig. 8-3a is

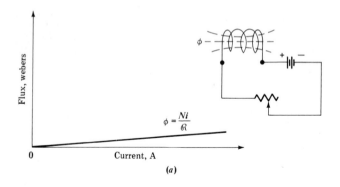

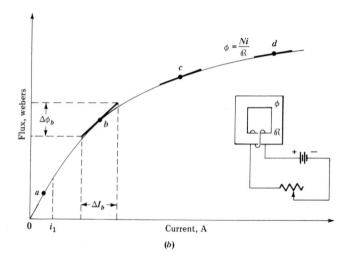

Figure 8-3 Magnetization curves: (*a*) air core; (*b*) ferromagnetic core.

constant over the entire range of current. Thus, the ratio $d\phi/di$ does not change and the inductance L is constant.

The inductor represented by the magnetization curve in Fig. 8-3*b* is called a nonlinear inductor because the slope of the curve and hence the inductance have different values when operated at different currents. The slope of the curve at any point on the curve may be determined graphically by drawing a tangent line to the curve at that point, as shown in Fig. 8-3*b*; the ratio $\Delta\phi/\Delta i$ is the slope at that point. For example, the slope at point *b* is

$$\left.\frac{d\phi}{di}\right|_{\text{point } b} = \frac{\Delta\phi_b}{\Delta I_b}$$

At higher values of current the slope is less, resulting in lower values of inductance. At relatively low values of current (between 0 and i_1 in Fig. 8-3b), the magnetization curve is a straight line; the slope and hence the inductance is constant over that region, and the inductance for that range of currents is called the *normal inductance of the inductor. The nameplate on an inductor always lists the normal inductance.*

Unless otherwise specified, it will be assumed that the inductors referred to in discussions and in problems throughout the text are operated in their linear regions (normal inductance).

INDUCTORS IN SERIES AND IN PARALLEL

The equivalent inductance of two or more inductors in series is determined by

$$L_{eq,S} = L_1 + L_2 + L_3 + \cdots + L_n \tag{8-13}$$

The equivalent inductance of two or more inductors in parallel is determined by

$$\frac{1}{L_{eq,P}} = \frac{1}{L_1} + \frac{1}{L_2} + \frac{1}{L_3} + \cdots + \frac{1}{L_n} \tag{8-14}$$

Note: Equations (8-13) and (8-14) are valid only if no magnetic coupling exists between coils. Magnetic coupling is present whenever the flux caused by current in one coil passes through the window of another coil. See Chaps. 18 and 19 for a discussion of magnetic coupling and its effect on the current in an electric circuit.

Example 8-2 Given the following ideal inductors (zero resistance): 2.5 H, 6.3 H, and 5.2 H. Determine (*a*) the equivalent inductance if connected in series; (*b*) the equivalent inductance if connected in parallel.

Solution

(*a*) $L_{eq,S} = L_1 + L_2 + L_3$

$L_{eq,S} = 2.5 + 6.3 + 5.2 = 14$ H

(*b*) $\dfrac{1}{L_{eq,P}} = \dfrac{1}{2.5} + \dfrac{1}{6.3} + \dfrac{1}{5.2}$

$L_{eq,P} = 1.33$ H

Derivation of Equation (8-13)

Applying Kirchhoff's voltage law to the circuit in Fig. 8-4a,

$$v_{\text{gen}} = v_{L1} + v_{L2} + v_{L3} + \cdots + v_{Ln}$$

$$v_{\text{gen}} = L_1 \frac{di}{dt} + L_2 \frac{di}{dt} + L_3 \frac{di}{dt} + \cdots + L_n \frac{di}{dt}$$

$$v_{\text{gen}} = (L_1 + L_2 + L_3 + \cdots + L_n) \frac{di}{dt}$$

$$v_{\text{gen}} = L_{\text{eq}, S} \frac{di}{dt}$$

where $L_{\text{eq}, S} = L_1 + L_2 + L_3 + \cdots + L_n$

Derivation of Equation (8-14)

$$v_L = L \frac{di}{dt}$$

$$di = \frac{v_L \, dt}{L}$$

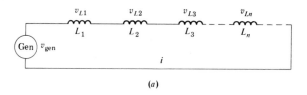

(a)

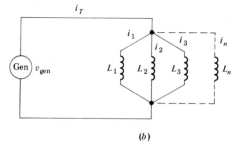

(b)

Figure 8-4 (a) Inductors in series; (b) inductors in parallel.

Integrating,

$$i = \frac{1}{L} \int v_L \, dt$$

Applying Kirchhoff's current law to the circuit in Fig. 8-4b

$$i_T = i_1 + i_2 + i_3 + \cdots + i_n$$

$$i_T = \frac{1}{L_1} \int v_1 \, dt + \frac{1}{L_2} \int v_2 \, dt + \frac{1}{L_3} \int v_3 \, dt + \cdots + \frac{1}{L_n} \int v_n \, dt$$

However, for a parallel circuit,

$$v_1 = v_2 = v_3 = v_n = v_{\text{gen}}$$

Therefore,

$$i_T = \left(\frac{1}{L_1} + \frac{1}{L_2} + \frac{1}{L_3} + \cdots + \frac{1}{L_n} \right) \int v_{\text{gen}} \, dt$$

$$i_T = \frac{1}{L_{\text{eq}, P}} \int v_{\text{gen}} \, dt$$

where $\quad \dfrac{1}{L_{\text{eq}, P}} = \dfrac{1}{L_1} + \dfrac{1}{L_2} + \dfrac{1}{L_3} + \cdots + \dfrac{1}{L_n}$

-5   TIME CONSTANT OF AN RL CIRCUIT

The equation for the current curves in Figs. 8-1c and 8-5c is†

$$i_L = \frac{E_{\text{bat}}}{R} (1 - \varepsilon^{-(R/L)t}) \qquad\qquad (8\text{-}15)$$

where i_L = instantaneous current, A
E_{bat} = battery voltage, V
R = resistance of circuit, Ω
L = inductance of circuit, H
t = time, s
ε = 2.718 (the base for natural logarithms)

† A complete analysis of the response of an *RL* circuit to a driving voltage is given in Chap. 27.

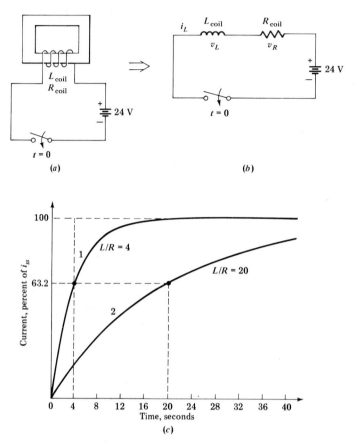

Figure 8-5 Comparison of current versus time curves for *LR* circuits with different time constants.

The value of the $\varepsilon^{-(R/L)t}$ term gets smaller with increasing time. Hence, as the time approaches infinity, the $\varepsilon^{-(R/L)t}$ term approaches zero. After an infinite time has elapsed, Eq. (8-15) reduces to the steady-state equation

$$i_{ss} = \frac{E_{bat}}{R}$$

The equation for the voltage drop across the inductance component of the coil may be obtained by applying Kirchhoff's voltage law to the equivalent circuit shown in Fig. 8-5*b* and solving for v_L. Thus,

$$E_{bat} = v_L + v_R$$

$$E_{bat} = v_L + i_L R$$

Solving for v_L,

$$v_L = E_{bat} - i_L R \tag{8-16}$$

Substituting Eq. (8-15) into Eq. (8-16),

$$v_L = E_{bat} - \frac{RE_{bat}}{R}(1 - \varepsilon^{-(R/L)t})$$

$$\boxed{v_L = E_{bat}\varepsilon^{-(R/L)t}} \qquad\qquad (8\text{-}17)$$

When $t = (0+)$,

$$i_L\Big|_{t=(0+)} = \frac{E_{bat}}{R}(1 - \varepsilon^0) \qquad\qquad v_L\Big|_{t=(0+)} = E_{bat}\varepsilon^0$$

$$i_L\Big|_{t=(0+)} = 0 \qquad\qquad v_L\Big|_{t=(0+)} = E_{bat}$$

As t increases, the exponential factor $\varepsilon^{-(R/L)t}$ gets smaller; the voltage v_L in Eq. (8-17) gets smaller, and the current i_L in Eq. (8-15) gets larger. As t approaches infinity, $\varepsilon^{-(R/L)t}$ approaches zero, causing v_L and i_L to approach their steady-state values:

$$i_L\Big|_{t\to\infty} = \frac{E_{bat}}{R}(1 - \varepsilon^{-\infty}) \qquad\qquad v_L\Big|_{t\to\infty} = E_{bat}\varepsilon^{-\infty}$$

$$i_L\Big|_{t\to\infty} = \frac{E_{bat}}{R} \qquad\qquad v_L\Big|_{t\to\infty} = 0$$

Thus, regardless of the values of L and R, the total time required to reach the absolute steady state will be the same, infinite.

However, a comparison of the current versus time curves for circuits with different ratios of inductance to resistance shows that *the initial rise in current takes a longer time in those circuits that have higher L/R ratios.* This is shown in Fig. 8-5c, where $L/R = 4$ for curve 1 and 20 for curve 2. The L/R ratio is called the *time constant* of an *LR* circuit; it is equal to the time in seconds that it takes for the voltage across an inductor, and the current in the inductor, to have completed 63.2 percent of its total change.

$$\boxed{\tau = \frac{L}{R}} \qquad\qquad (8\text{-}18)$$

where τ = time constant, s
L = inductance, H
R = resistance, Ω

Verification of the 63.2 percent change can be accomplished by substituting $t = (L/R)$ in Eqs. (8-15) and (8-17), and evaluating. Thus,

$$i_L = \frac{E_{bat}}{R}(1 - \varepsilon^{-(R/L)t}) \qquad\qquad v_L = E_{bat}(\varepsilon^{-(R/L)t})$$

$$i_L = \frac{E_{bat}}{R}(1 - \varepsilon^{-(R/L)(L/R)}) \qquad\qquad v_L = E_{bat}(\varepsilon^{-(R/L)(L/R)})$$

$$i_L = \frac{E_{bat}}{R}(1 - \varepsilon^{-1}) \qquad\qquad v_L = E_{bat}(\varepsilon^{-1})$$

$$i_L = \frac{E_{bat}}{R}(1 - 0.368) \qquad\qquad v_L = E_{bat}(0.368)$$

$$i_L = \frac{E_{bat}}{R}(0.632)$$

Note that i_L changed from an initial value of zero amperes to 63.2 percent of its final value; v_L changed from an initial value of E_{bat} to 36.8 percent E_{bat}, a change of 63.2 percent.

In five time constants (5τ), the current will have risen to

$$i_L = \frac{E_{bat}}{R}(1 - \varepsilon^{-(R/L)\ 5(L/R)}) = \frac{E_{bat}}{R}(1 - \varepsilon^{-5})$$

$$i_L = \frac{E_{bat}}{R}(1 - 0.0067) = 0.993\frac{E_{bat}}{R}$$

The voltage v_L will have dropped to

$$v_L = E_{bat}(\varepsilon^{-(R/L)\ 5(L/R)}) = E_{bat}(\varepsilon^{-5})$$

$$v_L = E_{bat}(0.0067)$$

Hence, for all practical purposes, after an elapsed time equal to five time constants, the current to the inductor, and the voltage across it, may be considered to be at steady state.

Figure 8-6 illustrates a method that can be used to closely approximate the graph of the transient behavior of an *LR* circuit. Draw four equally spaced horizontal lines to represent 0, 33.3 percent, 66.6 percent, and 100 percent of the steady-state current. Draw six equally spaced vertical lines to

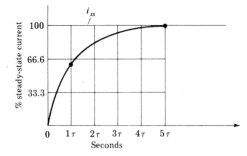

Figure 8-6 Graphical method for approximating the transient behavior of an *LR* circuit.

represent 0, 1τ, 2τ, 3τ, 4τ, and 5τ. Plot the points that represent the current for one time constant and for five time constants (shown with dots), and then approximate the curve.

The time constant for an *LR* circuit whose parameters are not known can be obtained graphically from an oscilloscope recording or *xy* plot of i_L versus time. This is accomplished by drawing a tangent line to the current curve at $t = 0$, as shown in Fig. 8-7. Then, from the point where the tangent line intersects the i_{ss} line, drop a perpendicular to the time axis. The point of intersection with the time axis is the time constant in seconds. Thus, ΔT in Fig. 8-7 is the time constant.

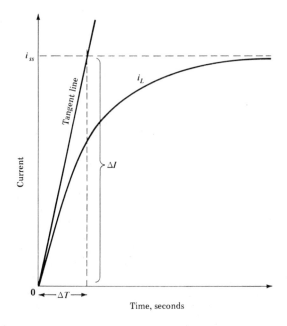

Figure 8-7 Determining the time constant of an *LR* circuit from an oscilloscope recording or *xy* plot.

Justification for the Graphical Determination of Time Constants

The slope of the current curve at $t = 0$ may be obtained algebraically by taking the derivative of Eq. (8-15) with respect to time, and evaluating it at $t = (0+)$. Thus,

$$\frac{di}{dt} = \frac{d}{dt}\left[\frac{E_{bat}}{R}(1 - \varepsilon^{-(R/L)t})\right]$$

$$\frac{di}{dt} = \frac{E_{bat}}{R}\left(\frac{R}{L}\varepsilon^{-(R/L)t}\right)$$

$$\left.\frac{di}{dt}\right|_{t=(0+)} = \frac{E_{bat}}{R}\frac{R}{L} = \frac{E_{bat}}{L}$$

The slope of the current curve at $t = 0$, obtained graphically from Fig. 8-7, is

$$\text{Slope}\Big|_{t=(0+)} = \frac{\Delta I}{\Delta T} = \frac{i_{ss}}{\Delta T} = \frac{E_{bat}/R}{\Delta T} = \frac{E_{bat}}{R\,\Delta T}$$

The slope obtained by the derivative must be equal to the slope obtained graphically. Thus,

$$\frac{E_{bat}}{L} = \frac{E_{bat}}{R\,\Delta T}$$

Solving for ΔT,

$$\Delta T = \frac{L}{R}$$

Example 8-3 The time constant of an *LR* circuit was determined to be 2.4 ms from an oscilloscope recording. An ohmmeter measurement of the circuit indicates the circuit resistance to be 2000 Ω. Determine the circuit inductance.

Solution

$$\tau = \frac{L}{R}$$

$$0.0024 = \frac{L}{2000}$$

$$L = 4.8 \text{ H}$$

Example 8-4 A coil whose resistance and inductance are 2.0 Ω and 8.0 Ω, respectively, is connected to a 12-V battery and a switch. Determine (*a*) the steady-state current;

(b) the time constant; (c) the current after one time constant has elapsed; (d) the current after 20 time constants have elapsed.

Solution

(a) $i_{ss} = \dfrac{E_{bat}}{R} = \dfrac{12}{2} = 6 \text{ A}$

(b) $\tau = \dfrac{L}{R} = \dfrac{8}{2} = 4 \text{ s}$

(c) $\left. i \right|_{t=1\tau} = 0.632(6) = 3.9 \text{ A}$

(d) After 20τ, which is $20 \times 4 = 80$ s, the circuit is at essentially steady state.

$i_{ss} \approx 6 \text{ A}$

6   ENERGY STORAGE IN AN INDUCTOR

It takes electrical energy, supplied by the driver, to establish the flux around a current-carrying conductor. All this energy is stored in the magnetic field as *magnetic energy*; none is expended. When the current is decreased, the flux surrounding the conductors is decreased, causing the stored energy to be released.

The energy stored in the magnetic field is distinct from the I^2Rt energy losses in the conductors, which is expended as heat energy. Thus, when analyzing the energy relationships in an inductor, it is convenient to make an *equivalent-circuit model*, showing the inductance and the resistance as separate lumped values. This is shown in Fig. 8-8. Figure 8-8a shows the actual inductor, and Fig. 8-8b shows the equivalent-circuit model where an ideal (pure) inductor and an ideal (pure) resistor take the place of the actual inductor.

The energy stored in the magnetic field of an inductor, at some instant of time t_k, is proportional to the self-inductance of the inductor and the square of the current at that instant. Expressed mathematically,

$$\boxed{W_{\phi k} = \tfrac{1}{2}Li_k^2} \qquad (8\text{-}19)$$

where $W_{\phi k}$ = energy accumulated in the inductor at time t_k, J
   i_k = current at time t_k, A
   L = inductance, H

The total energy accumulated in the magnetic field of the inductor at steady state is

$$W_{\phi,ss} = \tfrac{1}{2}LI_{ss}^2 \qquad \text{joules}$$

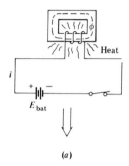

(a)

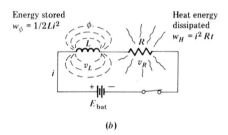

(b)

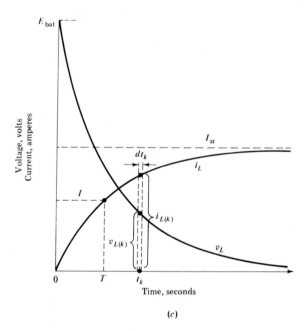

(c)

Figure 8-8 Energy relationships in an inductor: (a) lumped inductor; (b) equivalent circuit; (c) current and voltage characteristics.

This amount of energy will remain stored in the field until the circuit is opened or the current is changed by modifying the circuit.

Derivation of Equation (8-19)

At any instant of time, the instantaneous *rate of transfer* of energy from the battery to the ideal inductor is equal to the instantaneous voltage drop across the inductor multiplied by the instantaneous current.

$$p = v_L i_L \quad \text{watts or joules per second}$$

Referring to Fig. 8-8c, the increment of energy dw_k absorbed by the magnetic field of the inductor during the infinitesimal increment of time dt_k, at time t_k, is equal to the instantaneous power at time t_k multiplied by increment dt_k. That is,

$$dw_k = p_k \, dt_k \quad \text{joules}$$

In terms of voltage and current,

$$dw_k = v_{Lk} i_{Lk} \, dt_k \quad \text{joules}$$

where v_{Lk} = instantaneous voltage across the inductance at time t_k, V
i_{Lk} = instantaneous current at time t_k, A
dt_k = infinitesimal increment of time associated with the instant t_k

The total energy accumulated, from the instant the switch is closed until time T, may be determined by summing all the small increments of energy from time-zero to time T. Thus, using calculus,

$$W = \int_{t=0}^{t=T} v_L i_L \, dt \tag{8-20}$$

Expressing v_L in terms of i_L,

$$v_L = L \frac{di_L}{dt} \tag{8-21}$$

Substituting Eq. (8-21) into Eq. (8-20),

$$W = \int_{t=0}^{t=T} L \frac{di_L}{dt} i_L \, dt$$

$$W = L \int_{t=0}^{t=T} i_L \, di_L$$

Changing the limits of integration,

$$W = L \int_{i=0}^{i=I} i_L \, di_L$$

Current I corresponds to time T, as shown in Fig. 8-8c. Integrating,

$$W = \tfrac{1}{2}Li_L^2 \Big|_{i_L=0}^{i_L=I}$$

$$W = \tfrac{1}{2}LI^2$$

Example 8-5   The shunt-field circuit of a certain 30-hp DC motor has a resistance of 55.0 Ω and an inductance of 120 H. Calculate (*a*) the accumulated energy stored in the magnetic field when the switch is closed and the current attains its steady-state value. Assume the driving voltage is 240 V DC.

Solution

$$I_{ss} = \frac{E}{R} = \frac{240}{55} = 0.363 \text{ A}$$

$$W_{\phi,ss} = \tfrac{1}{2}LI_{ss}^2$$

$$W_{\phi,ss} = \tfrac{1}{2}(120)(4.363)^2$$

$$W_{\phi,ss} = 1142 \text{ J}$$

Example 8-6   A coil is connected in series with a 24-V battery and a switch as shown in Fig. 8-9a. The inductance and resistance of the coil are 6.0 H and 2.0 Ω, respectively. Determine (*a*) the time constant; (*b*) the steady-state current; (*c*) the current at one time constant; (*d*) the voltage drop caused by the inductance at one time constant; (*e*) the energy stored in the magnetic field when the current reaches steady state; (*f*) *the rate of expenditure* of heat energy at $t = 1$ time constant; (*g*) the *rate of accumulation* of magnetic energy at $t = 1$ time constant; (*h*) the magnetic energy accumulated 80 s after the switch was closed; (*i*) the *rate of expenditure* of heat energy at steady state; (*j*) the *rate of accumulation* of magnetic energy at steady state.

Solution
Figure 8-9b illustrates the v_L and i_L curves for the circuit.

(*a*) $\tau = \dfrac{L}{R} = \dfrac{6}{2} = 3.0 \text{ s}$

(*b*) $i_{ss} = \dfrac{E_{bat}}{R} = \dfrac{24}{2.0} = 12 \text{ A}$

(*c*) $i_{L,1\tau} = 12(0.632) = 7.58 \text{ A}$

(*d*) $v_{L,1\tau} = 0.368(24) = 8.83 \text{ V}$

(*e*) $W_{\phi,ss} = \tfrac{1}{2}Li_{ss}^2 = \tfrac{1}{2}(6)(12)^2 = 432 \text{ J}$

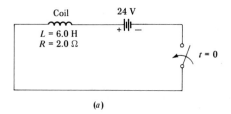

(a)

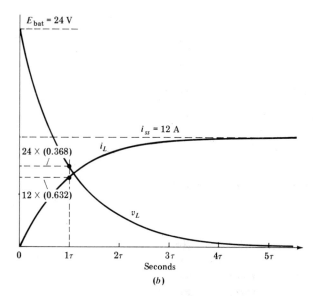

(b)

Figure 8-9 Circuit and characteristics curves for Example 8-6.

(f) $P_{\text{heat, }1\tau} = i^2 R = (7.584)^2(2) = 115.0$ W

(g) $P_{\phi, 1\tau} = v_L i_L = 8.832(7.584) = 66.98$ W

(h) $80 \text{ s} = \dfrac{80}{3.0} = 26.7$ time constants

Hence, current is at steady state.

$W_{\phi, \text{ss}} = \frac{1}{2}Li_{\text{ss}}^2 = 432$ J   as in part e

(i) $P_{\text{heat, ss}} = i_{\text{ss}}^2 R$

$P_{\text{heat, ss}} = (12)^2(2) = 288$ W

(j) $P_{\phi, \text{ss}} = v_L i_L = 0(2) = 0$ W

When the current reaches its steady-state value, the flux will no longer be changing, and the induced voltage will be zero.

Although energy to the amount of $\frac{1}{2}Li^2$ is required to establish the magnetic field, once the current attains its steady-state value, no additional energy is required to hold the flux at that level. However, there is a steady heat loss associated with the resistance of the coil and the connecting wires to the amount of i^2Rt. If the resistance is extremely small, such as in superconducting materials, the i^2Rt losses will be negligible.

Example 8-7   An inductor whose resistance and inductance are 2.0 Ω and 6.0 H, respectively, is connected in series with a 10-Ω resistor, a switch, and a 12-V battery as shown in Fig. 8-10a. Another switch is connected across the 10-Ω resistor. If both switches are closed, the steady-state circuit current is

$$I_{ss} = \frac{E_{bat}}{R}$$

$$I_{ss} = \frac{12}{2.0} = 6.0 \text{ A}$$

The energy accumulated in the magnetic field of the inductor at steady state is

$$W_{\phi, ss} = \tfrac{1}{2}LI_{ss}^2$$

$$W_{\phi, ss} = \tfrac{1}{2}(6)(6)^2$$

$$W_{\phi, ss} = 108 \text{ J}$$

Opening switch 2 introduces an additional 10 Ω in series with the inductor. This causes the current to change to a lower value, resulting in the release of some stored energy.
The new steady-state current is

$$I_{ss} = \frac{E_{bat}}{R}$$

$$I_{ss} = \frac{12}{2 + 10} = 1 \text{ A}$$

The energy remaining in the magnetic field of the inductor, at the new steady-state current, is

$$W_{\phi, ss} = \tfrac{1}{2}LI_{ss}^2$$

$$W_{\phi, ss} = \tfrac{1}{2}(6)(1)^2$$

$$W_{\phi, ss} = 3 \text{ J}$$

Thus the total energy released from the magnetic field after switch 2 was opened is

$$W_{released} = 108 - 3 = 105 \text{ J}$$

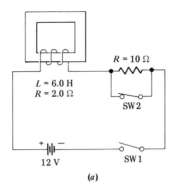

(a)

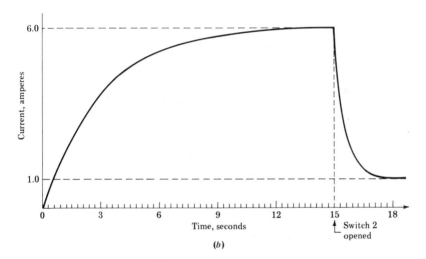

(b)

Figure 8-10   Circuit and current characteristics for Example 8-7.

A sketch showing the behavior of the current for the two conditions is shown in Fig. 8-10b. Note that the time constant of the circuit with switch 2 closed was

$$\tau = \frac{L}{R} = \frac{6}{2} = 3 \text{ s}$$

but the time constant for the circuit with switch 2 open is

$$\tau = \frac{L}{R} = \frac{6}{2 + 10} = 0.5 \text{ s}$$

The difference in time constants accounts for the slow buildup of current to 6 A and the rapid decrease to 1 A as shown in Fig. 8-10b.

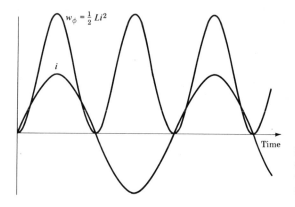

Figure 8-11 Current curve and energy curve associated with a sinusoidal current in an indicator.

Any change in the magnitude of the current in an inductive circuit will cause the magnetic field to absorb or release energy, depending on whether the current is increasing or decreasing, respectively. Hence, when an alternating current is in an inductor, the inductor will store energy during the quarter-cycle in which the current is increasing and will release energy during the next quarter-cycle when the current is decreasing. Figure 8-11 shows the current curve and the associated energy curve for a sinusoidal current in an inductor. The energy stored in the inductor at the instant the alternating current reaches its maximum value is

$$W = \tfrac{1}{2}LI^2_{\max}$$

8-7   ADVERSE EFFECTS CAUSED BY OPENING AN INDUCTIVE CIRCUIT

When a circuit containing appreciable inductance, such as the field windings of a generator, is opened, the inductive effect prevents the current from instantaneously dropping to zero; the current arcs across the switch, burning the switch blades. Unless adequately protected, such arcing and burning of switches and contacts can cause a serious maintenance problem. The severe arcing and burning that occur when the switch is opened are caused by the sudden release of energy that was stored in the magnetic field.

The *collapse of the magnetic field*, brought about by opening the circuit, generates a voltage of self-induction, which in accordance with Lenz's law is in a direction to *oppose the decrease* in current.

If, when the switch is opened, the voltage of self-induction is considerably higher than the voltage rating of the insulation, the insulation may be damaged.

Figure 8-12 illustrates the behavior of the current and the induced voltage e_L in an inductive circuit when the switch is closed and opened. At

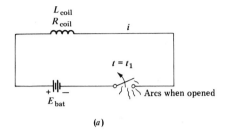

(a)

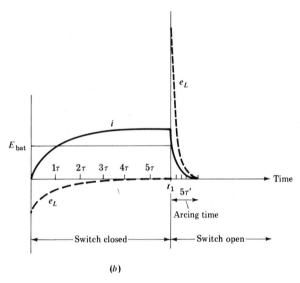

(b)

Figure 8-12 Behavior of the current and induced voltage in an inductive circuit when the switch is closed and opened.

the instant the switch is closed, the current is zero, and the induced voltage will be equal in magnitude but opposite in direction to the battery voltage. The current rises with time, approaching its steady-state value in five time constants, at which time the opposing voltage of self-induction approaches zero.

When the circuit is opened, the rapid collapse of magnetic flux induces a voltage in a direction to *oppose the decrease in current.* Hence, the voltage of self-induction will be in the same direction as the driving voltage; it will act as a voltage rise instead of a voltage drop!

The time constant when the switch is opened is much shorter than the time constant when the switch was closed. Opening the switch adds the resistance of the electric arc in series with the resistance of the circuit. When closing the circuit,

$$\tau = \frac{L_{coil}}{R_{coil}}$$

when opening the circuit,

$$\tau' = \frac{L_{coil}}{R_{coil} + R_{arc}}$$

The shorter time constant causes a faster drop in current and flux. Hence *the voltage induced when opening the switch is greater than the voltage induced when the circuit was closed.*

Although the resistance of the arc is not constant (nonlinear), the curves shown in Fig. 8-12b are representative of the behavior of the induced voltage when closing and opening a switch.

To prevent burning and arcing at the contacts of a switch or circuit breaker, a diode or varistor is sometimes permanently connected across the terminals of coils with high inductance as shown in Fig. 8-13. The solid lines indicate the path of current through the circuit when the switch is closed, and the broken lines indicate the path of current when the switch is opened. The self-inductance of the coil prevents an instantaneous change in the direction of the current. Hence the direction of the current through the coil at the instant the switch is opened will be the same as the direction it had when the switch was closed. When the switch is opened, the magnetic energy will be dissipated as heat in the resistance of the closed loop formed by the coil and the discharge element.

If a diode is used as the discharge element (Fig. 8-13a), almost all the magnetic energy will be dissipated in the resistance of the coil conductors; because of its low forward resistance, the I^2R dissipation in the diode will be insignificant. If a varistor is used as the discharge element (Fig. 8-13b), some of the energy will be dissipated in the varistor and some in the resistance of the coil. However, because of different time constants, the current will take a longer time to decay if the discharge element is a diode than if it is a varistor.

$$\tau_{diode\ loop} = \frac{L_{coil}}{R_{coil}}$$

$$\tau_{varistor\ loop} = \frac{L_{coil}}{R_{coil} + R_{varistor}}$$

Figure 8-13c shows how a *freewheeling diode* may be used to prevent the current in an inductive circuit from going to zero even though the driving

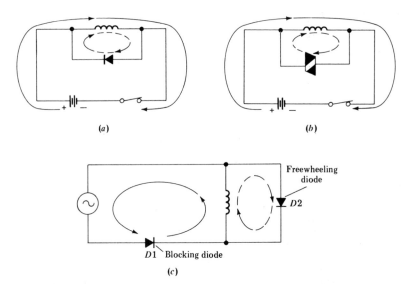

Figure 8-13 (a) Diode used as a discharge element; (b) varistor as a discharge element; (c) freewheeling diode used to maintain the current in a coil.

voltage is sinusoidal. Diode $D1$ is used as a rectifier, permitting only *one half-cycle of the generator current* to appear in the coil. During the half-cycle that diode $D1$ blocks the generator current, the coil current is continued through diode $D2$, sustained by the release of some of the stored energy in the magnetic field. To be effective, the time constant of the discharge loop must be much longer than the period of the alternating-current wave.

SUMMARY OF FORMULAS

$$v_L = N \frac{d\phi}{dt}$$

$$\phi = \frac{Ni_L}{\mathcal{R}}$$

$$L = \frac{N^2}{\mathcal{R}}$$

$$v_L = L \frac{di_L}{dt}$$

$$i_L = \frac{1}{L} \int_{t=0}^{t=T} v_L \, dt$$

$$L = N \frac{d\phi}{di_L}$$

$$v_L = E_{bat} (\varepsilon^{-(R/L)t})$$

$$i_L = \frac{E_{bat}}{R} (1 - \varepsilon^{-(R/L)t})$$

$$\tau = \frac{L}{R}$$

$$L_{eq,S} = L_1 + L_2 + L_3 + \cdots + L_n$$

$$\frac{1}{L_{eq,P}} = \frac{1}{L_1} + \frac{1}{L_2} + \frac{1}{L_3} + \cdots + \frac{1}{L_n}$$

$$W_\phi = \tfrac{1}{2}Li_L^2 \qquad P_\phi = v_L i_L$$

PROBLEMS

8-1 A coil of 50 turns has an inductance of 10 H and a resistance of 2.0 Ω. (a) What is the reluctance of the magnetic circuit; (b) what battery voltage should be applied to the coil in order to obtain a flux of 0.32 Wb?

8-2 A certain 40-turn coil has an inductance of 6.0 H. Determine the new inductance if 10 turns are added to the coil.

8-3 A DC field coil of a certain generator has 400 turns and a resistance of 200 Ω. The reluctance of the magnetic circuit is 1000 A · t/Wb. Determine (a) the flux in the magnetic circuit if the voltage drop across the coil is 40 V; (b) the inductance of the coil.

8-4 Determine the equivalent inductance of the following series-connected inductors: 5 H, 1 H, 100 H, and 12 H.

8-5 Using the same inductors in Prob. 8-4, determine the equivalent inductance if connected in parallel.

8-6 Determine the equivalent inductance of the following series-connected inductors: 2 H, 4 H, 6 H, and 8 H.

8-7 Using the same inductors in Prob. 8-6, determine the equivalent inductance if connected in parallel.

8-8 A 10-H 16-Ω coil is connected in series with a 4.0-Ω resistor, a 120-V battery, and a switch. Sketch the circuit and determine (a) the time constant of the coil; (b) the time constant of the circuit; (c) the steady-state current; (d) the current at $t = $ one time constant after the switch is closed; (e) the voltage drop across the coil at $t = $ one time constant; (f) the voltage across the coil at steady state.

8-9 An 8.0-H 4.0-Ω coil is connected in series with a 2.0-Ω resistor, a 6.0-V battery, and a switch. Sketch the circuit and determine (a) the initial current when the switch is closed; (b) the final current; (c) the voltage drop across the coil at $t = (0+)$; (d) the current in the 2.0-Ω resistor at $t = $ one time constant; (e) the voltage across the 2.0-Ω resistor at $t = $ one time constant; (f) if the 2.0-Ω resistor is to be replaced by another that would cause the time constant of the circuit to be 1/3 s, determine the resistance value of the new resistor.

8-10 A 10-Ω resistor, a 24-V battery, and a switch are connected in series with a coil whose resistance and inductance are 20 Ω and 6.0 H, respectively. Sketch the circuit and determine (a) the current at the instant the switch is closed ($t = 0+$); (b) the steady-state current; (c) the circuit time constant; (d) the current in the circuit and the voltage across the 10-Ω resistor at $t = 1\tau$; (e) the energy stored in the magnetic field at steady state.

8-11 A 6.0-H coil whose resistance is 12 Ω is connected in series with a 24-Ω resistor, a 144-V battery, and a switch. The switch is closed at $t = 0$. Sketch the circuit and determine (a) the time constant of the coil; (b) the time constant of the circuit; (c) the current at $t = (0+)$; (d) the steady-state current; (e) the energy stored in the magnetic field at steady state; (f) the heat power dissipated by the coil at $t = \tau_{\text{circ}}$; (g) the heat power dissipated by the circuit at $t = \tau_{\text{circ}}$; (h) the rate of storage of energy in the magnetic field at $t = \tau_{\text{circ}}$; (i) the new value of inductance if the number of turns of wire are doubled.

8-12 A coil whose resistance and inductance are 3.0 Ω and 10 H, respectively, is connected in series with a 2.0-Ω resistor, a 60-V battery, and a switch. Sketch the circuit. Determine (a) the time constant of the circuit; (b) the current at $t = (0+)$; (c) the steady-state current; (d) the energy stored in the magnetic field at steady state; (e) the heat power dissipated by the coil at an elapsed time equal to one time constant; (f) the rate of storage of energy at $t = $ one time constant; (g) the new time constant for the circuit if the number of turns of wire in the coil are tripled (assume the diameter of the coil and the size of the wire are not changed).

8-13 A 6.0-Ω 4.0-H coil is connected in series with a 2.0-Ω resistor, a switch, and a 48-V 0-Hz generator. Sketch the circuit and determine (a) the circuit time constant; (b) the current at $t = (0+)$; (c) the voltage across the coil at $t = (0+)$; (d) the steady-state current; (e) the current at $t = 1\tau$; (f) the voltage across the resistor at steady state; (g) the voltage across the coil at steady state; (h) the energy stored in the magnetic field at steady state; (i) the voltage induced in the coil at $t = 1\tau$; (j) the rate of storage of energy in the magnetic field at $t = 1\tau$; (k) the rate of expenditure of heat energy in the coil at $t = 1\tau$.

CHAPTER 9

INTRODUCTION TO THE SINUSOIDAL SYSTEM

In the study and application of electric circuits one will come across voltages and currents that vary with time in a periodic manner. Examples of such periodic waves are shown in Fig. 9-1. Rectangular waves may be found in DC generators and electronic circuits; sawtooth waves are used for sweep circuits in oscilloscopes; triangular waves are associated with the changing magnetic field passing through the window of a DC armature coil.

The sinusoidal wave shown in Fig. 9-1d has the greatest application and is the easiest to work with mathematically. Furthermore, as will be shown in Chap. 25, nonsinusoidal periodic waves can be broken down into a series of sine waves of different frequencies whose sum will equal the original periodic wave. Thus an understanding of the sinusoidal behavior of electric circuits will enable one to analyze nonsinusoidal periodic current and voltage waves found in electronic circuits and electrical machinery. These techniques are also useful in the analysis of mechanical vibrations and other variables that behave in a periodic manner.

9-1 NOMENCLATURE OF PERIODIC WAVES

The nomenclature of periodic waves is illustrated in Fig. 9-1 and defined below.

Cycle: A cycle is one complete set of values of a periodic wave.

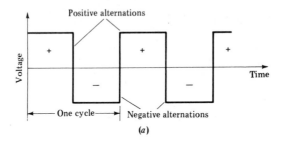

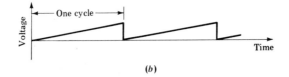

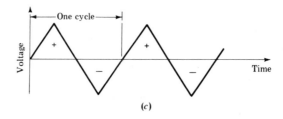

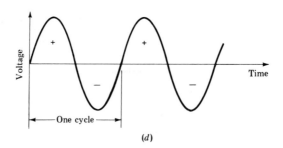

Figure 9-1 Examples of periodic voltage waves: (*a*) rectangular; (*b*) sawtooth; (*c*) triangular; (*d*) sinusoidal.

Frequency: The frequency (f) of a periodic wave is the number of cycles completed per second. One cycle per second is one hertz (abbreviated Hz).

Period: The period T of a periodic wave is the time required to complete one cycle. The period is generally expressed in seconds and is the reciprocal of the frequency.

$$T = \frac{1}{f}$$

(9-1)

For example, the time required to complete one cycle of a 60-Hz periodic wave is $\frac{1}{60}$ s.

Alternation: The positive and negative half-cycles of periodic waves that have alternately positive and negative values are called alternations. The periodic waves in Fig. 9-1a, c, and d have alternations. The wave in Fig. 9-1b is periodic but has no alternations.

9-2   EQUATIONS OF CONTINUOUS SINUSOIDAL CURRENT AND VOLTAGE WAVES

The general mathematical expressions for the continuous sinusoidal wave shown in Fig. 9-2 are

$$y = A \sin x \tag{9-2}$$

and

$$y = A \cos x \tag{9-3}$$

The choice of equation is dependent on the starting point or zero abscissa point of the continuous wave. The sine function may be used if the wave has its zero value at $x = 0$, whereas the cosine function may be used if the wave has its maximum value at $x = 0$.

Using Fig. 9-2 and Eq. (9-2) as a reference, the equation for the continuous voltage wave in Fig. 9-3 is

$$e = E_m \sin \alpha \tag{9-4}$$

where E_m = maximum (peak) value of the wave. Angle alpha (α) may be expressed in terms of degrees or radians.

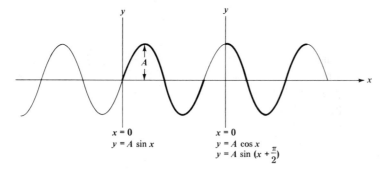

Figure 9-2   Example of a continuous sinusoidal wave.

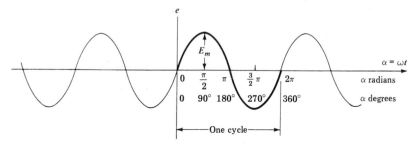

Figure 9-3   Example of a continuous sinusoidal voltage wave.

The radian is the unit for angular measure commonly used in the analysis and design of electrical circuits, machines, and controls. The relationship between degrees and radians is

$$2\pi \text{ radians} = 360°$$

or

$$1 \text{ radian} = \frac{360°}{2\pi} = 57.30°$$

$$\pi = 3.142$$

The superscript ( )^R will be used in equations to denote an angle in radians. Thus 2.5 radians is written 2.5^R.

Although Eq. (9-4) expresses the sinusoidal voltage wave of Fig. 9-3 as a function of angle α, many electrical relationships require the sine wave to be expressed in terms of time. Angle α is related to time t in the following manner:

$$\alpha = \omega t \tag{9-5}$$

where α = angle, radians (rad)
  t = elapsed time, seconds (s)
  ω = angular velocity, radians per second (rad/s)

The angular velocity of a sine wave may be obtained by dividing the angle spanned in one cycle by the period of the wave. Since the angle spanned in one cycle is 2π radians,

$$\omega = \frac{2\pi}{T} \quad \text{rad/s} \tag{9-6}$$

Since

$$T = \frac{1}{f}$$

$$\boxed{\omega = 2\pi f} \quad \text{rad/s} \tag{9-7}$$

Substituting Eqs. (9-5) and (9-7) into Eq. (9-4) results in the most frequently used expressions for a sinusoidal voltage wave, *the time-domain equations.*

$$e = E_m \sin \omega t \tag{9-8}$$

$$e = E_m \sin 2\pi f t \tag{9-9}$$

where e = instantaneous value of voltage at the instant of time t
      E_m = maximum value of the voltage wave, V
      f = frequency, Hz
      t = time, s, that has elapsed since the wave crossed the ωt axis going in the positive direction

Note: ωt and $2\pi f t$ are expressed in radians; to convert to degrees multiply by 57.3 (one radian = 57.2958°).

In a similar manner, the equation for a continuous sinusoidal current wave is

$$i = I_m \sin 2\pi f t \tag{9-10}$$

where i = instantaneous value of current at the instant of time t
      I_m = maximum value of the current wave, A
      f = frequency, Hz
      t = time, s, that has elapsed since the wave crossed the ωt axis going in the positive direction

Example 9-1   A certain sinusoidal voltage is expressed mathematically by

$$e = 0.40 \sin 377 t \quad \text{V}$$

Determine (a) the frequency of the wave; (b) the period; (c) the value of the voltage when the elapsed time is 0.01 s.

Solution

(a) $\omega = 2\pi f$

   $377 = 2\pi f$

    $f = 60$ Hz

(b) $T = \dfrac{1}{f}$

   $T = \dfrac{1}{60} = 0.0167$ s

(c) $e = 0.4 \sin(377 \times 0.01)^{\text{R}}$

   $e = 0.4 \sin(3.77)^{\text{R}}$

   $e = 0.4 \,(-0.588) = -0.235$ V

Or, using degrees,

$e = 0.4 \sin[377 \times 0.01 \times 57.3]^{\circ}$

$e = 0.4 \sin(216^{\circ})$

$e = 0.4 \,(-0.588) = -0.235$ V

Example 9-2   (a) What is the equation of a sinusoidal voltage wave that has a maximum value of 200 V and a frequency of 50 Hz? (b) What is the value of voltage 0.0025 s after the wave crosses the ωt axis going in the positive direction?

Solution

(a) $e = E_m \sin 2\pi f t$

   $e = 200 \sin 2\pi 50 t = 200 \sin 314.16 t$

(b) $e = 200 \sin(314.16 \times 0.0025)^{\text{R}}$

   $e = 200 \sin(0.7854)^{\text{R}}$

   $e = 200 \,(0.707) = 141$ V

Or, using degrees instead of radians,

$e = 200 \sin(0.7854 \times 57.3)^{\circ}$

$e = 200 \sin 45^{\circ}$

$e = 200 \,(0.707) = 141$ V

Waves of voltage and current are continuous. They do not stop after one cycle is completed but continue to repeat as long as the generator is operating. Hence, when the switch is closed, connecting the generator to a load, the instantaneous value of the voltage applied to the load depends on the instantaneous value of the voltage wave at the time of closing. *The instant that the switch blades make contact is arbitrarily considered as "time-zero," and the elapsed time is measured from that instant.* Figure 9-4 illustrates "time-zero" for three different points on the same continuous voltage wave and indicates the corresponding equations. The equations differ only by the angle of displacement of the "time-zero" line from the intersection of the wave with the time axis as it ascends in the positive direction. The equations representing the three different instants of switch closure are:
For Fig. 9-4b,

$$e = E_m \sin \omega t$$

For Fig. 9-4c,

$$e = E_m \sin(\omega t + \beta)$$

For Fig. 9-4d

$$e = E_m \sin(\omega t - \beta)$$

The angle β (beta), called the *phase angle or phase*, is measured along the $\alpha = \omega t$ line *from* a point of zero voltage with positive slope *to* the "time-zero" line. If the measurement is toward the right (Fig. 9-4c), the angle is positive; if the measurement is toward the left (Fig. 9-4d), tha angle is negative. If the "time-zero" line is coincident with the point of zero voltage with positive slope, as shown in Fig. 9-4b, the phase angle is zero. An easy way to determine the sign of β is to imagine yourself "riding" the wave on the up-swing, and at the instant it shoots above the time axis (ωt axis) jump off and walk to the "time-zero" line. If you walk to the right (positive direction), β is positive; if you walk to the left (negative direction), β is negative.

Example 9-3 For the voltage wave and time-zero corresponding to Fig. 9-4d, assume angle $\beta = 30°$, $f = 60$ Hz, and $E_m = 100$ V. Determine the voltage 0.0080 s after the switch is closed.

Solution

$$e = E_{max} \sin(\omega t - \beta)$$

$$e = 100 \sin(2\pi 60t - 30°)$$

$$e = 100 \sin[2\pi 60(0.008)(57.3) - 30]°$$

$$e = 100 \sin(142.81)° = 60 \text{ V}$$

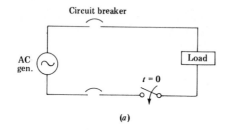

(a)

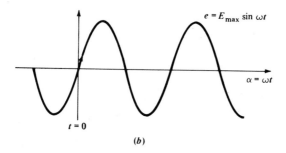

$$e = E_{max} \sin \omega t$$

$$\alpha = \omega t$$

$t = 0$

(b)

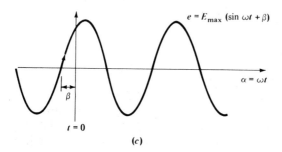

$$e = E_{max} (\sin \omega t + \beta)$$

$$\alpha = \omega t$$

β

$t = 0$

(c)

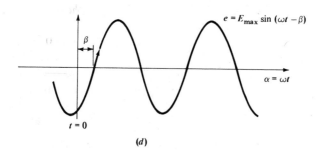

$$e = E_{max} \sin (\omega t - \beta)$$

$$\alpha = \omega t$$

β

$t = 0$

(d)

Figure 9-4 Examples of "time-zero" at different points on a sinusoidal voltage wave.

Example 9-4 The equation for a sinusoidal current wave is

$$i = 25 \sin\left(208t + \frac{\pi}{3}\right)$$

Determine (a) the frequency; (b) the period; (c) the current at $t = 0.012$ s.

Solution

(a) $\omega = 208$ rad/s

$$2\pi f = 208$$

$$f = \frac{208}{2\pi} = 33.10 \text{ Hz}$$

(b) $T = \dfrac{1}{f} = \dfrac{1}{33.1} = 0.030$ s

(c) $i = 25 \sin[208(0.012) + \pi/3]^{\text{R}}$

$\quad\quad i = 25 \sin[3.543]^{\text{R}}$

$\quad\quad i = 25\,(-0.3909) = -9.77$ A

9-3 ## ZERO HERTZ (DIRECT CURRENT)

Direct current or voltage may be considered as the special case of a sinusoidal current wave or sinusoidal voltage wave whose frequency is at the lower limit of zero hertz. This is explained with the aid of Fig. 9-5a, which shows an ideal generator supplying a sinusoidal driving voltage to a resistor load. The ideal generator has its E_m held constant, but the frequency is adjustable from zero hertz to infinite hertz. Figure 9-5b shows representative voltage waves of different frequencies *with the switch closed at the instant the respective voltage wave attains its maximum value.*

As the frequency of the voltage wave gets smaller, approaching 0 Hz in the limit, its period approaches infinity $(T = 1/f)$; that is, it takes an infinite time to complete one cycle. This is illustrated in Fig. 9-5b, where the 0-Hz voltage wave is shown to have a constant magnitude for all finite values of time.

The equations representative of all the voltage waves in Fig. 9-5b are

$$e = E_m \cos 2\pi f t$$

$$\tag{9-11}$$

$$e = E_m \sin(2\pi f t + 90°)$$

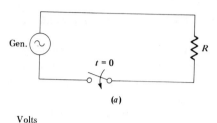

(a)

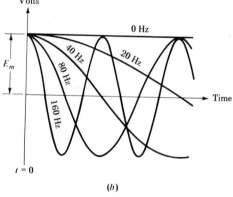

(b)

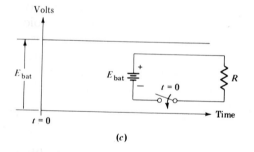

(c)

Figure 9-5 (a) Sinusoidal generator supplying a resistor load; (b) representative voltage waves of different frequencies; (c) battery-voltage curve.

Substituting 0 Hz for f in equation set (9-11),

$$e = E_m \cos 2\pi 0 t = E_m$$

$$e = E_m \sin(2\pi 0 t + 90°) = E_m \tag{9-12}$$

Thus, at a frequency of 0 Hz, $e = E_m$ for all finite time.

Comparison of the battery-voltage curve in Fig. 9-5c with the 0-Hz voltage of Fig. 9-5b indicates that a *DC voltage (or a direct current) may be considered as the special case of a sinusoidal voltage (or sinusoidal current) whose frequency is 0 Hz.*

Thus, from here on, the text develops the theory of electric circuits for the general case of any frequency, and zero is substituted for f when the response to a DC driver is desired.

4 EFFECTIVE VALUE OF SINUSOIDAL CURRENT AND VOLTAGE WAVES (RMS VALUE)

A sinusoidal generator delivers pulsating energy in a manner similar to a reciprocating pump delivering water; and just as the rate of flow of water is expressed in gallons or liters per minute rather than gallons or liters per stroke, so the measurement of the average rate of transfer of electrical energy in *joules per second* is preferred over the energy per cycle. *One joule per second is equal to one watt of electrical power.*

A battery or DC generator delivers energy to a resistor at a constant rate; this is in contrast to the pulsating energy delivered by a sinusoidal generator. For example, a sinusoidal current that has a peak value of I_m generates less heat in a given time, when passing through a resistor, than does a corresponding direct current of the same but constant I_m magnitude. For this reason, the DC ampere is used as the unit of measurement when determining the effectiveness of energy delivery by a periodic current.

Alternating-current ammeters and voltmeters, nameplate data of AC apparatus, and the capital letters, such as I, E, and V, used in electrical texts and other publications always indicate current and voltage as equivalent DC values (called rms), unless otherwise specified.

An experimental method for determining the equivalent DC value of a sinusoidal current is shown in Figs. 9-6a and b. The diagrams illustrate two circuits, each possessing identical resistance, one connected to a sinusoidal generator and the other to a DC generator. Wattmeters are used to measure the average heat power in each circuit. If the generator voltage to each circuit is adjusted so that the average heat power delivered to each resistor is the same, the amperes of direct current will be equal to the peak value of the sinusoidal wave divided by $\sqrt{2}$. That is,

$$I_{DC} = \frac{I_{m\sim}}{\sqrt{2}} = 0.707 I_{m\sim} \qquad\qquad (9\text{-}13)$$

where I_{DC} = equivalent DC value of the *sinusoidal* current
 $I_{m\sim}$ = maximum value of the *sinusoidal* current

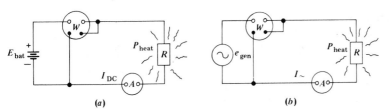

(a)   *(b)*

Figure 9-6 Circuits for determining the equivalent DC value of a sinusoidal current.

The equivalent DC value of a sinusoidal wave is also called the *effective value*, the *root-mean-square value*, or the *rms value*.

To reduce the number of subscripts used in the text, and to conform with standard practice, the letter symbols for rms current and rms voltage will be capital letters in lightface italic type. From here on, throughout the text,

> I means I_{rms} or I_{DC} as applicable
> E means E_{rms} or E_{DC} as applicable
> V means V_{rms} or V_{DC} as applicable

Thus

$$I = \frac{I_{m\sim}}{\sqrt{2}} \tag{9-14}$$

$$V = \frac{V_{m\sim}}{\sqrt{2}} \tag{9-15}$$

The average power delivered to a resistor by a sinusoidal driving voltage, in terms of the *rms current to the resistor and the rms voltage across the resistor is*

$$P_{av\sim} = V_R I_R \tag{9-16}$$

$$P_{av\sim} = I_R^2 R \tag{9-17}$$

$$P_{av\sim} = \frac{V_R^2}{R} \tag{9-18}$$

The rms current to a resistor in terms of the rms voltage across the resistor may be determined from Ohm's law:

$$I_R = \frac{V_R}{R} \tag{9-19}$$

It should be noted that the rms value of an alternating voltage or the rms value of an alternating current is only its equivalent DC value; although the rms values are very useful for calculating current, voltage, and average power, one must not lose sight of the fact that the actual current is an alternating one.

Example 9-5 The equation of a certain sinusoidal voltage wave is given by $e = 120 \sin 230t$. Determine (*a*) the maximum value of the wave; (*b*) the rms value; (*c*) the frequency; (*d*) the instantaneous voltage 0.01 s after the wave passes the zero-voltage value going in the positive direction; (*e*) the period of the wave.

Solution

(*a*) $E_m = 120$ V

(*b*) $E = 120 \, (0.707) = 84.84$ V

(*c*) $\omega = 230 = 2\pi f$ rad/s

$$f = \frac{230}{2\pi} = 36.62 \text{ Hz}$$

(*d*) $e = 120 \sin[230(0.01)]^R$

$\quad\ e = 120 \sin(2.3)^R$

$\quad\ e = 120 \, (0.7457) = 89.5$ V

(*e*) $T = \dfrac{1}{f} = \dfrac{1}{36.6} = 0.027$ s

Example 9-6 A 240-V 25-Hz sinusoidal generator is connected to a 20-Ω resistor through a circuit breaker and an ammeter as shown in Fig. 9-7a. Determine (*a*) the ammeter reading; (*b*) the period of the wave; (*c*) the average power; (*d*) the equation for the current wave if the circuit breaker is closed 30° after the wave crosses the time axis going in the positive direction; (*e*) the instantaneous current when the elapsed time is 0.01 s; (*f*) the ammeter reading if the generator is replaced by a 240-V battery.

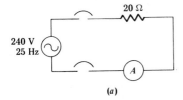

(*a*)

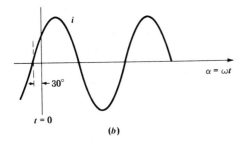

(*b*)

Figure 9-7 Circuit for Example 9-6.

Solution

(a) $I_R = \dfrac{V_R}{R} = \dfrac{240}{20} = 12$ A

(b) $T = \dfrac{1}{f} = \dfrac{1}{25} = 0.040$ s

(c) $P_{av} = I_R^2 R = (12)^2(20) = 2880$ W

or $P_{av} = V_R I_R = (240)(12) = 2880$ W

or $P_{av} = \dfrac{V_R^2}{R} = \dfrac{(240)^2}{20} = 2880$ W

(d) The current wave is sketched in Fig. 9-7b.

$\quad I_m = 12\sqrt{2} = 16.97$ A

$\quad \omega = 2\pi f = 2\pi(25) = 157$ rad/s

$\quad \beta = 30°$

$\quad i = 16.97 \sin(157t + 30°)$

(e) $i = 16.97 \sin[(157 \times 0.01)^R + 30°]$

Changing radians to degrees,

$\quad i = 16.97 \sin[(1.57 \times 57.3) + 30]°$

$\quad i = 16.97 \sin(119.96)° = 14.7$ A

or, using radians,

$\quad i = 16.97 \sin\left(1.57 + \dfrac{30}{57.3}\right)^R$

$\quad i = 16.97 \sin(2.09)^R = 14.7$ A

(f) $I_R = \dfrac{E_{bat}}{R} = \dfrac{240}{20} = 12$ A

Derivation of Equation (9-13)

$\quad I_{DC} = \dfrac{I_{m\sim}}{\sqrt{2}}$

If the circuits of Fig. 9-6a and b are adjusted to deliver the same average heat power to identical resistors, then

$$P_{DC} = P_{av\sim} \tag{9-20}$$

Since the direct current is nonvarying, the average power is the same as the power at any instant of time. This is shown in Fig. 9-8a. Thus,

$$P_{DC} = I_{DC}^2 R \tag{9-21}$$

Figure 9-8b shows the graphs of the sinusoidal current, and the resultant pulsating power supplied to the resistor by the AC generator in Fig. 9-6b. Note that there are two power pulses for each cycle of current. The average value of the power wave may be determined by summing all the instantaneous $i^2 R$ values over one cycle and dividing by 2π. Thus, using calculus,

$$P_{av\sim} = \int_0^{2\pi} \frac{i^2 R \, d\alpha}{2\pi}$$

$$P_{av\sim} = \frac{R}{2\pi} \int_0^{2\pi} i^2 \, d\alpha \tag{9-22}$$

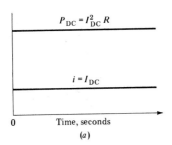

(a)

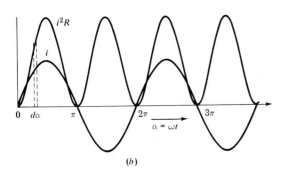

(b)

Figure 9-8 (a) Power curve for DC current; (b) power curve for a sinusoidal current.

Substituting Eqs. (9-21) and (9-22) into Eq. (9-20),

$$I_{DC}^2 R = \frac{R}{2\pi} \int_0^{2\pi} i^2 \, d\alpha$$

$$I_{DC}^2 = \frac{1}{2\pi} \int_0^{2\pi} i^2 \, d\alpha \qquad (9\text{-}23)$$

$$I_{DC} = \sqrt{\frac{1}{2\pi} \int_0^{2\pi} i^2 \, d\alpha} \qquad (9\text{-}24)$$

The equation for the sinusoidal current is

$$i = I_m \sin(\omega t) = I_m \sin \alpha \qquad (9\text{-}25)$$

Substituting Eq. (9-25) into Eq. (9-24), squaring both sides, and simplifying,

$$I_{DC}^2 = \frac{1}{2\pi} \int_0^{2\pi} (I_m \sin \alpha)^2 \, d\alpha$$

$$I_{DC}^2 = \frac{I_m^2}{2\pi} \int_0^{2\pi} \sin^2 \alpha \, d\alpha$$

Integrating and evaluating the limits,

$$I_{DC}^2 = \frac{I_m^2}{2\pi} \left[\tfrac{1}{2}(\alpha - \sin \alpha \cos \alpha)\right]_0^{2\pi}$$

$$I_{DC}^2 = \frac{I_m^2}{2\pi} \left[\tfrac{1}{2}(2\pi)\right]$$

Solving for I_{DC},

$$I_{DC} = \frac{I_{m\sim}}{\sqrt{2}} = 0.707 \, I_{m\sim} \qquad (9\text{-}26)$$

Solving Eq. (9-24) for I_{DC} required taking the square root of the mean value (average value) of the squares of the alternating current. Hence the DC value of the alternating-current wave is commonly referred to as the root-mean-square or rms value.

It should be noted that the same procedure outlined in this derivation may be used to obtain the rms value of any periodic wave. Thus, for the general case,

$$I_{rms} = \sqrt{\frac{1}{2\pi} \int_0^{2\pi} [f(\alpha)]^2 \, d\alpha} \qquad (9\text{-}27)$$

If the equation of the periodic wave is expressed as a function of time, rather than as a function of α, Eq. (9-27) may be expressed as

$$I_{rms} = \sqrt{\frac{1}{T} \int_0^T [f(t)]^2 \, dt} \tag{9-28}$$

where T represents the period of the wave in seconds.

AVERAGE VALUE OF A SINUSOIDAL WAVE

The average value of any periodic wave may be obtained by dividing one cycle of the wave in n equal parts to form n narrow vertical areas. Assign positive signs to the areas above the ωt axis, and minus signs to the areas below the ωt axis. Adding all the areas above and below the axis, taking into account their respective positive and negative signs, and dividing by n will result in the average value for the wave.

Inspection of the sine wave in Fig. 9-9a shows the area above the axis to be equal to the area below the axis. Hence the *average value of the sine wave is zero.*

$$I_{av\sim} = 0$$

However, if the sine wave is rectified using diodes as shown in Fig. 9-9b, the output of the rectifier will be the wave shown in Fig. 9-9c. The average value for the rectified sine wave is

$$I_{av\,\cap} = \frac{2}{\pi} I_{m\sim} = 0.637 \, I_{m\sim} \tag{9-29}$$

where $I_{av\,\cap}$ = average value of the *rectified* sine wave
$I_{m\sim}$ = maximum value of the sine wave

Note: The maximum value of the sinusoidal wave is equal to the maximum value of the rectified sine wave.

$$I_{av\sim} = 0$$

$$I_{av\,\cap} = 0.637 \, I_{m\sim}$$

$$I_{rms\sim} = 0.707 \, I_{m\sim}$$

$$I_{rms\,\cap} = 0.707 \, I_{m\sim}$$

The rms value of a sine wave and the rms value of a full-wave rectified sine wave are equal.

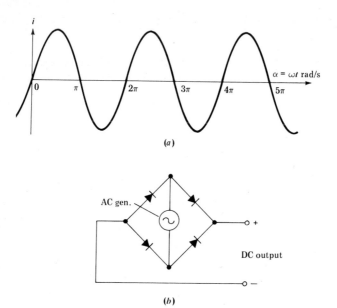

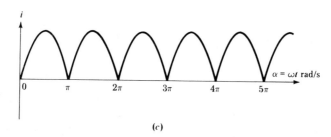

Figure 9-9 (a) Sine wave; (b) full-wave rectifier; (c) rectified sine wave.

Derivation of Equation (9-29)

Referring to Fig. 9-10, the average value of a rectified sine wave is equal to the sum of all the differential areas $i \, d\alpha$ divided by 2π. Using calculus,

$$I_{av} = \int_{\alpha=0}^{\alpha=2\pi} \frac{i \, d\alpha}{2\pi} \tag{9-30}$$

The equation representing the rectified sine wave may be written in two parts:

For $\alpha = 0$ to π,

$$i = I_m \sin \alpha \tag{9-31}$$

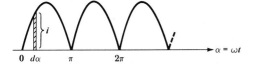

Figure 9-10 Rectified sine wave showing differential area.

For $\alpha = \pi$ to 2π,

$$i = -I_m \sin \alpha \qquad (9\text{-}32)$$

The equation has two components because the negative half-cycle of the sine wave was "flipped" when the wave was rectified. Substituting Eqs. (9-31) and (9-32) into Eq. (9-30),

$$I_{av\, \infty} = \frac{1}{2\pi} \left(\int_{\alpha = 0}^{\alpha = \pi} I_m \sin \alpha \, d\alpha + \int_{\alpha = \pi}^{\alpha = 2\pi} - I_m \sin \alpha \, d\alpha \right)$$

Integrating and simplifying,

$$I_{av\, \infty} = \frac{I_m}{2\pi} \left(- \cos \alpha \Big|_0^\pi + \cos \alpha \Big|_\pi^{2\pi} \right)$$

$$I_{av\, \infty} = \frac{I_m}{2\pi} (2 + 2)$$

$$I_{av\, \infty} = \frac{2I_{m\sim}}{\pi} \qquad (9\text{-}33)$$

It should be noted that the same procedure outlined in this derivation may be used to obtain the average value of any periodic wave. Thus, for the general case,

$$I_{av} = \int_0^{2\pi} \frac{f(\alpha) \, d\alpha}{2\pi} \qquad (9\text{-}34)$$

$$I_{av} = \int_0^T \frac{f(t) \, dt}{T} \qquad (9\text{-}35)$$

9-6   ADDITION OF SINUSOIDAL VARIABLES

The resultant wave formed by the addition of two or more sine waves of the same frequency will be another sine wave of the same frequency. This is true regardless of the relative magnitudes and phase angles of the component waves. Figure 9-11 shows two sine waves and their additive resultant.

The resultant wave is a sine wave whose frequency is the same as the frequency of the component waves, but whose magnitude and phase angle differ from those of the components.

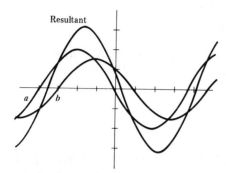

Resultant

a *b*

Figure 9-11 Two sine waves and their additive resultant.

Figure 9-12 shows two sinusoidal generators of the same frequency connected in series with a 20-Ω resistor and a switch. The voltage equations indicate the values of the respective generator voltages for the instant the switch is closed ($t = 0$), and for all elapsed time thereafter. When the switch is closed, the voltage impressed across the resistor will be ($e_1 + e_2$), and the current to the resistor will be

$$i = \frac{v_R}{R} = \frac{e_1 + e_2}{R} \qquad (9\text{-}36)$$

where $e_1 = 80 \sin(377t + 60°)$

$e_2 = 40 \sin(377t - 30°)$

Thus,

$$i = \frac{80 \sin(377t + 60°) + 40(\sin 377t - 30°)}{20} \qquad (9\text{-}37)$$

The current at the instant the switch is closed ($t = 0$) is

$$i\big|_{t=0} = \frac{80 \sin 60° + 40 \sin -30°}{20}$$

$$i\big|_{t=0} = \frac{69.28 - 20}{20} = 2.46 \text{ A}$$

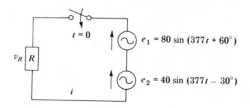

$t = 0$

$e_1 = 80 \sin (377t + 60°)$

$v_R\ R$

$e_2 = 40 \sin (377t - 30°)$

i

Figure 9-12 Two sinusoidal generators of the same frequency connected in series with a resistor.

The current, after an elapsed time of 0.02 s, is

$$\left.i\right|_{t=0.02} = \frac{80 \sin[377(0.02)(57.3) + 60]° + 40 \sin[377(0.02)(57.3) - 30]°}{20}$$

$$\left.i\right|_{t=0.02} = \frac{59.41 + 26.79}{20} = 4.31 \text{ A}$$

Equation (9-37) provides a means for obtaining the current at some elapsed time t. In order to obtain the rms or maximum value of the current wave, one must use trigonometric identities, which is at best a tedious process, or resort to a much simpler method called *phasor addition*.

9-7 PHASOR REPRESENTATION OF SINUSOIDAL VARIABLES

Mechanical quantities such as force, velocity, and acceleration are known as vector quantities. Whenever these quantities are to be added or subtracted, the direction in which each is acting must be considered. For example, if two equal and parallel forces pull on a wagon, as shown in Fig. 9-13a, the resultant pull is equal to $2F$; but if the two forces make an angle of $\theta°$ with each other, as shown in Fig. 9-13b, the resultant pull is less than $2F$. The resultant may be determined graphically by the parallelogram of forces shown in the diagram.

A similar technique, using *phasors* instead of vectors, may be used to simplify the addition or subtraction of two or more *sinusoidal voltages* (or

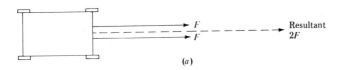

(a)

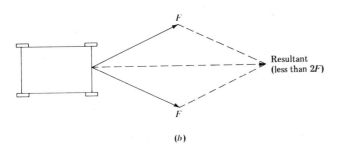

(b)

Figure 9-13 Resultant force due to (a) parallel forces; (b) nonparallel forces.

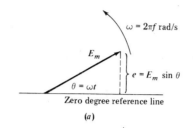

(a)

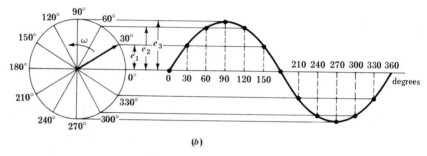

(b)

Figure 9-14 Development of a sinusoidal voltage wave by a revolving voltage phasor.

currents), all having the same frequency and the same or different E_m and phase angles.

A phasor, shown in Fig. 9-14a, is a directed line segment that rotates counter-clockwise around the origin at a constant angular velocity of ω radians per second. The length of the line segment is scaled in proportion to the peak value of the wave. Note that the positive direction of phasor rotation is the same as that for angular measure (counterclockwise). The angle θ is the instantaneous displacement angle in radians or degrees from the zero degree reference line.

The instantaneous value of the generated voltage is represented by the projection of the phasor on a vertical plane and may be determined by the product of E_m and the sine of the instantaneous angle of phase displacement.

$$e = E_m \sin \omega t$$

Figure 9-14b illustrates how the revolving phasor defines a sinusoidal voltage wave. The phasor, revolving counterclockwise at a uniform angular velocity, is shown at the instant it passes through the 30° position. At that instant the generated emf has a value e_1. At 60° it has a value e_2, at 90° a value e_3, etc. If these values are plotted as ordinates against an abscissa laid out in degrees, the result is a sine wave of voltage. Thus a sine wave can be represented by a phasor, and the value, or height, of the wave at any instant of time can be determined by the length of the "vertical shadow" cast by the phasor at that particular instant.

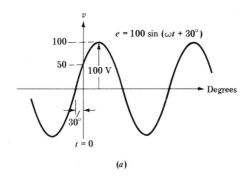

(a)

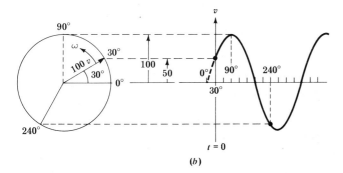

(b)

Figure 9-15 (a) Sine wave; (b) recreation of the sine wave by the vertical projection of its corresponding phasor.

Figure 9-15a illustrates a sinusoidal voltage wave whose equation, based on the given time-zero, is

$$e = 100 \sin(\omega t + 30°) \tag{9-38}$$

The corresponding voltage phasor, shown in Fig. 9-15b, has a length proportional to 100 V, and an angle at time-zero of $+30°$. The projection of this phasor on the vertical plane at time-zero indicates an instantaneous magnitude of $100 \sin 30°$ or 50 V; this corresponds to the time-zero value of the wave shown in Fig. 9-15a. As time elapses, the phasor rotates from its time-zero position, causing the magnitude of its projection on the vertical plane to change. The projection at any instant of time will be $e = 100 \sin(\omega t + 30°)$, where ωt is the increment increase in angle due to the elapsed time.

A plot of the vertical projection of the phasor versus its angular position, for all values of elapsed time, re-creates the sinusoidal voltage wave. *Thus the phasor represents the sine wave for every instant of time.* When the elapsed time is such that $(\omega t \times 57.3 + 30)° = 270°$, the projection on the vertical will be $100 \sin 270° = -100$ V. Similarly, if the elapsed time is such that

$(\omega t \times 57.3 + 30)° = 1470°$ (four revolutions of the phasor) the voltage at that instant would be $100 \sin 1470° = 100 \sin 30° = +50$ V.

When all the sinusoidal voltages and currents in a given circuit are at the same frequency, the angular velocity of each phasor is the same. Under such conditions, the phasors of the system are fixed in their relative positions with respect to one another as they rotate about the origin, and the fact that they rotate may be neglected. Hence, *for purposes of calculations and analysis, all phasors of the same frequency are "frozen" in their time-zero positions.*

The letter symbol representing a "frozen" phasor is printed in boldface type such as **E** and **I**. Expressed in polar form,

$$\mathbf{E} = E_{max}\underline{/\alpha°}$$

$$\mathbf{I} = I_{max}\underline{/\beta°}$$

E_{max} = maximum value of the voltage wave†
I_{max} = maximum value of the current wave†
$\underline{/\alpha°}$ = phase angle of the voltage phasor at $t = 0$
$\underline{/\beta°}$ = phase angle of the current phasor at $t = 0$

Thus the phasor representing the sinusoidal voltage wave of Eq. (9-38), expressed in polar form, is

$$\mathbf{E} = 100\underline{/30°}\text{ V}$$

Consider the two voltage waves given in Fig. 9-12,

$$e_1 = 80 \sin(377t + 60°)$$

$$e_2 = 40 \sin(377t - 30°)$$

The corresponding "frozen" voltage phasors are

$$\mathbf{E}_1 = 80\underline{/60°}\text{ V}$$

$$\mathbf{E}_2 = 40\underline{/-30°}\text{ V}$$

The two phasors and the corresponding sine waves are shown in Fig. 9-16. The sum of the two voltage waves may be obtained by graphical addition of wave e_1 and wave e_2 as shown in Fig. 9-16b, and the phasor representing the resultant voltage wave may be obtained by the parallelogram method as

---

† As will be seen in Chap. 12, E_{rms} and I_{rms} are often used in place of E_{max} and I_{max} in phasor calculations.

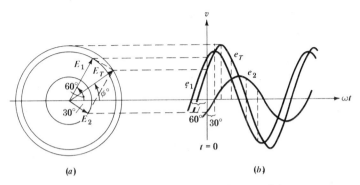

Figure 9-16 Two phasors, their additive resultant, and the corresponding sine waves.

shown in Fig. 9-16a. In the parallelogram method, $\mathbf{E}_1$ and $\mathbf{E}_2$ are laid off to scale (magnitude and angle) and the diagonal of the parallelogram is drawn to form the resultant phasor. The projection of the resultant phasor on the vertical plane, as it rotates counterclockwise, will result in the e_T wave shown in Fig. 9-16b.

9-8 **PHASOR ADDITION**

Phasor addition of sinusoidal voltages and phasor addition of sinusoidal currents may be accomplished by using any one of the following methods:

1. Graphically, by using the parallelogram method.
2. Graphically, by using the tip-to-tail method.
3. Algebraically, by resolving the phasors into horizontal and vertical components, combining like terms, and then obtaining the resultant.

A comparison of the three methods will be made by using the circuit shown in Fig. 9-17a. The sine waves represented by voltages $e_1, e_2,$ and e_3 are shown in Fig. 9-17b. The equations for these voltage waves, and their corresponding phasors, are

$$e_1 = 100 \sin(\omega t + 45°) \qquad \mathbf{E}_1 = 100\underline{/45°}$$

$$e_2 = 50 \sin(\omega t - 30°) \qquad \mathbf{E}_2 = 50\underline{/-30°}$$

$$e_3 = 60 \sin \omega t \qquad \mathbf{E}_3 = 60\underline{/0°}$$

The phasors are shown drawn to scale in Fig. 9-17c.

Parallelogram Method

To use the parallelogram method, start the phasor diagram by drawing any two of the three phasors; the angles should be marked off with a protractor,

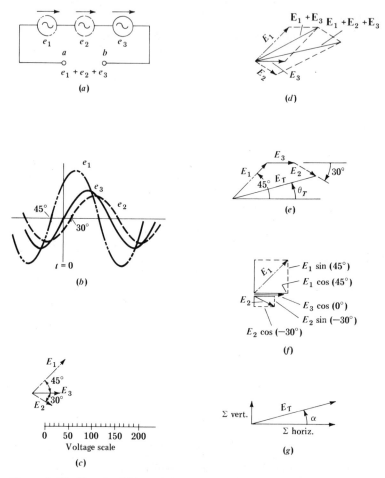

Figure 9-17 Phasor addition of sinusoidal voltages: (a) circuit;
(b) voltage waves; (c) phasors drawn to scale; (d) parallelogram
method; (e) tip-to-tail method; (f), (g) algebraic method.

and the length of the lines should be proportional to the peak values of the
voltages. This is shown in Fig. 9-17d, where E_1 and E_3 are the first two phasors
to be entered. Complete the parallelogram and draw the diagonal. The
diagonal is the phasor representing $(E_1 + E_3)$. Use this diagonal with the
remaining phasor E_2 to form another parallelogram. The diagonal of the
second parallelogram is $E_1 + E_2 + E_3$. The resultant voltage, magnitude,
and phase angle are obtained by applying the scale and the protractor to
the second diagonal. Thus,

$$E_T = 180\underline{/14.7°}\ V$$

Tip-To-Tail Method

To use the tip-to-tail method, enter the phasors on the phasor diagram, tip-to-tail, as shown in Fig. 9-17e. A line drawn from the origin to the tip of the last phasor is the resultant voltage.

$$\mathbf{E}_T = \mathbf{E}_1 + \mathbf{E}_2 + \mathbf{E}_3 = 180\underline{/14.7^\circ}\ \text{V}$$

Algebraic Method

The algebraic method provides a very accurate and convenient means for summing two or more phasors. Each phasor is broken down into horizontal and vertical components, as shown in Fig. 9-17f. The horizontals are summed, the verticals are summed, and the magnitude of the resultant voltage is obtained by taking the square root of the sum of the squares of the horizontals and the squares of the verticals. Thus,

$$E_T = \sqrt{\left(\sum \text{horiz.}\right)^2 + \left(\sum \text{vert.}\right)^2}$$

the phase angle of the resultant voltage is

$$\alpha = \tan^{-1} \frac{\sum \text{vert.}}{\sum \text{horiz.}}$$

Figure 9-17f and g illustrates the application of the algebraic method to obtain the summation of the voltages in the example problem. The corresponding computations follow:

Phasor	Horizontal component	Vertical component
$100\underline{/45^\circ}$	$100 \cos 45^\circ = 70.71$	$100 \sin 45^\circ = 70.71$
$50\underline{/-30^\circ}$	$50 \cos -30^\circ = 43.30$	$50 \sin -30^\circ = -25.00$
$60\underline{/0^\circ}$	$60 \cos 0^\circ = 60.00$	$60 \sin 0^\circ = 0.00$
Sums	174.01	45.71

$$E_T = \sqrt{(174.01)^2 + (45.71)^2}$$

$$E_T = 180\ \text{V}$$

$$\alpha = \tan^{-1} \frac{45.71}{174.0} = 14.72^\circ$$

Thus,

$$\mathbf{E}_T = 180\underline{/14.71^\circ}\ \text{V}$$

If terminals a and b in Fig. 9-17a are connected to a 20-Ω resistor, the rms current in the resistor and the heat power dissipated will be

$$I = \frac{E_T}{R} = \frac{180}{20} = 9.0 \text{ A} \qquad P = I^2R = (9.0)^2(20) = 1620 \text{ W}$$

9-9   **PHASOR ADDITION IN THE 0-Hz SYSTEM**

If the frequency of a sinusoidal generator is caused to approach 0 Hz, the angular velocity of the phasors representing the generator current and the generator voltage also approaches zero. That is,

$$\omega = 2\pi f \Big|_{f \to 0} = 0$$

Since a phasor is defined as a directed line segment that rotates counterclockwise around the origin at a constant angular velocity of ω radians per second, at 0 Hz, the line segments representing current and voltage will not rotate but will be fixed in position for all finite time. Thus, at 0 Hz, the current and voltage phasors are reduced to simple vectors as shown in Fig. 9-18b, and the voltage phasor is

$$E_{\text{bat}} = 24\underline{/0°} \text{ V}$$

However, as a general rule, when making calculations in a 0-Hz (DC) system, the 0° angle is omitted, the phasor diagram is omitted, and the voltage is

$$E_{\text{bat}} = 24 \text{ V}$$

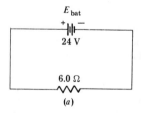

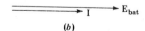

Figure 9-18   Phasor representation at 0 Hz.

SUMMARY OF FORMULAS

$$T = \frac{1}{f} \qquad \alpha = \omega t \qquad \omega = 2\pi f$$

$$1 \text{ rad} = 57.30° \qquad \pi = 3.142$$

$$I = I_{rms} = 0.7071_{max \sim} = 0.7071_{max \, \infty} \qquad I_{av \, \infty} = 0.6371_{m \sim}$$

$$P_{av \sim} = V_R I_R = I_R^2 R = \frac{V_R^2}{R} \qquad I_{av \sim} = 0$$

PROBLEMS

9-1 (a) Write the equation of a voltage wave that has a maximum value of 300 V and a frequency of 25 Hz; (b) what is the instantaneous value of voltage 10 s after the wave passes the zero position going in the positive direction?

9-2 The equation of a certain current wave is $i = 32 \sin 375t$. Determine (a) the maximum value of the wave; (b) the frequency; (c) the period; (d) the instantaneous value of current 0.060 s after the wave passes the zero position going in the positive direction.

9-3 The equation of the generated emf of an alternator is $e = 100 \sin 157t$. Determine (a) the maximum instantaneous voltage; (b) the rms value of voltage; (c) the average value of voltage; (d) the frequency; (e) the period of the wave; (f) the instantaneous value of voltage when $t = 0.020$ s.

9-4 The equation of a voltage wave is $e = 100 \sin 188t$ and that of its corresponding current wave is $i = 20 \sin[188t + (\pi/2)]$. Determine (a) the frequency of the voltage wave; (b) the frequency of the current wave; (c) the maximum value of the voltage; (d) the equivalent DC value of the current wave; (e) the magnitude of the current when the voltage is zero; (f) the phase angle between the two waves; (g) sketch the two waves on one set of coordinate axes.

9-5 The equation of a current wave passing through a 10-Ω resistor is $i = 2.08 \sin 1000t$. Calculate (a) the average power; (b) the maximum instantaneous power; (c) the power at an elapsed time of 0.001 s; (d) if the current passes through a full-wave rectifier, determine the rms value and the average value of the output current.

9-6 A 2000-Ω resistor is connected across a 240-V 60-Hz generator. (a) Calculate the current; (b) calculate the power.

9-7 Repeat Prob. 9-6 using a 30-Hz generator.

9-8 A certain incandescent lamp draws 100 W when connected to a 120-V 60-Hz generator. (a) Calculate the resistance of the lamp; (b) calculate the current; (c) write the equation of the current wave.

9-9 A 440-V 60-Hz voltage wave, when impressed across a certain resistor, dissipates energy at the rate of 10.0 kW. Determine (a) the rms value of current; (b) the maximum value of current; (c) the maximum instantaneous value of power; (d) the average value of power; (e) the resistance of the load.

9-10 Repeat Prob. 9-9 for a 440-V 0-Hz system (direct current).

9-11 (a) Write the equations for the current and voltage waves in Fig. 9-19. Determine (b) the current at $t = 0.1$ s; (c) the rms value of the current wave; (d) the average value of the current wave; (e) the average value if the current wave is rectified.

9-12 (a) Write the time-domain equations corresponding to the sinusoidal waves in Fig. 9-20. Determine (b) the current and voltage at $t = 0$ s; (c) the rms voltage; (d) the average value of the voltage wave; (e) the average value of the rectified voltage wave.

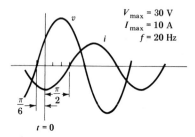

Figure 9-19 Current and voltage waves for Prob. 9-11.

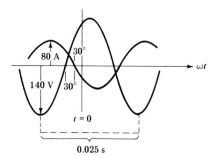

Figure 9-20 Current and voltage waves for Prob. 9-12.

9-13 Two 60-Hz generators are connected in series and the voltage wave of one is displaced 25° away from that of the other. The amplitude of each voltage wave is 339.5 V. (a) Draw the phasor diagram; (b) calculate the magnitude of the resultant voltage; (c) find what angle the resultant makes with respect to the others; (d) sketch the three voltage waves; (e) write the equations for the three waves.

9-14 Two 240-V DC generators are connected in series, and their voltages are in phase opposition. Draw the phasor diagram and calculate the resultant voltage.

9-15 Repeat Prob. 9-13 using 60-Hz generators.

9-16 The resultant magnitude of two 10-kHz voltage waves connected in series is 325 V. One of the component waves has a magnitude of 120 V and is displaced from the resultant by 25°. Draw the phasor diagram and calculate the magnitude and phase displacement of the other component.

9-17 Repeat Prob. 9-16 assuming that the displacement angle between the 120-V component and the 325-V resultant is 160°.

9-18 Three 90-Hz voltage waves are displaced from each other by 120°. The respective maximum voltages are 100, 115, and 130 V. Assuming that the three generators are connected in series, draw the phasor diagram and calculate the resultant voltage.

9-19 Determine the resultant of the following voltage phasors. The voltages are in series and the indicated angle is measured from the 0° reference line; 120 V at 30°; 85 V at −45°; 200 V at 0°; 191 V at 90°; 74 V at 120°. All operate at 400 Hz.

9-20 Figure 9-21 shows two generators in series supplying current to a 12-Ω resistor. (a) Determine the frequency; (b) write the equations of e_1 and e_2 for the given time-zero; (c) determine the rms value of e_1; (d) sketch the corresponding phasor diagram; (e) determine the resultant E_{max} across the 12-Ω resistor; (f) determine the power drawn by the resistor at $t = 0$ s.

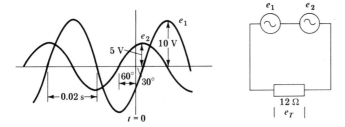

Figure 9-21   Current and voltage waves for Prob. 9-20.

9-21   The voltage equations for three generators connected in series are

$$e_1 = 25 \sin\left(600t + \frac{\pi}{6}\right)$$

$$e_2 = 60 \sin(600t + 54°)$$

$$e_3 = 20 \sin(600t - 20°)$$

Determine (a) the rms value of the e_1 wave; (b) the frequency of the e_1 wave; (c) the period of the e_1 wave; (d) sketch the three waves on a common time axis using the same time-zero; (e) sketch the corresponding phasor diagram to scale, and determine the magnitude and phase angle of the resultant voltage by tip-to-tail addition; (f) repeat part e using trigonometry.

9-22   Two generators e_1 and e_2 are connected in series and supply current to a 10-Ω resistor. The generator voltages are

$$e_1 = 12 \sin 125.6t$$

$$e_2 = 6 \sin\left(125.6t + \frac{\pi}{3}\right)$$

(a) Sketch the two voltage waves on a common time axis and enter the time-zero line; (b) draw the corresponding phasor diagram; (c) calculate the maximum instantaneous voltage across the resistor; (d) calculate the rms current in the resistor; (e) calculate the frequency of the current wave; (f) find the current at $t = 0.01$ s.

9-23   A series connection of three 20-Hz sinusoidal generators supplies power to a 40-Ω resistor load. The phasor voltages are

$$\mathbf{E}_1 = 100\underline{/45°} \qquad \mathbf{E}_2 = 100\underline{/-135°} \qquad \mathbf{E}_3 = 100\underline{/90°}$$

(a) Sketch the three voltage waves on a common time axis and indicate time-zero; (b) sketch the corresponding phasor diagram and construct the resultant voltage phasor; determine (c) the phasor expression and the time-domain equation for the voltage across the resistor; (d) the rms voltage across the resistor; (e) the rms current; (f) the average heat power drawn by the resistor; (g) the current at an elapsed time of 0.006 s from time-zero.

9-24 Three series-connected generators supply power to a 100-Ω resistor load. The time-domain equations representing the three generators are

$$e_1 = 36 \sin(377t + 56°)$$

$$e_2 = 41 \sin(377t + 76°)$$

$$e_3 = 86 \sin(377t - 54°)$$

(*a*) Sketch the circuit diagram and the phasor diagram; determine (*b*) the resultant rms voltage across the resistor; (*c*) the average power drawn by the resistor.

9-25 Two 60-Hz sinusoidal generators are connected in series and the resultant voltage is connected across a 10-Ω resistor. Generator 1 has an rms value of 100 V and a phase angle of 30°. Generator 2 has an rms value of 200 V and a phase angle of $-60°$. Determine (*a*) the rms voltage across the resistor; (*b*) the rms current through the resistor; (*c*) the equations for each generator voltage as functions of time; (*d*) the phasor diagram showing $\mathbf{E}_1$, $\mathbf{E}_2$, $\mathbf{V}_R$, $\mathbf{I}_R$; (*e*) the current at $t = 0.004$ s; (*f*) the total heat energy expended during two cycles of the current wave.

9-26 Two 50-Hz generators are connected in series and supply energy to a 10-Ω resistor. The phasors representing the generator voltages are

$$\mathbf{E}_1 = 50\underline{/30°} \qquad \mathbf{E}_2 = 100\underline{/0°}$$

(*a*) Sketch the corresponding sine waves and indicate time-zero; (*b*) sketch the corresponding phasor diagram and indicate the resultant phasor; (*c*) calculate the amplitude and phase angle of the resultant voltage; (*d*) determine the rms current and the average heat power dissipated; (*e*) find how much *heat energy* is expended during three cycles of the current wave.

CHAPTER 10

BEHAVIOR OF $R, L,$ AND C IN A SINUSOIDAL SYSTEM

Depending on the frequency of oscillation, the sinusoidal variation of current and voltage can have a profound effect on the behavior of the circuit elements.

10-1 EFFECT OF FREQUENCY ON THE CURRENT TO A RESISTOR

Figure 10-1a shows a sinusoidal generator connected in series with a switch and a 20-Ω resistor. The voltage v_R across the resistor is equal to the generator voltage:

$$v_R = 100 \sin(2\pi f t + 30°)$$

The current through the resistor, as determined by Ohm's law, is

$$i_R = \frac{v_R}{R} = \frac{100 \sin(2\pi f t + 30°)}{20} = 5 \sin(2\pi f t + 30°)$$

Note that both i_R and v_R have the same frequency and are in phase, as shown in Fig. 10-1b. There is *no angular difference between the current in a resistor and the voltage across it*. Expressed as phasors,

$$\mathbf{V}_R = 100\underline{/30°} \text{ V} \qquad \mathbf{I}_R = 5\underline{/30°} \text{ A}$$

The corresponding phasor diagram is shown in Fig. 10-1c.

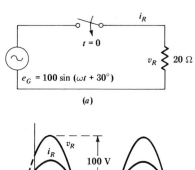

(a)

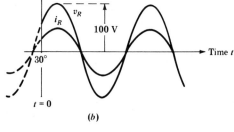

(b)

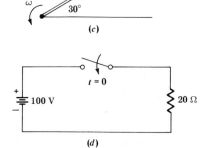

(c)

(d)

Figure 10-1 (a) Sinusoidal generator connected to a lumped resistance of 20 Ω; (b) current and voltage waves; (c) phasor diagram; (d) 100-V 0-Hz source connected to the 20-Ω resistor.

Figure 10-1d shows a 100-V 0-Hz (DC) supply connected to the same 20-Ω resistor. The current as determined from Ohm's law is

$$I = \frac{E_{bat}}{R}$$

$$I = \frac{100}{20} = 5 \text{ A}$$

Example 10-1 A 258-Ω resistor is connected to a sinusoidal generator whose voltage is expressed by $e_G = 126 \sin 1131t$. Determine (a) the rms voltage; (b) the rms current; (c) the time-domain equation for the current wave; (d) the instantaneous power delivered to the resistor at $t = 0.04$ s; (e) the average heat power expended by the resistor; (f) the heat energy expended in 10 s of operation.

Solution
The circuit is similar to that shown in Fig. 10-1*a*.

(*a*) $E_G = 126/\sqrt{2} = 89.1$ V

(*b*) $I_R = \dfrac{V_R}{R} = \dfrac{89.1}{258} = 0.345$ A

(*c*) $i = 0.345\sqrt{2} \sin 1131t = 0.4879 \sin 1121t$

(*d*) $P_{inst} = vi$

$P_{inst} = 126 \sin 1131t \times 0.4879 \sin 1131t$

$P|_{t=0.04\,s} = 126 \sin(1131 \times 0.04)^R \times 0.4879 \sin(1131 \times 0.04)^R$

$P|_{t=0.04\,s} = 119.90 \times 0.4642 = 55.6$ W

(*e*) $P_{av} = VI = (89.1)(0.345) = 30.74$ W

(*f*) $W = P_{av} \times t = 30.74 \times 10 = 307.4$ J

10-2 EFFECTIVE RESISTANCE

A phenomenon called *skin effect* causes a conductor to offer a higher resistance to an alternating current than it offers to a direct current. The resistance offered by a conductor to direct current is called *DC resistance* or *ohmic resistance*, and the resistance offered by the same conductor to an alternating current is called the *AC resistance* or *effective resistance*. The skin effect, explained in Sec. 10-7, is more pronounced at higher frequencies and is appreciably greater if the conductor is enclosed in a metallic sheath or pipe or is wound around a ferromagnetic core.

A table of multiplying factors for converting DC resistance to 60-Hz AC resistance is given in App. 4. The table shows the effect of 60 Hz on the resistance of conductors with different cross-sectional areas, with and without metallic sheaths.

Example 10-2 The ohmic resistance of a certain length of 1750-kcmil copper conductor is 0.012 Ω. Determine the effective resistance at 60 Hz if the conductor is enclosed in metallic pipe.

Solution
From App. 4, the multiplying factor for 1750-kcmil copper conductor enclosed in metallic pipe is 1.67. Hence the effective resistance at 60 Hz is

$R_{60\,Hz} = 0.012(1.67) = 0.020$ Ω

10-3   **EFFECT OF FREQUENCY ON THE CURRENT TO AN INDUCTOR**

As previously postulated, the inductance of a circuit serves to delay the buildup or decrease in current but does not prevent or in any way limit the change. Hence, when a 0-Hz voltage is applied to an inductor, the current, although delayed, is limited only by the resistance of the circuit. *However, if the circuit is operating at a frequency higher than 0 Hz, the current will be limited in magnitude as well as delayed in time.* The limitation imposed on the amplitude of the current wave, for a given value of inductance, is a direct result of the frequency of reversal of the applied voltage. Current waves at the lower frequencies have the same direction for longer periods of time and therefore attain proportionately higher amplitudes before decreasing than do current waves at the higher frequencies.

If there is no inductance in the circuit, then, regardless of the frequency, the current will not be delayed and its magnitude will be limited solely by the resistance of the circuit and the magnitude of the driving voltage. Doubling the inductance doubles the time delay and, assuming negligible circuit resistance, results in only half the value of current for the same frequency.

For example, if the time constant of an *LR* circuit is 2 s, it would require five time constants (10 s) to attain essentially "full current" at 0 Hz. However, if a 60-Hz driving voltage is applied to the circuit, the voltage will rise from zero to its peak value in one quarter-cycle, then decrease and reverse. Since one quarter-cycle of a 60-Hz wave is $\frac{1}{4} \times \frac{1}{60} = \frac{1}{240}$ s, the current could not possibly reach the same value that would be attained with a DC driver of the same voltage amplitude.

Figure 10-2 shows the effect of different sinusoidal frequencies on the current through an "ideal" inductor. The ideal generator in Fig. 10-2a has its E_{max} held constant, but its frequency is adjustable from 0 Hz to infinity. Representative voltage waves from 0 Hz to 160 Hz are shown in Fig. 10-2b; the switch is closed at the instant the respective voltage wave is at its peak value. The current waves that correspond to the different frequencies are shown in Fig. 10-2c.

The effect of frequency on the current through an inductor is quite similar to the effect of a continually reversing torque applied to a flywheel. In the case of the flywheel, more rapid reversals of the applied torque will result in less motion in each direction. Similarly, in the case of the inductor, higher frequencies of reversal of an applied voltage will result in less current in each direction, as shown in Fig. 10-2c.

Since inductors have a greater "choking effect" on currents at higher frequencies than on currents at lower frequencies, they are often used in rectifier filter circuits to reduce the ripple. Figure 10-3 illustrates the application of an inductor in a full-wave rectifier circuit. Inductors used for this purpose are called *chokes*.

Large current-limiting inductors, called *current-limiting reactors*, are used to protect alternating-current generators and other large AC apparatus

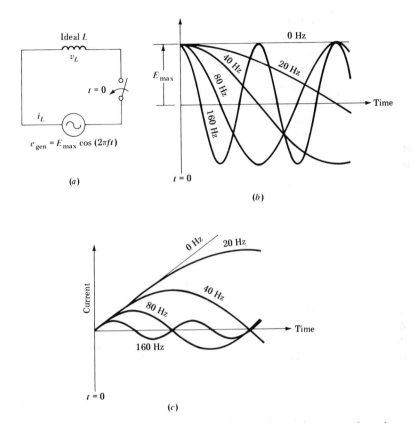

Figure 10-2 Effect of different sinusoidal frequencies on the current through an ideal inductor: (*a*) circuit; (*b*) representative voltage waves of different frequencies; (*c*) current waves that correspond to the voltages in (*b*).

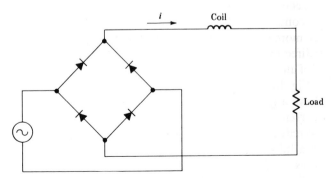

Figure 10-3 Application of an inductor to smooth the ripple in a rectifier circuit.

from damage by heavy short-circuit currents. When used for this purpose, the conductors are embedded in concrete to withstand the tremendous magnetic forces that tend to squeeze the turns together.

10-4 TIME LAG OF A SINUSOIDAL CURRENT IN A PURE INDUCTOR

Figure 10-4a shows a sinusoidal current in a pure inductor, and the resultant voltage drop across it. As indicated, the current through the inductor lags the voltage drop across the inductor by 90°. The 90° phase relationship between the i_L and v_L waves is explained by using Faraday's law as expressed in Eq. (10-1):

$$v_L = L \frac{di}{dt} \tag{10-1}$$

As indicated in Eq. (10-1) and shown in Fig. 10-4a, the voltage-drop wave has its maximum value at the instant that the rate of change of current di/dt is the greatest, and occurs when the current wave crosses the time axis. Likewise, the voltage developed across the coil is zero at the instant the rate of change of current is zero, and occurs when the current attains its peak value.

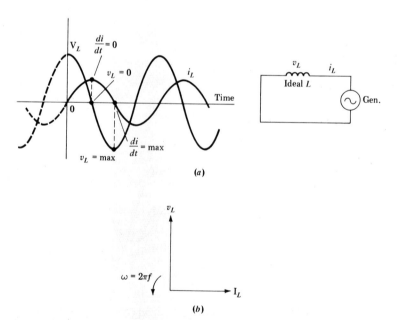

(a)

(b)

Figure 10-4   Time lag of a sinusoidal current in a pure inductor.

The equations representing the sinusoidal current wave and the sinusoidal voltage-drop wave in Fig. 10-4a are

$$i_L = I_{L,\text{max}} \sin 2\pi f t$$

$$v_L = V_{L,\text{max}} \sin(2\pi f t + 90°)$$

Expressing the current and voltage in polar form,

$$\mathbf{I}_L = I_{L,\text{max}} \underline{/0°}$$

$$\mathbf{V}_L = V_{L,\text{max}} \underline{/90°}$$

The phasor diagram for these sinusoids is shown in Fig. 10-4b. *Note that the current through the inductor lags the voltage drop across it by 90°.*

10-5   INDUCTIVE REACTANCE

In circuits *with sinusoidal drivers*, the ratio of the rms voltage across an ideal inductor to the rms current through it is called the inductive reactance of the inductor. Expressed mathematically,

$$X_L = \frac{V_L}{I_L}$$

where X_L = inductive reactance, Ω
        V_L = rms voltage across inductance, V
        I_L = rms current in inductance, A

Expressed in Ohm's-law form,

$$I_L = \frac{V_L}{X_L}$$

The inductive reactance may also be calculated from the known values of frequency and inductance by means of the following formula:

$$\boxed{X_L = 2\pi f L} \tag{10-2}$$

where f = frequency of driving voltage, Hz
        L = inductance, H

As indicated in Eq. (10-2), the inductive reactance offered by a given inductor decreases with decreasing frequency. If the frequency of the driving voltage is 0 Hz (DC), the inductive reactance will be zero; thus, discounting the initial time delay (five time constants) at 0 Hz, the inductance will offer no opposition to the current. *Equation (10-2) applies only to circuits containing sinusoidal drivers and linear elements.*

Derivation of Equation (10-2)

Referring to Fig. 10-4, the equation for the sinusoidal current through an ideal inductor is

$$i_L = I_{L,\max} \sin 2\pi ft \tag{10-3}$$

From Faraday's law, the voltage drop across the inductor is

$$v_L = L\frac{di_L}{dt} \tag{10-4}$$

Substituting Eq. (10-3) into Eq. (10-4) and performing the indicated operations,

$$v_L = L\frac{d}{dt}I_{L,\max} \sin 2\pi ft$$

$$v_L = 2\pi f LI_{L,\max} \cos 2\pi ft \tag{10-5}$$

$$V_{L,\max} = 2\pi f LI_{L,\max}$$

Expressed in Ohm's-law form,

$$I_{L,\max} = \frac{V_{L,\max}}{2\pi f L} \tag{10-6}$$

In terms of rms values

$$I_L = \frac{V_L}{2\pi f L} = \frac{V_L}{X_L}$$

where $X_L = 2\pi f L$

Note: Equations (10-3) and (10-5) are sine and cosine functions, respectively, indicating a 90° phase relationship between the two.

Example 10-3 A 10-H inductor of negligible resistance is connected to a 60-Hz sinusoidal driver whose voltage is 120 V rms. Calculate (*a*) the inductive reactance; (*b*) the rms current; (*c*) the current if the rms voltage remains the same but the frequency is changed to 10 Hz; (*d*) the polar form for the current and voltage in part *b*; (*e*) the equations for the voltage drop across the inductance, and the current to the inductance, as functions of time in part *b*; (*f*) repeat parts *a* and *b* for a frequency of 0 Hz (DC).

Solution

(a) $X_L = 2\pi f L = 2\pi(60)(10) = 3768 \ \Omega$

(b) $I_{rms} = \dfrac{V_{rms}}{X_L} = \dfrac{120}{3768} = 0.0318 \text{ A}$

(c) $X_L = 2\pi(10)(10) = 628 \ \Omega$

$I_{rms} = \dfrac{120}{628} = 0.191 \text{ A}$

(d) Assigning the driving voltage as the reference phasor at $0°$,

$\mathbf{E} = 120\underline{/0°}$

Since the current lags the voltage across the inductor by $90°$,

$\mathbf{I} = 0.0318\underline{/-90°}$

(e) $v_L = 120\sqrt{2} \sin 2\pi60t = 169.7 \sin 377t$

$i_L = 0.0318\sqrt{2} \sin(2\pi60t - 90°) = 0.045 \sin(377t - 90°)$

Note that the frequency of the current wave is the frequency of the driver, and the current wave lags the voltage wave by $90°$.

(f) $X_L = 2\pi f L = 2\pi(0)10 = 0$

$I = \dfrac{V}{X_L} = \dfrac{120}{0} = \infty$

If the frequency of the driving voltage is zero (direct current), the inductance will not limit the current. Hence, when used in DC systems, a coil must be designed with adequate internal resistance, by choice of conductor size, or connected in series with an external resistor of such magnitude as to prevent excessive current drain from the generator or battery.

0-6 **RATE OF RISE OF CURRENT THROUGH AN IDEAL INDUCTOR AT TIME-ZERO**

The rate of rise of current through an ideal inductor, at the instant the switch to the circuit is closed, depends on the value of the voltage wave at time-zero. If all the voltage waves have maximum values at time-zero, as shown in Fig. 10-2b, then the rate of rise of current will be the same for all frequencies including 0 Hz (DC). This is illustrated in Fig. 10-2c for the respective drivers and time-zero shown in Fig. 10-2b; changing the frequency changes the amplitude of the current wave but does not change its initial slope.

For the conditions shown in Fig. 10-2c, the rate of rise (or slope) of the current curve at $t = 0$ is expressed by

$$\frac{di_L}{dt}\bigg|_{t=(0+)} = \frac{V_{L,\,max}}{L} \tag{10-7}$$

The 0-Hz curve shown in Fig. 10-2c illustrates the behavior of a hypothetical circuit containing an *ideal inductance*, no resistance, and a DC driving voltage. Under such conditions, the current will rise continuously in a linear manner with time. With no resistance there will be no limit; it will continue to rise forever.

Derivation of Equation (10-7)

The slope of the current wave at time-zero may be obtained by substituting Eq. (10-6) into Eq. (10-3), taking the derivative of the resulting equation with respect to time, and then evaluating it at $t = 0+$. Thus,

$$i_L = \frac{V_{L,\,max}}{2\pi f L} \sin 2\pi f t \tag{10-8}$$

$$\frac{di_L}{dt} = \frac{d}{dt} \frac{V_{L,\,max}}{2\pi f L} \sin 2\pi f t$$

$$\frac{di_L}{dt} = \frac{2\pi f V_{L,\,max} \cos 2\pi f t}{2\pi f L} = \frac{V_{L,\,max} \cos 2\pi f t}{L}$$

At time-zero,

$$\frac{di_L}{dt}\bigg|_{t=(0+)} = \frac{V_{L,\,max}}{L}$$

To obtain the expression for the current wave when the frequency is 0 Hz (DC), substitute zero for the frequency in Eq. (10-8) and evaluate. Thus,

$$i_L = \frac{V_{L,\,max} \sin 2\pi 0 t}{2\pi 0 L} = \frac{0}{0} \tag{10-9}$$

Using L'Hospital's rule to evaluate indeterminate equation (10-9),

$$i_L = \frac{(d/df)(V_{L,\,max} \sin 2\pi f t)}{(d/df)(2\pi f L)} = \frac{2\pi t V_{L,\,max} \cos 2\pi f t}{2\pi L}$$

At a frequency of 0 Hz,

$$i_L\big|_{f=0} = \frac{V_{L,\,max}}{L} t \tag{10-10}$$

Equation (10-10) represents a straight line that passes through the point $t = 0$, $i = 0$, and whose slope is $V_{L,\,max}/L$.

Example 10-4   An ideal 12-V DC source is connected in series with an ideal 4.0-H inductor and a switch. Determine (*a*) the current 8.0 s after the switch is closed; (*b*) the steady-state current.

Solution

(*a*) $i_L\big|_{f=0} = \dfrac{V_L}{L}\,t = \dfrac{12}{4} \times 8 = 24$ A

(*b*) $i_{ss} = \dfrac{E_{bat}}{X_L} = \dfrac{12}{2\pi(0)4} = \infty$

10-7   **SKIN EFFECT**

A phenomenon called *skin effect* causes conductors to offer a higher resistance to an alternating current than they offer to a direct current. In a DC system, the current is distributed uniformly over the entire cross section of the conductor. However, with alternating current, the density of the current is greatest at the skin (or wall) of the conductor and has its lowest value at the center. A comparison of the current distribution in the same size conductor for DC and high-frequency AC is shown in Fig. 10-5*a*.

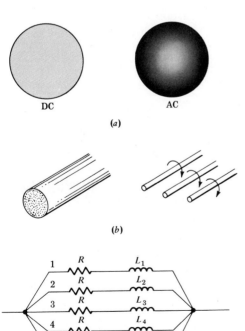

DC   AC

(*a*)

(*b*)

(*c*)

Figure 10-5   (*a*) Comparison of current distribution in the same size conductor for DC and high-frequency AC; (*b*) simulated conductor composed of an infinite number of identical filaments; (*c*) equivalent circuit for the paralleled filaments.

To explain this skin-effect phenomenon, imagine the conductor to be made up of an infinite number of identical parallel filaments of equal resistance, as shown in Fig. 10-5b. When carrying current, the center filament is encircled by more magnetic flux than is a filament at the wall. Thus the rate of change of flux about the center filament is greater than the rate of change of flux about the wall filament. The higher $d\phi/dt$ at the center of the conductor means a higher inductive reactance in the center filament, and hence less current in that filament.

An equivalent circuit for the conductor, illustrating the paralleled filaments, is shown in Fig. 10-5c; filaments 1 and 5 represent the wall filaments, filament 3 is the center filament, and filaments 2 and 3 are located between the center and the wall. Because of the differences in flux encirclements,

$$X_{L3} > X_{L2} > X_{L1} \qquad X_{L3} > X_{L4} > X_{L5}$$

resulting in

$$I_3 < I_2 < I_1 \qquad I_3 < I_4 < I_5$$

Since all filaments of the conductor do not carry equal shares of the current, the "resistance" offered to alternating current appears greater than the resistance offered to direct current. At very high frequencies (several million hertz) the current in the central section is so small that there is no need for solid conductors. For this reason, copper tubing is sometimes used as a substitute for solid-copper conductors in high-frequency applications.

However, at normal power-line frequencies (60 Hz) the skin effect for conductors smaller than 250,000 cmil area (250 kcmil) and not enclosed in metallic pipe or raceway is negligible. A 250-kcmil conductor has a diameter of 0.575 in (14.6 mm).

10-8   NONLINEAR INDUCTORS

The hypothetical curves shown in Fig. 10-6 illustrate the effect of a nonlinear magnetic core material on the inductance of a coil. For values of current ranging from 0 to i_a, the slope of the magnetization curve is essentially constant, and has its greatest value. This is the *linear region*. The corresponding inductance is constant, and is called the *normal or rated inductance*. See Sec. 8-3.

$$L_n = N \frac{d\phi}{di}\bigg|_{i \leq i_a} \tag{10-11}$$

where L_n = normal inductance, H.

For values of current in the knee region (i_a to i_b), the slope of the magnetization curve decreases with increasing current. Thus the inductance, which is a function of the slope, also decreases with increasing current.

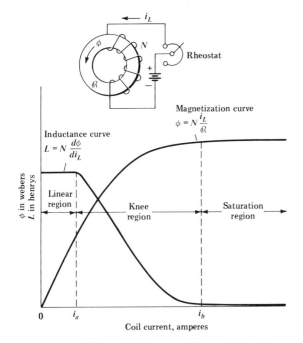

Figure 10-6 Flux and inductance characteristics of an inductor wound on a ferromagnetic core.

When operating in the saturation region (current excursions greater than i_b) the slope of the magnetization curve will be very small, and the inductance will have a very small value.

Since the self-inductance of a coil is a measure of its ability to oppose a change in the current through it, the greatest opposition to a changing current will occur in the linear region. Less but variable opposition to the same rate of change of current will occur in the knee region, and essentially little opposition to the same rate of change of current will occur in the saturation region.

Figure 10-7 shows oscilloscope recordings of current versus time for an inductor connected to a sinusoidal driving voltage. For sinusoidal current excursions in the linear region the inductance is constant. Hence the current is sinusoidal and the normal inductance L_N can be used to determine the current. That is,

$$X_L = 2\pi f L_n \tag{10-12}$$

$$I_L = \frac{V_L}{X_L} \tag{10-13}$$

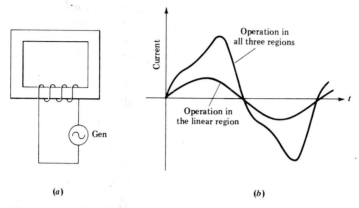

(a)　　　　　　　　　　　　　(b)

Figure 10-7 Recordings of current versus time of an iron-core inductor operating in the linear region, and in all three regions, when driven by a sinusoidal voltage.

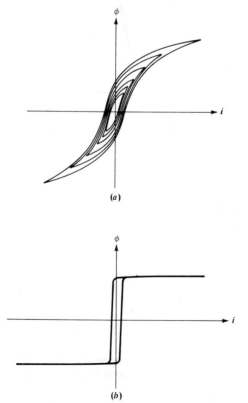

(a)

(b)

Figure 10-8 Hysteresis loops: (a) iron-core transformer; (b) saturable reactor.

However, *for sinusoidal current excursions that extend beyond the linear region, the inductance is not constant, the current is not sinusoidal, and the concept of inductive reactance does not apply.* Thus, when operating in the nonlinear region, Eqs. (10-12) and (10-13) have no significance and must not be used.

A series of overlapping hysteresis loops for the iron core of a transformer operating at different current ranges, with a sinusoidal driver, is shown in Fig. 10-8*a*. Note how increased current excursions cause the ferromagnetic core to operate farther into the saturation region.

Figure 10-8*b* shows a rectangular hysteresis loop for a *saturable reactor*. When saturation occurs, it is sudden and complete. The saturable reactor is used in magnetic amplifiers and other control applications.

10-9 EFFECT OF FREQUENCY ON THE CURRENT TO A CAPACITOR

Figure 10-9*a* shows an ideal sinusoidal generator whose frequency can be varied from 0 Hz to infinity, while keeping its voltage amplitude E_{max}

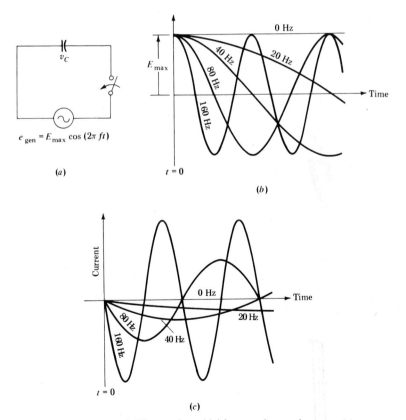

Figure 10-9 Effect of different sinusoidal frequencies on the current to a capacitor: (*a*) circuit; (*b*) representative voltage waves of different frequencies; (*c*) current waves that correspond to the voltages in (*b*).

constant. A few representative voltage waves of equal amplitude but different frequencies are shown in Fig. 10-9b.

As developed in Chap. 6, the current to a capacitor is proportional to the rate of change of voltage across it. That is,

$$i_C = C \frac{dv_C}{dt} \tag{10-14}$$

where dv_C/dt is the rate of change of voltage or slope of the voltage curve. Examination of the respective slopes of the sinusoidal voltage waves in Fig. 10-9b indicates that waves of higher frequency have greater (steeper) slopes and hence will cause greater currents to a capacitor than will waves of lower frequency.

The steady-state sinusoidal current waves shown in Fig. 10-9c are caused by the corresponding voltage waves in Fig. 10-9b. Note that higher frequencies cause higher current amplitudes and lower frequencies cause lower current amplitudes. As the frequency approaches zero, the amplitude of the current wave approaches zero.

10-10   TIME LEAD OF A SINUSOIDAL CURRENT TO A PURE CAPACITOR

Figure 10-10b shows the sinusoidal current to a pure capacitor, and the sinusoidal voltage across it. As indicated, the current to the capacitor leads the voltage across it by 90°. The 90° phase relationship between the i_C and v_C waves may be explained with the aid of Eq. (10-14).

As indicated in Eq. (10-14) and shown in Fig. 10-10b, the current wave has its maximum value at the instant that the rate of change of voltage dv_C/dt is the greatest, and occurs when the voltage wave crosses the time axis. Likewise, the current wave has zero value at the instant the rate of change of voltage is zero, and occurs when the voltage attains its peak value.

The equations representing the sinusoidal current wave and the sinusoidal voltage wave in Fig. 10-10b, for the given time-zero, are

$$i_C = I_{C(max)} \sin(2\pi f t + 90°)$$

$$v_C = V_{C(max)} \sin 2\pi f t$$

Expressing the current and voltage waves in polar form,

$$\mathbf{I}_C = I_{C(max)} \underline{/90°}$$

$$\mathbf{V}_C = V_{C(max)} \underline{/0°}$$

The phasor diagram corresponding to the waves in Fig. 10-10b is shown in Fig. 10-10c. Note that the current phasor leads the voltage phasor by 90°.

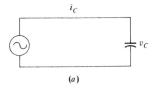

(a)

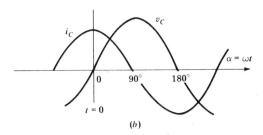

(b)

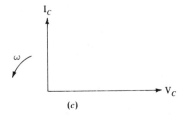

(c)

Figure 10-10   Time lead of a sinusoidal current to a pure capacitor.

10-11   CAPACITIVE REACTANCE

In circuits *with sinusoidal drivers*, the ratio of the rms voltage across a capacitor to the rms current to the capacitor is called the capacitive reactance of the capacitor. Expressed mathematically,

$$X_C = \frac{V_C}{I_C} \qquad (10\text{-}15)$$

where X_C = capacitive reactance, Ω.
Expressed in Ohm's-law form

$$I_C = \frac{V_C}{X_C} \qquad (10\text{-}16)$$

The capacitive reactance may also be calculated from the known values of frequency and capacitance by means of the following formula:

$$X_C = \frac{1}{2\pi f C}$$

(10-17)

where f = frequency of driving voltage, Hz
C = capacitance, F

As indicated in Eq. (10-17), the capacitive reactance offered by a given capacitor increases with decreasing frequency. If the frequency of the driving voltage is 0 Hz (DC), the capacitive reactance will be infinite and the capacitor will completely block the current (discounting the initial charging period).

Equation (10-17) applies only to circuits containing sinusoidal drivers and linear elements.

Derivation of Equation (10-17)

Figure 10-9a shows an ideal capacitor connected to a sinusoidal generator of constant amplitude but adjustable frequency. To compare the effects of all frequencies, including 0 Hz (DC), all driving voltages are assumed to have their maximum values at $t = 0$, as shown in Fig. 10-9b. The equation that represents the family of generator voltage waves shown in Fig. 10-9b is

$$e_{gen} = E_{max} \cos 2\pi f t$$

(10-18)

Applying Kirchhoff's voltage law to Fig. 10-9a,

$$v_C = e_{gen}$$

Substituting into Eq. (10-18),

$$v_C = E_{max} \cos 2\pi f t$$

(10-19)

From Eq. (10-14),

$$i_C = C \frac{dv_C}{dt}$$

(10-20)

Substituting Eq. (10-19) into Eq. (10-20) and performing the indicated operations,

$$i_C = C \frac{d}{dt} E_{max} \cos 2\pi f t$$

$$i_C = -2\pi f C E_{max} \sin 2\pi f t$$

(10-21)

The amplitude of the current wave in Eq. (10-21) is

$$I_{C(max)} = 2\pi f C E_{max}$$

In terms of rms values, obtained by dividing both sides by $\sqrt{2}$,

$$\frac{I_{C(max)}}{\sqrt{2}} = 2\pi f C \frac{E_{max}}{\sqrt{2}}$$

$$I_C = 2\pi f C E_{gen}$$

From Kirchhoff's voltage law, $V_C = E_{gen}$. Hence

$$I_C = 2\pi f V_C$$

$$I_C = \frac{V_C}{1/2\pi f C} = \frac{V_C}{X_C}$$

where

$$X_C = \frac{1}{2\pi f C}$$

Example 10-5 A 250-μF capacitor is connected across a 120-V 60-Hz system. Determine (*a*) the capacitive reactance; (*b*) the steady-state current; (*c*) the maximum instantaneous value of energy stored in the capacitor; (*d*) repeat parts *a*, *b*, and *c* for a frequency of 6 Hz; (*e*) repeat parts *a*, *b*, and *c* for a 120-V DC driver.

Solution

(*a*) $X_C = \dfrac{1}{2\pi f C} = \dfrac{1}{2\pi(60)250 \times 10^{-6}} = 10.61\ \Omega$

(*b*) $I = \dfrac{V}{X_C} = \dfrac{120}{10.61} = 11.31$ A

(*c*) $W = \frac{1}{2}CV^2 = \frac{1}{2}(250 \times 10^{-6})(120\sqrt{2})^2 = 3.6$ J

(*d*) $X_C = \dfrac{1}{2\pi f C} = \dfrac{1}{2\pi(6)(250 \times 10^{-6})} = 106.1\ \Omega$

$I = \dfrac{V}{X_C} = \dfrac{120}{106.1} = 1.131$ A

$W = \frac{1}{2}CV^2 = \frac{1}{2}(250 \times 10^{-6})(120\sqrt{2})^2 = 3.6$ J

(e) $X_C = \dfrac{1}{2\pi f C} = \dfrac{1}{2\pi(0)(250)10^{-6}} = \infty\ \Omega$

$I = \dfrac{V}{X_C} = \dfrac{120}{\infty} = 0\ \text{A}$

$W = \tfrac{1}{2}CV^2 = \tfrac{1}{2}(250 \times 10^{-6})(120)^2 = 1.8\ \text{J}$

10-12 ENERGY LOSS IN CAPACITORS

Although most capacitors closely approach the condition of pure capacitance, the dielectrics are not perfect and do permit a certain amount of electron leakage. If the leakage is excessive, it may cause overheating and result in the destruction of the capacitor. In addition, an alternating voltage applied to the capacitor causes a cyclic stressing of the dielectric that results in the development of heat. This loss in energy, called *dielectric hysteresis loss*, increases with increased frequency.

The small value of energy loss in a capacitor causes its phase angle to deviate from a true 90°, generally a small fraction of 1° for a ceramic capacitor and 1° to 2° for a paper capacitor. In most cases, however, the minute leakage current with its attendant energy loss is negligible, and for all such conditions the phase angle of the capacitor may be considered as 90°.

SUMMARY OF FORMULAS

$X_L = 2\pi f L \qquad X_C = \dfrac{1}{2\pi f C}$

$I_R = \dfrac{V_R}{R} \qquad I_L = \dfrac{V_L}{X_L} \qquad I_C = \dfrac{V_C}{X_C}$

PROBLEMS

10-1 A 25-Hz 120-V generator is connected to a 1568-Ω resistor. Sketch the circuit and determine (a) the rms current; (b) the average heat power expended; (c) the time-domain equations for the current and voltage waves.

10-2 A 154-Ω resistor is connected to a sinusoidal generator whose voltage is expressed by $e = 75 \sin 1000t$. Sketch the circuit and determine (a) the frequency; (b) the rms current; (c) the instantaneous power at $t = 0.01$ s; (d) the average heat power expended; (e) the energy expended in 56 h of operation.

10-3 The ohmic resistance of a certain length of 2000-kcmil aluminum conductor is 0.041 Ω. Determine the resistance at 60 Hz if the conductor is enclosed in a metallic sheath.

10-4 A 12-H ideal inductor is connected across a 240-V 25-Hz generator. Determine (a) the inductive reactance; (b) the rms current.

10-5 A 1.2-H ideal inductor is connected across a 208-V 60-Hz driver. Determine (a) the inductive reactance; (b) the rms current; (c) the peak instantaneous energy stored in the magnetic field.

10-6 An ideal 3-H inductor is connected to a sinusoidal generator whose voltage is expressed by $e = 100 \sin 400\,t$. Sketch the circuit and determine (a) the inductive reactance of the inductor; (b) the rms current; (c) the equation for the current wave; (d) the energy stored in the magnetic field at the instant the voltage wave goes through zero; (e) the rate of storage of energy in the field at the instant the voltage across the coil is zero; (f) if the AC generator is replaced by a 100-V battery, what will be the current at $t = 12$ s?

10-7 A 240-V 60-Hz 1200-rpm generator is connected across an inductive reactance of 2.0 Ω. Sketch the circuit and determine (a) the rms current; (b) the peak value of the sinusoidal current; (c) the inductance; (d) sketch on the same time axis the waves of driving voltage and current, indicating a time-zero; (e) write the equations for the waves for the given time-zero; (f) determine the energy stored in the field when the voltage across the coil is zero.

10-8 A 0.010-H ideal inductor is connected to a sinusoidal generator whose voltage is expressed by $v_{gen} = 50 \sin 10{,}584t$. Sketch the circuit and determine (a) the inductive reactance; (b) the rms current; (c) sketch the phasor diagram and the associated waves for the current and driving voltage; (d) write the equation for the current wave; (e) determine the steady-state current if the generator is replaced by a 200-V battery.

10-9 A 1000-μF capacitor is connected in series with a 120-V 60-Hz generator. Sketch the circuit and determine (a) the capacitive reactance; (b) the rms current; (c) sketch the phasor diagram, and the corresponding current and voltage waves on a common time axis.

10-10 A 400-μF capacitor is connected to a 208-V 25-Hz driver. Sketch the circuit and determine (a) the capacitive reactance; (b) the steady-state current; (c) the maximum instantaneous energy stored in the capacitor; (d) calculate the capacitive reactance and the steady current if the 25-Hz driver is replaced by a 208-V battery.

10-11 A 240-V 60-Hz sinusoidal driver is connected to a 500-μF capacitor. Sketch the circuit and determine (a) the capacitive reactance; (b) the steady-state current; (c) the maximum instantaneous voltage across the capacitor; (d) the maximum instantaneous charge in the capacitor; (e) the maximum instantaneous energy stored in the capacitor.

10-12 A 100-μF capacitor is connected to a generator whose voltage is expressed by $e_{gen} = 4.6 \sin 40{,}678t$. Determine (a) the capacitive reactance; (b) the rms currrent.

10-13 A 650-nF capacitor is connected to a 180-V 55-kHz generator. Determine (a) X_C; (b) the rms current; (c) the maximum instantaneous charge in the capacitor; (d) the maximum instantaneous energy stored in the capacitor.

10-14 A 0.020-F capacitor is connected to a sinusoidal generator. If the current to the capacitor is $i = 10 \sin 30t$, determine (a) rms voltage across the capacitor; (b) the time-domain equation for the voltage in part a; (c) the charge in coulombs accumulated when $t = 0.08$ s; (d) the maximum instantaneous energy stored in the capacitor during the current alternations; (e) the steady-state current if the AC generator is replaced by a 200-V battery.

CHAPTER 11

SOLVING CIRCUIT PROBLEMS WITH COMPLEX ALGEBRA

Complex algebra provides a relatively simple but powerful tool for obtaining the steady-state solutions of electric-circuit and electrical-machine problems involving sinusoidal drivers. It simplifies the mathematical manipulation of sinusoidal quantities and has the advantage of including the magnitudes and phase angles of current, voltage, and impedance in all related equations.

Complex algebra is associated with a two-dimensional plane, called a complex plane, whose horizontal axis is a real-number line and whose vertical axis is an imaginary-number line. Both axes must have the same scale (volts, amperes, ohms, etc.) and the divisions per unit length must be the same.

11-1 COMPLEX PLANE

To develop the complex plane, *we define j as an operator that rotates a directed line segment 90° in the counterclockwise (ccw) direction.* Multiplying line segment 5 in Fig. 11-1a by the j operator rotates it 90° ccw as shown in Fig. 11-1b. By extension, each successive multiplication by the j operator rotates the line segment an additional 90° ccw. Thus, as shown in Fig. 11-1c,

To rotate 90° multiply by j

To rotate 180° multiply by j^2

To rotate 270° multiply by j^3

To rotate 360° multiply by j^4

To rotate $(m \times 90°)$ multiply by j^m

As indicated in Fig. 11-1c, multiplying a directed line segment by j^2 rotates it 180°, thus reversing the direction of the directed line segment. In effect, multiplying by j^2 is the same as multiplying by -1. It therefore follows that

$$j^2 = -1$$

$$j = \sqrt{-1}$$

Since $\sqrt{-1}$ cannot be determined, it is called an *imaginary number*.

Figure 11-2 shows a directed line segment of length A making an angle of $\theta°$ with the real axis. Its position in the complex plane may be conveniently expressed as a *complex number*. This complex number in *polar form* is

$$\mathbf{A} = A\underline{/\theta°} \tag{11-1}$$

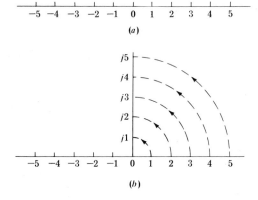

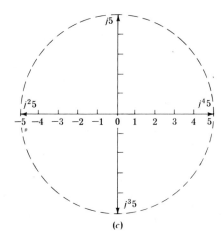

Figure 11-1 (*a*) Line segment; (*b*) effect of multiplying a line segment by the *j* operator; (*c*) effect of multiplying a line segment by exponential powers of the *j* operator.

The horizontal and vertical components produced by the projections of the directed line segment on the real and imaginary axes constitute the *real and imaginary components of the complex number that defines the directed line segment.* Thus line segment **A** in Fig. 11-2 may be defined by two numbers, a real number u on the real axis and an imaginary number jv on the imaginary axis; this is called the *rectangular form for the complex number* and may be written as

$$\mathbf{A} = u \text{ and } jv$$

or simply

$$\mathbf{A} = u + jv \qquad (11\text{-}2)$$

Note that the + sign in Eq. (11-2) means "and." *Imaginary numbers cannot be added to real numbers, and vice versa.*

The relationship between the polar form and the rectangular form may be obtained from the geometry of the right triangle formed by the directed line segment. Thus, referring to Fig. 11-2,

$$A = \sqrt{u^2 + v^2} \qquad (11\text{-}3)$$

$$\theta = \tan^{-1} \frac{v}{u} \qquad (11\text{-}4)$$

$$u = A \cos \theta \qquad (11\text{-}5)$$

$$v = A \sin \theta \qquad (11\text{-}6)$$

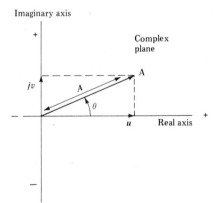

Figure 11-2   A directed line segment expressed in complex numbers.

Heavy roman letters (called boldfaced type) are used to designate complex numbers, and standard type is used to represent the lengths of the respective line segments. Thus **A** represents a complex number and A represents the length of the line segment; A is also called the *absolute value, the magnitude, or the modulus* of the complex number. Angle θ is called the *argument or phase* of the complex number.

When entering complex numbers that are in *polar form* into the complex plane, angles are measured *from* the positive real axis; a positive angle is a counterclockwise measurement, and a negative angle is a clockwise measurement. The lengths of all line segments representing voltage phasors are laid off to one scale, and the lengths of all line segments representing current phasors are laid off to another scale.

When entering complex numbers that are in rectangular form into the complex plane, use the real and imaginary components to establish the coordinates of the endpoint of the respective line segments, and then draw a line from each endpoint to the origin.

The complex plane is a natural extension of the phasor diagram, and as such can be used to facilitate the solutions of problems involving sinusoidal currents and voltages.

Example 11-1 Express the polar form of the voltage $\mathbf{V} = 50\underline{/36.87°}$ in rectangular form.

Solution
From Eqs. (11-5) and (11-6),

$$u = 50 \cos 36.87° = 40$$

$$v = 50 \sin 36.87° = 30$$

$$\mathbf{V} = u + jv = (40 + j30)$$

The complex number representing this voltage is shown plotted in the complex plane in Fig. 11-3.

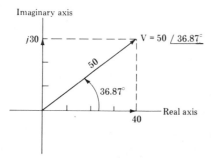

Figure 11-3 A plot of the complex number that represents the voltage in Example (11-1).

Exponential Form

Substituting Eqs. (11-5) and (11-6) into Eq. (11-2),

$$\mathbf{A} = A\cos\theta + jA\sin\theta$$

or

$$\mathbf{A} = A(\cos\theta + j\sin\theta) \tag{11-7}$$

Equation (11-7) is called the *trigonometric* form of the complex number $\mathbf{A}$. Substituting Euler's formula [Eq. (11-8)],

$$\varepsilon^{j\theta} = \cos\theta + j\sin\theta \tag{11-8}$$

into Eq. (11-7) results in the *exponential form* for the complex number

$$\mathbf{A} = A\varepsilon^{j\theta} \tag{11-9}$$

The exponential form has certain advantages that will be evidenced in specific problems later.

Conjugate of a Complex Number

The conjugate of a complex number is its mirror image in the complex plane; it is the reflection of the directed line segment about the real axis, as shown in Fig. 11-4. An asterisk on a complex number indicates it to be the conjugate. Thus, $\mathbf{A}^*$ is the conjugate of $\mathbf{A}$. Referring to Fig. 11-4,

$$\mathbf{A} = A\underline{/\theta} = u + jv$$

$$\mathbf{A}^* = A\underline{/-\theta} = u - jv$$

The conjugate of a complex number is used in phasor power calculations.

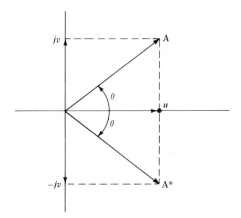

Figure 11-4 Conjugate of a complex number.

11-2 **ADDITION AND SUBTRACTION OF COMPLEX NUMBERS**

The sum of two or more complex numbers may be obtained graphically by tip-to-tail addition in the complex plane, or by complex algebra. When using complex algebra, *the addition or subtraction can be done only in rectangular form.*

Example 11-2 Determine the resultant voltage of three sinusoidal generators connected in series, whose voltages are $\mathbf{E}_1 = 100\underline{/42°}$, $\mathbf{E}_2 = 60\underline{/-36°}$, $\mathbf{E}_3 = 50\underline{/140°}$.

Solution
The resultant voltage $\mathbf{E}_T$ is

$$\mathbf{E}_T = \mathbf{E}_1 + \mathbf{E}_2 + \mathbf{E}_3$$

Converting the polar form to rectangular form,

$$\mathbf{E}_1 = 100 \cos 42° + j100 \sin 42° = 74.31 + j66.91$$

$$\mathbf{E}_2 = 60 \cos -36° + j60 \sin -36° = 48.54 - j35.27$$

$$\mathbf{E}_3 = 50 \cos 140° + j50 \sin 140° = -38.30 + j32.14$$

$$\mathbf{E}_T = \mathbf{E}_1 + \mathbf{E}_2 + \mathbf{E}_3 = 84.55 + j63.79$$

Returning to polar form,

$$\mathbf{E}_T = 106\underline{/37.0°} \text{ V}$$

The component voltages and the resultant voltage are drawn in the complex plane in Fig. 11-5a. Figure 11-5b shows the tip-to-tail addition.

Example 11-3 Given $\mathbf{A} = 35\underline{/40°}$. Determine $\mathbf{A} + \mathbf{A}^*$.

Solution

$$\mathbf{A} = 35\underline{/40°} = 26.81 + j22.50$$

$$\mathbf{A}^* = 35\underline{/-40°} = 26.81 - j22.50$$

$$\mathbf{A} + \mathbf{A}^* = 53.6 + j0 = 53.62$$

Thus the sum of a complex number and its conjugate is a real number; the imaginary terms add up to zero.

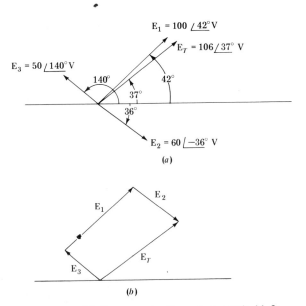

$E_1 = 100 \underline{/42°}\,\text{V}$

$E_T = 106 \underline{/37°}\,\text{V}$

$E_3 = 50 \underline{/140°}\,\text{V}$

$140°$

$42°$

$37°$

$36°$

$E_2 = 60 \underline{/-36°}\,\text{V}$

(a)

E_2

E_1

E_T

E_3

(b)

Figure 11-5 (a) Component voltages in Example 11-2;
(b) tip-to-tail addition.

1-3 MULTIPLICATION AND DIVISION OF COMPLEX NUMBERS

The multiplication and division of complex numbers may be done in rectangular, trigonometric, exponential, and polar form. The trigonometric form is rarely used, however, and will not be discussed.

Rectangular Form

Multiplication in rectangular form uses the ordinary rules of algebra and is explained in the following example.

Example 11-4 Obtain the product $\mathbf{A} \cdot \mathbf{B}$

where $\mathbf{A} = 4 + j3$

  $\mathbf{B} = 5 + j6$

Solution

$$\mathbf{A} \cdot \mathbf{B} = (4 + j3)(5 + j6)$$

Multiplying through,

$$\mathbf{A} \cdot \mathbf{B} = 20 + j24 + j15 + j^2 18$$

Collecting real and imaginary terms (note $j^2 = -1$),

$$\mathbf{A} \cdot \mathbf{B} = (20 - 18) + j(24 + 15)$$

$$\mathbf{A} \cdot \mathbf{B} = 2 + j39$$

$$\mathbf{A} \cdot \mathbf{B} = 39.05\underline{/87.06°}$$

Division in rectangular form can be accomplished only by multiplying numerator and denominator by the conjugate of the denominator. This operation, called *rationalizing the denominator*, is shown in the following example.

Example 11-5 Evaluate $\mathbf{A}/\mathbf{B}$

where $\mathbf{A} = 4 + j3$

$$\mathbf{B} = 5 + j6$$

Solution

$$\frac{\mathbf{A}}{\mathbf{B}} = \frac{4 + j3}{5 + j6} = \frac{(4 + j3)(5 - j6)}{(5 + j6)(5 - j6)}$$

$$\frac{\mathbf{A}}{\mathbf{B}} = \frac{20 - j24 + j15 - j^2 18}{25 - j30 + j30 - j^2 36}$$

$$\frac{\mathbf{A}}{\mathbf{B}} = \frac{38 - j9}{61} = \frac{39.05\underline{/-13.32°}}{61}$$

$$\frac{\mathbf{A}}{\mathbf{B}} = 0.64\underline{/-13.32°}$$

Exponential Form

Multiplication and division in exponential form are relatively simple compared with the same operations in rectangular form. This is indicated in the following examples.

Example 11-6 Use the exponential form to evaluate the product $(4 + j3)(5 + j6)$

Solution

$$\mathbf{A} = 4 + j3 = 5\varepsilon^{j36.87°}$$

$$\mathbf{B} = 5 + j6 = 7.81\varepsilon^{j50.19°}$$

$\mathbf{A} \cdot \mathbf{B} = (5\varepsilon^{j38.87°})(7.81\varepsilon^{j50.19°})$

$\mathbf{A} \cdot \mathbf{B} = (5)(7.81)\varepsilon^{j(36.87° + 50.19°)} = 39.05\varepsilon^{j87.06°}$

Converting to polar form,

$\mathbf{AB} = 39.05\underline{/87.06°}$

Example 11-7 Use exponential form to evaluate
$(5\underline{/36.87°})/(7.81\underline{/50.19°})$.

Solution

$$\frac{\mathbf{A}}{\mathbf{B}} = \frac{5\varepsilon^{j36.87°}}{7.81\varepsilon^{j50.19°}} = 0.640\varepsilon^{j(36.87° - 50.19°)}$$

$$\frac{\mathbf{A}}{\mathbf{B}} = 0.64\varepsilon^{-j13.32°}$$

Converting to polar form,

$$\frac{\mathbf{A}}{\mathbf{B}} = 0.64\underline{/-13.32°}$$

Thus, to multiply complex numbers that are in exponential form, multiply the moduli and add the angles (algebraically); to divide complex numbers, divide the moduli as indicated and subtract the denominator angle from the numerator angle.

Polar Form

Multiplication and division in polar form is a natural extension of the operation in exponential form, and is the preferred method. The polar method is justified by the exponential method, and the same procedure is used.

Example 11-8 Using the polar form, evaluate **AB** and **A/B**

where $\mathbf{A} = 4 + j3 = 5\underline{/36.87°}$

   $\mathbf{B} = 5 + j6 = 7.81\underline{/50.19°}$

Solution

$\mathbf{A} \cdot \mathbf{B} = (5\underline{/36.87°})(7.81\underline{/50.19°})$

$\mathbf{A} \cdot \mathbf{B} = (5)(7.81)\underline{/36.87° + 50.19°}$

$\mathbf{A} \cdot \mathbf{B} = 39.05\underline{/87.06°}$

Similarly,

$$\frac{\mathbf{A}}{\mathbf{B}} = \frac{5/36.87°}{7.81/50.19°} = \frac{5}{7.81} \underline{/36.87° - 50.19°}$$

$$\frac{\mathbf{A}}{\mathbf{B}} = 0.64\underline{/-13.32°}$$

It should be noted that *multiplication and division of complex numbers* (when more than two complex numbers are involved) *is much easier to do in polar form than in rectangular form,* and has the advantage of providing immediate information about the magnitude and phase angles of components, products, and quotients.

Example 11-9 Simplify the following using polar form:

$$\mathbf{X} = \frac{(40/30°)(60/20°)}{(70/-40°)(35/70°)}$$

Solution

$$\mathbf{X} = \frac{(40)(60)}{(70)(35)} \underline{/30° + 20° + 40° - 70°}$$

$$\mathbf{X} = 0.98/20°$$

Converting to rectangular form,

$$\mathbf{X} = (0.92 + j0.34)$$

Although Example 11-9 may be worked in rectangular form, the work will be tedious, messy, and subject to greater opportunity for error.

Example 11-10 This example is a repetition of Example 11-9 but is worked out in rectangular form in order to show the tediousness of this method.

Solution

$$40/30° = (34.64 + j20) \qquad 60/20° = (56.38 + j20.52)$$

$$70/-40° = (53.62 - j45) \qquad 35/70° = (11.97 + j32.89)$$

$$\mathbf{X} = \frac{(34.64 + j20)(56.38 + j20.52)}{(53.62 - j45)(11.97 + j32.89)}$$

$$\mathbf{X} = \frac{1953.06 + j710.81 + j1127.60 + j^2410.40}{641.83 + j1763.56 - j538.65 - j^21480.05}$$

$$\mathbf{X} = \frac{1542.66 + j1838.41}{2121.88 + j1224.91}$$

Rationalizing the denominator,

$$X = \frac{(1542.66 + j1838.41)(2121.88 - j1224.91)}{(2121.88 + j1224.91)(2121.88 - j1224.91)}$$

$$X = \frac{3,273,339.40 - j1,889,619.66 + j3,900,885.41 - j^2 2,251,886.79}{4,502,374.73 + 1,500,404.51}$$

$$X = \frac{5,525,226.19 + j2,011,265.75}{6,002,779.24}$$

$$X = (0.92 + j0.34) = 0.98\underline{/20°}$$

Examples 11-9 and 11-10 indicate the relative ease of multiplying and dividing complex numbers when they are in polar form compared with the same operation in rectangular form.

SUMMARY OF FORMULAS

$$\mathbf{A} = (u + jv) = A\underline{/\theta} = A\varepsilon^{j\theta}$$

$$\mathbf{A}^* = (u - jv) = A\underline{/-\theta} = A\varepsilon^{-j\theta}$$

$$A = \sqrt{u^2 + v^2} \qquad \theta = \tan^{-1}\left(\frac{v}{u}\right)$$

$$u = A\cos\theta \qquad v = A\sin\theta$$

PROBLEMS

11-1 Express the following complex numbers in rectangular form: (a) $50\underline{/80°}$; (b) $30\underline{/20°}$; (c) $120\underline{/180°}$; (d) $70\underline{/-90°}$; (e) $150\underline{/45°}$.

11-2 Express the following complex numbers in polar form: (a) $(3 + j7)$; (b) $(6 - j8)$; (c) $(0 + j4)$; (d) $(j3 - 5)$; (e) $(-2 - j6)$.

11-3 Determine the conjugate of the following complex numbers (a) $3\underline{/70°}$; (b) $5\underline{/-40°}$; (c) $(2 + j5)$; (d) $(-5 - j6)$; (e) $(4 + j0)$.

11-4 Perform the indicated operations and express the results in polar form:
(a) $(3 + j4) + (5 - j7) + (3 - j6) - (10 - j8)$
(b) $(5 + j9) - (6 + j15) + (4 - j3) + (6 - j5)$
(c) $3\underline{/30°} + 5\underline{/-50°} - 7\underline{/20°} + (6 + j5)$

11-5 Perform the indicated operations and express the results in polar form:

(a) $\dfrac{(3 + j9)(5 + j10)(9 - j7)}{5(3 + j6)}$

(b) $\dfrac{(17 + j9)(5 - j8)}{(4 + j7)(8 - j4)}$

11-6 Perform the indicated operations and express the result in polar form:

(a) $\dfrac{(13 + j19)(5 - j10)^{*}52\varepsilon^{-j13.4°}}{(14\underline{/30°})(17\underline{/50°})^{*}}$

(b) $\left[\dfrac{(15 - j7)(3 + j2)^{*}}{(4 + j6)^{*}(3\underline{/70°})}\right]^{*}$

CHAPTER 12

TWO-WIRE SYSTEMS WITH SINUSOIDAL DRIVERS

Two-wire systems with sinusoidal drivers represent a general class of circuits that include series circuits, parallel circuits, series-parallel circuits, and other circuit configurations that use a *single* sinusoidal driving voltage. These single-phase circuits may have any combination of R, L, and C, and any frequency, including 0 Hz (DC).

12-1   SERIES CIRCUIT

Figure 12-1 represents the *general case* of a series circuit containing lumped values of resistance, inductance, and capacitance. Applying Kirchhoff's voltage law,

$$e_{gen} = v_L + v_R + v_C \tag{12-1}$$

where e_{gen} = any sinusoidal driving voltage

v_L = instantaneous voltage drop across the inductance *due to the rate of change of current* in the inductance

v_C = instantaneous voltage drop across the capacitor *due to the accumulation of electric charge caused by the current to the capacitor*

v_R = instantaneous voltage drop across the resistor *due to the current in the resistor*

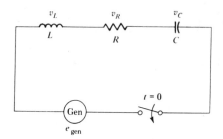

Figure 12-1   General case of a series circuit containing lumped values of R, L, and C.

Expressing Eq. (12-1) in terms of the common current at any instant of time,

$$e_{\text{gen}} = L\frac{di}{dt} + iR + \frac{1}{C}\int i\,dt \qquad\qquad (12\text{-}2)$$

Equation (12-2) is an integrodifferential equation that, when completely solved, describes the behavior of the current in the respective circuit; it describes the current for all time, from the instant the switch is closed to and through the steady-state condition.

Figure 12-2 illustrates some of the possible behaviors of the circuit current for different driving voltages and different combinations of R, L, and

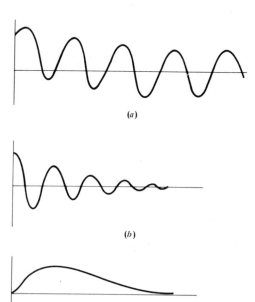

Figure 12-2   Some of the possible current curves for different driving voltages and different combinations of R, L, and C.

C. Note that the current curves have a transient section (which decays in five time constants), followed by the steady-state section. A mathematical determination of the transient section requires the solution of a differential equation, and will be discussed in Chaps. 26 and 27. By contrast, the mathematical solution of the steady-state section is relatively simple.

2-2   STEADY-STATE ANALYSIS OF THE SERIES CIRCUIT WITH A SINUSOIDAL OR DC DRIVER

The application of a sinusoidal driving voltage to a circuit containing linear elements will result in a steady-state sinusoidal current. The transient decays to insignificance after five time constants, leaving only the steady-state current. Figure 12-3 shows a series circuit containing linear elements of *R*, *L*, and *C* connected to a sinusoidal driving voltage. Above each element is the corresponding phasor diagram and below each element are the corresponding voltage and current waves. *The characteristics of the individual circuit elements remain the same regardless of how they are connected in the circuit, series, parallel, or any other combination.*

At any instant of time, the current in any one element of the series circuit must be identical to that in the other elements. Hence the voltage waves v_L, v_R, v_C that appear across the three respective elements in Fig. 12-3 may be sketched on the same time axis with respect to the *common* current wave. This is shown in Fig. 12-4*a*.

Similarly, the voltage phasors $\mathbf{V}_L$, $\mathbf{V}_R$, and $\mathbf{V}_C$ may be sketched on the same phasor diagram with respect to the common current phasor. This is shown in Fig. 12-4*b*.

The phasors shown in Fig. 12-4*b* are shown again in Fig. 12-5, "frozen" in different positions to reflect different time-zero references. Although the phase angles of the corresponding current and voltage phasors in Figs. 12-4*b*, 12-5*a*, and 12-5*b* are different, *the angles measured between the corresponding*

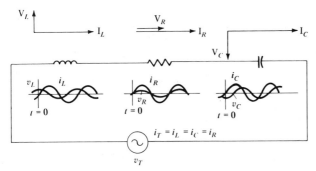

Figure 12-3   Series circuit containing *R*, *L*, and *C* connected to a sinusoidal driver.

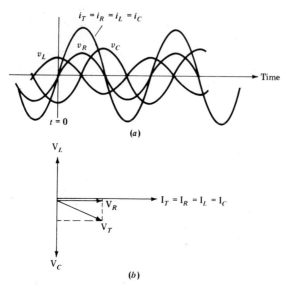

(a)

(b)

Figure 12-4 (a) Voltage-drop wave across each element of an *RLC* series circuit drawn with respect to the common current wave; (b) corresponding voltage phasors drawn with respect to the common current phasor.

phasors are the same; phasor V_L leads phasor I_T by $90°$, phasor V_C lags phasor I_T by $90°$, and phasor V_R is in phase with phasor I_T.

The *magnitude* of the rms voltage drop across the inductance is equal to the product of the rms current and the inductive reactance.

$$V_L = I_T X_L$$

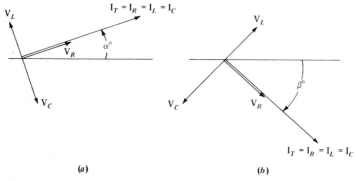

(a)

(b)

Figure 12-5 Phasors of Fig. 12-4b "frozen" in different positions to reflect different time-zero references.

Since the phasor representing the voltage drop across an ideal inductance leads the current through it by $90°$, if the phase angle of the current is $\alpha°$, as shown in Fig. 12-5a, then the phase angle for the voltage phasor must be $\alpha° + 90°$. Thus, expressed as a phasor in polar form,

$$\mathbf{V}_L = V_L\underline{/\alpha° + 90°} = I_T X_L\underline{/\alpha° + 90°} \tag{12-3}$$

Extracting the current phasor,

$$\mathbf{V}_L = (I_T\underline{/\alpha°})(X_L\underline{/90°}) \tag{12-4}$$

Substituting,

$$I_T\underline{/\alpha°} = \mathbf{I}_T \tag{12-5}$$

$$X_L\underline{/90°} = jX_L \tag{12-6}$$

Equation (12-4) becomes

$$\mathbf{V}_L = \mathbf{I}_T(jX_L) \tag{12-7}$$

The factor jX_L is the *inductive reactance expressed in terms of complex algebra.*

Similarly, the *magnitude* of the rms voltage drop across the capacitor is

$$V_C = I_T X_C$$

Since the phasor representing the voltage drop across an ideal capacitor lags the current to it by $90°$, if the phase angle of the current is $\alpha°$, as shown in Fig. 12-5a, then the phase angle of the voltage phasor must be $\alpha° - 90°$. Thus, expressed in polar form,

$$\mathbf{V}_C = V_C\underline{/\alpha° - 90°} = I_T X_C\underline{/\alpha° - 90°} \tag{12-8}$$

Extracting the current phasor,

$$\mathbf{V}_C = (I_T\underline{/\alpha°})(X_C\underline{/-90°})$$

Substituting,

$$I_T\underline{/\alpha°} = \mathbf{I}_T$$

$$X_C\underline{/-90°} = -jX_C$$

Equation (12-8) becomes

$$\mathbf{V}_C = \mathbf{I}_T(-jX_C) \tag{12-9}$$

The factor $-jX_C$ is the *capacitive reactance expressed in terms of complex algebra.*

Note: When using complex algebra in series circuits, parallel circuits, or any other type of circuit:

Inductive reactance is always $+jX_L$
Capacitive reactance is always $-jX_C$

The magnitude of the voltage drop across the resistor is

$$V_R = I_R R$$

Since the phasor representing the voltage drop across an ideal resistor is in phase with the current throughout it, if the phase angle of the current is $\alpha°$, as shown in Fig. 12-5a, then the phase angle of the voltage phasor must be $\alpha°$. Thus, expressed in polar form,

$$\mathbf{V}_R = V_R\underline{/\alpha°} = I_T R\underline{/\alpha°} \tag{12-10}$$

Extracting the current phasor,

$$\mathbf{V}_R = (I_T\underline{/\alpha°})(R\underline{/0°})$$

Substituting,

$$I_T\underline{/\alpha°} = \mathbf{I}_T$$

$$R\underline{/0°} = R$$

$$\mathbf{V}_R = \mathbf{I}_T R \tag{12-11}$$

Resistance R is always a *real number* when expressed in terms of complex algebra.

If a series circuit contains only linear elements and sinusoidal drivers, the equation representing Kirchhoff's voltage law may be expressed in terms of the voltage phasors; that is, *the phasor sum of the driving voltages is equal to the phasor sum of the voltage drops.* Thus, for the circuit in Fig. 12-3,

$$\mathbf{V}_T = \mathbf{V}_R + \mathbf{V}_L + \mathbf{V}_C \tag{12-12}$$

Substituting Eqs. (12-7), (12-9), and (12-11) into Eq. (12-12),

$$\mathbf{V}_T = \mathbf{I}_T R + \mathbf{I}_T(jX_L) + \mathbf{I}_T(-jX_C) \qquad (12\text{-}13)$$

Factoring out the common current phasor,

$$\mathbf{V}_T = \mathbf{I}_T(R + jX_L - jX_C) \qquad (12\text{-}14)$$

Defining,

$$\boxed{\mathbf{Z}_S = R + jX_L - jX_C} \qquad (12\text{-}15)$$

$$\mathbf{V}_T = \mathbf{I}_T\mathbf{Z}_S \qquad (12\text{-}16)$$

$\mathbf{Z}_S$ is called the *complex input impedance of the series circuit*; it is a measure of the overall opposition offered by a series circuit to the current *when the circuit is connected to a sinusoidal driving voltage.* Input impedance, also called *driving-point impedance,* is measured at the input terminals of the circuit.

It should be noted that the *resistance and reactance components of the input impedance are series-connected elements.*

A plot of Eq. (12-15) in the complex plane is called an *impedance diagram,* and is shown in Fig. 12-6. *The impedance diagram is not a phasor diagram*; phasors represent sinusoidally varying quantities, and resistance, reactance, and impedance are *constants* that do not vary sinusoidally. For this reason, sinusoidal variables should be plotted in a phasor diagram and impedances should be plotted in an impedance diagram.

From the geometry of the impedance diagram in Fig. 12-6, the input impedance of the series circuit is

$$\boxed{\mathbf{Z}_S = Z_S\underline{/\theta_S}} \qquad (12\text{-}17)$$

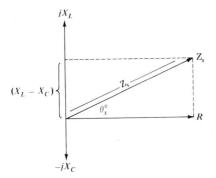

Figure 12-6   Impedance diagram.

The magnitude of $\mathbf{Z}_S$ is

$$Z_S = \sqrt{R^2 + (X_L - X_C)^2} \qquad (12\text{-}18)$$

The phase angle of $\mathbf{Z}_S$ is

$$\theta_S = \tan^{-1}\frac{X_L - X_C}{R} \qquad (12\text{-}19)$$

12-3 RMS PHASORS

Alternating-current ammeters and voltmeters and the nameplates of electrical apparatus indicate rms values unless otherwise specified. Hence, when solving problems involving sinusoidal currents and voltages, it is convenient to use *rms phasors*. An rms phasor uses rms values instead of the respective maximum values for the magnitude of the phasor. Thus, given a sinusoidal voltage of 100 V rms and a phase angle of 40°, the phasor would be written as

$$\mathbf{E} = 100\underline{/40^\circ}$$

The corresponding time-domain equation is

$$e = 100\sqrt{2}\,\sin(2\pi f t + 40^\circ)$$

The rms phasor is used in almost all literature pertaining to the analysis and solution of problems in electric circuits and electrical machinery.

Example 12-1 A coil whose resistance and inductive reactance are 10.0 Ω and 8.0 Ω, respectively, is connected in series with a 3.0-Ω capacitive reactance. If the circuit current is $30\underline{/20^\circ}$ A, as shown in Fig. 12-7a, determine the magnitude and phase angle of the driving voltage.

Solution
The equivalent series circuit is shown in Fig. 12-7b. Applying Kirchhoff's voltage law,

$$\mathbf{V}_T = \mathbf{V}_R = \mathbf{V}_L = \mathbf{V}_c$$

$$\mathbf{V}_T = \mathbf{I}_T R + \mathbf{I}_T(jX_L) + \mathbf{I}_T(-jX_C)$$

$$\mathbf{V}_T = (30\underline{/20^\circ})(10) + (30\underline{/20^\circ})(j8) + (30\underline{/20^\circ})(-j3)$$

$$\mathbf{V}_T = 30\underline{/20^\circ}\,(10 + j8 - j3)$$

$$\mathbf{V}_T = (30\underline{/20^\circ})(11.18\underline{/26.57^\circ})$$

$$\mathbf{V}_T = 335.4\underline{/46.57^\circ}\ \text{V}$$

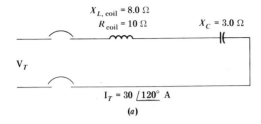

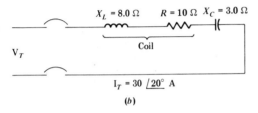

Figure 12-7   Circuit for Example 12-1.

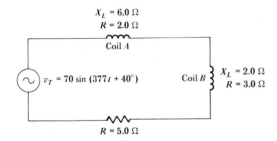

Figure 12-8   Circuit for Example 12-2.

Example 12-2   Two coils and a resistor are connected in series with a sinusoidal driver as shown in Fig. 12-8. Determine (a) the driving voltage in polar form; (b) the input impedance; (c) the circuit current; (d) the voltage drop across each component in the circuit; (e) the equation for the current as a function of time.

Solution

(a) $\mathbf{V}_T = \dfrac{70}{\sqrt{2}}\underline{/40^\circ} = 49.50\underline{/40^\circ}$ V

(b) $\mathbf{Z}_S = (2 + j6) + (3 + j2) + 5$

$\mathbf{Z}_S = 10 + j8 = 12.81\underline{/38.66^\circ}\ \Omega$

(c) $\mathbf{I}_T = \dfrac{\mathbf{V}_T}{\mathbf{Z}_S} = \dfrac{49.50\underline{/40^\circ}}{12.81\underline{/38.66^\circ}} = 3.86\underline{/1.34^\circ}$ A

(d) Voltage drop across coil A

$$\mathbf{V}_{\text{coil }A} = \mathbf{I}_T \mathbf{Z}_{\text{coil }A}$$

$$\mathbf{V}_{\text{coil }A} = (3.86\underline{/1.34°})(2 + j6) = (3.86\underline{/1.34°})(6.32\underline{/71.57°})$$

$$\mathbf{V}_{\text{coil }A} = 24.40\underline{/72.91°} \text{ V}$$

$$\mathbf{V}_{\text{coil }B} = \mathbf{I}_T \mathbf{Z}_{\text{coil }B} = (3.86\underline{/1.34°})(3 + j2) = (3.86\underline{/1.34°})(3.61\underline{/33.69°})$$

$$\mathbf{V}_{\text{coil }B} = 13.93\underline{/35.03°} \text{ V}$$

$$\mathbf{V}_R = \mathbf{I}_T R = (3.86\underline{/1.34°})(5) = 19.3\underline{/1.34°} \text{ V}$$

(e) $i = 3.86\sqrt{2} \sin(377t + 1.34°)$

Example 12-3 For the circuit shown in Fig. 12-9a, (a) determine the current, and then (b) sketch the phasor diagram showing the net driving voltage phasor and the current phasor.

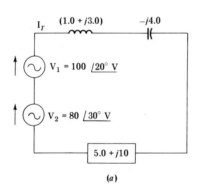

(a)

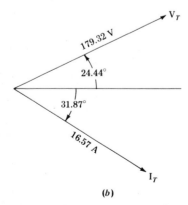

(b)

Figure 12-9 Circuit and phasor diagram for Example 12-3.

Solution

(a) Applying Kirchhoff's voltage law,

$$80\underline{/30°} + 100\underline{/20°} = I_T(1 + j3) + I_T(-j4) + I_T(5 + j10)$$

$$(69.28 + j40) + (93.97 + j34.20) = I_T(6 + j9)$$

$$(163.25 + j74.20) = I_T(6 + j9)$$

$$(179.32\underline{/24.44°}) = I_T(10.82\underline{/56.31°})$$

$$I_T = \frac{179.32\underline{/24.44°}}{10.82\underline{/56.31°}} = 16.57\underline{/-31.87°} \text{ A}$$

(b) The phasor diagram is shown in Fig. 12-9b. The net driving voltage is the phasor sum of the two series-connected driving voltages.

Example 12-4 A coil whose resistance and inductance are 60.0 Ω and 0.20 H, respectively, is connected to a 120-V 60-Hz generator. Determine (a) the inductive reactance; (b) the input impedance; (c) the steady-state current; (d) the steady-state current if the 60-Hz generator is replaced by a 120-V battery.

Solution

(a) $X_L = 2\pi f L = 2\pi(60)(0.2) = 75.4 \ \Omega$

(b) $Z_S = (60 + j75.4) = 96.36\underline{/51.49°} \ \Omega$

(c) $I_T = \dfrac{V_T}{Z_S} = \dfrac{120\underline{/0°}}{96.36\underline{/51.49°}} = 1.25\underline{/-51.49°} \text{ A}$

(d) With a battery applied, $f = 0$.

$$X_L = 2\pi(0)(0.2) = 0 \ \Omega$$

$$Z_S = R + jX_L = 60 + j0 = 60 \ \Omega$$

$$I_{DC} = \frac{V_{DC}}{Z_S} = \frac{120}{60} = 2 \text{ A}$$

12-4 **VOLTAGE-DIVIDER EQUATION FOR SERIES-CONNECTED IMPEDANCES**

A simple voltage divider consists of two or more impedances in series, connected to a driving voltage as shown in Fig. 12-10. The current in the series circuit is

$$I_T = \frac{V_T}{Z_T} \tag{12-20}$$

where $Z_T = Z_1 + Z_2 + \cdots + Z_k + \cdots + Z_n$

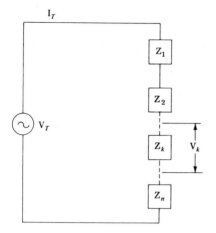

Figure 12-10 Simple voltage divider.

The voltage drop across impedance Z_k is

$$\mathbf{V}_k = \mathbf{I}_T \mathbf{Z}_k \tag{12-21}$$

Substituting Eq. (12-20) into Eq. (12-21) results in the *voltage-divider equation*

$$\mathbf{V}_k = \mathbf{V}_T \frac{\mathbf{Z}_k}{\mathbf{Z}_T} \tag{12-22}$$

Example 12-5 Using the voltage divider rule, determine the rms voltage across the 5.0-Ω resistor in Fig. 12-8.

Solution

$$\mathbf{Z}_T = (2 + j6) + (3 + j2) + 5$$

$$\mathbf{Z}_T = (10 + j8) = 12.81\underline{/38.66°}\ \Omega$$

$$\mathbf{V}_T = \frac{70}{\sqrt{2}}\underline{/40°} = 49.50\underline{/40°}\ \text{V}$$

$$\mathbf{V}_{5\Omega} = (49.50\underline{/40°})\frac{5\underline{/0°}}{12.81\underline{/38.66°}}$$

$$\mathbf{V}_{5\Omega} = 19.3\underline{/1.34°}\ \text{V}$$

12-5   PARALLEL CIRCUIT

Figure 12-11 represents the *general case* of a parallel circuit containing lumped values of resistance, inductance, and capacitance. Applying Kirchhoff's current law,

$$i_T = i_R + i_L + i_C \qquad \text{(12-23)}$$

In terms of the common driving voltage, which is the voltage across each parallel element,

$$i_T = \frac{v_T}{R} + \frac{1}{L} \int v_T \, dt + C \frac{dv_t}{dt} \qquad \text{(12-24)}$$

Equation (12-24) is an integrodifferential equation that, when completely solved, describes the behavior of the current in the respective circuit; it describes the current for all time, from the instant the switch is closed, to and through the steady-state condition.

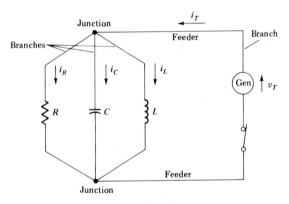

Figure 12-11   General case of a parallel circuit containing lumped values of *R*, *L*, and *C*.

12-6   STEADY-STATE ANALYSIS OF THE PARALLEL CIRCUIT WITH A SINUSOIDAL OR DC DRIVER

The application of a sinusoidal driving voltage to a circuit containing linear elements will result in a steady-state sinusoidal current. The transient decays to insignificance after five time constants, leaving only the steady-state sinusoidal component.

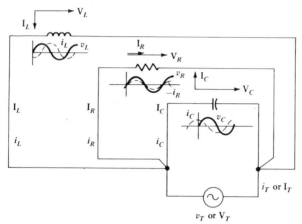

Figure 12-12 Parallel circuit containing R, L, and C connected to a sinusoidal driver.

Figure 12-12 shows a parallel circuit containing linear elements of R, L, and C connected to a sinusoidal driving voltage. Above each element is its corresponding phasor diagram, and below each element are the corresponding voltage and current waves.

At any instant of time, the voltage across any one element of a parallel circuit must be identical to that across each of the other elements. Hence the current waves i_L, i_R, and i_C that appear in the three elements may be sketched on the same time axis with respect to the common voltage wave. This is shown in Fig. 12-13a.

Similarly, the current phasors I_L, I_R, and I_C may be sketched on the same phasor diagram with respect to the common voltage phasor. This is shown in Fig. 12-13b.

The phasors shown in Fig. 12-13b are shown again in Fig. 12-14, frozen in different positions to reflect different time-zero references. Although the

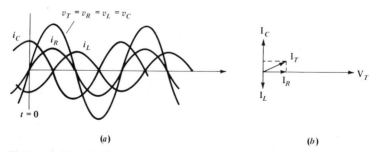

(a)　　　　　　　　　　　　(b)

Figure 12-13 (a) Current waves to each element of an RLC parallel circuit drawn with respect to the common voltage wave; (b) corresponding current phasors drawn with respect to the common voltage phasor.

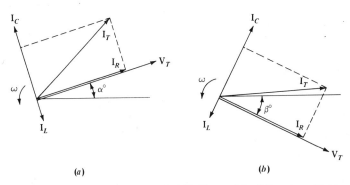

Figure 12-14 Phasors of Fig. 12-13b "frozen" in different positions to reflect different time-zero references

phase angles of the corresponding current and voltage phasors in Figs. 12-13b, 12-14a, and 12-14b are different, the *angles measured between the corresponding phasors are the same*; phasor $\mathbf{I}_L$ lags phasor $\mathbf{V}_T$ by 90°, phasor $\mathbf{I}_C$ leads phasor $\mathbf{V}_T$ by 90°, and phasor $\mathbf{I}_R$ is in phase with phasor $\mathbf{V}_T$.

If the parallel circuit contains only linear elements and a sinusoidal driver, as shown in Fig. 12-12, the equation representing Kirchhoff's current law may be expressed in terms of the current phasors; that is, *the phasor sum of the currents going to a junction is equal to the phasor sum of the currents leaving the junction.* Thus, for the circuit in Fig. 12-12,

$$\mathbf{I}_T = \mathbf{I}_R + \mathbf{I}_L + \mathbf{I}_C \qquad (12\text{-}25)$$

Applying Ohm's law to each of the parallel branches in Fig. 12-12,

$$\mathbf{I}_L = \frac{\mathbf{V}_T}{jX_L} \qquad \mathbf{I}_C = \frac{\mathbf{V}_T}{-jX_C} \qquad \mathbf{I}_R = \frac{\mathbf{V}_T}{R} \qquad (12\text{-}26)$$

Substituting the Ohm's-law relationships into Eq. (12-25),

$$\mathbf{I}_T = \frac{\mathbf{V}_T}{R} + \frac{\mathbf{V}_T}{jX_L} + \frac{\mathbf{V}_T}{-jX_C} \qquad (12\text{-}27)$$

Factoring out the common voltage phasor,

$$\mathbf{I}_T = \mathbf{V}_T\left(\frac{1}{R} + \frac{1}{jX_L} + \frac{1}{-jX_C}\right) \qquad (12\text{-}28)$$

Defining

$$\boxed{\frac{1}{\mathbf{Z}_P} = \left(\frac{1}{R} + \frac{1}{jX_L} + \frac{1}{-jX_C}\right)} \tag{12-29}$$

$$\mathbf{Z}_P = \frac{1}{1/R + 1/jX_L + 1/-jX_C} \tag{12-30}$$

Substituting Eq. (12-29) into Eq. (12-28) and simplifying,

$$\boxed{\mathbf{I}_T = \frac{\mathbf{V}_T}{\mathbf{Z}_P}} \tag{12-31}$$

$\mathbf{Z}_P$ is called the *complex input impedance of the parallel RLC circuit*. It is a measure of the overall opposition offered by a parallel circuit to the current *when the circuit is connected to a sinusoidal driving voltage.*

12-7   ADMITTANCE METHOD FOR PARALLEL-CIRCUIT CALCULATIONS

The admittance method for parallel-circuit calculations was developed in order to avoid awkward equations like Eq. (12-30). The admittance method is the recommended method for solving parallel-circuit problems and is particularly advantageous when the circuit includes three or more paralleled branches.

Referring to Fig. 12-15, the total current supplied by the sinusoidal generator is

$$\mathbf{I}_T = \mathbf{I}_1 + \mathbf{I}_2 + \mathbf{I}_3 + \cdots + \mathbf{I}_n \tag{12-32}$$

From Ohm's law,

$$\mathbf{I}_1 = \frac{\mathbf{V}_T}{\mathbf{Z}_1}$$

$$\mathbf{I}_2 = \frac{\mathbf{V}_T}{\mathbf{Z}_2}$$

$$\mathbf{I}_3 = \frac{\mathbf{V}_T}{\mathbf{Z}_3}$$

$$\vdots \quad \vdots$$

$$\mathbf{I}_n = \frac{\mathbf{V}_T}{\mathbf{Z}_n}$$

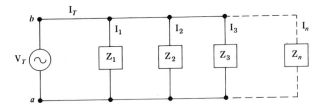

Figure 12-15 Paralleled impedances.

Substituting the Ohm's-law relationships into Eq. (12-32),

$$I_T = \frac{V_T}{Z_1} + \frac{V_T}{Z_2} + \frac{V_T}{Z_3} + \cdots + \frac{V_T}{Z_n}$$

Factoring out the common driving voltage,

$$I_T = V_T\left(\frac{1}{Z_1} + \frac{1}{Z_2} + \frac{1}{Z_3} + \cdots + \frac{1}{Z_n}\right) \tag{12-33}$$

Defining the reciprocal of impedance to be the admittance **Y**,

$$\boxed{\frac{1}{Z} = Y} \tag{12-34}$$

Substituting Eq. (12-34) into Eq. (12-33),

$$I_T = V_T(Y_1 + Y_2 + Y_3 + \cdots + Y_n) \tag{12-35}$$

Defining,

$$Y_T = (Y_1 + Y_2 + Y_3 + \cdots + Y_n) \tag{12-36}$$

Equation (12-35) becomes

$$\boxed{I_T = V_T Y_T} \tag{12-37}$$

Admittances Y_1, Y_2, $Y_3, \ldots, Y_n$ are the individual admittances of the parallel branches, and Y_T is the total circuit admittance, also called the *input admittance or driving-point admittance. The unit of admittance is the* siemen,†

---

† Previously called the mho.

and its letter symbol is S. The admittance of a circuit may be considered as a measure of the ease with which a circuit can conduct a *sinusoidal current.* Thus circuits with higher values of admittance will have higher values of current.

The admittance of a circuit or branch, expressed in rectangular form, is

$$\mathbf{Y} = Y\underline{/\alpha^\circ} = G + jB$$

The real part of the admittance is called the *conductance,* and the imaginary part is called the *susceptance.* The unit for each is the siemen. It should be noted that the *conductance and susceptance components of an admittance are paralleled elements.*

12-8   SPECIAL-CASE FORMULA FOR TWO IMPEDANCES IN PARALLEL

A simple special-case formula that *applies to only two impedances in parallel* is derived from Eq. (12-36) as follows:

$$\mathbf{Y}_{P2} = \mathbf{Y}_A + \mathbf{Y}_B$$

$$\mathbf{Z}_{P2} = \frac{1}{\mathbf{Y}_{P2}} = \frac{1}{\mathbf{Y}_A + \mathbf{Y}_B} = \frac{1}{1/\mathbf{Z}_A + 1/\mathbf{Z}_B} = \frac{1}{(\mathbf{Z}_A + \mathbf{Z}_B)/\mathbf{Z}_A \mathbf{Z}_B}$$

$$\boxed{\mathbf{Z}_{P2} = \frac{\mathbf{Z}_A \mathbf{Z}_B}{\mathbf{Z}_A + \mathbf{Z}_B}} \qquad\qquad (12\text{-}38)$$

where $\mathbf{Z}_{P2}$ is the equivalent impedance of two paralleled impedances.

Example 12-6   Referring to Fig. 12-15, Assume $\mathbf{Z}_1 = 2.8\underline{/-56^\circ}$, $\mathbf{Z}_2 = 1.9\underline{/79^\circ}$, $\mathbf{Z}_3 = 1.5\underline{/-63^\circ}$, and $\mathbf{V}_T = 100\underline{/0^\circ}$. Determine (*a*) the input admittance of the circuit; (*b*) the steady-state current supplied by the generator; (*c*) the input impedance.

Solution

(*a*) $\mathbf{Y}_1 = \dfrac{1}{2.8\underline{/-56^\circ}} = 0.3571\underline{/56^\circ} = (0.1997 + j0.2961)$ S

$\mathbf{Y}_2 = \dfrac{1}{1.9\underline{/79^\circ}} = 0.5263\underline{/-79^\circ} = (0.1004 - j0.5166)$ S

$\mathbf{Y}_3 = \dfrac{1}{1.5\underline{/-63^\circ}} = 0.6667\underline{/63^\circ} = (0.3027 + j0.5940)$ S

$\mathbf{Y}_T = \mathbf{Y}_1 + \mathbf{Y}_2 + \mathbf{Y}_3$

$\mathbf{Y}_T = (0.1997 + j0.2961) + (0.1004 - j0.5166) + (0.3027 + j0.5940)$

$\mathbf{Y}_T = 0.6028 + j0.3735 = 0.7091\underline{/31.78^\circ}$ S

(b) $\mathbf{I}_T = \mathbf{V}_T \mathbf{Y}_T = (100\underline{/0°})(0.7091\underline{/31.78°})$

$\mathbf{I}_T = 70.91\underline{/31.78°} \text{ A}$

(c) $\mathbf{Z}_T = \dfrac{1}{\mathbf{Y}_T} = \dfrac{1}{0.7091\underline{/31.78°}}$

$\mathbf{Z}_T = 1.41\underline{/-31.78°} \ \Omega$

The input impedance may also be determined from Ohm's law,

$$\mathbf{Z}_T = \frac{\mathbf{V}_T}{\mathbf{I}_T} = \frac{100\underline{/0°}}{70.91\underline{/31.78°}} = 1.41\underline{/-31.78°} \ \Omega$$

12-9   SERIES-TO-PARALLEL CONVERSION

In many circuit applications it becomes necessary to convert the series-connected parameters of the impedance coil shown in Fig. 12-16a to equivalent parallel-connected parameters shown in Fig. 12-16c. To be equivalent, the input impedance of the parallel circuit must equal the input impedance of the series circuit it replaces. Thus,

$$\mathbf{Z}_P = \mathbf{Z}_S = R_S + jX_S$$

Converting $\mathbf{Z}_p$ to admittance form,

$$\mathbf{Y}_P = \frac{1}{\mathbf{Z}_P} = \frac{1}{R_S + jX_S}$$

Rationalizing the denominator and expanding into rectangular form,

$$\mathbf{Y}_P = \frac{R_S - jX_S}{R_S^2 + X_S^2} = \frac{R_S}{R_S^2 + X_S^2} - \frac{jX_S}{R_S^2 + X_S^2}$$

As indicated in the preceding equation and shown in Fig. 12-16b, the input admittance consists of a conductance component in parallel with a susceptance component.

$$\mathbf{Y}_P = G - jB$$

where

$$G = \frac{R_S}{R_S^2 + X_S^2} \qquad \text{and} \qquad -jB = \frac{-jX_S}{R_S^2 + X_S^2}$$

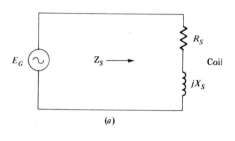

(a)

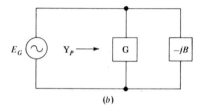

(b)

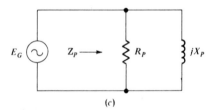

(c)

Figure 12-16 Series-to-parallel conversion.

Converting the respective admittance components in Fig. 12-16b to impedance form results in the equivalent parallel parameters shown in Fig. 12-16c.

$$\frac{1}{G} = \frac{R_S^2 + X_S^2}{R_S}$$

$$\frac{1}{-jB} = \frac{j(R_S^2 + X_S^2)}{X_S}$$

Thus

$$R_P = \frac{R_S^2 + X_S^2}{R_S} \qquad \frac{R_S^2 + X_S^2}{X_S} = X_P \qquad (12\text{-}39)$$

Although derived from the series-connected components of an impedance coil, equation set (12-33) may also be used to convert a series-connected RC branch to its equivalent parallel parameters.

Example 12-7 Assume R_S, X_S, and E_G in Fig. 12-16a are 3.0 Ω, 4.0 Ω, and 100 V, respectively. Determine (a) the circuit current; (b) the equivalent parallel circuit parameters; (c) the feeder current to the equivalent parallel circuit.

Solution

(a) $I_T = \dfrac{V_T}{Z_T} = \dfrac{100/\underline{0°}}{3.0 + j4.0} = 20/\underline{-53.1}$ A

(b) $R_P = \dfrac{R_S^2 + X_S^2}{R_S} = \dfrac{(3)^2 + (4)^2}{3} = 8.33$ Ω

$X_P = \dfrac{R_S^2 + X_S^2}{X_S} = \dfrac{(3)^2 + (4)^2}{4} = 6.25$ Ω

(c) Using Fig. 12-16c as a guide,

$$I_T = I_{R_P} + I_{X_P} = \frac{E_G}{R_P} + \frac{E_G}{jX_P}$$

$$I_T = \frac{100/\underline{0°}}{8.33} + \frac{100/\underline{0°}}{6.25/\underline{90°}} = (12.0 - j16.0) = 20/\underline{-53.1°}\ \text{A}$$

Since the current supplied by the generator to the series circuit is identical to that supplied to the parallel circuit, *the two circuits are equivalent as viewed from the generator terminals.*

12-10   CURRENT-DIVIDER EQUATION FOR PARALLEL-CONNECTED ADMITTANCES

A simple formula for determining the current drawn by any one of many parallel-connected admittances can be derived from Fig. 12-17. The total current supplied to the parallel circuit is

$$I_T = V_T Y_T \tag{12-40}$$

where $Y_T = Y_1 + Y_2 + \cdots + Y_k + \cdots + Y_n$

The current to admittance Y_k is

$$I_k = V_T Y_k \tag{12-41}$$

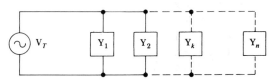

Figure 12-17 Circuit for determining the current-divider equation.

Solving Eq. (12-40) for $\mathbf{V}_T$ and substituting it into Eq. (12-41) results in the current-divider equation

$$\boxed{\mathbf{I}_k = \mathbf{I}_T \frac{\mathbf{Y}_k}{\mathbf{Y}_T}}$$

(12-42)

Example 12-8 Using the current-divider equation, determine the current in impedance $\mathbf{Z}_2$ of Example 12-6.

Solution

$$\mathbf{I}_2 = \mathbf{I}_T \frac{\mathbf{Y}_2}{\mathbf{Y}_T}$$

$$\mathbf{I}_2 = (70.91\underline{/31.78°}) \frac{0.5263\underline{/-79°}}{0.7091\underline{/31.78°}}$$

$$\mathbf{I}_2 = 52.63\underline{/-79.00°})\ \text{A}$$

Example 12-9 A 350-μF capacitor, a 10.0-Ω resistor, and a coil whose inductance and resistance are 0.030 H and 5 Ω, respectively, are connected in parallel and supplied by a 120-V 60-Hz generator. Sketch the circuit and determine (*a*) the steady-state current through each branch; (*b*) the steady-state line current; (*c*) the input impedance; (*d*) the input admittance; (*e*) the steady-state circuit current if the sinusoidal generator is replaced by a 120-V battery.

Solution

The circuit diagram is shown in Fig. 12-18.

(*a*) $X_C = \dfrac{1}{2\pi(60)(350 \times 10^{-6})} = 7.58\ \Omega$

$X_L = 2\pi(60)(0.03) = 11.31\ \Omega$

$\mathbf{Z}_{\text{coil}} = (5 + j11.31) = 12.37\underline{/66.15°}\ \Omega$

$\mathbf{I}_C = \dfrac{\mathbf{V}_T}{-jX_C} = \dfrac{120\underline{/0°}}{7.58\underline{/-90°}} = 15.83\underline{/90°}\ \text{A}$

$\mathbf{I}_{\text{coil}} = \dfrac{\mathbf{V}_T}{\mathbf{Z}_{\text{coil}}} = \dfrac{120\underline{/0°}}{12.37\underline{/66.15°}} = 9.70\underline{/-66.15°}\ \text{A}$

$\mathbf{I}_R = \dfrac{\mathbf{V}_T}{R} = \dfrac{120\underline{/0°}}{10\underline{/0°}} = 12\underline{/0°}\ \text{A}$

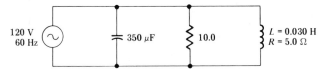

Figure 12-18   Circuit for Example 12-9.

(b) $\mathbf{I}_T = \mathbf{I}_C + \mathbf{I}_{coil} + \mathbf{I}_R$

$\mathbf{I}_T = 15.83\underline{/90°} + 9.70\underline{/-66.15°} + 12\underline{/0°}$

$\mathbf{I}_T = (0 + j15.83) + (3.92 - j8.87) + (12 + j0)$

$\mathbf{I}_T = 15.92 + j6.96 = 17.38\underline{/23.60°}$ A

(c) From Ohm's law,

$$\mathbf{Z}_{in} = \frac{\mathbf{V}_T}{\mathbf{I}_T} = \frac{120\underline{/0°}}{17.38\underline{/23.60°}} = 6.90\underline{/-23.60°} \ \Omega$$

(d) $\mathbf{Y}_{in} = \dfrac{1}{\mathbf{Z}_{in}} = \dfrac{1}{6.90\underline{/-23.60°}} = 0.1448\underline{/23.60°}$ S

(e) With a DC driver, $f = 0$ Hz

$$X_L = 2\pi(0)L = 0 \ \Omega \qquad \mathbf{Z}_{coil} = R + jX_L = 5 + j0 = 5 \ \Omega$$

$$X_C = \frac{1}{2\pi(0)C} = \infty \ \Omega$$

$$I_C = \frac{120}{\infty} = 0 \ A$$

$$I_{coil} = \frac{120}{5} = 24 \ A$$

$$I_R = \frac{120}{10} = 12 \ A$$

$$I_T|_{f=0} = 24 + 12 = 36 \ A$$

12-11   INPUT IMPEDANCE OF SERIES-PARALLEL CIRCUITS

The input impedance, also called driving-point impedance, of a series-parallel circuit is the impedance as viewed from the generator terminals of the circuit. Determination of the input impedance is accomplished through the use of admittance-impedance conversions and network-reduction formulas, as appropriate. To reduce the risk of error, all admittance-to-impedance conversions and vice versa should be done in polar form, and all calculations should be carried out to at least four decimal places.

Example 12-10 Determine (a) the input impedance of the series-parallel circuit in Fig. 12-19a; (b) the circuit current; (c) the voltage drop across Z_1; (d) the voltage drop across Z_2.

Solution
(a) *Start with that part of the circuit farthest from the driver*, combining elements as you progress toward the input terminals. Thus the paralleled impedances Z_2 and Z_3 are combined to form impedance Z_{cd}. Impedance Z_{cd} is then added to Z_1 to obtain the

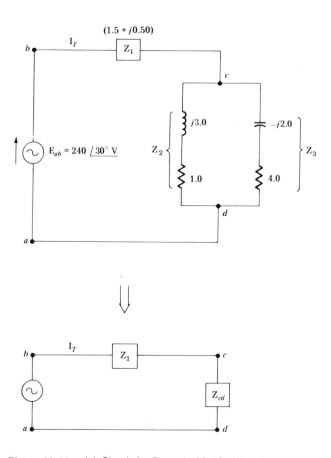

Figure 12-19 (a) Circuit for Example 12-10; (b) reduced circuit.

input impedance, as shown in Fig. 12-19b. Using the special-case formula for two paralleled impedances,

$$\mathbf{Z}_{cd} = \frac{\mathbf{Z}_2 \mathbf{Z}_3}{\mathbf{Z}_2 + \mathbf{Z}_3} = \frac{(1 + j3)(4 - j2)}{(1 + j3) + (4 - j2)}$$

$$\mathbf{Z}_{cd} = \frac{4 - j2 + j12 - j^2 6}{5 + j1} = \frac{10 + j10}{5 + j1} = \frac{14.14/45°}{5.10/11.31°} = 2.77/33.69°$$

$$\mathbf{Z}_{cd} = 2.31 + j1.54 \ \Omega$$

$$\mathbf{Z}_{in} = (1.5 + j0.5) + (2.31 + j1.54) = (3.81 + j2.04) \ \Omega$$

$$\mathbf{Z}_{in} = 4.32/28.17° \ \Omega$$

(b) $\mathbf{I} = \dfrac{\mathbf{V}}{\mathbf{Z}_{in}} = \dfrac{240/30°}{4.32/28.17°} = 55.56/1.83° \ \text{A}$

(c) $\mathbf{V}_{Z1} = \mathbf{I}\mathbf{Z}_1 = (55.56/1.83°)(1.5 + j0.5)$

$\quad \mathbf{V}_{Z1} = (55.56/1.83°)(1.58/18.44°) = 87.79/20.27° \ \text{V}$

(d) $\mathbf{V}_{Z2} = \mathbf{I}\mathbf{Z}_{cd} = (55.56/1.83°)(2.77/33.69°)$

$\quad \mathbf{V}_{Z2} = 153.9/35.52° \ \text{V}$

Example 12-11 For the circuit shown in Fig. 12-20a, determine (a) the input impedance of the series-parallel combination; (b) the ammeter reading; (c) the equation for the current wave as a function of time; (d) the input impedance and steady-state current if the AC generator is replaced with a 240-V battery.

Solution
Using the special-case formula for two parallel impedances,

(a) $\mathbf{Z}_{bc} = \dfrac{(5 + j7)5}{(5 + j7) + 5} = 3.52/19.47° = (3.32 + j1.17) \ \Omega$

$\quad \mathbf{Z}_{in} = (3.32 + j1.17) + (3 + j1) = 6.68/18.95° \ \Omega$

(b) $\mathbf{I} = \dfrac{\mathbf{V}}{\mathbf{Z}_{in}} = \dfrac{240/0°}{6.68/18.95°} = 35.92/{-18.95°} \ \text{A}$

The ammeter indicates 35.92 A.

(c) $i = 35.92\sqrt{2} \sin(2\pi 25t - 18.95°)$

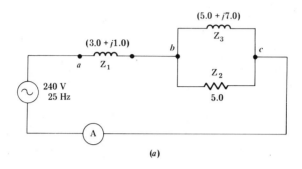

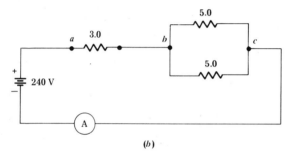

Figure 12-20   Circuits for Example 12-11.

(*d*) At steady-state DC, the inductive reactances are zero and the circuit is equivalent to that shown in Fig. 12-20*b*. Therefore,

$$R_{\text{in}} = 3 + 2.5 = 5.5 \ \Omega \qquad I_{\text{DC}} = \frac{240}{5.5} = 43.64 \text{ A}$$

Example 12-12   Determine the input impedance of the circuit shown in Fig. 12-21*a*. The circuit impedances are

$$\mathbf{Z}_1 = (0.054 + j0) \ \Omega \qquad \mathbf{Z}_2 = (1 + j2) \ \Omega$$

$$\mathbf{Z}_3 = (0 - j4) \ \Omega \qquad \mathbf{Z}_4 = (3 - j3) \ \Omega$$

$$\mathbf{Z}_5 = (1 + j3) \ \Omega \qquad \mathbf{Z}_6 = 0.278\underline{/-56.3°} \ \Omega$$

Solution
Combine $\mathbf{Z}_2, \mathbf{Z}_3$, and $\mathbf{Z}_4$ to form a simplified circuit as shown in Fig. 12-21*b*. Then combine $\mathbf{Z}_A$ and $\mathbf{Z}_5$ to form the circuit in Fig. 12-21*c*. The final reduction is shown in

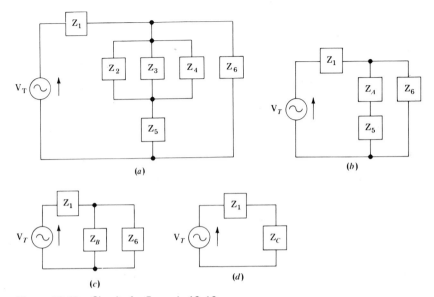

Figure 12-21 Circuits for Example 12-12.

Fig. 12-21*d*. Combining $\mathbf{Z}_2$, $\mathbf{Z}_3$, $\mathbf{Z}_4$,

$$\mathbf{Z}_2 = 1 + j2 = 2.24\underline{/63.43°}\ \Omega$$

$$\mathbf{Z}_3 = 0 - j4 = 4\underline{/-90°}\ \Omega$$

$$\mathbf{Z}_4 = 3 - j3 = 4.24\underline{/-45°}\ \Omega$$

$$\mathbf{Y}_2 = \frac{1}{\mathbf{Z}_2} = \frac{1}{2.24\underline{/63.43°}} = 0.4464\underline{/-63.43°} = (0.1997 - j0.3993)\ \text{S}$$

$$\mathbf{Y}_3 = \frac{1}{\mathbf{Z}_3} = \frac{1}{4\underline{/-90°}} = 0.2500\underline{/90°} = (0 + j0.2500)\ \text{S}$$

$$\mathbf{Y}_4 = \frac{1}{\mathbf{Z}_4} = \frac{1}{4.24\underline{/-45°}} = 0.2357\underline{/45°} = (0.1667 + j0.1667)\ \text{S}$$

$$\mathbf{Y}_A = \mathbf{Y}_2 + \mathbf{Y}_3 + \mathbf{Y}_4$$

$$\mathbf{Y}_A = (0.3664 + j0.0174) = 0.3668\underline{/2.72°}\ \text{S}$$

$$\mathbf{Z}_A = \frac{1}{\mathbf{Y}_A} = \frac{1}{0.3668\underline{/2.72°}} = 2.726\underline{/-2.72°} = (2.723 - j0.1294)\ \Omega$$

$$\mathbf{Z}_B = \mathbf{Z}_A + \mathbf{Z}_5 = (2.723 - j0.1294) + (1 + j3) = 4.701\underline{/37.64°}\ \Omega$$

Combining $\mathbf{Z}_B$ and $\mathbf{Z}_6$,

$$\mathbf{Z}_C = \frac{\mathbf{Z}_B \mathbf{Z}_6}{\mathbf{Z}_B + \mathbf{Z}_6} = \frac{(4.701\underline{/37.64°})(0.278\underline{/-56.3°})}{(4.701\underline{/37.64°}) + (0.278\underline{/-56.3°})}$$

$$\mathbf{Z}_C = \frac{1.307\underline{/-18.66°}}{4.69\underline{/34.25°}} = 0.2787\underline{/-52.91°} = (0.1680 - j0.2223)\ \Omega$$

$$\mathbf{Z}_{\text{input}} = \mathbf{Z}_C + \mathbf{Z}_1 = (0.1680 - j0.2223) + (0.054 + j0) = 0.3142\underline{/-45.04°}\ \Omega$$

Example 12-13 Determine (a) the potential difference between points B and D in Fig. 12-22a; (b) the voltmeter reading; (c) the voltmeter reading (DC voltmeter) if the generator is replaced by a 250-V battery, as shown in Fig. 12-22b.

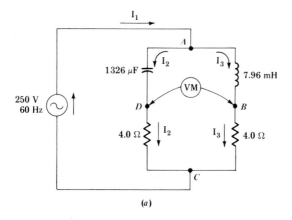

(a)

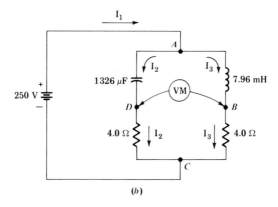

(b)

Figure 12-22 Determining the potential difference between any two points in a circuit.

Solution

(*a*) Using the potential-difference equation developed in Sec. 5-3, and traveling path *DCB*,

$$\mathbf{V}_{DB} = (\textstyle\sum \text{driving voltages}) - (\textstyle\sum \text{voltage drops})$$

$$\mathbf{V}_{DB} = 0 - (4\mathbf{I}_2 + (-1)(4)\mathbf{I}_3) \tag{12-43}$$

The current in each of the two branches may be determined by calculating the impedance of each branch and then applying Ohm's law. Thus,

$$X_L = 2\pi f L = 2\pi(60)(7.96 \times 10^{-3}) = 3.00 \ \Omega$$

$$X_C = \frac{1}{2\pi 60(1326 \times 10^{-6})} = 2.00 \ \Omega$$

The current in branch *ADC* is

$$\mathbf{I}_2 = \frac{250\underline{/0°}}{(4 - j2)} = \frac{250\underline{/0°}}{4.47\underline{/-26.57°}} = 55.90\underline{/26.57°} \ \text{A}$$

The current in branch *ABC* is

$$\mathbf{I}_3 = \frac{250\underline{/0°}}{(4 + j3)} = \frac{250\underline{/0°}}{5\underline{/36.87°}} = 50.0\underline{/-36.87°} \ \text{A}$$

Substituting the values for $\mathbf{I}_2$ and $\mathbf{I}_3$ into Eq. (12-43),

$$\mathbf{V}_{DB} = -[4(55.90\underline{/26.57°}) - 4(50\underline{/-36.87°})]$$

$$\mathbf{V}_{DB} = -223.6\underline{/26.57°} + 200\underline{/-36.87°}$$

$$\mathbf{V}_{DB} = (-199.99 - j100.01) + (160.00 - j120)$$

$$\mathbf{V}_{DB} = (-39.99 - j220.01) = 223.6\underline{/-100.3°}$$

(*b*) The voltmeter will indicate 223.6 V.

(*c*) At 0 Hz (DC),

$$X_L = 2\pi f L = 2\pi(0)(7.96 \times 10^{-3}) = 0$$

$$X_C = \frac{1}{2\pi f C} = \frac{1}{2\pi 0(1326 \times 10^{-6})} = \infty$$

The current in branch *ADC* is

$$\mathbf{I}_2 = \frac{250\underline{/0°}}{4 + j\infty} = \frac{250}{\infty\underline{/90°}} = 0 \ \text{A}$$

The current in branch ABC is

$$\mathbf{I}_3 = \frac{250\underline{/0^\circ}}{4 + j0} = \frac{250}{4} = 62.5 \text{ A}$$

Traveling path DCB,

$$\mathbf{V}_{DB} = \left(\sum \text{ driving voltages}\right) - \left(\sum \text{ voltage drops}\right)$$

$$\mathbf{V}_{DB} = 0 - [4\mathbf{I}_2 + (-1)(4\mathbf{I}_3)] = -(4 \times 0 - 4 \times 62.5) = 250 \text{ V}$$

SUMMARY OF FORMULAS

Series Circuit

$$\mathbf{Z}_T = \mathbf{Z}_1 + \mathbf{Z}_2 + \mathbf{Z}_3 + \cdots + \cdots + \mathbf{Z}_n \qquad \mathbf{I}_T = \frac{\mathbf{V}_T}{\mathbf{Z}_T}$$

$$\mathbf{Z}_S = R + j(X_L - X_C) = Z_S\underline{/\theta_S}$$

$$Z_S = \sqrt{R^2 + (X_L - X_C)^2} \qquad \theta_S = \tan^{-1}\frac{X_L - X_C}{R}$$

$$\mathbf{V}_k = \mathbf{V}_T\frac{\mathbf{Z}_k}{\mathbf{Z}_T}$$

Parallel Circuit

$$\mathbf{I}_T = \mathbf{V}_T\mathbf{Y}_T = \frac{\mathbf{V}_T}{\mathbf{Z}_T} \qquad \mathbf{I}_k = \mathbf{I}_T\frac{\mathbf{Y}_k}{\mathbf{Y}_T}$$

$$\mathbf{Y}_T = \mathbf{Y}_1 + \mathbf{Y}_2 + \mathbf{Y}_3 + \cdots + \mathbf{Y}_n = G + jB = Y\underline{/\alpha^\circ}$$

$$\mathbf{Y} = \frac{1}{\mathbf{Z}} \qquad \mathbf{Z}_{P2} = \frac{\mathbf{Z}_A\mathbf{Z}_B}{\mathbf{Z}_A + \mathbf{Z}_B}$$

Series-to-Parallel Conversion

$$R_P = \frac{R_S^2 + X_S^2}{R_S} \qquad X_P = \frac{R_S^2 + X_S^2}{X_S}$$

Potential Difference between Two Points

$$\mathbf{V}_{AB} = \left(\sum \text{ driving voltages}\right) - \left(\sum \text{ voltage drops}\right)$$

PROBLEMS

Series Circuit

12-1 A 6.0-Ω resistor and a 4.0-Ω inductive reactance are connected in series with a sinusoidal generator. If the generator voltage is given as $e = 20 \sin 157t$ V, determine (a) the complex impedance; (b) the circuit current in polar form; (c) the frequency of the alternating current.

12-2 A coil, a capacitor, and a resistor are connected in series and supplied by a 1.5-kHz generator. The capacitive reactance is 4.0 Ω, the resistor is 4.0 Ω, and the coil has a resistance of 3.0 Ω and an inductive reactance of 4.0 Ω. If the current in the circuit is $10/\underline{20°}$ A, determine (a) the complex impedance, (b) the phasor voltage drop across the coil.

12-3 Determine the resistance, inductance, or capacitance values of series-connected elements that will draw a current of $20/\underline{-30°}$ A from a 60-Hz sinusoidal generator whose voltage is $100/\underline{0°}$ V.

12-4 A current of $(2.0 + j3.0)$ A is determined to be in a series circuit consisting of a sinusoidal generator and the following impedances: $\mathbf{Z}_1 = 4.0/\underline{0°}$, $\mathbf{Z}_2 = (6.0 + j5.0)$ $\mathbf{Z}_3 = -j2.0$. Sketch the circuit and calculate (a) the circuit impedance; (b) the voltage drop (in polar form) across each component; (c) the generator voltage.

12-5 A certain series circuit is composed of two generators, a 4.0-Ω resistor, a 10.0-Ω inductive reactance, and a 3.0-Ω capacitive reactance. If the generator voltages are $60 \sin (1000t + 50°)$ and $100 \sin (1000t + 30°)$, determine (a) the complex impedance; (b) the total driving voltage in polar form; (c) the magnitude and phase angle of the circuit current.

12-6 A 100-V AC generator is connected in series with a $3.61/\underline{33.69°}$-Ω impedance, a 5.0-Ω capacitive reactance, and a 2.0-Ω resistor. Sketch the circuit and determine the reading of a clip-on ammeter hooked over one of the connecting wires.

12-7 A 240-V 400-Hz generator supplies a series connection of three ideal circuit elements. The elements are a 2.0-Ω resistor, a 3.0-Ω inductive reactance, and a 4.0-Ω capacitive reactance. Sketch the circuit and determine (a) the complex impedance; (b) the rms current; (c) the voltage drop across the inductance; (d) the steady-state current if the driver is replaced by a 240-V battery; (e) the steady-state voltage across the capacitor for the conditions in part d.

12-8 A high-impedance rms voltmeter is used to measure the voltage drop across each of three series-connected ideal circuit elements. If the rms readings are 40 V, 25 V, and 60 V for V_L, V_R, and V_C, respectively, determine the equation for the voltage wave representing the driving voltage whose frequency is 20 Hz.

12-9 A 6000-Hz 200-V generator is connected in series with a capacitive reactance of 6.0 Ω and a coil whose resistance and inductive reactance are 4.0 Ω and 2.0 Ω, respectively. Sketch the circuit and determine (a) the circuit impedance; (b) the circuit current; (c) the phasor diagram; (d) the rms voltage across the capacitor; (e) the rms voltage across the coil; (f) the maximum instantaneous energy stored in the capacitor; (g) the maximum instantaneous charge stored in the capacitor.

Parallel Circuit

12-10 A coil of 30 mH, a resistor of 10 Ω, and a capacitance of 350 μF are connected in parallel and supplied by a 120-V 60-Hz generator. Sketch the circuit and determine (a) the current through each element; (b) the total line current; (c) the circuit impedance; (d) the phase angle between the feeder current and the driving voltage.

12-11 A parallel circuit consisting of a 4.0-Ω resistor, a 2.0-Ω capacitive reactance, and a 6.0-Ω inductive reactance are fed from a 30-Hz 240-V generator. Sketch the circuit and determine (a) the circuit admittance; (b) the circuit impedance; (c) the feeder current; (d) the time-domain equations for the driving voltage and the feeder current; (e) the phasor diagram (drawn to scale) showing branch currents, feeder current, and driving voltage.

12-12   A 400-V 50-Hz generator supplies current to a parallel circuit consisting of a 4.0-Ω resistor, a 5.0-Ω inductive reactance, and a 3.0-Ω capacitive reactance. Sketch the circuit and determine (*a*) the input admittance; (*b*) the input impedance; (*c*) the feeder current; (*d*) the time-domain equation for the feeder current; (*e*) the steady-state current if the sinusoidal generator is replaced by a 400-V battery.

12-13   A parallel circuit consisting of a 2.0-Ω resistor, a 4.0-H inductor, and a 0.0070-F capacitor is connected to a sinusoidal driver. The equation for the driving voltage is $v = 100 \sin (30t + 40°)$. Sketch the circuit and determine (*a*) the frequency of the supply voltage; (*b*) the inductive reactance; (*c*) the rms current drawn by the inductor; (*d*) the equation for the current wave to the inductor; (*e*) the instantaneous energy stored in the capacitor at $t = 0.1$ s; (*f*) the rate of expenditure of heat energy at $t = 0.1$ s.

12-14   A parallel circuit consisting of a 5.0-Ω capacitive reactance, a 4.0-Ω resistor, and a 2.0-Ω inductive reactance is connected to a 240-V 60-Hz generator. Sketch the circuit and determine (*a*) the circuit admittance; (*b*) the current to each component; (*c*) the feeder current; (*d*) the values of inductance and capacitance; (*e*) the maximum instantaneous energy stored in the magnetic field.

12-15   Two parallel-connected impedances are fed by a sinusoidal voltage wave expressed by $v_t = 170 \sin (377t + 60°)$. The impedances are $2.9/\!-16°$ Ω and $8.5/\underline{10.5°}$ Ω. Determine the rms current in each impedance and the rms feeder current.

12-16   A branch consisting of a 200-μF capacitor in series with a 5.0-Ω resistor is connected in parallel with a $(10 + j4.0)$-Ω impedance. The generator voltage is 100 sin 377t. Determine (*a*) the current in the series branch; (*b*) the current in the $(10 + j4.0)$-Ω branch; (*c*) the feeder current.

12-17   Determine the input impedance of a parallel circuit consisting of a 2.0-Ω resistor branch, a $(3.0 + j8.0)$-Ω branch and a branch containing a 4.0-Ω capacitive reactance.

12-18   A 240-V 60-Hz generator supplies energy to a parallel circuit consisting of a 5.0-Ω capacitive reactance, a 2.0-Ω resistor, and a $(0.10 + j0.40)$-S admittance. Sketch the circuit and determine the feeder current.

12-19   A 400-V 50-Hz generator supplies current to a parallel circuit consisting of a $(2.0 + j3.0)$-Ω impedance and a $(3.0 + j4.0)$-Ω impedance. Sketch the circuit and determine (*a*) the input admittance; (*b*) the feeder current; (*c*) the time-domain equation for the current in the $(3.0 + j4.0)$-Ω impedance.

12-20   A branch consisting of a 200-μF capacitor in series with a 5.0-Ω resistor is connected in parallel with a coil whose resistance and inductance are 10 Ω and 0.0106 H, respectively. The generator voltage is 100 sin 377t V. Sketch the circuit and determine (*a*) the current to the capacitor; (*b*) the current in the coil; (*c*) the feeder current; (*d*) the steady-state feeder current if the sinusoidal generator is replaced by a 12-V battery.

12-21   A 450-V 60-Hz source supplies a phasor current of $5.0/\!-60°$ A to an impedance. Sketch the circuit and determine (*a*) the impedance in ohms; (*b*) a set of series-connected circuit elements that can be used to construct the impedance (express the elements in henrys, ohms, and/or farads as applicable); (*c*) a set of parallel-connected circuit elements that can be used to make an equivalent impedance.

12-22   (*a*) Determine the resistance, inductance, and/or capacitive values of parallel-connected circuit elements that draw a total of $20/\!-30°$ A from a 60-Hz sinusoidal generator whose driving voltage is $100/\underline{0°}$ V; (*b*) repeat part *a* for series-connected elements.

12-23   Determine the ammeter readings for the circuit shown in Fig. 12-23.

12-24   A feeder current of 250 A is supplied to the following paralleled impedances:

$Z_1 = 5/\underline{20°}$ Ω $Z_3 = 10/\underline{15°}$ Ω
$Z_2 = 6/\underline{0°}$ Ω $Z_4 = 8/\!-60°$ Ω

Sketch the circuit and determine the current drawn by Z_3.

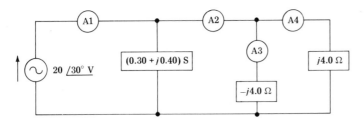

Figure 12-23 Circuits for Prob. 12-23.

12-25 A DC generator supplies a total of 100 A to the following paralleled resistors.

$R_1 = 2\,\Omega$ $R_3 = 5\,\Omega$ $R_5 = 4\,\Omega$
$R_2 = 3\,\Omega$ $R_4 = 7\,\Omega$ $R_6 = 9\,\Omega$

Sketch the circuit and determine the current in R_1 and R_6.

12-26 A $40\underline{/30°}$ Ω impedance is connected in series with the following paralleled group of impedances:

$Z_1 = 3\underline{/60°}\ \Omega$ $Z_3 = 8\underline{/60°}\ \Omega$
$Z_2 = 5\underline{/40°}\ \Omega$ $Z_4 = 4\underline{/-25°}\ \Omega$

If the voltage drop across the $40\underline{/30°}$-Ω impedance is $200\underline{/0°}$ V, determine (a) the current in Z_3, (b) the voltage drop across the paralleled section.

Series-Parallel Circuit

12-27 Determine the input impedance for the circuit shown in Fig. 12-24. The circuit parameters are $Z_1 = (3 + j4)$, $Z_2 = 9.8\underline{/-78°}$, $Z_3 = 18.5\underline{/21.8°}$. The 400-Hz sinusoidal generator has an rms voltage of 100 V.

12-28 For the circuit section shown in Fig. 12-25, $Z_1 = 4\underline{/50°}$, $Z_2 = 6\underline{/40°}$, $Z_3 = 5\underline{/30°}$, $Z_4 = 2\underline{/20°}$, $Z_5 = 6\underline{/60°}$. Determine (a) the voltmeter reading; (b) the ammeter reading, (c) the time-domain equation for the current in Z_1 (assume a 60-Hz driver).

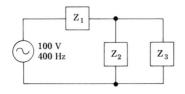

Figure 12-24 Circuit for Prob. 12-27.

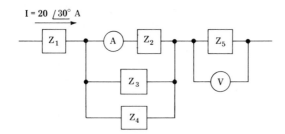

Figure 12-25 Circuit for Prob. 12-28.

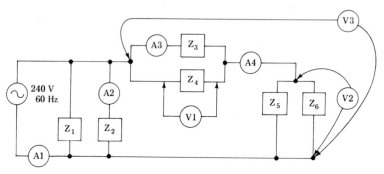

Figure 12-26 Circuit for Prob. 12-30.

12-29 A certain series-parallel circuit contains an impedance Z_1 in series with a parallel combination of Z_2 and Z_3. The sinusoidal driving voltage is $240\underline{/30°}$ V. The impedances are $Z_1 = (1 - j1)$, $Z_2 = (1 + j1)$, $Z_3 = (1 - j1)$. Sketch the circuit and determine (a) the driving-point impedance; (b) the driving-point admittance; (c) the current supplied by the generator; (d) the voltage drop across Z_1.

12-30 Determine the ammeter and voltmeter reading for the circuit shown in Fig. 12-26. The 60-Hz generator supplies 240 V to the circuit, and the impedances are $Z_1 = 3 + j2$, $Z_2 = 4 + j1$, $Z_3 = 2 - j5$, $Z_4 = 3 + j6$, $Z_5 = 4 + j7$, $Z_6 = 2 + j3$.

12-31 The voltage across a certain parallel section of a series-parallel circuit is $120\underline{/60°}$ V. The parallel section consists of a $(2.0 + j3.0)$-Ω branch in parallel with a $(4.0 - j8.0)$-Ω branch. Sketch the circuit and determine (a) the current in each branch; (b) the total current to the parallel section.

12-32 An inductive reactance of 1.0 Ω is connected in series with two parallel-connected branches whose respective impedances are $(4.0 + j3.0)$ and $(2.0 - j3.0)$ Ω. The circuit is supplied by a 450-V 25-Hz generator. Sketch the circuit and determine (a) the input impedance; (b) the current supplied by the generator; (c) the current in the $(2.0 - j3.0)$-Ω impedance; (d) the voltage drop across the $(4.0 + j3.0)$-Ω impedance; (e) the voltage drop across the 1.0-Ω inductive reactance; (f) sketch the phasor diagram showing all component voltage drops, component

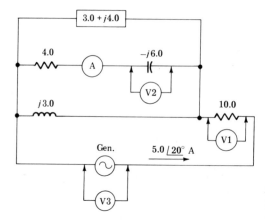

Figure 12-27 Circuit for Prob. 12-33.

currents, driving voltage, and input current; (g) write the time-domain equation for the current through the (4.0 + j3.0)-Ω impedance; (h) assuming the sinusoidal generator is replaced by a 450-V battery, sketch the circuit to reflect the new conditions, and solve for the steady-state current supplied by the battery.

12-33 (a) Determine the voltmeter and ammeter readings for the circuit shown in Fig. 12-27; (b) repeat part a assuming the sinusoidal generator is replaced by a battery, and the battery current is 5.0 A.

CHAPTER 13

POWER IN A TWO-WIRE SYSTEM WITH SINUSOIDAL DRIVERS

Power in an electric circuit is defined as the time rate of transfer of energy and is expressed in joules per second or watts. At any instant of time the instantaneous rate of transfer of energy in a circuit is equal to the instantaneous value of voltage across the circuit multiplied by the instantaneous value of current.

$$p = vi$$

where $v =$ instantaneous voltage, V
$i =$ instantaneous current, A

Multiplying the instantaneous values of voltage with the corresponding instantaneous values of current, and plotting the product versus time, results in a power or volt-ampere wave.

13-1   ACTIVE POWER

The wave representing the product of the sinusoidal current and voltage waves for a pure resistor circuit is shown in Fig. 13-1a. It is a pulsating power wave whose frequency is double that of the driving voltage. The average value of this wave, taken over a period of one cycle, is a measure of the rate of transfer of energy from the generator to the resistor. The

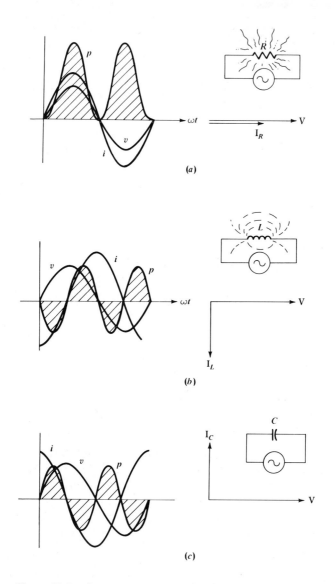

Figure 13-1   Power waves in (*a*) pure resistor circuit; (*b*) pure inductive circuit; (*c*) pure capacitive circuit.

energy delivered to the resistor is released in the form of heat energy. It is *unidirectional*; that is, the heat energy does not reconvert to electrical energy and flow back to the generator. Therefore, it does work, and the energy is said to *be active*. The average rate of transfer of this unidirectional energy is called *active power*, *average power*, *true power*, *actual power*, or *power*.

The active power supplied to a resistor by a sinusoidal driver is numerically equal to the rms voltage across the resistor, multiplied by the rms current through it. Thus,

$$P_R = V_R I_R \qquad (13\text{-}1)$$

The active power may be expressed in joules per second, volt-amperes, or *watts*. The watt is the unit generally used for the volt-amperes of active power. From Ohm's law,

$$I_R = \frac{V_R}{R} \qquad V_R = I_R R$$

Substituting the Ohm's-law relationships into Eq. (13-1) provides two additional power equations:

$$P_R = \frac{V_R^2}{R} \qquad (13\text{-}2)$$

$$P_R = I_R^2 R \qquad (13\text{-}3)$$

13-2   REACTIVE POWER

The wave representing the product of the current wave and the voltage wave for a pure inductive circuit is shown in Fig. 13-1*b*. It is an *alternating power* wave whose frequency is double that of the driving voltage. Referring to Fig. 13-1*b*, when the power wave is positive the generator is storing energy in the magnetic field of the inductance, and when the power wave is negative the stored energy is returned to the generator. Thus the average value of the power wave taken over a period of one cycle is zero.

$$P_{L,\,av} = 0$$

The volt-amperes represented by the product of the rms voltage across the inductor times the rms current through the inductor is called the *reactive power drawn by the inductor*. Thus,

$$Q_L = V_L I_L \qquad (13\text{-}4)$$

where Q_L is the reactive power drawn by an inductor. Reactive power may be expressed in joules per second, reactive volt-amperes, or *vars*. The var (volt-amperes reactive) is the unit generally used in reactive power measurement. From Ohm's law,

$$I_L = \frac{V_L}{X_L} \qquad V_L = IX_L$$

Substituting the Ohm's-law relationships into Eq. (13-4) provides two additional reactive-power equations:

$$Q_L = \frac{V_L^2}{X_L} \tag{13-5}$$

$$Q_L = I_L^2 X_L \tag{13-6}$$

The reactive power drawn by an inductor is called lagging volt-amperes or lagging vars because it is associated with a lagging current.

Alternating-current motors, transformers, and other apparatus that make use of magnetic fields draw lagging vars from a generator.

The wave representing the product of the current wave and the voltage wave for a pure capacitive circuit is shown in Fig. 13-1c. Like that for the inductor, the average value of the power wave for the capacitor, taken over a period of one cycle, is zero.

$$P_{C,\text{av}} = 0$$

Similarly, the volt-amperes represented by the product of the rms voltage across the capacitor and the rms current to the capacitor is called the *reactive power drawn by the capacitor*. Thus,

$$Q_C = V_C I_C \tag{13-7}$$

where Q_C is the reactive power drawn by a capacitor and may be expressed in joules per second, reactive volt-amperes, or vars. From Ohm's law,

$$I_C = \frac{V_C}{X_C} \qquad V_C = I_C X_C$$

Substituting the Ohm's-law relationships into Eq. (13-7) provides two additional reactive-power equations:

$$Q_c = \frac{V_c^2}{X_c} \tag{13-8}$$

$$Q_c = I_c^2 X_c \tag{13-9}$$

The reactive power drawn by a capacitor is called leading volt-amperes or leading vars because it is associated with a leading current.

13-3 ACTIVE AND REACTIVE POWER IN AN ELECTRIC CIRCUIT

The current, voltage, and power waves for a circuit containing resistance as well as reactance are shown in Fig. 13-2a. The active power is equal to the average value of the power wave and may be obtained by averaging the negative and positive areas with due regard to their sign.

Since neither the inductor nor the capacitor draws active power, the total active power expended in either the series or the parallel circuit is that drawn by the resistor alone. Hence the total active power is

$$P_T = V_R I_R \tag{13-10}$$

With reference to the phasor diagrams of Fig. 13-2b and c,

For the parallel circuit	For the series circuit
$I_R = I_T \cos \theta_T$	$V_R = V_T \cos \theta_T$
$V_R = V_T$	$I_R = I_T$

Substituting into Eq. (13-10),

$P_T = V_T I_T \cos \theta_T$	$P_T = V_T I_T \cos \theta_T$

Note that the equation for determining the active power drawn by a series circuit is identical to the equation for determining the active power drawn by a parallel circuit. Furthermore, the same equation applies to any series-parallel combination: this must be so, because any series-parallel combination can be reduced to the simple parallel circuit or simple series circuit shown in Fig. 13-2b or c, respectively. For example, the series branch shown

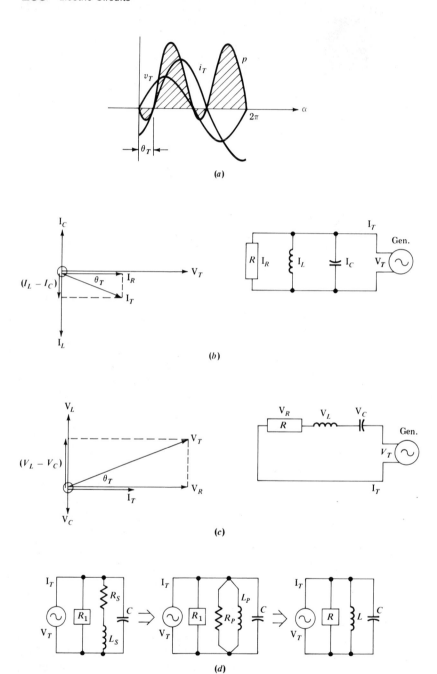

Figure 13-2 (*a*) Current voltage and power waves for a circuit containing *R*, *L*, and *C*; (*b*) parallel circuit; (*c*) series circuit; (*d*) converting a series branch into an equivalent parallel branch.

in Fig. 13-2*d* can be converted into an equivalent parallel section using the procedure outlined in Example 12-7 of Chap. 12, and the resulting paralleled resistors are combined to form the single resistor R. The current I_T in the reduced circuit is identical to the current I_T in the original series-parallel circuit.

Thus the active power drawn by any two-wire circuit *from a sinusoidal generator* is

$$P_T = V_T I_T \cos \theta_T \qquad (13\text{-}11)$$

where V_T = driving voltage at the input terminals of the circuit, V
  I_T = current at the input terminals of the circuit, A
  P_T = active power supplied to the circuit, W
  θ_T = phase angle, measured *from* the current $\mathbf{I}_T$ *to* the voltage phasor $\mathbf{V}_T$. A counterclockwise measurement is a positive angle, and a clockwise measurement is a negative angle.

The reactive power drawn by the circuit in Fig. 13-2*b* and *c* is a combination of lagging vars and leading vars. A comparison of the power wave for the inductance in Fig. 13-1*b* with the power wave for the capacitance in Fig. 13-1*c* shows that the reactive power drawn by the inductance is positive when the reactive power drawn by the capacitance is negative, and vice versa.

Reactive power supplied by a generator to a load is considered positive if the load is inductive, and negative if the load is capacitive.† In those applications where the load is composed of inductive and capacitive components, the net reactive power supplied by the generator is expressed by

$$Q_T = |Q_L| - |Q_C| \qquad (13\text{-}12)$$

Substituting Eqs. (13-4) and (13-7) into Eq. (13-12),

$$Q_T = V_L I_L - V_C I_C \qquad (13\text{-}13)$$

---

† *IEEE Standard Dictionary of Electrical and Electronic Terms*, IEEE STD 100–1977, 2d ed., John Wiley & Sons, Inc., New York, 1977.

With reference to the phasor diagrams in Fig. 13-2b and c.

For the parallel circuit	For the series circuit
$V_L = V_C = V_T$	$I_L = I_C = I_T$

Substituting into Eq. (13-13) and factoring,

$Q_T = V_T(I_L - I_C)$	$Q_T = I_T(V_L - V_C)$
From Fig. 13-2b,	From Fig. 13-2c
$I_L - I_C = I_T \sin \theta_T$	$V_L - V_C = V_T \sin \theta_T$
Hence,	Hence,
$Q_T = V_T I_T \sin \theta_T$	$Q_T = V_T I_T \sin \theta_T$

Note that the equation for determining the total reactive power drawn by a series circuit is identical to the equation for determining the total reactive power drawn by a parallel circuit. Furthermore, the same equation may also be used for any series-parallel combination. Thus the reactive power drawn by any two-wire circuit from a *sinusoidal generator* is

$$Q_T = V_T I_T \sin \theta_T \qquad (13\text{-}14)$$

where V_T = driving voltage at the input terminals of the circuit, V
I_T = current at the input terminals of the circuit, A
Q_T = reactive power supplied to the circuit, var
θ_T = phase angle measured *from* the current phasor $\mathbf{I}_T$ to the voltage phasor $\mathbf{V}_T$

A counterclockwise measurement is a positive angle, and a clockwise measurement is a negative angle. Failure to consider the direction of angular measurement may introduce serious errors in reactive-power calculations.

13-4 APPARENT POWER

Examination of Eqs. (13-11) and (13-14) indicates that they may be represented as the two legs of a right triangle whose hypotenuse is $V_T I_T$. This is shown in Fig. 13-3a for an inductive load and is called a *power diagram*. The power diagram for a capacitive load is shown in Fig. 13-3b.

The *hypotenuse of the power diagram is called the apparent power, and is equal to the right-angle summation of the active and reactive components.* Thus, from the power diagram,

$$S_T = V_T I_T \tag{13-15}$$

$$S_T = \sqrt{P_T^2 + Q_T^2} \tag{13-16}$$

$$\theta_T = \cos^{-1} \frac{P_T}{S_T} \tag{13-17}$$

Note that angle θ_T is the angle between the $\mathbf{V}_T$ phasor and the $\mathbf{I}_T$ phasor. From Ohm's law

$$I_T = \frac{V_T}{Z_{\text{in}}} \quad \text{and} \quad V_T = I_T Z_{\text{in}}$$

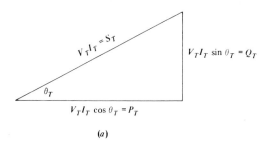

(a)

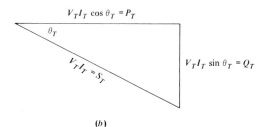

(b)

Figure 13-3 Power diagram: (a) lagging pf; (b) leading pf.

Figure 13-4   Power-factor meter.

Substituting the Ohm's-law relationships into Eq. (13-15),

$$S_T = \frac{V_T^2}{Z_{\text{in}}}$$

(13-18)

$$S_T = I_T^2 Z_{\text{in}}$$

(13-19)

where Z_{in} = input impedance of the two-wire circuit.

Since Q_T, P_T, and S_T are products of volts times amperes, it is correct to express each in terms of volt-amperes (VA). The volt-amperes of active power are generally called watts, and the volt-amperes of reactive power are generally called vars. However, there is no special designator for the volt-amperes of apparent power. The volt-ampere rating of electrical apparatus, unless otherwise specified, is always given as apparent power.

13-5   POWER FACTOR

The ratio of the active power drawn by a circuit to its total volt-ampere input is called the power factor (pf) of the circuit. Thus the power factor measured at the terminals of the circuit is

$$pf_T = \frac{P_T}{S_T}$$

(13-20)

The power factor of a circuit, whether leading or lagging, may be measured with a power-factor meter, shown in Fig. 13-4. The power factor may have values ranging between zero and one. At the high end of the scale are such circuit elements as resistors, incandescent lamps, electric ranges, and similar heating equipment that draw only active power; for these circuits, the apparent power and the active power are the same, and the power factor is unity (1). Approaching the low end of the scale are such circuit elements as capacitors and inductors. A ceramic capacitor draws essentially zero active power; hence its power factor is almost zero. Power factor is sometimes expressed in percent, with unity power factor equal to 100 percent.

Substituting Eqs. (13-11) and (13-15) into Eq. (13-20), and simplifying,

$$pf_T = \frac{V_T I_T \cos \theta_T}{V_T I_T}$$

$$\boxed{pf_T = \cos \theta_T}$$

(13-21)

or

$$\boxed{\theta_T = \cos^{-1}(pf_T)}$$

Thus, as indicated in Eqs. (13-21), the power factor is equal to the cosine of the phase angle between the V_T and I_T phasors. Angle θ_T is called the power factor angle.

The power factor for a *simple series circuit* and for a simple parallel circuit may be determined from the circuit resistance and the input impedance. Thus, referring to Fig. 13-2*b* and *c*,

For the simple parallel circuit

$$pf_T = \frac{P_T}{S_T} = \frac{V_T^2/R}{V_T^2/Z_{in}}$$

$$pf_T = \frac{Z_{in}}{R}$$

For the simple series circuit

$$pf_T = \frac{P_T}{S_T} = \frac{I_T^2 R}{I_T^2 Z_{in}}$$

$$pf_T = \frac{R}{Z_{in}}$$

However, regardless of the circuit arrangement, series, parallel, or series-parallel combination, the power factor can always be determined from Eq. (13-20) or Eq. (13-21).

The power factor of a circuit, system, or motor is a measure of its effectiveness in utilizing the apparent power that it draws from the generator. Power factor must not be confused with efficiency. For example, some of the active power drawn by an electric motor is lost in heat because of the $I^2 R$ effect in the conductors, hysteresis and eddy-current losses in the iron, bearing friction, and windage. Hysteresis loss is a heat loss caused by the alternating movement of magnetic domains in the iron, and eddy-current loss is a heat loss caused by an alternating current induced in the iron. Both are generated by the alternating magnetic field of the motor coils.

As a result of these internal losses, the useful output power of the motor is less than the input power.

$$P_{out} = P_{in} - P_{losses}$$

The efficiency of a motor is the ratio of the useful active power output from the motor shaft to the total active power input to the motor. That is,

$$\text{Efficiency} = \frac{P_{\text{out}}}{P_{\text{in}}} \qquad\qquad (13\text{-}22)$$

The shaft output power of a motor is generally expressed in horsepower (hp); this must be converted to watts for use in Eq. (13-22). The conversion factor is

$$1 \text{ hp} = 746 \text{ W}$$

The power factor of a motor is *the ratio of the active power drawn by the motor to the apparent power drawn by the motor.*

$$pf = \frac{P_{\text{in}}}{S_{\text{in}}} \qquad\qquad (13\text{-}23)$$

Note the difference between Eq. (13-23) and Eq. (13-22).

The nameplate rating of a motor specifies the current that it will draw at rated voltage, rated frequency, and rated temperature when delivering rated shaft horsepower to a load.

Example 13-1 A 20-Ω resistor, a 50-μF capacitor, and a coil whose inductance and resistance are 50 mH and 8.0 Ω, respectively, are connected in parallel and supplied by a 440-V 60-Hz system as shown in Fig. 13-5a. Calculate (*a*) the current to each branch; (*b*) the feeder current; (*c*) the system active power; (*d*) the system reactive power; (*e*) the system apparent power; (*f*) the system power factor and power-factor angle.

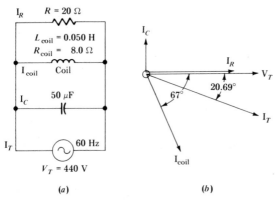

(*a*)   (*b*)

Figure 13-5 Circuit diagram and phasor diagram for Example 13-1.

Solution

(a) $\mathbf{I}_R = \dfrac{\mathbf{V}_T}{R} = \dfrac{440\underline{/0°}}{20\underline{/0°}} = 22\underline{/0°}\ \text{A}$

$\mathbf{Z}_{\text{coil}} = R + jX_L = 8 + j(2\pi)(60)(0.05)$

$\mathbf{Z}_{\text{coil}} = 8 + j18.85 = 20.48\underline{/67.0°}\ \Omega$

$\mathbf{I}_{\text{coil}} = \dfrac{\mathbf{V}_T}{\mathbf{Z}_{\text{coil}}} = \dfrac{440\underline{/0°}}{20.48\underline{/67.0°}} = 21.48\underline{/-67°}\ \text{A}$

$X_C = \dfrac{1}{2\pi f C} = \dfrac{1}{2\pi 60(50 \times 10^{-6})} = 53.05\ \Omega$

$\mathbf{I}_C = \dfrac{\mathbf{V}_C}{-jX_C} = \dfrac{440\underline{/0°}}{53.05\underline{/-90°}} = 8.29\underline{/90°}\ \text{A}$

(b) $\mathbf{I}_T = \mathbf{I}_R + \mathbf{I}_{\text{coil}} + \mathbf{I}_C$

$\mathbf{I}_T = 22\underline{/0°} + 21.48\underline{/-67.0°} + 8.29\underline{/90°}$

$\mathbf{I}_T = (22 + j0) + (8.39 - j19.77) + (0 + j8.29) = 32.49\underline{/-20.69°}\ \text{A}$

(c) The phasor diagram is shown in Fig. 13-5b.

$P_T = V_T I_T \cos\!\measuredangle \quad = 440(32.49)\cos(+20.69°)$

$P_T = 13{,}373.62\ \text{W}$

(d) $Q_T = V_T I_T \sin\!\measuredangle \quad = 440\,(32.49)\sin(+20.69°)$

$Q_T = 5050.80\ \text{var}$

(e) $S_T = \sqrt{P_T^2 + Q_T^2} = \sqrt{(13{,}373.62)^2 + (5050.80)^2}$

$S_T = 14{,}295.60\ \text{V A}$

(f) $pf = \dfrac{P_T}{S_T} = \dfrac{13{,}373.62}{14{,}295.60} = 0.9355$

$\theta_T = \cos^{-1} 0.9355 = 20.7°$

Example 13-2 A coil whose resistance and inductance are 40 Ω and 0.20 H, respectively, is supplied by a 120-V DC *generator*. Calculate (a) the circuit impedance; (b) the circuit current; (c) the total active power; (d) the circuit power factor and the power-factor angle; (e) the circuit var; (f) the phasor diagram.

Solution

The circuit diagram is shown in Fig. 13-6a. Direct current is considered the special case of alternating current with the frequency approaching 0 Hz. Thus,

(a) $X_L = 2\pi f L = 2\pi(0)(0.2) = 0 \; \Omega$

$$Z_T = \sqrt{R^2 + X_L^2} = \sqrt{(40)^2 + 0^2} = 40 \; \Omega$$

(b) $I_T = \dfrac{V_T}{Z_T} = \dfrac{120}{40} = 3 \; \text{A}$

(c) $P_T = I_R^2 R = (3)^2 40 = 360 \; \text{W}$

(d) $S_T = V_T I_T = 120 \times 3 = 360 \; \text{VA}$

$$pf_T = \dfrac{P_T}{S_T} = \dfrac{360}{360} = 1$$

$$\theta_T = \cos^{-1}(pf) = \cos^{-1}(1) = 0°$$

(e) $Q_T = V_T I_T \sin \theta_T = 120 \times 3 \times \sin 0° = 0 \; \text{var}$

(f) The phasor diagram is shown in Fig. 13-6b.

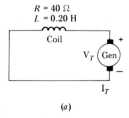

(a)

(b)

Figure 13-6 Circuit diagram and phasor diagram for Example 13-2.

3-6   QUADERGY

The energy that is locked in the system, cycling back and forth between the generator and an inductor, or the generator and a capacitor, is called *quadergy* and is equal to

$$K = Qt \qquad (13\text{-}24)$$

where K = quadergy, var · s
$\qquad Q$ = reactive power, var
$\qquad t$ = time, s

The time t must equal one or more complete periods or must be quite long compared with the time of one period. If the reactive power is in kilovars and the time is in hours, the quadergy is expressed in kilovar-hours (kvar · h).

3-7   POWER-FACTOR IMPROVEMENT

A generator feeding power to a load, at a power factor other than unity, must supply a greater value of kVA than would be otherwise supplied if the load operated at unity power factor. For example, the 16-kW load in Fig. 13-7a is operating at 0.8 pf lagging and draws an apparent power of

$$S = \frac{P}{\text{pf}} = \frac{16}{0.8} = 20 \text{ kVA}$$

If the same 16-kW load were operating at unity power factor, it would draw

$$S = \frac{P}{\text{pf}} = \frac{16}{1} = 16 \text{ kVA}$$

The reactive power drawn by the 20-kVA load at 0.8 pf lagging is

$$S_T = \sqrt{P_T^2 + Q_T^2}$$

$$20 = \sqrt{(16)^2 + Q_T^2}$$

$$Q_T = 12 \text{ kvar lagging}$$

If a 12-kvar capacitor is connected in parallel with the load as shown in Fig. 13-7b, the total kilovars supplied by the generator will be

$$Q_T = Q_L - Q_C$$

$$Q_T = 12 - 12 = 0$$

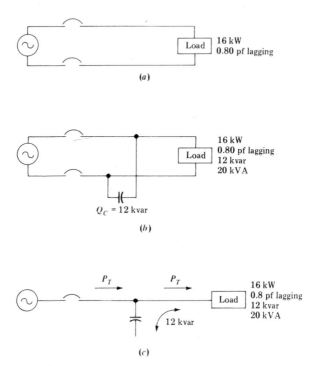

Figure 13-7 (a) System with a lagging pf load;
(b) capacitor used to modify the system pf; (c) one-line
diagram of system.

Under such conditions, where $Q_L = Q_C$, no quadergy is supplied by the generator; the quadergy that was previously supplied to the inductor by the generator is now supplied by the capacitor. Thus, in AC circuits containing inductance and capacitance, *inductance is considered to be a consumer of quadergy, and capacitance is considered to be a source of quadergy.* The one-line diagram in Fig. 13-7c is an abbreviated drawing that indicates the energy or power flow between connected components. Although it is not a circuit diagram (no return lines are shown), it provides all the significant information about the circuit. One-line diagrams are particularly useful in three-phase systems where one line represents three or four conductors that feed many wye-connected and delta-connected loads.

Installing capacitors for the purpose of relieving the generator of its requirement to supply volt-amperes for the lagging power needs of an inductive circuit reduces the apparent power demand on the generator and improves the circuit power factor. If the capacitor rating is large enough to supply all the quadergy drawn by the inductance of the circuit, the power factor will be raised to unity; if it supplies only part of the quadergy, the power factor will not be unity but will be of a higher value than before the capacitor was installed.

Although power-factor improvement (also called power-factor correction) may be accomplished by the addition of series- or parallel-connected capacitors, the parallel connection has the advantage of not affecting the current through the other parallel-connected branches.

Another advantage realized by power-factor improvement is the reduction in line current, and hence a reduction in I^2R losses, IR drops, and IX_L drops in the distribution lines. Excessive voltage drops in the connecting lines have an adverse effect on the performance of electrical apparatus; motors develop less torque, heaters take a longer time to provide the desired heat energy, fluorescent lights flicker, and incandescent lamps are not as bright. When used for pf improvement, *properly sized capacitors raise the voltage at the point of their installation by reducing the voltage drops in the lines. Hence, for maximum benefit they should be installed at the load rather than at the generator.*

Example 13-3 A 220-V, $1\frac{1}{2}$-hp, single-phase, 60-Hz induction motor draws 7.6 A when operating at rated conditions. The efficiency at rated conditions is 85 percent. Calculate (a) the kVA input; (b) the kW input; (c) the power factor; (d) the capacitance of a parallel-connected capacitor that will cause the system to operate at unity power factor: (e) the current supplied by the generator after the capacitor is installed; (f) the motor current after the capacitor is installed; (g) the capacitor current.

Solution
The circuit diagram and phasor diagram are shown in Fig. 13-8a and b, respectively.

(a) $S_T = V_T I_T = 220(7.6) = 1672 \text{ VA} = 1.672 \text{ kVA}$

(b) Eff. $= \dfrac{P_{out}}{P_{in}}$ $1 \text{ hp} = 746 \text{ W}$

$0.85 = \dfrac{1.5\,(746)}{P_{in}}$

$P_{in} = 1316.5 \text{ W}$

(c) $\text{pf}_T = \dfrac{P_T}{S_T} = \dfrac{1316.5}{1672} = 0.787$

(d) $\theta_T = \cos^{-1} \text{pf} = \cos^{-1} 0.787 = 38.1°$

$Q_T = V_T I_T \sin \theta_T = 220\,(7.6) \sin 38.1° = 1030.8 \text{ var}$

To restore the pf to unity, the capacitor must supply all the vars. Therefore, $Q_C = 1030.8$ var.

$$Q_C = \frac{V_C^2}{X_C}$$

$$1030.8 = \frac{(220)^2}{X_C}$$

$$X_C = 46.95 \ \Omega$$

$$X_C = \frac{1}{2\pi f C}$$

$$46.95 = \frac{1}{2\pi(60)C}$$

$$C = 0.0000565 \ \text{F} = 56.5 \ \mu\text{F}$$

(e) After the capacitor is installed, as shown in Fig. 13-8c, the power factor is unity and the apparent power is equal to the active power. Thus, at pf $= 1$,

$$S_T = P_T = V_T I_T$$

$$1316.5 = 220 \ I_T$$

$$I_T = 5.98 \ \text{A}$$

(f) $I_{\text{motor}} = 7.6 \ \text{A}$.

(g) $\mathbf{I}_C = \dfrac{\mathbf{V}_T}{-jX_C}$

$$\mathbf{I}_C = \frac{220/\underline{0°}}{46.95/\underline{-90°}}$$

$$\mathbf{I}_C = 4.69/\underline{90°} \ \text{A}$$

The phasor diagram corresponding to the circuit in Fig. 13-8c is shown in Fig. 13-8d. As demonstrated in this example, the current to the motor is unaffected by the addition of the parallel-connected capacitor. *The current through each branch of a simple parallel circuit is independent of the current through the other branches and is unaffected by the addition of other parallel-connected elements.*

Example 13-4 A 12-kVA load *A* operating at 0.70 pf lagging and a 10-kVA load *B* operating at 0.80 pf lagging are connected in parallel and supplied by a 440-V 60-Hz generator. Determine (a) the total active power; (b) the total reactive power; (c) the total apparent power; (d) the system power factor; (e) the kvar rating of a parallel-connected capacitor required to raise the system power factor to unity; (f) repeat

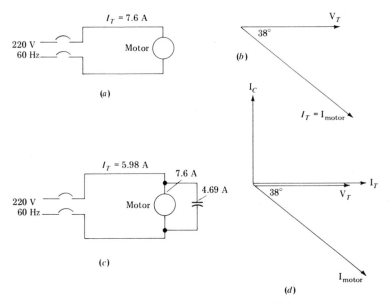

Figure 13-8 Circuit and phasor diagram for Example 13-3.

part (e) for a correction to 0.90 pf lagging; (g) the kVA and power factor of each load for the corrected conditions in parts e and f.

Solution
The circuit diagram is shown in Fig. 13-9a.

(a) $P_T = P_A + P_B = 12 \times 0.7 + 10 \times 0.8 = 16.4 \text{ kW}$

(b) $\theta_A = \cos^{-1} 0.7 = 45.57°$

$\theta_B = \cos^{-1} 0.8 = 36.87°$

$Q_A = S_A \sin \theta_A = 12 \sin 45.57° = 8.57 \text{ kvar}$

$Q_B = S_B \sin \theta_B = 10 \sin 36.87° = 6.00 \text{ kvar}$

$Q_T = Q_A + Q_B = 8.57 + 6.00 = 14.57 \text{ kvar}$

(c) $S_T = \sqrt{P_T^2 + Q_T^2}$

$S_T = \sqrt{(16.4)^2 + (14.57)^2} = 21.94 \text{ kVA}$

(d) $\text{pf}_T = \dfrac{P_T}{S_T} = \dfrac{16.4}{21.94} = 0.747$

$\theta_T = \cos^{-1} \text{pf}_T = 41.62°$

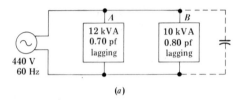

(a)

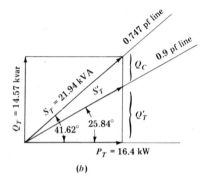

$P_T = 16.4$ kW

(b)

Figure 13-9 Circuit and phasor diagram for Example 13-4.

(e) To obtain unity pf, a capacitor must have a kvar rating equal to the lagging kilovars of the system. Thus,

$Q_C = Q_T = 14.57$ kvar

(f) Figure 13-9b shows the system power diagram. For a system power factor of 0.9 lagging, the new power-factor angle will be

$\theta'_T = \cos^{-1} 0.9 = 25.84°$

Regardless of the value to which the power factor has been corrected, the active power component for a given load will always remain the same. This must be so, because the capacitor draws little or no active power. Hence the total circuit power for all possible power-factor corrections will be

$P_T = 16.4$ kW

The value of Q'_T can be obtained from the power diagram in Fig. 13-9b by using the system active power and the new power-factor angle. Thus,

$$\tan 25.84° = \frac{Q'_T}{16.4}$$

$Q'_T = 7.94$ kvar

As indicated in the power diagram, the kvar rating of the capacitor required to obtain a pf of 0.90 lagging is

$$Q_C = Q_T - Q'_T$$

$$Q_C = 14.57 - 7.94 = 6.63 \text{ kvar}$$

(g) The kVA and power factor of each load are unaffected by the addition of parallel-connected capacitors. Hence the kVA and power factor of each load will remain unchanged: 12 kVA at 0.70 pf and 10 kVA at 0.80 pf.

13-8   ECONOMICS OF POWER-FACTOR IMPROVEMENT

The operation of electrical systems at low power factors necessitates larger power-plant facilities to deliver the same kW, i.e., bigger generators and associated apparatus. In addition, the higher currents drawn at low values of power factor cause greater voltage drops and higher I^2R losses in the connecting lines and apparatus, necessitating conductors of larger cross-sectional area. To encourage the alleviation of such adverse conditions, utility companies institute power-factor clauses in their rate structures, offering reductions in billing in return for a higher power factor at the load.

Power-factor improvement by the addition of capacitors is based on an engineering study of the economics of the particular installation. The rate of return on capacitor investment depends on the type of power-factor clause and the extent to which the power factor deviates from unity. Power-factor improvements up to 90 to 95 percent are generally economically justified.

13-9   PHASOR POWER

As indicated in Eq. (13-16) and illustrated in Fig. 13-3, *the apparent power is a right-angle summation of the active and reactive components.* Thus the components may be expressed in complex numbers and plotted in the complex plane as shown in Fig. 13-10a. Using complex algebra,

$$\mathbf{S}_T = P_T + jQ_T = S_T\underline{/\theta^\circ} \tag{13-25}$$

where $\mathbf{S}$ is defined as *phasor power* (previously called vector power or complex power) and θ is the power-factor angle. For circuits containing both leading and lagging vars,

$$\mathbf{S}_T = P_T + jQ_L - jQ_C \tag{13-26}$$

This is shown in Fig. 13-10b.

If the circuit is predominantly inductive, the net reactive component is positive, and if the circuit is predominantly capacitive, the net reactive component is negative.

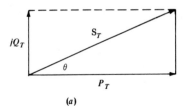

(a)

(b)

Figure 13-10 Phasor-power diagrams.

Phasor power may be obtained directly from the product of *the phasor voltage and the conjugate of the phasor current.*

$$\boxed{S_T = V_T I_T^*}$$

(13-27)

It is called phasor power because it is the product of two phasors. The reason for defining phasor power as $V_T I_T^*$ instead of $V_T I_T$ can be explained by reference to Fig. 13-11. From previously derived equations,

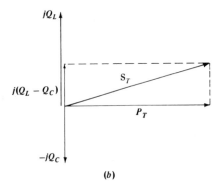

$$P_T = V_T I_T \cos\theta_T = V_T I_T \cos(\alpha° - \beta°)$$

(13-28)

$$Q_T = V_T I_T \sin\theta_T = V_T I_T \sin(\alpha° - \beta°)$$

(13-29)

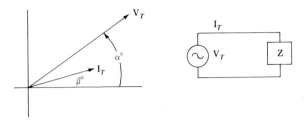

Figure 13-11 Circuit diagram and phasor diagram for justification of Eq. (13-27).

Applying Eq. (13-27) to the phasors in Fig. 13-11,

$$\mathbf{V}_T = V_T\underline{/\alpha^\circ}$$

$$\mathbf{I}_T = I_T\underline{/\beta^\circ}$$

$$\mathbf{I}_T^* = I_T\underline{/-\beta^\circ}$$

$$\mathbf{S}_T = \mathbf{V}_T\mathbf{I}_T^* = (V_T\underline{/\alpha^\circ})(I_T\underline{/-\beta^\circ}) = V_TI_T\underline{/(\alpha^\circ - \beta^\circ)}$$

$$\mathbf{S}_T = V_TI_T\cos(\alpha^\circ - \beta^\circ) + jV_TI_T\sin(\alpha^\circ - \beta^\circ) \qquad (13\text{-}30)$$

Note that the active and reactive components of power obtained through the use of the phasor-power equation are identical to the components obtained in Eqs. (13-28) and (13-29).

If we *erroneously* use $\mathbf{S}_T = \mathbf{V}_T\mathbf{I}_T$ instead of $\mathbf{V}_T\mathbf{I}_T^*$,

$$\left.\begin{aligned}\mathbf{S}_T &= (V_T\underline{/\alpha^\circ})(I_T\underline{/\beta^\circ}) \\[4pt] \mathbf{S}_T &= V_TI_T\underline{/\alpha^\circ + \beta^\circ} \\[4pt] \mathbf{S}_T &= V_TI_T\cos(\alpha^\circ + \beta^\circ) + jV_TI_T\sin(\alpha^\circ + \beta^\circ)\end{aligned}\right\} \quad \text{Wrong} \qquad (13\text{-}31)$$

Comparing the components of Eq. (13-31) with the correct components in Eqs. (13-28) and (13-29) indicates that Eq. (13-31) is *wrong* and must not be used to determine phasor power.

Example 13-5 The voltage across a circuit and the current to it are given by

$$\mathbf{V}_T = 140\underline{/30^\circ}\text{ V}$$

$$\mathbf{I}_T = 7.0\underline{/80^\circ}\text{ A}$$

(a) Using the phasor-power equation, determine the active power, reactive power, apparent power, and power factor; (b) determine the equivalent input impedance; (c) repeat parts a and b for a circuit that draws 10 A from a 120-V DC supply.

Solution

(a) $\mathbf{S}_T = \mathbf{V}_T \mathbf{I}_T^*$

$\qquad \mathbf{S}_T = (140\underline{/30°})(7\underline{/-80°})$

$\qquad \mathbf{S}_T = 980\underline{/-50°}$

$\qquad \mathbf{S}_T = 980 \cos(-50°) + j980 \sin(-50°)$

$\qquad \mathbf{S}_T = 629.93 - j750.72$

Thus,

$S_T = 980$ VA

$P_T = 629.93$ W

$Q_T = -750.72$ var

pf $= \cos(-50°) = 0.64$

The negative sign for the reactive power indicates a leading power factor (net capacitive effect).

(b) $\mathbf{Z}_{\text{in}} = \dfrac{\mathbf{V}_T}{\mathbf{I}_T} = \dfrac{140\underline{/30°}}{7.0\underline{/80°}} = 20\underline{/-50°}$ Ω

$\qquad \mathbf{Z}_{\text{in}} = 12.86 - j15.32$ Ω

Thus the equivalent impedance of the series circuit consists of a 12.86-Ω resistor and a 15.32-Ω capacitive reactance.

(c) In phasor form,

$\mathbf{V}_T = 120\underline{/0°}$ V

$\mathbf{I}_T = 10\underline{/0°}$ A

$\mathbf{S}_T = \mathbf{V}_T \mathbf{I}_T^* = (120\underline{/0°})(10\underline{/0°})^* = 1200\underline{/0°}$ VA

$\mathbf{S}_T = 1200 + j0$

$S_T = 1200$ VA

$P_T = 1200$ W

$Q_T = 0$ var

pf $= \dfrac{P_T}{S_T} = 1$

$\mathbf{Z}_{\text{in}} = \dfrac{\mathbf{V}_T}{\mathbf{I}_T} = \dfrac{120\underline{/0°}}{10\underline{/0°}} = 12\underline{/0°} = (12 + j0)$

Thus the equivalent input impedance of the circuit is 12 Ω resistive.

3-10   MAXIMUM POWER TRANSFER IN A SINUSOIDAL SYSTEM

As discussed in Sec. 4-14, in low-energy DC circuits, such as in very low level generators or amplifiers, the amount of energy transferred to the load is more important than the overall efficiency of the operation. In such circuits, maximum power will be transferred to the load if the resistance of the load is equal to the resistance of the generator. However, if the driver is sinusoidal, as shown in Fig. 13-12, *maximum power will be transferred to the load if the impedance of the load is equal to the conjugate of the generator impedance.* Referring to Fig. 13-12,

$$Z_T = R_G + R_{\text{load}} + jX_{\text{load}} + jX_G$$

where $R_G + jX_G$ = impedance of generator
$R_{\text{load}} + jX_{\text{load}}$ = impedance of load
$$Z_T = \sqrt{(R_G + R_{\text{load}})^2 + (X_{\text{load}} + X_G)^2}$$

The rms current is

$$I_T = \frac{E}{Z_T}$$

The power delivered to the load is

$$P_{\text{load}} = I_T^2 R_{\text{load}} = \left(\frac{E}{Z_T}\right)^2 R_{\text{load}}$$

$$P_{\text{load}} = \frac{E^2 R_{\text{load}}}{(R_G + R_{\text{load}})^2 + (X_{\text{load}} + X_G)^2} \tag{13-32}$$

Examination of Eq. (13-32) indicates that the power to the load will be maximized if

$$X_{\text{load}} = -X_G$$

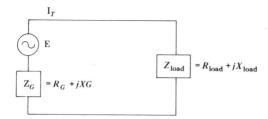

Figure 13-12 Circuit for deriving the maximum-power-transfer relationship.

Selecting X_{load} to be the negative of X_G, Eq. (13-32) becomes

$$P_{load} = \frac{E^2 R_{load}}{(R_G + R_{load})^2} \tag{13-33}$$

As proved before (see Sec. 4-14), Eq. (13-33) will be maximized if $R_{load} = R_G$. Hence, for maximum power to be transferred,

$$R_{load} = R_G$$

$$X_{load} = -X_G$$

Therefore,

$$\mathbf{Z}_{load} = R_G - jX_G$$

or

$$\mathbf{Z}_{load} = \mathbf{Z}_G^*$$

Example 13-6 Assume the generator voltage and generator impedance in Fig. 13-12 are $160\underline{/30°}$ V and $(4.0 + j6.0)$ Ω, respectively. Determine (a) the required load impedance so that maximum power transfer occurs; (b) load current; (c) apparent power, active power, and reactive power drawn by the load.

Solution

(a) $\mathbf{Z}_{load} = \mathbf{Z}_G^* = 4 - j6 = 7.21\underline{/-56.31°}$ Ω

(b) $\mathbf{I}_T = \dfrac{\mathbf{V}_T}{\mathbf{Z}_T} = \dfrac{\mathbf{V}_T}{\mathbf{Z}_G + \mathbf{Z}_{load}} = \dfrac{160\underline{/30°}}{(4+j6)+(4-j6)} = 20\underline{/30°}$ A

(c) $\mathbf{V}_{load} = \mathbf{I}_T \mathbf{Z}_{load} = (20\underline{/30°})(7.21\underline{/-56.31°})$

$\quad \mathbf{V}_{load} = 144.20\underline{/-26.31°}$ V

$\quad \mathbf{S}_{load} = \mathbf{V}_{load} \mathbf{I}_T^* = (144.20\underline{/-26.31°})(20\underline{/-30°})$

$\quad \mathbf{S}_{load} = 2884\underline{/-56.31°} = (1599.75 - j2399.63)$ VA

Thus,

Apparent power = 2884 VA

Active power = 1600 W

Reactive power = 2400 var (leading)

13-11 **MEASUREMENT OF ACTIVE POWER IN THE TWO-WIRE SYSTEM**

Active power may be measured directly with a wattmeter as shown in Fig. 13-13*a*. The current and potential coils shown in Fig. 13-13*a* refer to electromechanical wattmeters. Digital wattmeters (DWM) use electronic circuitry in place of current and potential coils but have the same terminal markings and are connected in the same manner as are electromechanical wattmeters. The two current terminals are connected in series with the load, and the two potential terminals are connected in parallel with the load. The ± markings, or spots of paint, on the wattmeter terminals are polarity marks which indicate that these terminals should connect to the same power line.

The wattmeter indicates a value equal to the product of the voltage across its potential terminals, the current through its current terminals, and the cosine of the angle between the voltage and current waves.

$$P = VI \cos \overset{\textstyle \mathbf{V}}{\underset{\textstyle \mathbf{I}}{\diagdown}}$$

If the angle between the voltage and current waves is large, the wattmeter will indicate a small value of active power, even though the magnitudes of

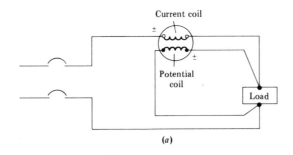

(*a*)

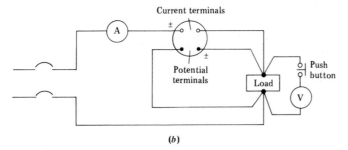

(*b*)

Figure 13-13 Circuits for the measurement of active power in the two-wire system.

current and voltage may be quite high. If electromechanical wattmeters are used, and the current and/or voltage rating of the wattmeter is exceeded, the respective current and voltage coils may overheat and possibly burn out. Hence for circuits of unknown current and voltage, using electromechanical wattmeters, an ammeter and a voltmeter should be used in conjunction with the wattmeter, as shown in Fig. 13-13b. The push button in series with the voltmeter enables the wattmeter to be read without adding the voltmeter load. The button is pushed to obtain the voltmeter reading.

13-12 CORRECTION OF ELECTROMECHANICAL WATTMETER INDICATIONS FOR INTERNAL LOSSES

As shown in Fig. 13-13a, the potential coil of a wattmeter acts as a small load in parallel with the load being measured. Hence the wattmeter indication will include the active power drawn by the potential coil. This loss may be determined by

$$P_{pc} = \frac{V_{pc}^2}{R_{pc}}$$
(13-34)

where P_{pc} = active power consumed by the potential coil of the wattmeter, W

V_{pc} = voltage across the potential coil, V

R_{pc} = resistance of the potential coil, generally printed on the scale card of the wattmeter, Ω

The active power supplied to the load is the wattmeter indication minus the losses in the potential coil.

$$P_{load} = P_{ind} - \frac{V_{pc}^2}{R_{pc}}$$
(13-35)

If the potential coil of the wattmeter is connected across the line, instead of across the load, the voltage across it will be greater than the load voltage by the amount of voltage drop in the current coil. The wattmeter would indicate the active power of the load plus that drawn by the resistance of the current coil. Since the resistance of the current coil is rarely given, this correction cannot always be made. Hence the circuits of Fig. 13-13a and b are the preferred connections.

Note: Digital wattmeter readings do not have to be corrected for instrument losses.

13-13 CURRENT TRANSFORMERS AND THEIR APPLICATION IN CURRENT AND POWER MEASUREMENT

In those AC applications where the current magnitudes exceed the ampere rating of the wattmeter or ammeter, a current transformer (called a C-T) is connected between the circuit and the instruments to provide proportionally smaller currents to the instruments. The meter readings are multiplied by a factor called the *C-T ratio* to obtain the actual values.

Figure 13-14*a* shows the general construction details of a laboratory-type current transformer. The primary and secondary coils are wound around an iron core, and the primary has several taps to provide different C-T ratios. (The principle of transformer action is discussed in Chap. 18.) The ampere rating of the current-transformer secondary must match the ampere ratings of the wattmeter and ammeter. For example, if the ampere rating of the watt-meter is 5 A, which is generally the case, the secondary of the current trans-former must have a 5-A rating. The C-T ratios are expressed as 300/5,

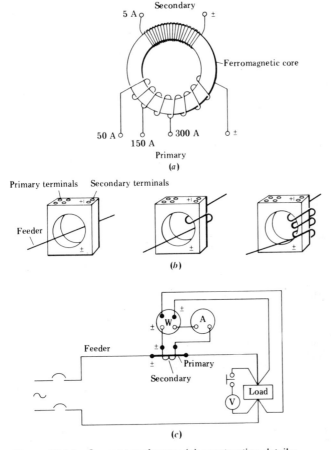

Figure 13-14 Current transformer: (*a*) construction details; (*b*) using the window of a C-T to obtain different ratios; (*c*) standard graphical symbol for a C-T and its application in current and power measurement.

150/5, 50/5, etc. The ratio is read as 300 A to 5 A, 150 A to 5 A, 50 A to 5 A, etc. Higher current ratios are obtained by passing the circuit feeder through the window of the current transformer as shown in Fig. 13-14*b*. *When using the window, the circuit feeder becomes the primary, and the primary terminals are not used.* The current transformer nameplate indicates the C-T ratio for one, two, or four conductors through the window. However, regardless of the primary used, the two secondary terminals are always connected to the meters.

Figure 13-14*c* shows the standard graphical symbol for a current transformer, and its application in current and power measurement. The primary of the current transformer (shown as a heavy line) is connected in series with the load circuit, and the secondary (shown by two loops) is connected in series with the ammeter and the current terminals of the watt-meter. Polarity marks on the current transformer help ensure correct connections. Current-transformer secondaries are provided with a short-circuiting switch to short the secondary terminals when there is no burden on the transformer; the instrument acts as the *burden* or load. With no burden on the secondary, a dangerously high voltage may appear across the secondary terminals; it may also result in excessive heating of the transformer iron, as well as introducing errors caused by permanent magnetization of the iron core.

In a DC system, the power factor is unity, the reactive power is zero, and the active power is equal to the apparent power. Hence most power

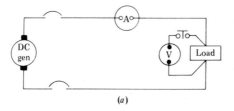

(*a*)

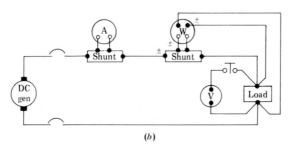

(*b*)

Figure 13-15 (*a*) Power measurement in a 0-Hz system; (*b*) using a shunt to extend the range of the instruments.

measurements in DC systems are generally made indirectly with an ammeter and a voltmeter as shown in Fig. 13-15a. The product of the two meter readings $(V \cdot I)$ is the power. If it is desirable to use a wattmeter in a DC system, the connections are identical to those shown in Fig. 13-13a and b.

For high-current applications in a DC system, current measurement is accomplished with a calibrated resistor of very low resistance, called a *shunt*, in parallel with a DC millivoltmeter as shown in Fig. 13-15b. Power measurement is accomplished with a shunt connected in parallel with the current terminals of the wattmeter. Each instrument and its associated shunt must be a matching set.

Current transformers are designed for use with sinusoidally varying currents and cannot be used in a DC system.

13-14 MEASUREMENT OF REACTIVE POWER IN THE TWO-WIRE SYSTEM

Reactive power may be measured directly with a varmeter. The varmeter is an ordinary wattmeter that uses a phase-shifting network to shift the phase angle of the voltage wave by 90°. Varmeters with internal phase shifters are connected into the circuit in the same manner as are wattmeters, as shown in Figs. 13-13 and 13-14c.

If used without a phase shifter, the wattmeter is calibrated in watts and indicates a value equal to

$$P_T = V_T I_T \cos\!\underset{\mathbf{I}_T}{\overset{\mathbf{V}_T}{\diagup}} \theta_T \qquad \text{W}$$

If used with a 90° lag phase shifter, the wattmeter is calibrated in reactive power and indicates a value equal to

$$Q_T = V_T I_T \cos\!\underset{\mathbf{I}_T}{\overset{(-j)(\mathbf{V}_T)}{\diagup}} \qquad \text{var} \qquad (13\text{-}36)$$

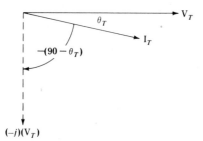

Figure 13-16 Phasor diagram illustrates the effect of a 90° lag phase shifter on the voltage phasor.

where $(-j)(\mathbf{V}_T)$ is the shifted voltage phasor. The $(-j)$ operator rotates the $\mathbf{V}_T$ phasor 90° in the clockwise direction as shown in Fig. 13-16.

From the phasor diagram, measuring the angle *from* phasor $\mathbf{I}_T$ to phasor $(-j\mathbf{V}_T)$, a clockwise measurement,

$$(-j)(\mathbf{V}_T)$$
$$= -(90° - \theta_T^\circ) = (\theta_T^\circ - 90°)$$
$$\mathbf{I}_T$$

Thus Eq. (13-36) becomes

$$Q_T = V_T I_T \cos(\theta_T^\circ - 90°) \tag{13-37}$$

From trigonometry,

$$\cos(\theta_T^\circ - 90°) = \sin \theta_T^\circ \tag{13-38}$$

Substituting Eq. (13-38) into Eq. (13-37) results in

$$Q_T = V_T I_T \sin \theta_T \qquad \begin{matrix} \mathbf{V}_T \\ \\ \mathbf{I}_T \end{matrix} \tag{13-39}$$

Equation (13-39) indicates that a wattmeter with a 90° lag phase shifter in its potential circuit indicates reactive power.

Example 13-7 Assume the ammeter, wattmeter, and voltmeter in Fig. 13-14c indicate 3.50 A, 200 W, and 226 V, respectively, and the C-T ratio is 500/5. If the wattmeter losses are negligible compared with the power drawn by the load, determine the apparent power, active power, reactive power, and power factor of the load.

Solution

$$S = VI = 226(3.5)(\tfrac{500}{5}) = 79,100 \text{ VA} = 79.1 \text{ kVA}$$

$$P = 200(\tfrac{500}{5}) = 20,000 \text{ W} = 20.0 \text{ kW}$$

$$S = \sqrt{P^2 + Q^2}$$

$$79,100 = \sqrt{(20,000)^2 + Q^2}$$

$$Q = 76,529.8 \text{ var} = 76.5 \text{ kvar}$$

$$\text{pf} = \frac{P}{S} = \frac{20,000}{79,100} = 0.25$$

UMMARY OF FORMULAS

$$P_R = V_R I_R = I_R^2 R = \frac{V_R^2}{R}$$

$$Q_L = V_L I_L = I_L^2 X_L = \frac{V_L^2}{X_L}$$

$$Q_C = V_C I_C = I_C^2 X_C = \frac{V_C^2}{X_C}$$

$$Q_T = Q_L - Q_C$$

$$P_T = V_T I_T \cos \theta_T \qquad Q_T = V_T I_T \sin \theta_T$$

$$S_T = V_T I_T = \sqrt{P_T^2 + Q_T^2} = I_T^2 Z_{in} = \frac{V_T^2}{Z_{in}}$$

$$pf_T = \frac{P_T}{S_T} = \cos \theta_T \qquad Eff. = \frac{P_{out}}{P_{in}}$$

$$K = Qt$$

$$S_T = (P_T + jQ_T) = V_T I_T^*$$

PROBLEMS

13-1   A parallel circuit consisting of a 3.0-Ω resistor, a 10-Ω capacitive reactance, and an ideal inductor whose inductive reactance is 6.0 Ω is connected to a 120-V 60-Hz source. Sketch the circuit and determine (a) the circuit impedance; (b) the current through each branch; (c) the feeder current; (d) the total active power input; (e) the total reactive power input; (f) the apparent power input; (g) the power factor.

13-2   A 120-V 60-Hz source supplies energy to a circuit consisting of a 6.0-μF capacitor in series with an inductor whose resistance and inductance are 500 Ω and 2.0 H, respectively. Determine (a) the circuit impedance; (b) the circuit current; (c) the active power; (d) the apparent power; (e) the reactive power; (f) the power factor.

13-3   A 440-V 60-Hz source supplies 20 kVA to a load whose power factor is 70 percent. Sketch the circuit and determine (a) kW; (b) kvars.

13-4   A 120-V 60-Hz generator supplies power to two parallel loads. One load draws 10 kVA at unity power factor, and the other draws 40 kVA at 0.60 pf lagging; Sketch the circuit and determine (a) the total active power; (b) the total reactive power; (c) the total apparent power; (d) the system power factor.

13-5 A capacitor, an electric resistance heater, and an impedance are connected in parallel to a 120-V 60-Hz system. The capacitor draws 50 var, the heater draws 100 W, and the impedance draws 269 VA at a power factor of 0.74 lagging. Sketch the circuit and determine (a) the system active power; (b) the system reactive power; (c) the system apparent power; (d) the system power factor.

13-6 A 250-V 30-Hz generator supplies power to a parallel circuit consisting of a 20-hp motor whose efficiency and power factor are 80 percent and 90 percent lagging, respectively, a 100-kW resistive-type heater, and some unknown impedance. The feeder current supplied by the generator is $648\underline{/23°}$ A. Sketch the circuit and determine (a) the current drawn by the heater; (b) the current drawn by the motor; (c) the current drawn by the unknown impedance; (d) the system active power; (e) the system reactive power; (f) the system apparent power; (g) the system power factor.

13-7 Two parallel-connected loads A and B are supplied by a 440-V 60-Hz generator. Load A draws an apparent power of 100 kVA at 0.80 pf lagging, and load B draws an apparent power of 70 kVA at unity pf. Sketch the circuit and determine (a) the system active power; (b) the system reactive power; (c) the system apparent power; (d) the system power factor; (e) the feeder current; (f) the kW · h of energy supplied by the generator in 20 min; (g) the kvar · h of quadergy supplied by the generator in 20 min.

13-8 An electric circuit draws an active power of 100 kW at 0.8 pf lagging from a 240-V 60-Hz source. Sketch the circuit and determine (a) the apparent power input; (b) the kvar input; (c) the capacitance of a parallel-connected capacitor required to adjust the system power factor to 95 percent.

13-9 A 240-V, 60-Hz, single-phase system supplies energy to a fully loaded 20-hp induction motor. The full-load efficiency and power factor of the motor are 85.5 percent and 74 percent respectively. Determine (a) the kW drawn by the motor; (b) the apparent power; (c) the reactive power; (d) the kilovar rating of a capacitor required to adjust the power factor to 0.90 lagging; (e) the current drawn by the capacitor; (f) the capacitance of the capacitor.

13-10 A 4.0-kW resistor is in parallel with a 10-hp, 0.82-pf, 76 percent efficient motor. The driving voltage is 230 V at 60 Hz, and the motor is operating at rated load. Sketch the circuit and determine (a) the circuit active power; (b) the circuit reactive power; (c) the circuit apparent power; (d) the power factor; (e) the kvars of capacitance required to eliminate the reactive component of current supplied by the generator.

13-11 A 2.0-kW heater is in parallel with a 5.0-hp, 240-V, 60-Hz induction motor whose efficiency and power factor are 0.85 and 0.72, respectively. The parallel combination is supplied by a 240-V 60-Hz system. Sketch the circuit and determine (a) the system active power; (b) the system reactive power; (c) the system apparent power; (d) the kVA rating of a capacitor required to adjust the system power factor to 0.80 leading; (e) the capacitance of the capacitor.

13-12 A 450-V 60-Hz generator supplies power to a 150 kVA 0.80-pf lagging load and a 100-kVA 0.75-pf lagging load. Sketch the circuit and determine (a) the total active power; (b) the total reactive power; (c) the total apparent power; (d) the overall system power factor; (e) the kilovar rating of a capacitor required to adjust the system power factor to 0.90 lagging; (f) the capacitance of the capacitor.

13-13 A 20-hp 450-V motor operating at rated load has an efficiency of 85 percent and a power factor of 0.76 lagging. The motor is in parallel with a 30-kVA load whose power factor angle is 35° lagging. The two loads are supplied from a 450-V 60-Hz source. Sketch the circuit and determine (a) the active power and reactive power drawn by the motor; (b) the active power and reactive power drawn by the 30-kVA load; (c) the total active power supplied by the system; (d) the total reactive power supplied by the system; (e) the total apparent power supplied by the system; (f) the kvars of capacitance that must be paralleled with the system in order to adjust the power factor to 0.90 lagging; (g) the capacitance of the capacitor.

13-14 The current and driving voltage to a single-phase motor are $50/-30°$ A and $100/0°$ V, respectively. Determine (a) the apparent power; (b) the active power; (c) the reactive power; (d) the power factor.

13-15 A 220-V 60-Hz generator supplies power to an $11/20°$ Ω impedance. Determine (a) the current; (b) the apparent power; (c) the active power; (d) the reactive power.

13-16 A voltage of $100/40°$ V appears across an impedance when it passes a current of $30/-50°$ A. Sketch the circuit and determine (a) the apparent power drawn by the impedance; (b) the active power; (c) the reactive power; (d) the power factor.

13-17 Oscilloscope recordings indicate that the 60-Hz voltage across an impedance and the current through it are $100/20°$ V and $10/60°$ A, respectively. Sketch the circuit and determine (a) the apparent power; (b) the active power; (c) the reactive power; (d) the power factor; (e) the equivalent series circuit parameters (R, L, C) that make up an equivalent impedance.

13-18 The voltage and current to a certain induction motor are 100 V and 30 A, respectively, and the phase angle between the current and the voltage is 40°. Sketch the phasor diagram and determine (a) the apparent power; (b) the active power; (c) the reactive power; (d) the power factor; (e) the equivalent impedance of the load.

13-19 Oscilloscope measurements of the voltage across an impedance and the current to it are $208/30°$ V and $40/10°$ A, respectively. Determine (a) the active and reactive power drawn by the impedance; (b) the power factor.

13-20 A certain impedance draws 300 W and 600 lagging var from a 100-V 25-Hz supply. Determine (a) the current; (b) the equivalent values of R and L that make up the impedance.

13-21 A 25-Hz generator supplies a total of 60 kW and 20 kvar to a circuit containing two impedances (A and B) in parallel. The driving voltage is $120/0°$ V and the current to A is $60/30°$ A. Sketch the circuit and determine (a) the apparent power drawn by A; (b) the apparent power drawn by B; (c) the current drawn by B; (d) the overall system power factor.

13-22 The phasor power supplied by a 240-V 25-Hz generator to a series connection of ideal circuit elements is $300/20°$ VA. Sketch the circuit and determine (a) the complex impedance of the series circuit; (b) the equivalent values of the circuit elements that comprise the series circuit.

13-23 Determine the required components (R, L, C) for a load that will maximize the transfer of power from a 120-V 60-Hz source. The source impedance is $5.0/30°$ Ω.

13-24 A generator whose voltage and internal impedance are $300/60°$ V and $(5.0 + j10)$ Ω, respectively, is connected to a load whose components are such as to provide maximum power transfer. Sketch the circuit and determine (a) the impedance of the load in polar form; (b) the active power drawn by the load.

13-25 A 25-Hz generator supplies power to an induction-motor load. Instrumentation is similar to that shown in Fig. 13-13. If the ammeter indicates 7.3 A, the voltmeter indicates 450 V, and the wattmeter indication (corrected for internal losses) is 2797 W, determine (a) the apparent power; (b) the active power; (c) the power factor; (d) the reactive power; (e) sketch the power diagram and calculate the var rating of a parallel-connected capacitor required to adjust the system power factor to 0.95 lagging; (f) determine the capacitance of the capacitor required in part d.

13-26 Assume the instrument indications for the circuit shown in Fig. 13-14 are wattmeter reading = 45 W, ammeter reading = 0.63 A, voltmeter reading = 122 V, C-T ratio is 5/1. The resistance of the wattmeter potential coil is 10,150 Ω. Determine (a) the apparent power; (b) the wattmeter losses; (c) the active power drawn by the load; (d) the power factor of load; (e) the reactive power drawn by the load; (f) sketch the power diagram and determine the kvars of capacitance required to adjust the system power factor to 0.80 lagging; (g) if the system frequency is 50 Hz, determine the capacitance of the capacitor required in part f.

CHAPTER 14

NETWORKS

An electric network is a combination of circuit elements and sources connected in series, parallel, series-parallel, or other circuit configurations. Although the series circuit, parallel circuit, and series-parallel circuit require only the simple application of Ohm's and Kirchhoff's laws for their solution, the determination of currents and voltage drops in circuits of greater complexity requires more involved techniques such as loop and node analysis.

14-1 TOPOLOGICAL FEATURES OF A NETWORK

The branches, nodes, and loops of a network are the topological features that determine the number of independent equations required for the complete solution of the network currents and voltages.

A *major node, shown in Fig. 14-1, is a junction where one branch divides into two or more branches.* Junctions B, C, and D are major nodes. All other junctions (A, E, F, and G) are *minor nodes.*

The branches are the connecting links between major nodes. The branches between node B and node C are branch $\mathbf{Z}_3$ and the series branch of $\mathbf{Z}_5 + \mathbf{Z}_6$.

A loop is a closed path formed by connected branches. The loops in Fig. 14-1 are loop $BECB$, loop $BCDGB$, and loop $BECDGB$.

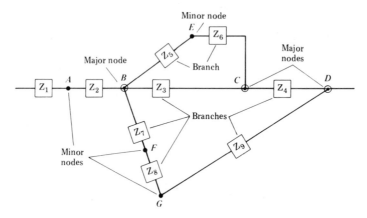

Figure 14-1   Topological features of a network.

14-2   ASSUMED DIRECTIONS OF DRIVING VOLTAGE AND CURRENT

The determination of the branch currents, and the voltage between any two junctions, in a simple series or parallel circuit is relatively easy. However, if the network is complicated and contains more than one source, the determination of branch currents and branch voltages requires careful attention to the relative directions of the currents and driving voltages in the system.

One method of indicating the direction of the driving voltage is through *generator arrows* as shown in Fig. 14-2a.

Another method uses a + mark in place of the point of the arrow as shown in Fig. 14-2b.

The phasor diagram and the associated voltage waves for the generator voltages are shown in Fig. 14-2c and d, respectively.

The equations representing the waves for the three driving voltages are

$$e_a = 30.0\sqrt{2}\, \sin(\omega t + 45°)$$

$$e_b = 100\sqrt{2}\, \sin(\omega t + 30°)$$

$$e_c = 70\sqrt{2}\, \sin(\omega t - 90°)$$

At $t = 0$, the instantaneous voltage developed by each generator is

$$e_a = 30\sqrt{2}\, \sin 45° = 30.0 \text{ V}$$

$$e_b = 100\sqrt{2}\, \sin 30° = 70.71 \text{ V}$$

$$e_c = 70\sqrt{2}\, \sin -90° = -98.99 \text{ V}$$

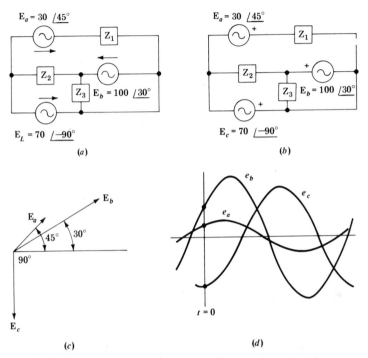

Figure 14-2 Method of indicating the relative directions of driving voltages: (a) arrows; (b) plus (+) mark; (c) phasor diagram; (d) associated voltage waves.

Thus, at time-zero, the generator terminal *corresponding to the point of the generator arrow* in Fig. 14-2a, or the + mark in Fig. 14-2b, is +30 V for generator a, +70.71 V for generator b, and −98.99 V for generator c.

4-3   VOLTAGE AND CURRENT SOURCES

Every generator, or battery, provides both current and voltage outputs to a load. If the voltage output is relatively constant over the allowable range of current, it is generally classified as a voltage source. If the current output is relatively constant over the allowable range of voltage, it is generally classified as a current source.

Voltage sources are used extensively in almost all circuit applications; current sources have significant usage in transistor-circuit analysis, node analysis of networks, Millman's theorem, and Norton's theorem.

Voltage sources such as batteries or generators have internal impedances, called the source impedance, that cause a change in the output voltage when the source is supplying current to a circuit. The source impedance of a battery or DC generator causes an internal IR drop, and the source impedance of a sinusoidal generator causes an internal $\mathbf{IZ}$ drop.

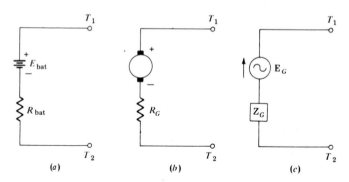

Figure 14-3 Equivalent circuit models for voltage sources.

Equivalent circuit models for voltage sources are shown in Fig. 14-3. Each model includes an *ideal constant-voltage generator or ideal constant-voltage battery in series with the source impedance. The ideal generator maintains a constant-voltage amplitude and waveshape regardless of the amount of current it supplies to the circuit.* The source impedance accounts for the change in voltage at the $T_1 T_2$ terminals of the generator when it supplies current to the circuit.

If the impedance of the voltage source is negligible compared with the input impedance of the circuit, the impedance of the source may be neglected without introducing any significant error. However, in those applications where the impedance of the source is not negligible, it must be accounted for in the circuit analysis.

Example 14-1 A 100-V DC source with an internal resistance of 2.0 Ω is connected through a switch to an 18-Ω load as shown in Fig. 14-4. Determine the voltage at terminals $T_1 T_2$ with the switch closed and with the switch open.

Solution
With the switch closed, the current in the circuit is

$$I = \frac{V}{R} = \frac{100}{2 + 18} = 5 \text{ A}$$

The voltmeter reads the IR drop across the 18-Ω resistor:

$$V_{T_1 T_2} = IR = 5 \times 18 = 90 \text{ V}$$

The internal voltage drop due to the resistance of the generator is

$$V_{\text{drop}} = IR_G = 5 \times 2 = 10 \text{ V}$$

With the switch open, there can be no current. Hence there can be no voltage drop across the internal resistance of the generator, and the voltmeter will read 100 V.

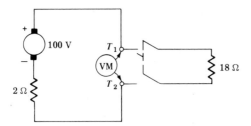

Figure 14-4 Circuit for Example 14-1.

Example 14-2 Determine the voltage at the source terminals $T_1 T_2$ in Fig. 14-5, with the switch closed and with the switch open. Assume a high-impedance voltmeter that draws insignificant current.

Solution
With the switch closed,

$$\mathbf{Z}_T = 10\underline{/30°} + 4\underline{/-47°}$$

$$\mathbf{Z}_T = (8.66 + j5.00) + (2.73 - j2.93)$$

$$\mathbf{Z}_T = (11.39 + j2.07) = 11.58\underline{/10.32°} \ \Omega$$

$$\mathbf{I} = \frac{\mathbf{V}}{\mathbf{Z}_T} = \frac{50\underline{/0°}}{11.58\underline{/10.32°}} = 4.32\underline{/-10.32°} \text{ A}$$

The voltmeter will read the magnitude of the IZ drop across the load.

$$V = IZ = 4.32 \times 4 = 17.28 \text{ V}$$

With the switch open, there can be no current. Hence there can be no voltage drop across the internal impedance of the generator, and the voltmeter will read 50 V.

A *current source,* shown in Fig. 14-6, *consists of an ideal constant-current generator in parallel with the source impedance. The ideal constant-current generator will deliver a constant amplitude and constant waveshape of current*

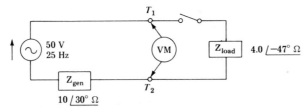

Figure 14-5 Circuit for Example 14-2.

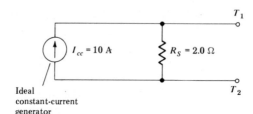

Ideal
constant-current
generator

Figure 14-6 Current source.

regardless of the circuit to which it is connected. The arrow on the symbol for the ideal constant-current generator indicates the direction of current.

The constant-current generator shown in Fig. 14-6 supplies a total of 10 A regardless of whether or not a load is connected to terminals $T_1 T_2$. Thus, in Fig. 14-6, all 10 A pass through the 2-Ω internal resistor. The voltage across the 2-Ω resistor is

$$V = 10 \times 2 = 20 \text{ V}$$

If the same current source is connected to a 4-Ω resistor, as shown in Fig. 14-7, the constant 10-A output is shared by the two paralleled resistors. The current drawn by each resistor may be determined by the current-divider equation:

$$\mathbf{Y_4} = \tfrac{1}{4} = 0.25 \text{ S}$$

$$\mathbf{Y_2} = \tfrac{1}{2} = 0.5 \text{ S}$$

$$\mathbf{Y_T} = \mathbf{Y_4} + \mathbf{Y_2} = 0.25 + 0.5 = 0.75 \text{ S}$$

$$\mathbf{I_4} = \frac{\mathbf{Y_4}}{\mathbf{Y_T}} \mathbf{I_T} = \frac{0.25}{0.75} (10) = \frac{1}{3} (10) \text{ A}$$

$$\mathbf{I_2} = \frac{\mathbf{Y_2}}{\mathbf{Y_T}} \mathbf{I_T} = \frac{0.50}{0.75} (10) = \frac{2}{3} (10) \text{ A}$$

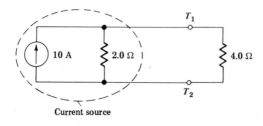

Current source

Figure 14-7 Current source connected to a resistor load.

The voltage across $T_1 T_2$ is equal to the IR drop across the 4-Ω resistor

$$V_{T_1 T_2} = IR = \tfrac{1}{3}(10)(4) = 13.3 \text{ V}$$

A current source is equivalent to a voltage source if, and only if, the respective open-circuit voltages are equal and the respective short-circuit currents are equal. To ensure equality, when converting from a voltage source to a current source, and vice versa, use Fig. 14-8a or b as appropriate, and the following source-conversion formula:

$$\boxed{\mathbf{I}_{cc} = \frac{\mathbf{E}_{cv}}{\mathbf{Z}_S}} \qquad (14\text{-}1)$$

where $\mathbf{I}_{cc}$ = constant current, A
   $\mathbf{V}_{cv}$ = constant voltage, V
   $\mathbf{Z}_S$ = source impedance (series for voltage source, parallel for current source), Ω

Equivalent voltage and current sources are equivalent at only the external terminals $T_1 T_2$; they produce identical terminal voltages at no load and identical terminal voltages and currents when connected to a load. However, on open circuit (no load connected to terminals $T_1 T_2$), the voltage source consumes no energy, whereas the current source consumes energy at the rate of $I^2 R$ W in the internal source impedance.

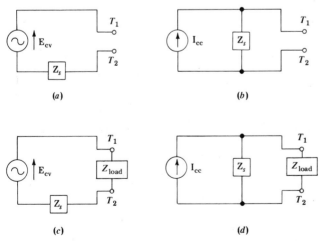

Figure 14-8   (a) Voltage source; (b) current source; (c) voltage source with load; (d) current source with load.

Derivation of Equation (14-1)

Figure 14-8c and d shows respective voltage and current sources connected to identical load impedances $\mathbf{Z}_{\text{load}}$. The two sources have identical impedances $\mathbf{Z}_S$. The current supplied to the load impedance by the respective sources are:

For the voltage source

$$\mathbf{I}_{\text{load}} = \frac{\mathbf{E}_{cv}}{\mathbf{Z}_S + \mathbf{Z}_{\text{load}}}$$

For the current source

Using the current-divider equation,

$$\mathbf{I}_{\text{load}} = \mathbf{I}_{cc} \frac{\mathbf{Y}_{\text{load}}}{\mathbf{Y}_S + \mathbf{Y}_{\text{load}}}$$

$$\mathbf{I}_{\text{load}} = \mathbf{I}_{cc} \frac{1/\mathbf{Z}_{\text{load}}}{1/\mathbf{Z}_S + 1/\mathbf{Z}_{\text{load}}}$$

$$\mathbf{I}_{\text{load}} = \frac{I_{cc}}{\mathbf{Z}_{\text{load}}} \frac{\mathbf{Z}_S \cdot \mathbf{Z}_{\text{load}}}{\mathbf{Z}_S + \mathbf{Z}_{\text{load}}}$$

For the two load currents to be equal,

$$\frac{\mathbf{E}_{cv}}{\mathbf{Z}_S + \mathbf{Z}_{\text{load}}} = \frac{\mathbf{I}_{cc}}{\mathbf{Z}_{\text{load}}} \frac{\mathbf{Z}_S \cdot \mathbf{Z}_{\text{load}}}{\mathbf{Z}_S + \mathbf{Z}_{\text{load}}}$$

Simplifying,

$$\mathbf{E}_{cv} = \mathbf{I}_{cc} \mathbf{Z}_S$$

or

$$\mathbf{I}_{cc} = \frac{\mathbf{E}_{cv}}{\mathbf{Z}_S}$$

Example 14-3 A 120-V voltage source with an internal resistance of 3.0 Ω is shown in Fig. 14-9a. Determine the parameters for an equivalent current source.

Solution
Using the source-conversion formula,

$$\mathbf{I}_{cc} = \frac{\mathbf{E}_{cv}}{\mathbf{Z}_S} = \frac{120}{3} = 40 \text{ A}$$

$\mathbf{Z}_S = 3.0$-Ω resistance in parallel with the constant-current generator. Figure 14-9b shows the equivalent current source.

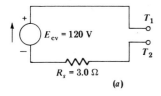

(a)

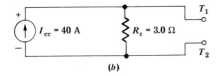

(b)

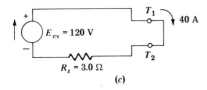

(c)

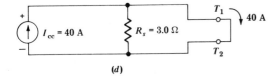

(d)

Figure 14-9 Circuits for Example 14-3, and for determining the equivalence of current and voltage sources.

The current source and the voltage source, shown in Fig. 14-9a and b, are equivalent at the $T_1 T_2$ terminals. This equivalence can be proved by calculating the open-circuit voltage across terminals $T_1 T_2$ for both sources, and then calculating the short-circuit current at terminals $T_1 T_2$ for both sources. If the open-circuit voltages are equal, and the short-circuit currents are equal, the sources are equivalent at the terminals. The calculations for the circuits in Fig. 14-9 follow.

(a) Open-Circuit Calculation, Voltage Source

With no load connected to the $T_1 T_2$ terminals of the voltage source (Fig. 14-9a), there can be no current and thus no IR drop. Hence the voltage across the $T_1 T_2$ terminals is 120 V.

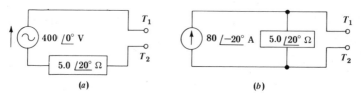

Figure 14-10 Circuits for Example 14-4: (*a*) equivalent voltage source; (*b*) equivalent source.

(b) Open-Circuit Calculation, Current Source

With no load connected to the $T_1 T_2$ terminals of the current source (Fig. 14-9*b*), the total 40 A of the constant-current generator will pass through the 3-Ω resistor, resulting in an IR drop equal to $40 \times 3 = 120$ V. Hence the voltage across the $T_1 T_2$ terminals is 120 V.

(c) Short-Circuit Calculation, Voltage Source

A very low impedance jumper (short circuit) connected across terminals $T_1 T_2$ of the voltage source (Fig. 14-9*c*) will cause a current at the terminals equal to $I = V/R = 120/3 = 40$ A.

(d) Short-Circuit Calculation, Current Source

A jumper (short circuit) connected across terminals $T_1 T_2$ of the current source (Fig. 14-9*d*) will cause all 40 A to pass through the jumper, bypassing the 3-Ω resistor. Hence the current at the terminals will be 40 A.

Calculations (*a*), (*b*), (*c*), and (*d*) prove the equivalence of the current and voltage sources in Fig. 14-9.

Example 14-4 A 400-V 60-Hz generator, shown in Fig. 14-10*a*, has an internal impedance of $5.0/\underline{20°}$ Ω. Determine the parameters for an equivalent current source.

Solution
Using the source-conversion formula,

$$\mathbf{I}_{cc} = \frac{\mathbf{E}_{cv}}{\mathbf{Z}_S} = \frac{400/\underline{0°}}{5/\underline{20°}} = 80/\underline{-20°} \text{ A}$$

The equivalent current source is shown in Fig. 14-10*b*.

14-4   LOOP-CURRENT ANALYSIS OF NETWORKS

Loop-current analysis, also called mesh-current analysis, is a method that uses Kirchhoff's voltage law to obtain a set of simultaneous equations that, when solved, can be used to determine the magnitudes and phase angles of the branch currents.

The circuit shown in Fig. 14-11a will be used to illustrate and develop the loop-analysis technique.

I_1 = current in Z_1

I_2 = current in Z_2

I_3 = current in Z_3

I_4 = current in Z_4

I_5 = current in Z_5

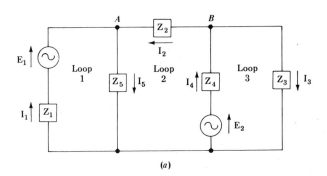

(a)

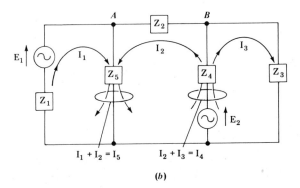

(b)

Figure 14-11 Circuits used to illustrate the loop-current analysis of networks.

Applying Kirchhoff's voltage law to each of the three loops in Fig. 14-11a,

For loop 1,

$$E_1 = I_1 Z_1 + I_5 Z_5$$

For loop 2, (14-2)

$$E_2 = I_4 Z_4 + I_2 Z_2 + I_5 Z_5$$

For loop 3,

$$E_2 = I_4 Z_4 + I_3 Z_3$$

Applying Kirchhoff's current law to node A and node B in Fig. 14-11a,

For node A,

$$I_5 = I_1 + I_2$$

(14-3)

For node B,

$$I_4 = I_2 + I_3$$

Substituting equation set (14-3) into equation set (14-2),

$$E_1 = I_1 Z_1 + (I_1 + I_2)Z_5$$

$$E_2 = I_2 Z_2 + (I_1 + I_2)Z_5 + (I_2 + I_3)Z_4$$ (14-4)

$$E_2 = I_3 Z_3 + (I_2 + I_3)Z_4$$

Rearranging equation set (14-4),

$$E_1 = (Z_1 + Z_5)I_1 + Z_5 I_2$$

$$E_2 = Z_5 I_1 + (Z_2 + Z_4 + Z_5)I_2 + Z_4 I_3$$ (14-5)

$$E_2 = (Z_3 + Z_4)I_3 + Z_4 I_2$$

Solving the three simultaneous equations in equation set (14-5) will determine the branch currents I_1, I_2, I_3; substituting these values into equation set (14-3) will determine the remaining branch currents I_4 and I_5. Figure 14-11b, along with equation set (14-3), shows how currents I_1, I_2, and I_3 combine to form the branch currents in impedances Z_4 and Z_5. The currents I_1, I_2, and I_3 in Fig. 14-11*b* are called the *loop currents* or *mesh currents*.

GUIDELINES FOR THE EFFICIENT WRITING OF LOOP EQUATIONS

The following guidelines will help minimize "bookkeeping" errors, avoid errors in sign, ensure the selection of independent loops, and permit writing the loop equations in the proper format for solution by determinants or by computer.

The network in Fig. 14-12*a* will be used as a guide for the development of the loop-current technique. Although this is a simple three-loop network, it is adequate for instructional purposes.

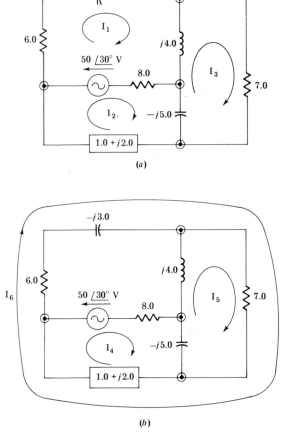

Figure 14-12   Circuits used in the preparation of guidelines for the efficient writing of loop equations.

Procedure

1. Convert all current sources, if any, to voltage sources. *If a current source is ideal (no parallel impedance), source conversion formulas cannot be used, and the procedure outlined in Sec. 14-4 should be followed.*
2. Circle all major nodes.
3. Determine the number of branches.
4. Determine the required number of independent loop equations from the following equation:

$$l = (b - n) + 1 \tag{14-6}$$

where l = number of independent loops

b = number of branches

n = number of major nodes

For Fig. 14-12*a*,

$n = 4$
$b = 6$

$l = (6 - 4) + 1 = 3$ loops or meshes

5. Select the loops, and enter on the network diagram, an arbitrarily assigned direction for each loop current. *Each loop must contain at least one impedance or one generator that is not in any of the other loops.* This is necessary so that all the loop equations will be independent equations. Although the choice of loop-current direction will not affect the solution, once chosen, they should not be changed for the duration of the problem. Changing the direction of the assumed loop currents while in the process of solving the problem will introduce an error. The arbitrary directions assigned to the loop currents in Fig. 14-12*a* are all clockwise.
6. Using Kirchhoff's voltage law, write the loop equations *while traveling in the assigned directions around the respective loops:*
   (*a*) When summing the driving voltages, assign a (+) sign to every driving voltage whose arrow or subscripts indicate that it is in the same direction as the assumed loop current; assign a minus (−) sign to driving voltages that are opposite in direction to the assumed loop current.
   (*b*) When summing voltage drops, assign a plus (+) sign to all *currents* in the **IZ** drops that are in the same direction as the assumed loop current; assign a minus (−) sign to all currents in the **IZ** drops that are opposite in direction to the assumed loop current.

Thus, for loop 1 in Fig. 14-12a,

$$+50\underline{/30°} = (+\mathbf{I}_1)(6) + (+\mathbf{I}_1)(-j3)$$
$$+ (+\mathbf{I}_1 - \mathbf{I}_3)(j4) + (+\mathbf{I}_1 - \mathbf{I}_2)(8) \quad (14\text{-}7)$$

Note that the **IZ** drop across $j4$ is caused by the difference between loop currents $\mathbf{I}_1$ and $\mathbf{I}_3$, and the **IZ** drop in the 8-Ω resistor is caused by the difference between loop currents $\mathbf{I}_1$ and $\mathbf{I}_2$.

For loop 2,

$$-50\underline{/30°} = (+\mathbf{I}_2 - \mathbf{I}_1)(8) + (+\mathbf{I}_2 - \mathbf{I}_3)(-j5) + (+\mathbf{I}_2)(1 + j2) \quad (14\text{-}8)$$

For loop 3,

$$0 = (+\mathbf{I}_3)(7) + (+\mathbf{I}_3 - \mathbf{I}_2)(-j5) + (+\mathbf{I}_3 - \mathbf{I}_1)(j4) \quad (14\text{-}9)$$

Rearranging the loop equations,

Loop 1

$$50\underline{/30°} = \mathbf{I}_1[14 + j1] + \mathbf{I}_2[-8] + \mathbf{I}_3[-j4] \quad (14\text{-}10)$$

Loop 2

$$-50\underline{/30°} = \mathbf{I}_1[-8] + \mathbf{I}_2[9 - j3] + \mathbf{I}_3[j5] \quad (14\text{-}11)$$

Loop 3

$$0 = \mathbf{I}_1[-j4] + \mathbf{I}_2[j5] + \mathbf{I}_3[7 - j1] \quad (14\text{-}12)$$

The format shown in Table 14-1 will enable the loop equations to be written *by inspection*, and at the same time, ensure that all current contributions to the voltage drops in each loop will be considered.

Table 14-1 Loop-current format

Loop	Driving voltage	Voltage drops				
1	[ ] = + [	]$\mathbf{I}_1$ + [	]$\mathbf{I}_2$ + [	]$\mathbf{I}_3$ + $\cdots$ + [	]$\mathbf{I}_n$
2	[ ] = + [	]$\mathbf{I}_1$ + [	]$\mathbf{I}_2$ + [	]$\mathbf{I}_3$ + $\cdots$ + [	]$\mathbf{I}_n$
3	[ ] = + [	]$\mathbf{I}_1$ + [	]$\mathbf{I}_2$ + [	]$\mathbf{I}_3$ + $\cdots$ + [	]$\mathbf{I}_n$
$\vdots$	$\vdots$	$\vdots$	$\vdots$	$\vdots$	$\vdots$
n	[ ] = + [	]$\mathbf{I}_1$ + [	]$\mathbf{I}_2$ + [	]$\mathbf{I}_3$ + $\cdots$ + [	]$\mathbf{I}_n$

Table 14-2 Loop equations for Fig. 14-12a

Loop	Loop equation
1	$[+(50\underline{/30°})] = [6 - j3 + j4 + 8]\mathbf{I}_1 + [-(8)]\mathbf{I}_2 + [-(j4)]\mathbf{I}_3$
2	$[-(50\underline{/30°})] = [-(8)]\mathbf{I}_1 + [1 + j2 + 8 - j5]\mathbf{I}_2 + [-(-j5)]\mathbf{I}_3$
3	$[\quad 0 \quad] = [-(j4)]\mathbf{I}_1 + [-(-j5)]\mathbf{I}_2 + [j4 + 7 - j5]\mathbf{I}_3$

The application of the loop-current format to the three loops in Fig. 14-12a is shown in Table 14-2. The procedure for entering the data is as follows:

1. Prepare a blank format for the three loops using Table 14-1 as a guide.
2. (a) Sum the driving voltages in loop 1 and enter it in the voltage column; if any driving voltage is opposite in direction to loop current $\mathbf{I}_1$, multiply that driving voltage by (-1).
   (b) Sum all the impedances in *loop 1* and enter it as the coefficient of current $\mathbf{I}_1$.
   (c) Sum all the impedances in *loop 1* that carries current $\mathbf{I}_2$, and enter it as the coefficient of current $\mathbf{I}_2$; if $\mathbf{I}_2$ is opposite in direction to loop current $\mathbf{I}_1$, multiply the coefficient by (-1).
   (d) Sum all the impedances in *loop 1* that carries current $\mathbf{I}_3$ and enter it as the coefficient of $\mathbf{I}_3$; if $\mathbf{I}_3$ is opposite in direction to loop current $\mathbf{I}_1$, multiply the coefficient by (-1).
3. Use a similar procedure for *loop 2*.
   (a) Sum the driving voltages in *loop 2* and enter it in the voltage column; if the driving voltage is opposite in direction to loop current $\mathbf{I}_2$, multiply the driving voltage by (-1).
   (b) Sum all the impedances that make up *loop 2* and enter it as the coefficient of current $\mathbf{I}_2$.
   (c) Sum all the impedances in *loop 2* that carries current $\mathbf{I}_1$ and enter it as the coefficient of current $\mathbf{I}_1$; if $\mathbf{I}_1$ is opposite in direction to loop current $\mathbf{I}_2$, multiply the coefficient by (-1).
   (d) Sum all the impedances in *loop 2* that carries current $\mathbf{I}_3$, and enter it as the coefficient of $\mathbf{I}_3$; if $\mathbf{I}_3$ is opposite in direction to loop current $\mathbf{I}_2$, multiply the coefficient by (-1).
4. Follow the same procedure for loop 3.

Although the description of the procedure is rather verbose, the actual filling out of the table is relatively rapid. Furthermore, the loop-current format makes it easy to recheck the equations to ensure accuracy before proceeding with the solution.

Simplifying the equations in Table 14-2,

$$50\underline{/30°} = (14 + j1)\mathbf{I}_1 + (-8)\mathbf{I}_2 + (-j4)\mathbf{I}_3$$

$$-50\underline{/30°} = (-8)\mathbf{I}_1 + (9 - j3)\mathbf{I}_2 + (j5)\mathbf{I}_3 \qquad (14\text{-}13)$$

$$0 = (-j4)\mathbf{I}_1 + (j5)\mathbf{I}_2 + (7 - j1)\mathbf{I}_3$$

If the current in only one branch of a network is to be determined, the calculations can be simplified by judicious selection of loops; the loops should be selected so that the branch of interest will be threaded by only one loop current. For example, if the current in the 8-Ω branch of Fig. 14-12a is to be determined, loop currents $\mathbf{I}_1$ and $\mathbf{I}_2$ would have to be calculated and the branch current determined from the phasor difference $\mathbf{I}_1 - \mathbf{I}_2$ or $\mathbf{I}_2 - \mathbf{I}_1$. However, if the loop selections are as shown in Fig. 14-12b only $\mathbf{I}_4$ need be calculated.

14-6 **SOLUTION OF NETWORK EQUATIONS BY DETERMINANTS**

The method of determinants is a straightforward systematic method for solving network equations through manipulation of its coefficients. To develop the technique, assume the following network equations were written in *standard loop-current* format, using Table 14-1 as a guide:

$$\mathbf{E}_{G1} = \mathbf{Z}_1\mathbf{I}_1 + \mathbf{Z}_2\mathbf{I}_2$$

$$\mathbf{E}_{G2} = \mathbf{Z}_3\mathbf{I}_2 + \mathbf{Z}_4\mathbf{I}_2$$

From this set of equations, an ordered array of impedances called the *impedance determinant is formed*:

$$\Delta_\mathbf{z} = \begin{vmatrix} \mathbf{Z}_1 & \mathbf{Z}_2 \\ \mathbf{Z}_3 & \mathbf{Z}_4 \end{vmatrix} \qquad (14\text{-}14)$$

Defining $\Delta_{\mathbf{z}1}$ to be $\Delta_\mathbf{z}$ with its first column replaced by the voltage column, and $\Delta_{\mathbf{z}2}$ to be $\Delta_\mathbf{z}$ with its second column replaced by the voltage column, then, provided that $\Delta_\mathbf{z}$ is not zero,

$$\mathbf{I}_1 = \frac{\Delta_{\mathbf{z}1}}{\Delta_\mathbf{z}} \qquad \Delta_\mathbf{z} \neq 0 \qquad (14\text{-}15)$$

$$\mathbf{I}_2 = \frac{\Delta_{\mathbf{z}2}}{\Delta_\mathbf{z}} \qquad \Delta_\mathbf{z} \neq 0 \qquad (14\text{-}16)$$

The method of solution outlined in Eqs. (14-15) and (14-16) is called Cramer's rule. Equation (14-14) is called a second-order determinant, or a 2 by 2 determinant, because it has 2 rows and 2 columns. The evaluation of a second-order determinant is accomplished by cross multiplication in the following manner:

$$\Delta_\mathbf{Z} = \begin{vmatrix} \mathbf{Z}_1 & \mathbf{Z}_2 \\ \mathbf{Z}_3 & \mathbf{Z}_4 \end{vmatrix} = \mathbf{Z}_1\mathbf{Z}_4 - \mathbf{Z}_3\mathbf{Z}_2 \tag{14-17}$$

Example 14-5   Using determinants, solve the following network equations:

$$10\underline{/0°} = 3.0\mathbf{I}_1 + j4.0\mathbf{I}_2$$

$$8.0\underline{/90°} = j7.0\mathbf{I}_1 + (3.0 + j4.0)\mathbf{I}_2$$

$$\Delta_\mathbf{Z} = \begin{vmatrix} 3 & j4 \\ j7 & (3 + j4) \end{vmatrix} = 3(3 + j4) - (j7)(j4) = (9 + j12) + 28$$

$$\Delta_\mathbf{Z} = 37 + j12 = 38.90\underline{/17.97°}$$

$$\Delta_{\mathbf{Z}1} = \begin{vmatrix} 10 & j4 \\ 8\underline{/90°} & (3 + j4) \end{vmatrix} = 10(3 + j4) - (j8)(j4) = (30 + j40) - j^2 32$$

$$\Delta_{\mathbf{Z}1} = 62 + j40 = 73.78\underline{/32.83°}$$

$$\Delta_{\mathbf{Z}2} = \begin{vmatrix} 3 & 10 \\ j7 & 8\underline{/90°} \end{vmatrix} = 3(j8) - 10(j7) = 24j - 70j$$

$$\Delta_{\mathbf{Z}2} = -46j = 46\underline{/-90°}$$

$$\mathbf{I}_1 = \frac{\Delta_{\mathbf{Z}1}}{\Delta_\mathbf{Z}} = \frac{73.78\underline{/32.83°}}{38.90\underline{/17.97°}} = 1.90\underline{/14.9°} \text{ A}$$

$$\mathbf{I}_2 = \frac{\Delta_{\mathbf{Z}2}}{\Delta_\mathbf{Z}} = \frac{46\underline{/-90°}}{38.90\underline{/17.97°}} = 1.18\underline{/-108°} \text{ A}$$

Three simultaneous equations provide a third-order impedance matrix. Thus,

$$\mathbf{E}_{G1} = \mathbf{Z}_1\mathbf{I}_1 + \mathbf{Z}_2\mathbf{I}_2 + \mathbf{Z}_3\mathbf{I}_3$$

$$\mathbf{E}_{G2} = \mathbf{Z}_4\mathbf{I}_1 + \mathbf{Z}_5\mathbf{I}_2 + \mathbf{Z}_6\mathbf{I}_3$$

$$\mathbf{E}_{G3} = \mathbf{Z}_7\mathbf{I}_1 + \mathbf{Z}_8\mathbf{I}_2 + \mathbf{Z}_9\mathbf{I}_3$$

$$\Delta_\mathbf{Z} = \begin{vmatrix} \mathbf{Z}_1 & \mathbf{Z}_2 & \mathbf{Z}_3 \\ \mathbf{Z}_4 & \mathbf{Z}_5 & \mathbf{Z}_6 \\ \mathbf{Z}_7 & \mathbf{Z}_8 & \mathbf{Z}_9 \end{vmatrix} \tag{14-18}$$

Evaluation of Third-Order Determinants

1. Circle all elements in the left column.
2. Assign a plus mark to the top left element in the array, and then assign alternate minus and plus marks to the remaining elements in the left-hand column as follows:

$$\Delta_Z = \begin{vmatrix} + & Z_1 & Z_2 & Z_3 \\ - & Z_4 & Z_5 & Z_6 \\ + & Z_7 & Z_8 & Z_9 \end{vmatrix} \qquad (14\text{-}19)$$

The circled elements are called *factors*.

3. Form *minor determinants, called minors,* by deleting (in turn) the row and column that contain each circled element. Thus,

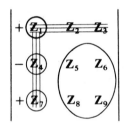   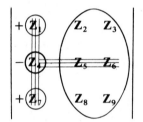

Minor of Z_1 Minor of Z_4

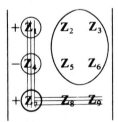

Minor of Z_7

4. The sum of the products of the signed factors with their corresponding minors is the determinant Δ_Z. Thus,

$$\Delta_Z = (+Z_1)\begin{vmatrix} Z_5 & Z_6 \\ Z_8 & Z_9 \end{vmatrix} + (-Z_4)\begin{vmatrix} Z_2 & Z_3 \\ Z_8 & Z_9 \end{vmatrix} + (+Z_7)\begin{vmatrix} Z_2 & Z_3 \\ Z_5 & Z_6 \end{vmatrix} \qquad (14\text{-}20)$$

$$\Delta_Z = Z_1(Z_5 Z_9 - Z_8 Z_6) - Z_4(Z_2 Z_9 - Z_8 Z_3) + Z_7(Z_2 Z_6 - Z_5 Z_3) \qquad (14\text{-}21)$$

Example 14-6 Determine (*a*) the rms current drawn by the $(1 + j2)$-Ω impedance using the loops selected in Fig. 14-12*a*; (*b*) repeat part *a* using the loops selected in Fig. 14-12*b*; (*c*) the rms voltage drop across the $(1 + j2)$-Ω impedance; (*d*) the phasor power, active power, and reactive power drawn by the $(1 + j2)$-Ω impedance, and the power factor of the impedance.

Solution

(*a*) The loop equations for Fig. 14-12*a*, as previously determined [equation set (14.13)], are

$$50\underline{/30^\circ} = (14 + j1)\mathbf{I}_1 + (-8)\mathbf{I}_2 + (-j4)\mathbf{I}_3$$

$$-50\underline{/30^\circ} = (-8)\mathbf{I}_1 + (9 - j3)\mathbf{I}_2 + (j5)\mathbf{I}_3$$

$$0 = (-j4)\mathbf{I}_1 + (j5)\mathbf{I}_2 + (7 - j1)\mathbf{I}_3$$

$$
\Delta_{\mathbf{z}} = \begin{array}{c} + \\ - \\ + \end{array}
\begin{vmatrix}
(14 + j1) & -8 & -j4 \\
-8 & (9 - j3) & j5 \\
-j4 & j5 & (7 - j1)
\end{vmatrix}
$$

$$\Delta_{\mathbf{z}} = +(14 + j1)[(9 - j3)(7 - j1) - (j5)(j5)] - (-8)[(-8)(7 - j1) - (j5)(-j4)]$$

$$+ (-j4)[(-8)(j5) - (9 - j3)(-j4)]$$

$$\Delta_{\mathbf{z}} = 676\underline{/-28.16^\circ}$$

The current in the $(1 + j2)$-Ω impedance is loop current $\mathbf{I}_2$.

$$
\Delta_{\mathbf{z}2} = \begin{array}{c} + \\ - \\ + \end{array}
\begin{vmatrix}
(14 + j1) & 50\underline{/30^\circ} & -j4 \\
-8 & -50\underline{/30^\circ} & j5 \\
-j4 & 0 & (7 - j1)
\end{vmatrix}
$$

$$\Delta_{\mathbf{z}2} = (14 + j1)[(-50\underline{/30^\circ})(7 - j1) - (0)(j5)] - (-8)[(50\underline{/30^\circ})(7 - j1)$$

$$-0(-j4)] + (-j4)[(50\underline{/30^\circ})(j5) - (-50\underline{/30^\circ})(-j4)]$$

$$\Delta_{\mathbf{z}2} = 1950\underline{/-148.53^\circ}$$

$$\mathbf{I}_2 = \frac{\Delta_{\mathbf{z}2}}{\Delta_{\mathbf{z}}} = \frac{1950\underline{/-148.53^\circ}}{676\underline{/-28.16^\circ}} = 2.88\underline{/-120.37^\circ} \text{ A}$$

The rms current is 2.88 A.

(b) Based on the loop selections in Fig. 14-12b, the current in the $(1 + j2)$ impedance is the phasor sum of loop current $\mathbf{I}_4$ and loop current $\mathbf{I}_6$. Using the standard loop-current format, the loop equations for Fig. 14-12b are

Loop

4    $[-50\underline{/30°}] = [8 - j5 + 1 + j2]\mathbf{I}_4 + [-(-j5)]\mathbf{I}_5 + [1 + j2]\mathbf{I}_6$

5    $[0] \quad = [-(-j5)]\mathbf{I}_4 + [j4 + 7 - j5]\mathbf{I}_5 + [7]\mathbf{I}_6$

6    $[0] \quad = [1 + j2]\mathbf{I}_4 + [7]\mathbf{I}_5 + [6 - j3 + 7 + 1 + j2]\mathbf{I}_6$

Simplifying,

$$-50\underline{/30°} = (9 - j3)\mathbf{I}_4 + (j5)\mathbf{I}_5 + (1 + j2)\mathbf{I}_6$$

$$0 = (j5)\mathbf{I}_4 + (7 - j1)\mathbf{I}_5 + (7)\mathbf{I}_6$$

$$0 = (1 + j2)\mathbf{I}_4 + (7)\mathbf{I}_5 + (14 - j1)\mathbf{I}_6$$

$$\Delta_{\mathbf{z}} = \begin{matrix} + \\ - \\ + \end{matrix} \begin{vmatrix} (9 - j3) & j5 & (1 + j2) \\ (j5) & (7 - j1) & 7 \\ (1 + j2) & 7 & (14 - j1) \end{vmatrix} = 676\underline{/-28.16°}$$

$$\Delta_{\mathbf{z4}} = \begin{matrix} + \\ - \\ + \end{matrix} \begin{vmatrix} (-50\underline{/30°}) & j5 & (1 + j2) \\ 0 & (7 - j1) & 7 \\ 0 & 7 & (14 - j1) \end{vmatrix} = 2619.58\underline{/-173.63°}$$

$$\Delta_{\mathbf{z6}} = \begin{matrix} + \\ - \\ + \end{matrix} \begin{vmatrix} (9 - j3) & j5 & -50\underline{/30°} \\ j5 & (7 - j1) & 0 \\ (1 + j2) & 7 & 0 \end{vmatrix} = 1188.46\underline{/-37.75°}$$

$$\mathbf{I}_6 = \frac{\Delta_{\mathbf{z6}}}{\Delta_{\mathbf{z}}} = \frac{1188.46\underline{/-37.75°}}{676\underline{/-28.16°}} = 1.76\underline{/-9.59°} \text{ A}$$

$$\mathbf{I}_4 = \frac{\Delta_{\mathbf{z4}}}{\Delta_{\mathbf{z}}} = \frac{2619.58\underline{/-173.63°}}{676\underline{/-28.16°}} = 3.88\underline{/-145.47°} \text{ A}$$

The current in the $(1 + j2)$-Ω impedance is

$$\mathbf{I} = \mathbf{I_4} + \mathbf{I_6} = 3.88\underline{/-145.47^\circ} + 1.76\underline{/-9.59^\circ}$$

$$\mathbf{I} = (-3.20 - j2.20) + (1.74 - j0.29)$$

$$\mathbf{I} = -1.46 - j2.49 = 2.886\underline{/-120.38^\circ} \text{ A}$$

$I_{\text{rms}} = 2.886$ A

A review of parts a and b of this example should prove that, although the choice of loops does not affect the solution, it does affect the amount of mathematical manipulation required to obtain the solution.

$\cdot$ (c) The voltage drop across the $(1 + j2)$ impedance is

$$\mathbf{V} = \mathbf{IZ}$$

$$\mathbf{V} = 2.886\underline{/-120.38^\circ} (1 + j2) = (2.886\underline{/-120.38^\circ})(2.236\underline{/63.43^\circ})$$

$$\mathbf{V} = 6.453\underline{/-56.95^\circ} \text{ V}$$

The rms voltage drop across the impedance is 6.45 V.

(d) The phasor power is

$$\mathbf{S} = \mathbf{VI^*} = (6.453\underline{/-56.95^\circ})(2.886\underline{/+120.38^\circ})$$

$$\mathbf{S} = 18.623\underline{/63.43^\circ} \text{ V A}$$

$$\mathbf{S} = (8.33 + j16.65) \text{ V A}$$

thus $\quad P = 8.3$ W

$\qquad Q = 16.6$ var lagging

$\qquad$ pf $= \cos 63.4^\circ = 0.45$ lagging

Note: The component loop currents that make up the current to an impedance must not be used in I^2R or I^2X formulas to calculate the respective active and reactive power drawn by that impedance; it is wrong. The phasor sum of the currents in an impedance is the actual current in that impedance, and is the only current that can be used in I^2R and I^2X calculations. Thus,

(Correct) $\quad Q = I^2X = (2.886)^2(2) = 16.6$ var

(Wrong) $\quad Q = I_4^2 X + I_6^2 X = (3.88)^2(2) + (1.76)^2(2) = 36.3$

(Correct) $\quad P = I^2R = (2.886)^2(1) = 8.33$ W

(Wrong) $\quad P = I_4^2 R + I_6^2 R = (3.88)^2(1) + (1.76)^2(1) = 18.5$

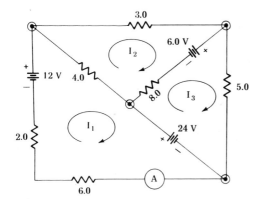

Figure 14-13   Circuit diagram for Example 14-7.

Example 14-7   Using loop analysis, determine the ammeter reading in Fig. 14-13.

Solution
Using the standard loop-current format of Table 14-1:

Loop 1

$$[+12 - 24] = [2 + 4 + 6]\mathbf{I}_1 + [-4]\mathbf{I}_2 + [0]\mathbf{I}_3$$

Loop 2

$$[-6] = [-4]\mathbf{I}_1 + [8 + 4 + 3]\mathbf{I}_2 + [-8]\mathbf{I}_3$$

Loop 3

$$[6 + 24] = [0]\mathbf{I}_1 + [-8]\mathbf{I}_2 + [8 + 5]\mathbf{I}_3$$

Simplifying,

$$-12 = (12)\mathbf{I}_1 + (-4)\mathbf{I}_2 + (0)\mathbf{I}_3$$

$$-6 = (-4)\mathbf{I}_1 + (15)\mathbf{I}_2 + (-8)\mathbf{I}_3$$

$$30 = (0)\mathbf{I}_1 + (-8)\mathbf{I}_2 + (13)\mathbf{I}_3$$

To determine the ammeter reading, solve for loop current $\mathbf{I}_1$

$$\Delta_{\mathbf{z}} = \begin{vmatrix} + & 12 & -4 & 0 \\ - & -4 & 15 & -8 \\ + & 0 & -8 & 13 \end{vmatrix}$$

$$\Delta_z = (12)[(15)(13) - (-8)(-8)] - (-4)[(-4)(13) - (-8)(0)]$$

$$+ 0[(-4)(-8) - (15)(0)]$$

$$\Delta_z = 1572 - 208 = 1364$$

$$\Delta_{z1} = \begin{vmatrix} + & \boxed{-12} & -4 & 0 \\ - & \boxed{-6} & 15 & -8 \\ + & \boxed{30} & -8 & 13 \end{vmatrix}$$

$$\Delta_{z1} = (-12)[(15)(13) - (-8)(-8)] - (-6)[(-4)(13) - (-8)(0)]$$

$$+ 30[(-4)(-8) - (15)(0)]$$

$$\Delta_{z1} = -1572 - 312 + 960 = -924$$

$$I_1 = \frac{\Delta_{z1}}{\Delta_z} = \frac{-924}{1364} = -0.68; \text{ ammeter reads 0.68 A}$$

Example 14-8 Using loop analysis, determine the ammeter reading in Fig. 14-14.

Solution
The number of loop equations, as determined by Eq. 14-6, is

$$l = (b - n) + 1 = (6 - 4) + 1 = 3$$

The selected loops are shown in Fig. 14-14. The loop equations are

Loop 1

$$[100\underline{/90°} - 100\underline{/30°}] = [-j2]I_1 + [0]I_2 + [0]I_3$$

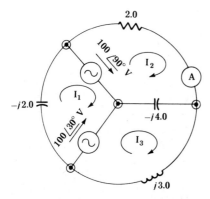

Figure 14-14 Circuit diagram for Example 14-8.

Loop 2

$$[-100\underline{/90°}] = [0]\mathbf{I}_1 + [2 - j4]\mathbf{I}_2 + [-(-j4)]\mathbf{I}_3$$

Loop 3

$$[100\underline{/30°}] = [0]\mathbf{I}_1 + [-(-j4)]\mathbf{I}_2 + [j3 - j4]\mathbf{I}_3$$

To determine the ammeter reading, solve for $\mathbf{I}_2$:

$$\Delta_\mathbf{z} = \begin{vmatrix} + & \boxed{-j2} & 0 & 0 \\ - & \boxed{0} & (2 - j4) & j4 \\ + & \boxed{0} & j4 & (-j1) \end{vmatrix}$$

$$\Delta_\mathbf{z} = (-j2)[(2 - j4)(-j1) - (j4)(j4)] + 0 + 0$$

$$\Delta_\mathbf{z} = (-j2)[12 - j2] = (2\underline{/-90°})(12.17\underline{/-9.46°})$$

$$\Delta_\mathbf{z} = 24.34\underline{/-99.46°}$$

$$\Delta_{\mathbf{z}2} = \begin{vmatrix} + & \boxed{-j2} & 100\underline{/90°} - 100\underline{/30°} & 0 \\ - & \boxed{0} & -100\underline{/90°} & j4 \\ + & \boxed{0} & 100\underline{/30°} & -j1 \end{vmatrix}$$

$$\Delta_{\mathbf{z}2} = (-j2)[(-100\underline{/90°})(-j1) - (100\underline{/30°})(j4)]$$

$$\Delta_{\mathbf{z}2} = (2\underline{/-90°})[-100 - 400\underline{/120°}]$$

$$\Delta_{\mathbf{z}2} = (2\underline{/-90°})[-100 - (-200 + j346.41)] = 721.11\underline{/-163.90°}$$

$$\mathbf{I}_2 = \frac{\Delta_{\mathbf{z}2}}{\Delta_\mathbf{z}} = \frac{721.11\underline{/-163.90°}}{24.34\underline{/-99.46°}} = 29.63\underline{/-64.4°} \text{ A}$$

The ammeter reads 29.6 A.

4-7   NODE-VOLTAGE ANALYSIS OF NETWORKS

Node-voltage analysis of networks is a method that uses Kirchhoff's current law to obtain a set of simultaneous equations that, when solved, will provide information concerning the magnitudes and phase angles of the voltages across each branch. *The node-voltage method has excellent applications in circuits that have common ground connections such as those used in electronic circuits.*

The circuit shown in Fig. 14-15a will be used to illustrate the node-voltage technique. *The number of node equations required is equal to one less than the number of major nodes.* Hence, when selecting the major nodes, omit the node that connects to the greatest number of branches. Thus in Fig. 14-15a, node 3 will not be used; it has the greatest number of branches and is generally the ground node.

When solving problems using node analysis, the node equations will be simplified if all voltage sources are converted to equivalent current sources, as shown in Fig. 14-15b. The voltage source consisting of constant voltage E_{cv1} and source impedance Z_{S1} is converted to constant current I_{cc1} in parallel with Z_{S1}:

$$I_{cc1} = \frac{E_{cv1}}{Z_{S1}}$$

Similarly,

$$I_{cc2} = \frac{E_{cv2}}{Z_{S2}}$$

The circuit shown in Fig. 14-15c is identical to that of Fig. 14-15b, but redrawn to highlight the three major nodes. *The voltage, or potential, at the common node is arbitrarily set at zero, and the voltage at each of the other nodes is measured with respect to the common node.* Thus, referring to Fig. 14-15c, V_1 is the voltage measured *from* node 1 *to* common, and V_2 is the voltage measured *from* node 2 *to* common. Voltage $(V_1 - V_2)$ is the voltage measured *from* node 1 *to* node 2, and $(V_2 - V_1)$ is the voltage measured *from* node 2 *to* node 1.

In accordance with Kirchhoff's current law, the phasor sum of the load currents at a node is equal to the phasor sum of the source currents that supply the node; load currents are currents in the impedances. The polarity of a source current is considered positive if it is directed toward the node and negative if it is directed away from the node. The polarity of all load currents is considered positive. Thus, referring to Fig. 14-15c,

At node 1,

$$I_{cc1} = I_{S1} + I_b + I_c \tag{14-22}$$

At node 2,

$$-I_{cc2} = I_{S2} + I_a + I_c \tag{14-23}$$

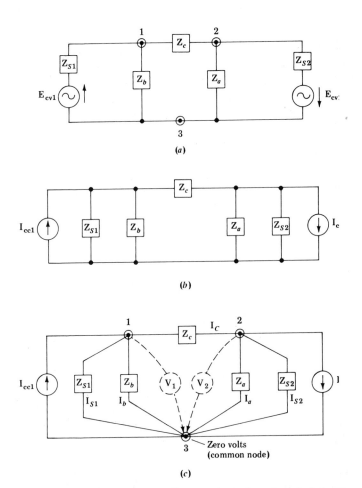

Figure 14-15 Circuits used to illustrate the node-voltage technique.

Expressing the currents in each load in terms of the voltage drops across the respective load impedances,

$$I_{cc1} = \frac{V_1 - 0}{Z_{S1}} + \frac{V_1 - 0}{Z_b} + \frac{V_1 - V_2}{Z_c} = \frac{V_1}{Z_{S1}} + \frac{V_1}{Z_b} + \frac{V_1 - V_2}{Z_c} \qquad (14\text{-}24)$$

$$-I_{cc2} = \frac{V_2 - 0}{Z_{S2}} + \frac{V_2 - 0}{Z_a} + \frac{V_2 - V_1}{Z_c} = \frac{V_2}{Z_{S2}} + \frac{V_2}{Z_a} + \frac{V_2 - V_1}{Z_c} \qquad (14\text{-}25)$$

Note: The potential of the node under consideration is always considered positive with respect to all other nodes. Thus, as viewed from node 1, the load current in Z_c is $(V_1 - V_2)/Z_c$, whereas when viewed from node 2, the load current in Z_c is $(V_2 - V_1)/Z_c$.

14-8   GUIDELINES FOR THE EFFICIENT WRITING OF NODE EQUATIONS

The following guidelines will help minimize "bookkeeping" errors, avoid errors in sign, and permit writing the node equations in the proper format for solution by determinants or by computer. The network in Fig. 14-16*a* will be used as a guide for the development of the nodal technique.

Procedure

1. Circle all major nodes.
2. Select the node with the greatest number of branches to be the common node.
3. Number the remaining nodes 1, 2, 3, etc.
4. Convert all voltage sources to equivalent current sources. *If a voltage source is ideal (no series impedance), source conversion formulas cannot be used, and the procedure outlined in Sec. 14-7 should be followed.*
5. Write Kirchhoff's current equations for each numbered node. Express each load current in terms of its respective node voltage and branch impedance. The polarity of the source current is assumed positive if it is directed toward the node, and negative if it is directed away from the node. The polarity of all load currents is positive.

Referring to Fig. 14-16*a*, node 5 is selected as the common node, and is arbitrarily set at zero volts. Converting the voltage sources to current sources,

$$\mathbf{I}_{cc1} = \frac{50/30°}{-j2} = \frac{50/30°}{2/-90°} = 25/120° \text{ A}$$

$$\mathbf{I}_{cc2} = \frac{75/20°}{j4} = \frac{75/20°}{4/90°} = 18.75/-70° \text{ A}$$

The revised circuit, showing the voltage sources replaced by equivalent current sources, is shown in Fig. 14-16*b*.

Applying Kirchhoff's current law to the four nodes:

For node 1

$$-25/120° = \frac{\mathbf{V}_1}{3} + \frac{\mathbf{V}_1 - \mathbf{V}_2}{-j2} + \frac{\mathbf{V}_1 - \mathbf{V}_4}{\div j5}$$

For node 2

$$25/120° = \frac{\mathbf{V}_2}{2} + \frac{\mathbf{V}_2 - \mathbf{V}_1}{-j2} + \frac{\mathbf{V}_2 - \mathbf{V}_3}{j3}$$

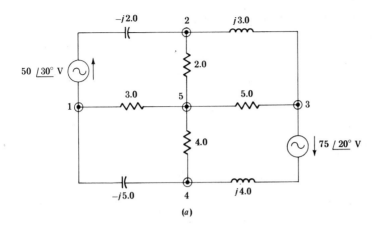

(a)

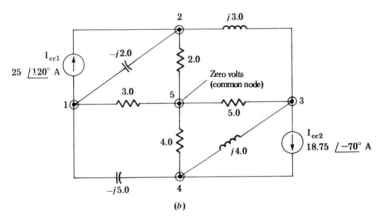

(b)

Figure 14-16 Circuits used in the preparation of guidelines for the efficient writing of node equations.

For node 3

$$-18.75\underline{/-70°} = \frac{V_3}{5} + \frac{V_3 - V_4}{j4} + \frac{V_3 - V_2}{j3}$$

For node 4

$$18.75\underline{/-70°} = \frac{V_4}{4} + \frac{V_4 - V_3}{j4} + \frac{V_4 - V_1}{-j5}$$

Rearranging the node equations,

$$(-25\underline{/120°}) = \left(-\frac{1}{j2} + \frac{1}{3} - \frac{1}{j5}\right)\mathbf{V}_1 + \frac{1}{j2}\mathbf{V}_2 + \frac{1}{j5}\mathbf{V}_4$$

$$(25\underline{/120°}) = \frac{1}{j2}\mathbf{V}_1 + \left(\frac{1}{2} - \frac{1}{j2} + \frac{1}{j3}\right)\mathbf{V}_2 + \left(-\frac{1}{j3}\right)\mathbf{V}_3$$

$$(-18.75\underline{/-70°}) = \left(-\frac{1}{j3}\right)\mathbf{V}_2 + \left(\frac{1}{5} + \frac{1}{j4} + \frac{1}{j3}\right)\mathbf{V}_3 + \left(-\frac{1}{j4}\right)\mathbf{V}_4$$

$$(18.75\underline{/-70°}) = \frac{1}{j5}\mathbf{V}_1 + \frac{-1}{j4}\mathbf{V}_3 + \left(\frac{1}{4} + \frac{1}{j4} - \frac{1}{j5}\right)\mathbf{V}_4$$

Note: The coefficients of the voltages in the preceding equations are the branch admittances.

The format shown in Table 14-3 will permit rapid writing of the node equations, ensure that all node voltages are considered, and eliminate or minimize bookkeeping errors.

The application of the node-voltage format to the four nodes in Fig. 14-16b is shown in Table 14-4. The procedure used to enter the data is as follows:

1. Prepare a blank format for the four nodes using Table 14-3 as a guide.
2. (a) Sum the current sources connected to node 1 and enter it in the source-current column; if a source current is directed away from node 1, multiply it by (-1).
   (b) Sum all *branch admittances* that connect to node 1, and enter it as the coefficient of $\mathbf{V}_1$.
   (c) Sum all branch admittances that connect to *both node 1 and node 2*, and enter it as the coefficient of $\mathbf{V}_2$.
   (d) Sum all branch admittances that connect to *both node 1 and node 3*, and enter it as the coefficient of $\mathbf{V}_3$.

Table 14-3 Node-voltage format

Node	Source current			Load currents		
1	[ ] = + [	]$\mathbf{V}_1$ − [	]$\mathbf{V}_2$ − [	]$\mathbf{V}_3 \cdots$ − [	]$\mathbf{V}_n$
2	[ ] = − [	]$\mathbf{V}_1$ + [	]$\mathbf{V}_2$ − [	]$\mathbf{V}_3 \cdots$ − [	]$\mathbf{V}_n$
3	[ ] = − [	]$\mathbf{V}_1$ − [	]$\mathbf{V}_2$ + [	]$\mathbf{V}_3 \cdots$ − [	]$\mathbf{V}_n$
⋮	⋮	⋮	⋮	⋮	⋮
n	[ ] = − [	]$\mathbf{V}_1$ − [	]$\mathbf{V}_2$ − [	]$\mathbf{V}_3 \cdots$ + [	]$\mathbf{V}_n$

(e) Sum all branch admittances that connect to *both node 1 and node 4*, and enter it as the coefficient of $\mathbf{V}_4$.
3. Use a similar procedure for node 2.
   (a) Sum the current sources connected to node 2 and enter it in the source-current column; if a source current is directed away from node 2, multiply it by (-1).
   (b) Sum all the branch admittances that connect to node 2, and enter it as the coefficient of $\mathbf{V}_2$.
   (c) Sum all the branch admittances that connect to *both node 2 and node 1* and enter it as the coefficient of $\mathbf{V}_1$.
   (d) Sum all the branch admittances that connect to *both node 2 and node 3* and enter it as the coefficient of $\mathbf{V}_3$.
   (e) Sum all the branch admittances that connect to *both node 2 and node 4*, and enter it as the coefficient of $\mathbf{V}_4$.
4. Follow a similar procedure for node 3 and node 4.

Although the description of the procedure is verbose, the actual filling out of the table is relatively simple and can be done by inspection. Using the format illustrated in Tables 14-3 and 14-4 makes it easy to recheck the equations to ensure accuracy before proceeding with the solution.

Example 14-9 Determine the ammeter reading in Fig. 14-17a using (a) node-voltage analysis, and (b) loop-current analysis.

Solution
(a) Node-voltage analysis: Use node-voltage analysis to determine the voltage drop across the 3-Ω resistor. Then use Ohm's law to determine the current. Converting the voltage sources to equivalent current sources,

$$\mathbf{I}_{cc2} = \tfrac{24}{4} = 6 \text{ A} \qquad \mathbf{I}_{cc1} = \tfrac{60}{6} = 10 \text{ A}$$

Table 14-4

Node	Equation
1	$[-25\underline{/120°}] = + \left[\dfrac{1}{-j2} + \dfrac{1}{3} + \dfrac{1}{-j5}\right]\mathbf{V}_1 - \left[\dfrac{1}{-j2}\right]\mathbf{V}_2 - [0]\mathbf{V}_3 - \left[\dfrac{1}{-j5}\right]\mathbf{V}_4$
2	$[25\underline{/120°}] = - \left[\dfrac{1}{-j2}\right]\mathbf{V}_1 + \left[\dfrac{1}{-j2} + \dfrac{1}{2} + \dfrac{1}{j3}\right]\mathbf{V}_2 - \left[\dfrac{1}{j3}\right]\mathbf{V}_3 - [0]\mathbf{V}_4$
3	$[-18.75\underline{/-70°}] = -[0]\mathbf{V}_1 - \left[\dfrac{1}{j3}\right]\mathbf{V}_2 + \left[\dfrac{1}{j3} + \dfrac{1}{5} + \dfrac{1}{j4}\right]\mathbf{V}_3 - \left[\dfrac{1}{j4}\right]\mathbf{V}_4$
4	$[18.75\underline{/-70°}] = - \left[\dfrac{1}{-j5}\right]\mathbf{V}_1 - [0]\mathbf{V}_2 - \left[\dfrac{1}{j4}\right]\mathbf{V}_3 + \left[\dfrac{1}{-j5} + \dfrac{1}{j4} + \dfrac{1}{4}\right]\mathbf{V}_4$

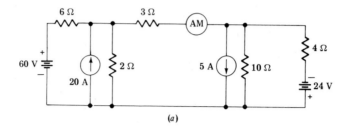

(a)

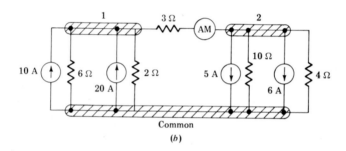

Common

(b)

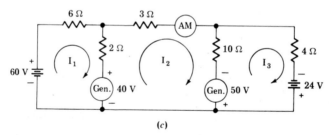

(c)

Figure 14-17 Circuits for Example 14-9: (a) original circuit; (b) modified for node analysis; (c) modified for loop analysis.

Substituting the equivalent current sources for the voltage sources results in Fig. 14-17b. The node-voltage equations for node 1 and node 2, expressed in the node-voltage format, are

Node 1

$$[10 + 20] = + [\tfrac{1}{6} + \tfrac{1}{2} + \tfrac{1}{3}]V_1 - [\tfrac{1}{3}]V_2$$

Node 2

$$[-5 - 6] = - [\tfrac{1}{3}]V_1 + [\tfrac{1}{3} + \tfrac{1}{10} + \tfrac{1}{4}]V_2$$

Simplifying,

$$30 = 1.0V_1 \quad -0.333V_2$$
$$-11 = -0.333V_1 + 0.683V_2$$

From this set of equations an ordered array of admittances, called the *admittance determinant*, is formed.

$$\Delta_Y = \begin{vmatrix} 1 & -0.333 \\ -0.333 & 0.683 \end{vmatrix} = (1)(0.683) - (-0.333)(-0.333)$$

$$\Delta_Y = 0.572$$

$$\Delta_{Y1} = \begin{vmatrix} 30 & -0.333 \\ -11 & 0.683 \end{vmatrix} = 30(0.683) - (-11)(-0.333) = 16.827$$

$$\Delta_{Y2} = \begin{vmatrix} 1 & 30 \\ -0.333 & -11 \end{vmatrix} = (1)(-11) - (-0.333)(30)$$

$$\Delta_{Y2} = -1.01$$

$$V_1 = \frac{\Delta_{Y1}}{\Delta_Y} = \frac{16.827}{0.572} = 29.42 \text{ V}$$

$$V_2 = \frac{\Delta_{Y2}}{\Delta_Y} = \frac{-1.01}{0.572} = -1.77 \text{ V}$$

$$V_{3\Omega} = V_1 - V_2 = 29.42 - (-1.77) = 31.19 \text{ V}$$

$$I = \frac{V_1 - V_2}{3} = \frac{31.19}{3} = 10.4 \text{ A}$$

(b) Loop-current analysis: Converting the current sources in Fig. 14-17a to equivalent voltage sources,

$$V_{cv1} = 20 \times 2 = 40 \text{ V}$$

$$V_{cv2} = 10 \times 5 = 50 \text{ V}$$

Substituting the equivalent voltage sources for the current sources results in Fig. 14-17c. The loop-current equations for loops 1, 2, and 3, expressed in the loop-current format, are

Loop 1

$$[60 - 40] = [6 + 2]I_1 + [-2]I_2 + [0]I_3$$

Loop 2

$$[40 + 50] = [-2]I_1 + [2 + 3 + 10]I_2 + [-10]I_3$$

Loop 3

$$[24 - 50] = [0]I_1 + [-10]I_2 + [10 + 4]I_3$$

Simplifying,

$$20 = (8)I_1 + (-2)I_2 + (0)I_3$$

$$90 = (-2)I_1 + (15)I_2 + (-10)I_3$$

$$-26 = (0)I_1 + (-10)I_2 + (14)I_3$$

To determine the ammeter reading, solve for I_2. Using determinants,

$$\Delta_z = \begin{vmatrix} +\,\boxed{8} & -2 & 0 \\ -\,\boxed{-2} & 15 & -10 \\ +\,\boxed{0} & -10 & 14 \end{vmatrix}$$

$$\Delta_z = 8[(15)(14) - (-10)(-10)] - (-2)[(-2)(14) - (-10)(0)] + 0$$

$$\Delta_z = 880 - 56 = 824$$

$$\Delta_{z2} = \begin{vmatrix} +\,\boxed{8} & 20 & 0 \\ -\,\boxed{-2} & 90 & -10 \\ +\,\boxed{0} & -26 & 14 \end{vmatrix}$$

$$\Delta_{z2} = 8[(90)(14) - (-26)(-10)] - (-2)[(20)(14) - (-26)(0)] + 0$$

$$\Delta_{z2} = 8000 + 560 = 8560$$

$$I_2 = \frac{\Delta_{z2}}{\Delta_z} = \frac{8560}{824} = 10.4 \text{ A}$$

PROBLEMS

Source Conversion

14-1 Convert a current source, consisting of a 60-A constant-current generator and a paralleled 2.0-Ω resistor, to an equivalent voltage source. Sketch the circuits.

14-2 Convert a voltage source, consisting of a 50-V constant-voltage generator in series with a 2.5-Ω resistor, to an equivalent current source. Sketch the circuits.

14-3 A 10-Ω resistor is connected to the terminals of a current source. The source consists of a 90-A constant-current generator and a 5.0-Ω paralleled resistor. (a) Sketch the circuit and determine the current to the 10-Ω resistor and the voltage across it; (b) replace the current source by an equivalent voltage source and determine the current to the 10-Ω resistor and the voltage across it. Sketch the circuits and show all work.

14-4 A 60-Hz 200-V sinsusoidal voltage source, whose internal impedance is $4/60°$ Ω, is connected to a $10/20°$-Ω load. (a) Determine the current to the load and the voltage across the load; (b) replace the voltage source by an equivalent current source and determine the current to the load and the voltage across the load. Sketch the circuits and show all work.

Loop Analysis

14-5 The loop currents and two impedances for the 60-Hz network shown in Fig. 14-18 are:

$$I_1 = 20/60° \text{ A} \qquad I_4 = 70/100° \text{ A}$$

$$I_2 = 50/30° \text{ A} \qquad Z_5 = 6.0/25° \text{ Ω}$$

$$I_3 = 10/210° \text{ A} \qquad Z_6 = 15/40° \text{ Ω}$$

Determine (a) the ammeter reading; (b) the voltmeter reading; (c) active, reactive, and apparent power drawn by Z_5; (d) series-connected parameters (henrys, farads, ohms) that make up Z_5.

14-6 (a) Write the loop equations for the network shown in Fig. 14-19; (b) determine the ammeter readings; (c) find the voltage drop across the 6.0-Ω resistor.

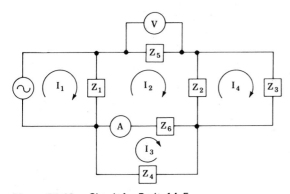

Figure 14-18   Circuit for Prob. 14-5.

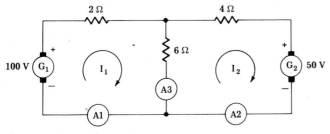

Figure 14-19   Circuit for Prob. 14-6.

14-7 (a) Write the loop equations for the network shown in Fig. 14-20; (b) determine the ammeter readings; (c) find the voltage drop across the (1.0 + j3.0)-Ω impedance.

14-8 (a) Write the loop equations for the network shown in Fig. 14-21; determine (b) the ammeter readings; (c) the voltage drop across the (5.0 + j6.0)-Ω impedance; (d) the active, reactive, and apparent power drawn by the (5.0 + j6.0)-Ω impedance.

14-9 Using loop analysis, determine the current in the 4.0-Ω resistor in Fig. 14-22.

14-10 (a) Convert the current source in Fig. 14-23 to a voltage source; (b) determine the current in the 4.0-Ω resistor.

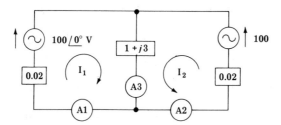

Figure 14-20   Circuit for Probs. 14-7 and 14-23.

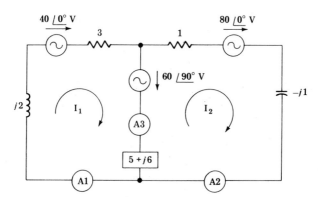

Figure 14-21   Circuit for Prob. 14-8.

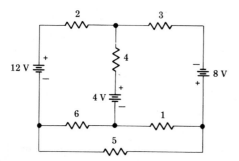

Figure 14-22   Circuit for Prob. 14-9.

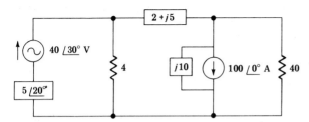

Figure 14-23   Circuit for Probs. 14-10 and 14-24.

14-11   (*a*) Write the loop equations for the network shown in Fig. 14-24: (*b*) calculate the loop currents and determine the ammeter readings.

14-12   Resketch Fig. 14-24, substituting a 10-V battery for generator *A* and a 6.0-V battery for generator *B*. Note the effect of 0 Hz on the capacitance and inductance in the network, and determine the ammeter readings at steady-state.

14-13   Determine the reading of the ammeter in Fig. 14-25. The loops should be selected so that only one loop current will pass through the ammeter.

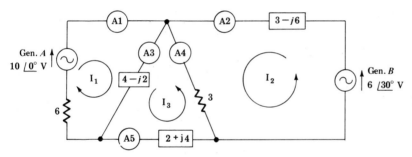

Figure 14-24   Circuit for Probs. 14-11, 14-12, and 14-25.

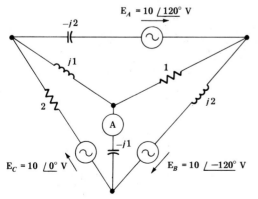

Figure 14-25   Circuit for Probs. 14-13 and 14-26.

14-14 Determine the ammeter reading in Fig. 14-26 and the power drawn by the 12-Ω resistor.

14-15 Determine the ammeter reading in Fig. 14-27.

14-16 Resketch Fig. 14-27 showing the 20-V and 90-V sinusoidal generators replaced by 20-V and 90-V batteries, respectively. Determine the ammeter reading at steady state.

14-17 Determine the ammeter and voltmeter reading in Fig. 14-28.

14-18 Resketch Fig. 14-28 assuming the 100-V sinusoidal generator is replaced by a 100-V battery. Determine the ammeter and voltmeter readings at steady state.

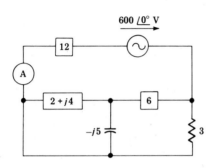

Figure 14-26   Circuit for Probs. 14-14 and 14-27.

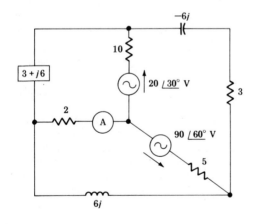

Figure 14-27   Circuit for Probs. 14-15 and 14-16.

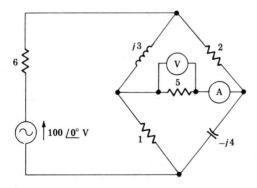

Figure 14-28   Circuit for Probs. 14-17 and 14-18.

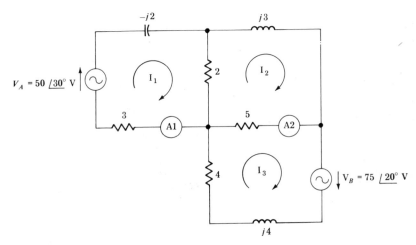

Figure 14-29 Circuit for Probs. 14-19 and 14-20.

14-19 Write the loop equations and determine the ammeter readings in Fig. 14-29.

14-20 Resketch Fig. 14-29 showing the 50-V and 75-V sinusoidal generators replaced by 50-V and 75-V batteries, respectively. Determine the ammeter readings at steady state.

Node Analysis

14-21 (*a*) Convert the voltage source in Fig. 14-30 to a current source; (*b*) write the node equations and determine the voltmeter reading.

14-22 Convert the voltage source in Fig. 14-31 to a current source, write the node equations, and determine the voltage across the 20-Ω resistor.

14-23 Using node analysis, determine the voltage across, and the current to, the (1 + *j*3)-Ω impedance in Fig. 14-20.

14-24 (*a*) Convert the voltage source in Fig. 14-23 to a current source; (*b*) write the node equations; (*c*) solve for the voltage drop across the (2 + *j*5)-Ω impedance and determine the current through it.

14-25 Using node analysis, determine the voltage drop across each of the three delta-connected impedances in Fig. 14-24.

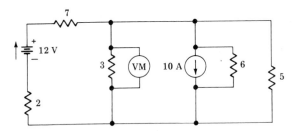

Figure 14-30 Circuit for Prob. 14-21.

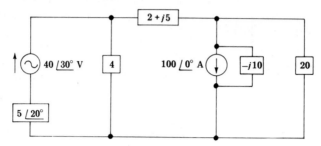

Figure 14-31   Circuit for Prob. 14-22.

14-26   Using node analysis, determine the voltage drop across each of the wye-connected impedances in Fig. 14-25.

14-27   (*a*) Using node analysis, determine the voltage drop across the 6-Ω resistor in Fig. 14-26; (*b*) determine the current through the 6-Ω resistor.

CHAPTER 15

NETWORK THEOREMS AND EQUIVALENT CIRCUITS

The network theorems and equivalent circuits introduced in this chapter provide additional techniques for simplifying the solution of complex circuit problems.

15-1 THEVENIN EQUIVALENT SECTION

Thevenin's theorem provides a mathematical method for *replacing a section of a network containing one or more generators and impedances with an equivalent circuit model that contains only one generator and one series-connected impedance.* All circuit elements must be linear and bilateral. The Thevenin equivalent section has the same current and voltage relationships at its output terminals as does the section it replaces. The Thevenin equivalent is useful when the arrangement and parameters of a section of the circuit containing one or more generators remain the same but another section is subject to change. If the circuit contains inductance and/or capacitance, a different Thevenin equivalent is required for each frequency.

The general procedure for calculating the Thevenin equivalent section is developed with the aid of Fig. 15-1a. In this figure, the section to the left of terminals $T_1 T_2$ is to be replaced by its Thevenin equivalent, as shown in Fig. 15-1b.

Note: Convert all current sources to voltage sources before calculating the Thevenin equivalent.

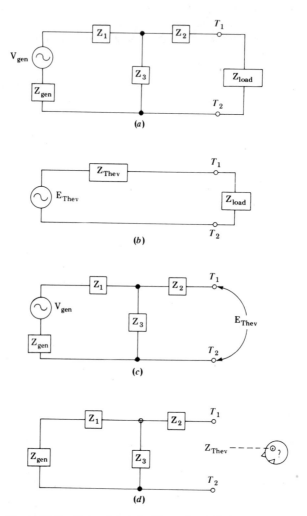

Figure 15-1 Determining the Thevenin equivalent section: (a) original circuit; (b) Thevenin equivalent model; (c) determining E_{Thev}; (d) determining Z_{Thev}.

Determination of E_{Thev}

To determine E_{Thev}, remove the load as shown in Fig. 15-1c and calculate the voltage drop across terminals $T_1 T_2$. This voltage drop is the Thevenin voltage E_{Thev}.

Determination of Z_{Thev}

To determine Z_{Thev}, disconnect the load and replace the generator with an impedance equal to its own internal impedance as shown in Fig. 15-1d.

Then calculate the impedance looking back into the circuit from the $T_1 T_2$ terminals. The calculated value is $\mathbf{Z}_{\text{Thev}}$.

Note: The Thevenin impedance is also called the *output impedance* of the replaced section.

Example 15-1 Determine the Thevenin equivalent section to the left of terminals $T_1 T_2$ in Fig. 15-2*a*.

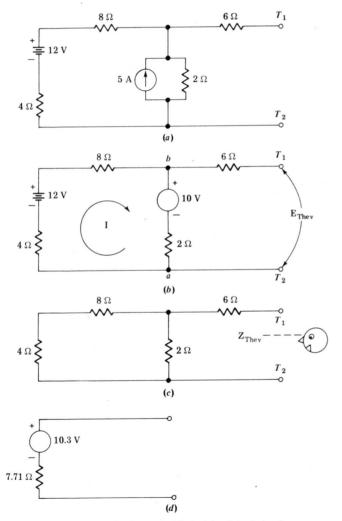

Figure 15-2 Circuits for Example 15-1: (*a*) original circuit; (*b*) determining $\mathbf{E}_{\text{Thev}}$; (*c*) determining $\mathbf{Z}_{\text{Thev}}$; (*d*) Thevenin equivalent circuit.

Solution

Converting the current source to a voltage source,

$$I_{cc} = \frac{E_{cv}}{Z_S}$$

$$5 = \frac{E_{cv}}{2}$$

$$E_{cv} = 10 \text{ V}$$

Substituting the voltage source for the current source results in Fig. 15-2*b*. The current in the loop formed by the two sources is determined by applying Kirchhoff's voltage law:

$$\sum \text{driving voltages} = \sum \text{voltage drops}$$

$$12 - 10 = 4I + 8I + 2I$$

$$I = 0.14 \text{ A}$$

Using the potential-difference equation along path $T_2 \, ab T_1$,

$$E_{Thev} = \left(\sum \text{driving voltages} \right) - \left(\sum \text{voltage drops} \right)$$

$$E_{Thev} = 10 - [(-0.14)2 + 0(6)] = 10.3 \text{ V}$$

Replacing the voltage sources by their internal impedances results in Fig. 15-2*c*. The equivalent parallel resistance of the two-resistor series branch and the 2-Ω branch is

$$R_{eq,P} = \frac{(8 + 4)(2)}{(8 + 4) + 2} = 1.71 \; \Omega$$

Adding the 6-Ω series branch,

$$Z_{Thev} = 1.71 + 6 = 7.71 \; \Omega$$

The completed Thevenin equivalent circuit is shown in Fig. 15-2*d*.

Example 15-2 Determine (*a*) the Thevenin equivalent for the section to the left of terminals $T_1 T_2$ in Fig. 15-3*a*; (*b*) the required load impedance for maximum power transfer to the load.

Solution

(*a*) To determine $\mathbf{E}_{Thev}$, remove the load, as shown in Fig. 15-3*b*, and calculate the voltage across the $T_1 T_2$ terminals. With the circuit open at terminals $T_1 T_2$, there will

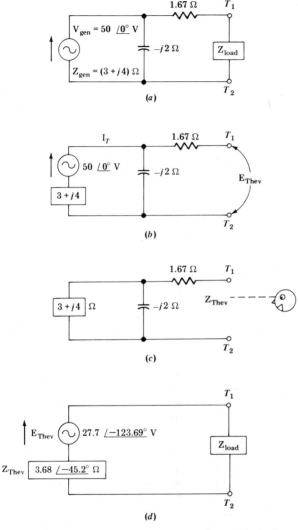

Figure 15-3 Circuits for Example 15-2: (a) original circuit; (b) determining E_{Thev}; (c) determining Z_{Thev}; (d) Thevenin equivalent circuit.

be no current through, and hence no voltage drop across, the 1.67-Ω resistor. Thus E_{Thev} will be equal to the voltage drop across the capacitor. Solving for current I_T,

$$I_T = \frac{V_{gen}}{Z_T} = \frac{50\underline{/0°}}{(3 + j4) - j2} = \frac{50\underline{/0°}}{3.61\underline{/33.69°}} = 13.85\underline{/-33.69°} \text{ A}$$

$$V_C = I_T jX_C = (13.85\underline{/-33.69°})(2\underline{/-90°}) = 27.7\underline{/-123.69°} \text{ V}$$

Thus $\mathbf{E}_{\text{Thev}} = 27.7/\!-\!123.69°$ V. The voltage across the capacitor may also be calculated by using the voltage-divider equation

$$\mathbf{V}_K = \mathbf{V}_T \frac{\mathbf{Z}_K}{\mathbf{Z}_T}$$

$$\mathbf{V}_C = \frac{(50/\!0°)(2/\!-\!90°)}{(3 + j4) - j2}$$

$$\mathbf{V}_C = 27.7/\!-\!123.69° \text{ V}$$

To determine $\mathbf{Z}_{\text{Thev}}$, replace the generator by its internal impedance, as shown in Fig. 15-3c, and calculate the impedance of the section looking back into the circuit from the $T_1 T_2$ terminals. The impedance of the paralleled branches is

$$\mathbf{Z}_{P2} = \frac{(3 + j4)(-j2)}{(3 + j4) + (-j2)} = \frac{8 - j6}{3 + j2} = 0.92 - j2.61$$

$$\mathbf{Z}_{\text{Thev}} = (0.92 - j2.61) + 1.67 = (2.59 - j2.61) = 3.68/\!-\!45.2° \ \Omega$$

The completed Thevenin equivalent for the section to the left of terminals $T_1 T_2$ is shown in Fig. 15-3d.

(b) Referring to Fig. 15-3d, for maximum power transfer to occur,

$$\mathbf{Z}_{\text{load}} = \mathbf{Z}_{\text{Thev}}^*$$

$$\mathbf{Z}_{\text{load}} = (3.68/\!-\!45.2°)^* = 3.68/\!45.2° \ \Omega$$

15-2   NORTON EQUIVALENT SECTION

The Norton equivalent section is similar to the Thevenin equivalent, except that *it consists of a constant-current source, called the Norton current, in parallel with an impedance called the Norton impedance.* This is shown in Fig. 15-4a. The Norton equivalent can be determined from the Thevenin equivalent, and vice versa, by means of the following relationships:

$$\mathbf{Z}_{\text{Nort}} = \mathbf{Z}_{\text{Thev}} \tag{15-1}$$

$$\mathbf{I}_{\text{Nort}} = \frac{\mathbf{E}_{\text{Thev}}}{\mathbf{Z}_{\text{Thev}}} \tag{15-2}$$

Like the Thevenin equivalent, the Norton equivalent is equivalent only at the output terminals $T_1 T_2$. That is, the voltage and current at the output terminals of the Norton equivalent section and the Thevenin equivalent section are identical to the voltage and current at the corresponding terminals of the section they replace.

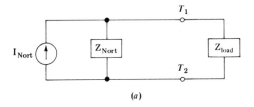

(a)

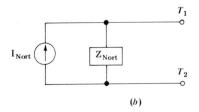

(b)

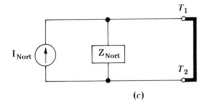

(c)

Figure 15-4 Norton equivalent section: (a) connected to a load; (b) open-circuited; (c) short-circuited.

The Norton current is a constant current that remains the same regardless of the impedance of the connected load; it remains the same if the terminals $T_1 T_2$ are open, as shown in Fig. 15-4b, or shorted as shown in Fig. 15-4c.

If the original circuit is frequency-dependent, that is, if it contains inductance and/or capacitance, the Norton equivalent will have to be recalculated for each change in frequency.

Example 15-3 The Thevenin equivalent for a certain circuit is $E_{Thev} = 20\underline{/30°}$ V, $Z_{Thev} = 5\underline{/20°}$ Ω. A load impedance of $6\underline{/40°}$ Ω is connected across the output terminals. Determine the load current and the voltage across the load using (a) the Thevenin equivalent; (b) the Norton equivalent. (c) Determine the active and reactive power delivered to the load.

Solution

$$Z_{Nort} = Z_{Thev} = 5\underline{/20°}\ \Omega$$

$$I_{Nort} = \frac{E_{Thev}}{Z_{Thev}} = \frac{20\underline{/30°}}{5\underline{/20°}} = 4\underline{/10°}\ A$$

The Thevenin equivalent section is shown in Fig. 15-5a, and the Norton equivalent section is shown in Fig. 15-5b.

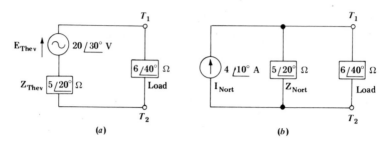

Figure 15-5 Circuits for Example 15-3: (*a*) Thevenin equivalent; (*b*) Norton equivalent.

(*a*) Thevenin equivalent

From Ohm's law,

$$\mathbf{I}_{load} = \frac{\mathbf{E}_{Thev}}{\mathbf{Z}_{Thev} + \mathbf{Z}_{load}}$$

$$\mathbf{I}_{load} = \frac{20\underline{/30°}}{(5\underline{/20°}) + (6\underline{/40°})}$$

$$\mathbf{I}_{load} = \frac{20\underline{/30°}}{(9.295 + j5.567)}$$

$$\mathbf{I}_{load} = \frac{20\underline{/30°}}{10.834\underline{/30.92°}}$$

$$\mathbf{I}_{load} = 1.846\underline{/-0.92°} \ \text{A}$$

$$\mathbf{V}_{load} = \mathbf{I}_{load}\,\mathbf{Z}_{load}$$

$$\mathbf{V}_{load} = (1.846\underline{/-0.92°})(6\underline{/40°})$$

$$\mathbf{V}_{load} = 11.076\underline{/39.08°} \ \text{V}$$

(*b*) Norton equivalent

Using the current-divider rule,

$$\mathbf{I}_{load} = \mathbf{I}_T\,\frac{\mathbf{Y}_{load}}{\mathbf{Y}_T}$$

$$\mathbf{Y}_{load} = \frac{1}{6\underline{/40°}} = 0.1667\underline{/-40°}$$

$$\mathbf{Y}_{load} = (0.1277 - j0.1071)$$

$$\mathbf{Y}_T = \frac{1}{5\underline{/20°}} + \frac{1}{6\underline{/40°}}$$

$$\mathbf{Y}_T = 0.2\underline{/-20°} + 0.1667\underline{/-40°}$$

$$\mathbf{Y}_T = (0.1879 - j0.0684)$$

$$+ \ (0.1227 - j0.1071)$$

$$\mathbf{Y}_T = (0.3156 - j0.1755)$$

$$\mathbf{Y}_T = 0.3611\underline{/-29.077°}$$

$$\mathbf{I}_{load} = (4\underline{/10°})\,\frac{0.1667\underline{/-40°}}{0.3611\underline{/-29.077°}}$$

$$\mathbf{I}_{load} = 1.846\underline{/-0.92°} \ \text{A}$$

$$\mathbf{V}_{load} = \mathbf{I}_{load}\,\mathbf{Z}_{load} = (1.846\underline{/-92°})(6\underline{/40°})$$

$$\mathbf{V}_{load} = 11.076\underline{/39.08°} \ \text{V}$$

A comparison of Fig. 15-5 with Fig. 14-10 indicates that the Thevenin equivalent section of a circuit is in effect an equivalent voltage source, and the Norton equivalent section is in effect an equivalent current source.

(c) $\mathbf{S} = \mathbf{V}_{load}\, \mathbf{I}^*_{load}$

$\mathbf{S} = (11.076\underline{/39.08°})(1.846\underline{/+0.92°})$

$\mathbf{S} = 20.45\underline{/40°} = 15.67 + j13.15$

$P = 15.7 \text{ W}$

$Q = 13.2 \text{ var}$

15-3 CONTROLLED SOURCES

The voltage and current sources previously discussed were *independent sources*; they involved ideal voltages and ideal currents, and were *not dependent* on currents or voltages elsewhere in the circuit. However, many of the voltages and currents generated within electronic devices and circuits *are dependent* on the currents and/or voltages elsewhere in the circuit. These sources are called *controlled sources* or *dependent sources*, and are represented in circuit diagrams by diamond-shaped symbols as shown in Fig. 15-6.

The procedure for loop or node analysis of circuits that have controlled sources is the same as was outlined for circuits with independent sources. However, the methods previously outlined for obtaining the Thevenin equivalent impedance of a network cannot always be used if controlled sources are present. If controlled sources are present, the Thevenin impedance should be obtained from the ratio of the open-circuit voltage at the $T_1 T_2$ terminals to the short-circuit current at the same terminals. That is,

$$\boxed{\mathbf{Z}_{Thev} = \frac{\mathbf{V}_{oc}}{\mathbf{I}_{sc}}} \tag{15-3}$$

where $\mathbf{V}_{oc}$ = voltage between terminals T_1 and T_2, assuming the terminals are open-circuited

$\mathbf{I}_{sc}$ = current through a "zero-ohm" jumper connected between terminals T_1 and T_2

Note 1: Equation (15-3) may be used with circuits containing both independent and dependent sources.

Note 2: $\mathbf{V}_{oc}$ *is* $\mathbf{E}_{Thev}$

Example 15-4 Determine the Thevenin equivalent of the circuit section to the left of terminals $T_1 T_2$ in Fig. 15-7a.

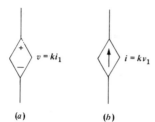

Figure 15-6 (a) Controlled voltage source; (b) controlled current source.

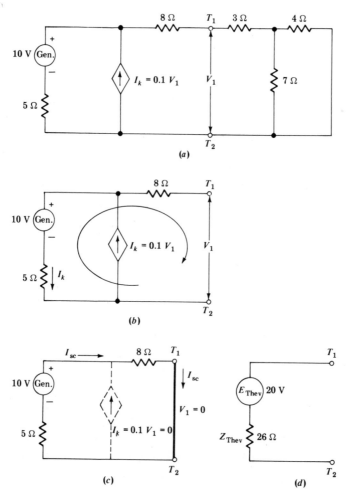

Figure 15-7 Circuits for Example 15-4: (a) original circuit; (b) open-circuit conditions; (c) short-circuit conditions; (d) Thevenin equivalent circuit.

Solution

The section to the left of terminals $T_1 T_2$ is shown in Fig. 15-7b and represents the open-circuit conditions. The open-circuit voltage V_{oc} is the voltage V_1 between terminals T_1 and T_2. This voltage may be determined by writing the potential-difference equation for the outside loop and solving for V_1. Thus, summing in the clockwise direction

$$V_1 = (\sum \text{driving voltages}) - (\sum \text{voltage drops})$$

$$V_1 = 10 - [(-I_k)5 + (0)8]$$

However, $I_k = 0.1\ V_1$
Substituting and solving for V_1

$$V_1 = 10 - [(-0.1\ V_1)5 + 0]$$

$$V_1 = 20\ V$$

Hence,

$$E_{\text{Thev}} = V_{oc} = 20\ V$$

To determine I_{sc}, calculate the current through a jumper that short-circuits terminals $T_1 T_2$, as shown in Fig. 15-7c. The voltage drop across the zero-ohm jumper is zero. Hence $V_1 = 0$ and $I_k = 0$. Applying Kirchhoff's voltage law to the loop, and recognizing that $V_1 = 0$,

$$10 = 8\ I_{sc} + 0 + 5\ I_{sc}$$

$$10 = 13\ I_{sc}$$

$$I_{sc} = 0.769\ A$$

$$Z_{\text{Thev}} = \frac{V_{oc}}{I_{sc}} = \frac{20}{0.769} = 26\ \Omega$$

The Thevenin equivalent section is shown in Fig. 15-7d.

Example 15-5 Determine the Thevenin equivalent of the section to the left of terminals $T_1 T_2$ in Fig. 15-8a.

Solution

The section to the left of terminals $T_1 T_2$ is shown in Fig. 15-8b with the load removed, and represents the open-circuit conditions. As indicated, $I_1 = 0$; hence $0.8 I_1 = 0$,

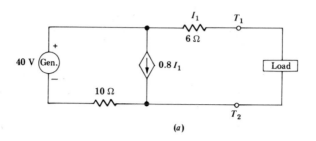

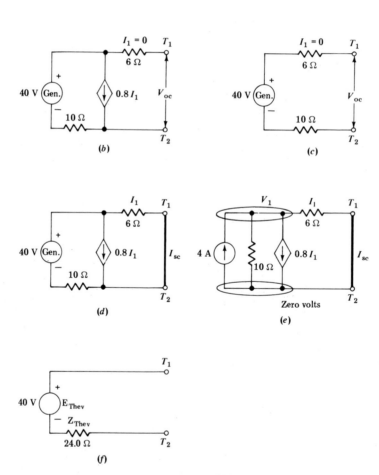

Figure 15-8 Circuits for Example 15-5: (*a*) original circuit; (*b*), (*c*) open-circuit conditions; (*d*), (*e*) short-circuit conditions; (*f*) Thevenin equivalent.

and Fig. 15-8b reduces to that shown in Fig. 15-8c. Applying the potential-difference equation to the reduced loop,

$$V_{oc} = (\sum \text{driving voltages}) - (\sum \text{voltage drops})$$

$$V_{oc} = 40 - [0(10) + 0(6)]$$

$$V_{oc} = 40 \text{ V}$$

Hence,

$$E_{Thev} = V_{oc} = 40 \text{ V}$$

To determine I_{sc}, short terminals $T_1 T_2$ as shown in Fig. 15-8d and solve for I_{sc}. As indicated, under short-circuit conditions $I_1 = I_{sc}$. Converting the independent voltage source to a current source, as shown in Fig. 15-8e, and then writing the node equation,

$$I_{cc} = \frac{E_{cv}}{Z_S} = \frac{40}{10} = 4 \text{ A}$$

at node V_1,

$$(4 - 0.8 I_1) = (\tfrac{1}{10} + \tfrac{1}{6})V_1$$

The voltage across the 6-Ω resistor, as determined by Ohm's law, is

$$V_1 = 6 I_1$$

Substituting into the previous equation and solving for I_1,

$$4 - 0.8 I_1 = (\tfrac{1}{10} + \tfrac{1}{6})6 I_1$$

$$I_1 = 1.67 \text{ A}$$

$$I_{sc} = I_1 = 1.67 \text{ A}$$

$$Z_{Thev} = \frac{V_{oc}}{I_{sc}} = \frac{40}{1.67} = 24.0 \ \Omega$$

The Thevenin equivalent section is shown in Fig. 15-8f.

15-4 SUPERPOSITION THEOREM

The superposition theorem has applications in circuits having two or more generators, and is particularly useful when the generators have different frequencies. The supersposition theorem can be applied only to circuits having linear bilateral elements.

To apply the superposition theorem, calculate the component currents for each branch using one generator at a time, with all other generators replaced by their respective impedances. Each branch current is the sum of the component currents contributed by each generator to that branch.

Example 15-6 A 12-V battery and a sinusoidal generator are connected in series with a coil, as shown in Fig. 15-9a. The inductance and resistance of the coil are 1.0 H and 6.0 Ω, respectively. The equation for the generator voltage wave is $e_{gen} = 30 \sin 10t$. Assume the resistance of the battery is negligible, and the inductance and resistance of the generator are 0.20 H and 0.10 Ω, respectively. Determine the steady-state circuit current.

Solution
To calculate the sinusoidal component of the current, replace the battery by its internal resistance (which is assumed to be zero in this problem). This is shown in Fig. 15-9b.

$$X_{L,\,coil} = \omega L_{coil} = 10(1) = 10\ \Omega \qquad X_{L,\,gen} = \omega L_{gen} = 10(0.2) = 2\ \Omega$$

$$\mathbf{Z}_T = \mathbf{Z}_{gen} + \mathbf{Z}_{coil} = (0.1 + j2) + (6 + j10) = 6.1 + j12$$

$$\mathbf{Z}_T = 13.46\underline{/63.05^\circ}\ \Omega$$

$$\mathbf{I} = \frac{\mathbf{E}_{gen}}{\mathbf{Z}_T} = \frac{30/\sqrt{2}\underline{/0^\circ}}{13.46\underline{/63.05^\circ}} = 1.58\underline{/-63.05^\circ}\ A$$

$$i_\sim = 1.58\sqrt{2}\sin(10t - 63.05^\circ)$$

$$i_\sim = 2.23\sin(10t - 63.05^\circ)$$

To calculate the DC component of current, replace the sinusoidal generator by its internal impedance, as shown in Fig. 15-9c. The inductance of the circuit has no effect on direct current at steady state. Hence,

$$I_{DC} = \frac{E_{bat}}{R_T} = \frac{12}{6 + 0.1} = 1.97\ A$$

or using phasors,

$$\mathbf{Z} = R + j2\pi f L = 6.1 + j2\pi 0(1.2)$$

$$\mathbf{Z} = 6.1\ \Omega$$

The total current in the circuit is

$$i_T = i_\sim + I_{DC}$$

$$i_T = 2.23\sin(10t - 63.05^\circ) + 1.97$$

A sketch of the steady-state current is shown in Fig. 15-9d.

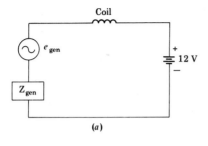

(a)

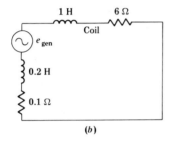

(b)

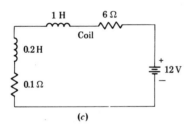

(c)

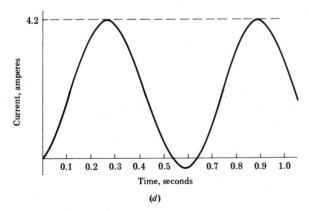

(d)

Figure 15-9 Circuits for Example 15-6: (a) original
circuit; (b) sinusoidal component; (c) 0-Hz component;
(d) graph of steady-state circuit current.

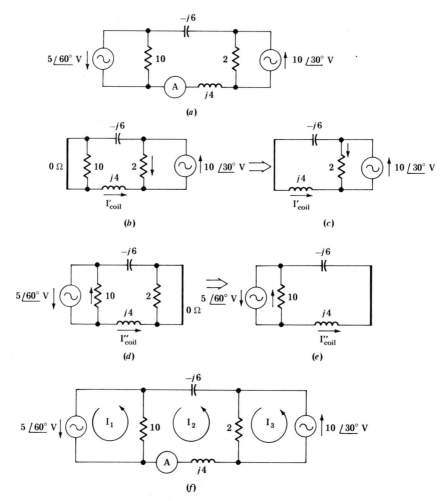

Figure 15-10 Circuits for Example 15-7: (*a*) original circuit; (*b*), (*c*), (*d*), (*e*) superposition method; (*f*) loop analysis.

Example 15-7 Determine the ammeter reading in Fig. 15-10*a* using (*a*) superposition; (*b*) loop analysis.

Solution

(*a*) Superposition method: Replacing the $5\underline{/60°}$ V generator by its impedance, which is indicated as $0\,\Omega$ for this problem, results in the circuit of Fig. 15-10*b* and *c*. The component current in the coil, as obtained from Ohm's law, is

$$\mathbf{I}'_{coil} = \frac{10\underline{/30°}}{(j4) + (-j6)} = \frac{10\underline{/30°}}{2\underline{/-90°}} = 5\underline{/120°} \text{ A}$$

Replacing the $10/30°$ generator by its impedance, which is indicated as $0 \, \Omega$ for this problem, results in the circuit shown in Fig. 15-10d and e. The component current in the coil, for this circuit, is

$$I''_{coil} = \frac{5/60°}{j4 + (-j6)} = \frac{5/60°}{2/-90°} = 2.5/150° \text{ A}$$

$$I_{coil} = I'_{coil} + I''_{coil}$$

$$I_{coil} = 5/120° + 2.5/150° = (-2.5 + j4.33) + (-2.17 + j1.25)$$

$$I_{coil} = (-4.67 + j5.58) = 7.27/129.9° \text{ A}$$

The ammeter will read 7.27 A.
(b) Using loop analysis and the loops in Fig. 15-10f,

Loop 1

$$[5/60°] = [10]I_1 + [(-10)]I_2 + [0]I_3$$

Loop 2

$$[0] = [-(10)]I_1 + [10 + j4 - j6 + 2]I_2 + [-(2)]I_3$$

Loop 3

$$[10/30°] = [0]I_1 + [-(2)]I_2 + [2]I_3$$

$$\Delta_z = \begin{vmatrix} + & 10 & -10 & 0 \\ - & -10 & (12 - j2) & -2 \\ + & 0 & -2 & 2 \end{vmatrix}$$

$$\Delta_z = 10[(12 - j2)(2) - (-2)(-2)] - (-10)[(-10)(2) - (-2)(0)] + 0$$

$$\Delta_z = (200 - j40) - 200 = -j40 = 40/-90°$$

$$\Delta_{z2} = \begin{vmatrix} + & 10 & 5/60° & 0 \\ - & -10 & 0 & -2 \\ + & 0 & 10/30° & 2 \end{vmatrix}$$

$$\Delta_{z2} = 10[(0)(2) - (10\underline{/30°})(-2)] - (-10)[(5\underline{/60°})(2) - (10\underline{/30°})(0)] + 0$$

$$\Delta_{z2} = 200\underline{/30°} + 100\underline{/60°} = (173.21 + j100) + (50 + j86.6)$$

$$\Delta_{z2} = (223.21 + j186.60) = 290.93\underline{/39.9°}$$

$$I_2 = \frac{\Delta_{z2}}{\Delta_z} = \frac{290.93\underline{/39.9°}}{40\underline{/-90°}} = 7.27\underline{/129.9°}\ A$$

The ammeter will read 7.27 A.

15-5 MILLMAN'S THEOREM

Millman's theorem provides a simple method for replacing parallel voltage sources such as those shown in Fig. 15-11a with an equivalent single voltage source shown in Fig. 15-11d. To obtain the Millman equivalent circuit:

1. Convert all parallel voltage sources to equivalent current sources, as shown in Fig. 15-11b.

   Note:

   $$I_k = \frac{E_k}{Z_k} \qquad Y_k = \frac{1}{Z_k}$$

2. Obtain the phasor sum of the currents supplied by the current sources:

   $$I_T = I_1 + I_2 + I_k + \cdots + I_N$$

3. Sum the admittances of the current sources:

   $$Y_T = Y_1 + Y_2 + Y_k + \cdots + Y_N$$

   The reduced circuit is shown in Fig. 15-11c. *Note:* I_T and Y_T in Fig. 15-11c represent an equivalent current source whose parameters are

   $$I_{cc} = I_T \tag{15-4}$$

   $$Z_S = \frac{1}{Y_T} \tag{15-5}$$

4. Use the source-conversion formula to obtain the Millman equivalent voltage source.

   $$E_{cv} = I_{cc} Z_S \tag{15-6}$$

   Substituting Eqs. (15-4) and (15-5) into Eq. (15-6),

   $$E_{cv} = \frac{I_T}{Y_T} \tag{15-7}$$

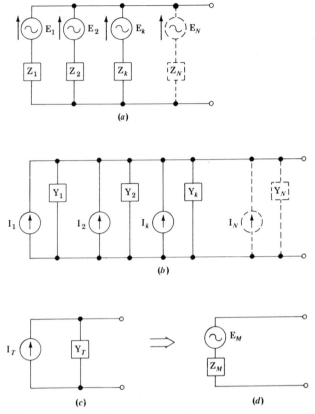

Figure 15-11 Determining the Millman equivalent circuit:
(*a*) original voltage sources; (*b*) equivalent current sources;
(*c*) reduced circuit; (*d*) Millman equivalent.

Relating Eqs. (15-5) and (15-7) to the Millman equivalent circuit shown in Fig. 15-11*d*,

$$\mathbf{Z}_M = \mathbf{Z}_S = \frac{1}{\mathbf{Y}_T} \tag{15-8}$$

$$\mathbf{E}_M = \mathbf{E}_{cv} = \frac{\mathbf{I}_T}{\mathbf{Y}_T} \tag{15-9}$$

Warning: Paralleling *low impedance* voltage sources whose open-circuit voltages differ considerably in magnitude, or differ in direction, may cause severe damage to generators and batteries; batteries with large amounts of available energy may explode.

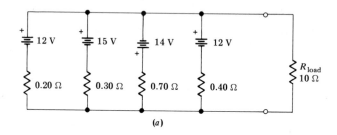

(a)

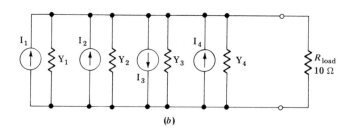

(b)

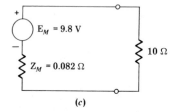

(c)

Figure 15-12 Circuits for Example 15-8: (a) original circuit;
(b) conversion to equivalent current sources; (c) Millman
equivalent.

Example 15-8 For the circuit shown in Fig. 15-12a, determine (a) the Millman equivalent of the four paralleled voltage sources; (b) the current to the load.

Solution
Converting the voltage sources to equivalent current sources results in Fig. 15-12b, where

$$I_1 = \frac{12}{0.20} = 60 \text{ A} \qquad I_3 = \frac{14}{0.70} = 20 \text{ A}$$

$$I_2 = \frac{15}{0.30} = 50 \text{ A} \qquad I_4 = \frac{12}{0.40} = 30 \text{ A}$$

Noting the direction of the current arrows,

$$I_T = I_1 + I_2 - I_3 + I_4$$

$$I_T = 60 + 50 - 20 + 30 = 120 \text{ A}$$

$$Y_1 = \frac{1}{0.20} = 5.000 \qquad Y_3 = \frac{1}{0.70} = 1.428$$

$$Y_2 = \frac{1}{0.30} = 3.333 \qquad Y_4 = \frac{1}{0.40} = 2.500$$

$$Y_T = Y_1 + Y_2 + Y_3 + Y_4$$

$$Y_T = 5.000 + 3.333 + 1.428 + 2.500 = 12.261 \text{ S}$$

From Eqs. (15-8) and (15-9),

$$Z_M = \frac{1}{Y_T} = \frac{1}{12.261} = 0.082 \ \Omega$$

$$E_M = \frac{I_T}{Y_T} = \frac{120}{12.261} = 9.8 \text{ V}$$

Figure 15-12c shows the Millman equivalent connected to the 10-Ω load.

$$I_{\text{load}} = \frac{9.8}{10 + 0.082} = 0.97 \text{ A}$$

Example 15-9 Determine the Millman equivalent for the circuit shown in Fig. 15-13a. The 60-Hz voltages E_1, E_2, and E_3 are $100\underline{/0°}$ V, $120\underline{/20°}$ V, and $110\underline{/-30°}$ V, respectively. Impedances Z_1, Z_2, and Z_3 are, respectively, $12\underline{/30°}$ Ω, $10\underline{/40°}$ Ω, $15\underline{/0°}$ Ω.

Solution
Converting the voltage sources to current sources results in Fig. 5-13b.

$$I_1 = \frac{V_1}{Z_1} = \frac{100\underline{/0°}}{12\underline{/30°}} = 8.33\underline{/-30°} = 7.21 - j4.17$$

$$I_2 = \frac{V_2}{Z_2} = \frac{120\underline{/20°}}{10\underline{/40°}} = 12\underline{/-20°} = 11.28 - j4.10$$

$$I_3 = \frac{V_3}{Z_3} = \frac{110\underline{/-30°}}{15\underline{/0°}} = 7.33\underline{/-30°} = 6.35 - j3.67$$

$$I_T = I_1 + I_2 + I_3 = (24.84 - j11.93) = 27.56\underline{/-25.66°} \text{ A}$$

$$Y_1 = \frac{1}{12/30°} = 0.0833/-30° = 0.0722 - j0.0417$$

$$Y_2 = \frac{1}{10/40°} = 0.1000/-40° = 0.0766 - j0.0643$$

$$Y_3 = \frac{1}{15/0°} = 0.0667/0° = 0.0667 + j0.000$$

$$Y_T = (0.2154 - j0.1059) = 0.2401/-26.186° \text{ S}$$

$$Z_M = \frac{1}{Y_T} = \frac{1}{0.2401/-26.186°} = 4.16/26.19° \text{ } \Omega$$

$$E_M = \frac{I_T}{Y_T} = \frac{27.56/-25.66°}{0.2401/-26.186°} = 115/0.53° \text{ V}$$

The Millman equivalent circuit for the three parallel voltage sources is shown in Fig. 15-13c.

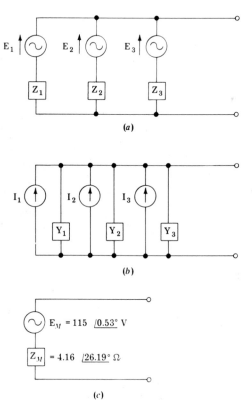

(a)

(b)

$E_M = 115 /0.53°$ V

$Z_M = 4.16 /26.19° \Omega$

(c)

Figure 15-13 Circuits for Example 15-9: (a) original circuit; (b) conversion to equivalent current sources; (c) Millman equivalent.

15-6 RECIPROCITY THEOREM

The reciprocity theorem applies to networks with linear bilateral elements and a single voltage source or a single current source. The reciprocity theorem states that if a voltage source in branch A of a network causes $\mathbf{I}_X$ amperes in branch B, shifting the source voltage (but not its impedance) to branch B will cause the same $\mathbf{I}_X$ amperes to appear in branch A. However, the currents in other branches will not remain the same.

The ratio of *input voltage* in branch A to the *output* current in branch B is called the *transfer impedance*:

$$Z_{tr} = \frac{\mathbf{V}_{in,\,A}}{\mathbf{I}_{out,\,B}} = \frac{\mathbf{V}_{in,\,B}}{\mathbf{I}_{out,\,A}} \qquad (15\text{-}10)$$

The transfer impedance is very useful for predicting the current in one branch of a network resulting from a voltage inserted in another branch.

Similarly, if a current source connected between nodes 1 and 2 causes a potential difference $\mathbf{V}_X$ between nodes 3 and 4, shifting the current source (but not its admittance) to nodes 3 and 4 will cause the same voltage $\mathbf{V}_X$ between nodes 1 and 2. However, the voltages between other nodes will not remain the same.

The ratio of *input current* between one set of nodes to *output voltage* between another set of nodes is called the *transfer admittance*.

$$Y_{tr} = \frac{\mathbf{I}_{in,\,1,\,2}}{\mathbf{V}_{out,\,3,\,4}} = \frac{\mathbf{I}_{in,\,3,\,4}}{\mathbf{V}_{out,\,1,\,2}} \qquad (15\text{-}11)$$

Note: Both Z_{tr} and Y_{tr} are frequency-dependent. Hence, new values must be calculated for different frequencies.

Example 15-10 For the circuit in Fig. 15-14, the transfer impedance between branches $(1 - 4)$ and $(3 - 5)$ is determined to be $28.4\underline{/39.59°}\,\Omega$ at 1000 Hz. (*a*) Calculate the current in the 7-Ω resistor if the generator voltage is 40 V at 1000 Hz; (*b*) repeat part *a* assuming generator terminals T_1 and T_2 are interchanged.

Solution

(*a*) $Z_{tr} = \dfrac{\mathbf{V}_{in}}{\mathbf{I}_{out}}$

$$28.4\underline{/39.59°} = \frac{40\underline{/10°}}{\mathbf{I}_{out}}$$

$$\mathbf{I}_{out} = 1.41\underline{/-29.59°}\ \text{A}$$

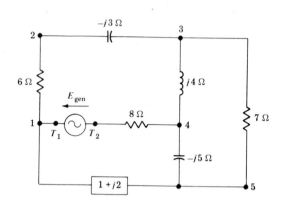

Figure 15-14 Circuit for Example 15-10.

(*b*) Reversal of the generator terminals reverses the generator polarity. Hence,

$$28.4\underline{/39.59°} = \frac{-40\underline{/10°}}{\mathbf{I}_{out}}$$

$$\mathbf{I}_{out} = -1.41\underline{/-29.59°} \text{ A}$$

Reversing the source voltage reverses the output current.

15-7   WYE-DELTA TRANSFORMATION

Circuits that are more complex than those composed of combinations of simple series and simple parallel sections may be reduced to less complex form by using wye-delta transformations. For example, the circuit in Fig. 15-15*a* is neither a parallel circuit, series circuit, nor series-parallel circuit. Hence the input impedance cannot be determined by using simple series or simple parallel techniques developed in previous chapters.

However, if a triangular section of the circuit (called a delta) is replaced by an equivalent wye section, as shown in Fig. 15-15*b*, the circuit is reduced to the simple series-parallel configuration shown in Fig. 15-15*c*. Wye and delta sections of networks are also called T and Π sections respectively.

Delta to Wye Conversion

Referring to Fig. 15-16, the three branch impedances for the *equivalent wye section* of a circuit may be obtained from the three branch impedances of the delta section by means of the following formulas:

$$\mathbf{Z}_1 = \frac{\mathbf{Z}_A \mathbf{Z}_B}{\mathbf{Z}_A + \mathbf{Z}_B + \mathbf{Z}_C}$$

$$\mathbf{Z}_2 = \frac{\mathbf{Z}_B \mathbf{Z}_C}{\mathbf{Z}_A + \mathbf{Z}_B + \mathbf{Z}_C} \qquad (15\text{-}12)$$

$$\mathbf{Z}_3 = \frac{\mathbf{Z}_C \mathbf{Z}_A}{\mathbf{Z}_A + \mathbf{Z}_B + \mathbf{Z}_C}$$

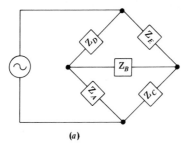

(a)

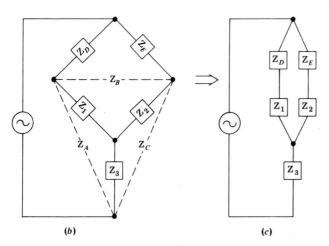

(b)   (c)

Figure 15-15   Replacing a delta section by an equivalent wye section.

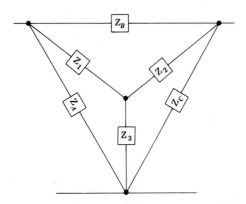

Figure 15-16   Circuit for equation sets (15-12) and (15-13).

Delta to Wye Conversion

Referring to Fig. 15-16, the three branch impedances for the *equivalent delta section* of a circuit may be obtained from the three branch impedances of the wye section by means of the following formulas:

$$\mathbf{Z}_A = \frac{\mathbf{Z}_1\mathbf{Z}_2 + \mathbf{Z}_2\mathbf{Z}_3 + \mathbf{Z}_3\mathbf{Z}_1}{\mathbf{Z}_2}$$

$$\mathbf{Z}_B = \frac{\mathbf{Z}_1\mathbf{Z}_2 + \mathbf{Z}_2\mathbf{Z}_3 + \mathbf{Z}_3\mathbf{Z}_1}{\mathbf{Z}_3} \qquad (15\text{-}13)$$

$$\mathbf{Z}_C = \frac{\mathbf{Z}_1\mathbf{Z}_2 + \mathbf{Z}_2\mathbf{Z}_3 + \mathbf{Z}_3\mathbf{Z}_1}{\mathbf{Z}_1}$$

Conversion of Balanced Sections

If all branches of a delta section have identical impedances, the section is called a balanced delta. Similarly, if all branches of a wye section have identical impedances, the section is called a balanced wye. The relationship between a branch of a balanced delta section and a branch of a balanced wye section is

$$\boxed{\mathbf{Z}_{\Delta,\,br} = 3\,\mathbf{Z}_{Y,\,br}} \qquad (15\text{-}14)$$

Equation (15-14) may be derived from any one equation in equation set (15-12) or any one equation in equation set (15-13).

Example 15-11 For the circuit in Fig. 15-15a, assume

$$\mathbf{Z}_A = 5.0\underline{/20°}\ \Omega \qquad \mathbf{Z}_B = 10\underline{/90°}\ \Omega \qquad \mathbf{Z}_C = 8.0\underline{/0°}\ \Omega$$

$$\mathbf{Z}_D = 6.0\underline{/-45°}\ \Omega \qquad \mathbf{Z}_E = 2.0\underline{/0°}\ \Omega$$

Convert the lower delta into an equivalent wye and then determine the input impedance of the circuit.

Solution

$$\mathbf{Z}_1 = \frac{\mathbf{Z}_A\mathbf{Z}_B}{\mathbf{Z}_A + \mathbf{Z}_B + \mathbf{Z}_C} = \frac{(5\underline{/20°})(10\underline{/90°})}{5\underline{/20°} + 10\underline{/90°} + 8\underline{/0°}}$$

$$\mathbf{Z}_1 = \frac{50\underline{/110°}}{4.70 + j1.71 + j10 + 8} = \frac{50\underline{/110°}}{12.7 + j11.71}$$

$$Z_1 = \frac{50\underline{/110°}}{17.27\underline{/42.68°}} = 2.90\underline{/67.32°}$$

$$Z_1 = (1.12 + j2.67)\ \Omega$$

$$Z_2 = \frac{Z_B Z_C}{Z_A + Z_B + Z_C}$$

$$Z_2 = \frac{(10\underline{/90°})(8\underline{/0°})}{17.27\underline{/42.68°}} = 4.63\underline{/47.32°} = (3.14 + j3.40)\ \Omega$$

$$Z_3 = \frac{Z_C Z_A}{Z_A + Z_B + Z_C} = \frac{(8\underline{/0°})(5\underline{/20°})}{17.27\underline{/42.68°}} = 2.32\underline{/-22.68°} = (2.14 - j0.89)\ \Omega$$

Thus, the circuit is reduced to that shown in Fig. 15-15c. To evaluate the parallel combination,

$$Z_1 + Z_D = (1.12 + j2.67) + 6\underline{/-45°}$$

$$Z_1 + Z_D = (1.12 + j2.67) + (4.24 - j4.24) = 5.59\underline{/-16.34°}\ \Omega$$

$$Z_2 + Z_E = (3.14 + j3.40) + 2\underline{/0°}$$

$$Z_2 + Z_E = 5.14 + j3.40 = 6.16\underline{/33.48°}\ \Omega$$

Using the special-case formula for two impedances in parallel,

$$Z_{P,\,2} = \frac{(5.59\underline{/-16.34°})(6.16\underline{/33.48°})}{5.59\underline{/-16.34°} + 6.16\underline{/33.48°}} = \frac{34.43\underline{/17.14°}}{10.66\underline{/9.86°}}$$

$$Z_{P,\,2} = 3.23\underline{/7.28°} = (3.20 + j0.41)\ \Omega$$

Hence,

$$Z_{\text{in}} = Z_{P,\,2} + Z_3 = (3.20 + j0.41) + (2.14 - j0.89)$$

$$Z_{\text{in}} = (5.34 - j0.48) = 5.36\underline{/-5.14°}\ \Omega$$

Example 15-12 Determine the input impedance of the circuit shown in Fig. 15-17a. Assume:

$$Z_1 = 2.0\underline{/30°}\ \Omega \qquad Z_3 = 6.0\underline{/90°}\ \Omega$$

$$Z_2 = 3.0\underline{/40°}\ \Omega \qquad Z_4 = 7.0\underline{/50°}\ \Omega$$

$$Z_5 = 10\underline{/-60°}\ \Omega$$

Solution
First convert the wye into an equivalent delta, as shown in Fig. 15-17b, and then determine the input impedance.

$$Z_A = \frac{Z_1 Z_2 + Z_2 Z_3 + Z_3 Z_1}{Z_2}$$

$$Z_A = \frac{(2\underline{/30°})(3\underline{/40°}) + (3\underline{/40°})(6\underline{/90°}) + (6\underline{/90°})(2\underline{/30°})}{3\underline{/40°}}$$

$$Z_A = \frac{33.62\underline{/117.49°}}{3\underline{/40°}} = 11.21\underline{/77.49°} \ \Omega$$

$$Z_B = \frac{Z_1 Z_2 + Z_2 Z_3 + Z_3 Z_1}{Z_3}$$

$$Z_B = \frac{33.62\underline{/117.49°}}{6\underline{/90°}} = 5.60\underline{/27.49°} \ \Omega$$

$$Z_C = \frac{Z_1 Z_2 + Z_2 Z_3 + Z_3 Z_1}{Z_1}$$

$$Z_C = \frac{33.62\underline{/117.49°}}{2\underline{/30°}} = 16.81\underline{/87.49°} \ \Omega$$

Combining parallel branches, using the special-case formula,

$$Z_{kl} = \frac{Z_C Z_5}{Z_C + Z_5} = \frac{(16.81\underline{/87.49°})(10\underline{/-60°})}{16.81\underline{/87.49°} + 10\underline{/-60°}}$$

$$Z_{kl} = \frac{168.1\underline{/27.49°}}{(5.74 + j8.13)} = \frac{168.1\underline{/27.49°}}{9.95\underline{/54.81°}} = 16.89\underline{/-27.32°} \ \Omega$$

$$Z_{ml} = \frac{Z_A Z_4}{Z_A + Z_4} = \frac{(11.21\underline{/77.49°})(7\underline{/50°})}{11.21\underline{/77.49°} + 7\underline{/50°}}$$

$$Z_{ml} = \frac{78.47\underline{/127.49°}}{(6.93 + j16.31)} = 4.43\underline{/60.51°} \ \Omega$$

The reduced circuit is shown in Fig. 15-17c.

$$Z_{kl} + Z_{ml} = 16.89\underline{/-27.32°} + 4.43\underline{/60.51°}$$

$$Z_{kl} + Z_{ml} = (15.01 - j7.75) + (2.18 + j3.86)$$

$$Z_{kl} + Z_{ml} = (17.19 - j3.90) = 17.62\underline{/-12.77°} \ \Omega$$

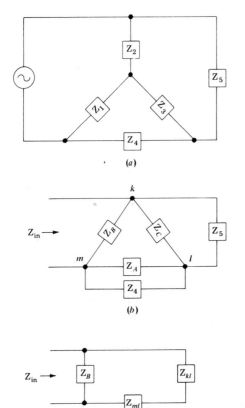

(a)

(b)

(c)

Figure 15-17 Circuit for Example 15-12.

Combining the two parallel branches in Fig. 15-17c,

$$Z_{in} = \frac{Z_B(Z_{kl} + Z_{ml})}{Z_B + (Z_{kl} + Z_{ml})} = \frac{(5.60\underline{/27.49°})(17.62\underline{/-12.77°})}{(5.60\underline{/27.49°} + 17.62\underline{/-12.77°})}$$

$$Z_{in} = \frac{98.67\underline{/14.72°}}{22.19\underline{/-3.38}} = 4.45\underline{/18.1°} \ \Omega$$

Proof of Wye-Delta Transformation

Equation set (15-12) and equation set (15-13) were derived by equating the input impedances of the wye section to the corresponding input impedances of the delta section. Referring to Fig. 15-18 and defining Z_{kl} to be the input impedance between terminals k and l,

$$Z_{kl,Y} = Z_1 + Z_2 \qquad Z_{kl,\Delta} = \frac{(Z_B)(Z_A + Z_C)}{Z_B + (Z_A + Z_C)}$$

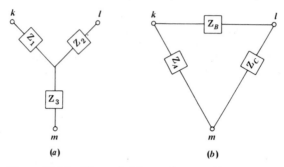

Figure 15-18 Wye and delta sections used in proving the wye-delta transformation.

Similarly,

$$\mathbf{Z}_{lm,Y} = \mathbf{Z}_2 + \mathbf{E}_3 \qquad \mathbf{Z}_{lm,\Delta} = \frac{\mathbf{Z}_C(\mathbf{Z}_B + \mathbf{Z}_A)}{\mathbf{Z}_C + (\mathbf{Z}_B + \mathbf{Z}_A)}$$

$$\mathbf{Z}_{mk,Y} = \mathbf{Z}_3 + \mathbf{Z}_1 \qquad \mathbf{Z}_{mk,\Delta} = \frac{\mathbf{Z}_A(\mathbf{Z}_B + \mathbf{Z}_C)}{\mathbf{Z}_A + (\mathbf{Z}_B + \mathbf{Z}_C)}$$

Equating the corresponding input impedances,

$$\mathbf{Z}_1 + \mathbf{Z}_2 = \frac{\mathbf{Z}_B(\mathbf{Z}_A + \mathbf{Z}_C)}{\mathbf{Z}_A + \mathbf{Z}_B + \mathbf{Z}_C} \tag{15-15}$$

$$\mathbf{Z}_2 + \mathbf{Z}_3 = \frac{\mathbf{Z}_C(\mathbf{Z}_B + \mathbf{Z}_A)}{\mathbf{Z}_A + \mathbf{Z}_B + \mathbf{Z}_C} \tag{15-16}$$

$$\mathbf{Z}_3 + \mathbf{Z}_1 = \frac{\mathbf{Z}_A(\mathbf{Z}_B + \mathbf{Z}_C)}{\mathbf{Z}_A + \mathbf{Z}_B + \mathbf{Z}_C} \tag{15-17}$$

The solution of Eqs. (15-15), (15-16), and (15-17) for $\mathbf{Z}_1$, $\mathbf{Z}_2$, $\mathbf{Z}_3$, $\mathbf{Z}_A$, $\mathbf{Z}_B$, and $\mathbf{Z}_C$ results in equation sets (15-12) and (15-13).

SUMMARY OF FORMULAS

$$\mathbf{Z}_{\text{Nort}} = \mathbf{Z}_{\text{Thev}} = \frac{\mathbf{V}_{oc}}{\mathbf{I}_{sc}} \qquad \mathbf{I}_{\text{Nort}} = \frac{\mathbf{E}_{\text{Thev}}}{\mathbf{Z}_{\text{Thev}}}$$

$$\mathbf{Z}_M = \frac{1}{\mathbf{Y}_T} \qquad \mathbf{Z}_{tr} = \frac{\mathbf{V}_{in}}{\mathbf{I}_{out}}$$

$$\mathbf{E}_M = \frac{\mathbf{I}_T}{\mathbf{Y}_T} \qquad \mathbf{Y}_{tr} = \frac{\mathbf{I}_{in}}{\mathbf{V}_{out}}$$

Wye-Delta Conversion

Balanced branches   *Unbalanced branches*

$Z_{\Delta,\,br} = 3Z_{Y,\,br}$ See equation sets (15-12) and (15-13)

ROBLEMS

Thevenin and Norton

15-1 Determine E_{Thev} and Z_{Thev} to the left of terminals $T_1 T_2$ in Fig. 15-19, and sketch the new circuit.

15-2 Determine E_{Thev} and Z_{Thev} to the left of terminals $T_1 T_2$ in Fig. 15-20, and sketch the new circuit.

15-3 Determine E_{Thev} and Z_{Thev} to the left of terminals $T_1 T_2$ in Fig. 15-21, and sketch the new circuit.

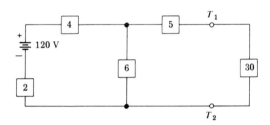

Figure 15-19   Circuit for Prob. 15-1.

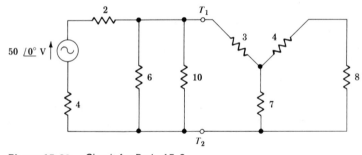

Figure 15-20   Circuit for Prob. 15-2.

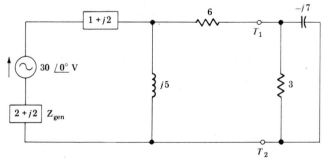

Figure 15-21   Circuit for Prob. 15-3.

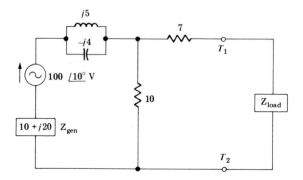

Figure 15-22 Circuit for Probs. 15-4 and 15-5.

15-4 (a) Determine the Thevenin equivalent section to the left of terminals $T_1 T_2$ in Fig. 15-22, and sketch the new circuit; (b) determine $\mathbf{Z}_{load}$ that will result in maximum power transfer; (c) calculate the active power drawn by $\mathbf{Z}_{load}$ in part b.

15-5 Replace the 100-V sinusoidal generator in Fig. 15-22 with a 100-V battery, and determine (a) the steady-state Thevenin equivalent section to the left of terminals $T_1 T_2$; (b) $\mathbf{Z}_{load}$ that will result in maximum power transfer; (c) power drawn by $\mathbf{Z}_{load}$ in part b.

15-6 Determine (a) the Thevenin equivalent section to the left of terminals $T_1 T_2$ in Fig. 15-23; (b) the ammeter and voltmeter readings; (c) the active, reactive, and apparent power drawn by the load; (d) the power factor of the load.

15-7 (a) Determine the Thevenin equivalent section to the left of terminals $T_1 T_2$ in Fig. 15-24; (b) sketch the new equivalent circuit and determine the ammeter and voltmeter readings; (c) the active and reactive power supplied to the load; (d) the power factor of the load.

15-8 (a) Replace the 450-V sinusoidal generator in Fig. 15-24 with a 450-V DC generator, and determine the Thevenin equivalent section to the left of terminals $T_1 T_2$; (b) sketch the new equivalent circuit and determine the ammeter and voltmeter readings; (c) determine the energy *stored* in the load at steady state.

15-9 (a) Determine the Thevenin equivalent section to the left of terminals $T_1 T_2$ in Fig. 15-25; (b) sketch the equivalent circuit and determine the ammeter and voltmeter readings.

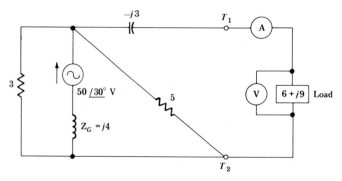

Figure 15-23 Circuit for Prob. 15-6.

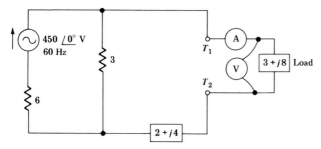

Figure 15-24   Circuit for Probs. 15-7 and 15-8.

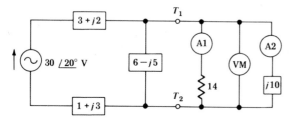

Figure 15-25   Circuit for Prob. 15-9.

15-10   The Thevenin equivalent voltage and Thevenin equivalent impedance of a circuit section are $60/30°$ V and $20/30°$ Ω, respectively. Determine the Norton equivalent section and sketch the circuit.

15-11   The Thevenin equivalent values of a circuit section are $Z_{Thev} = 5.0/20°$ Ω, $E_{Thev} = 100/60°$ V. Determine the Norton equivalent and sketch the circuit.

15-12   Determine the Thevenin equivalent of the circuit section to the left of terminals $T_1 T_2$ in Fig. 15-26.

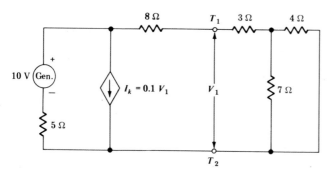

Figure 15-26   Circuit for Prob. 15-12.

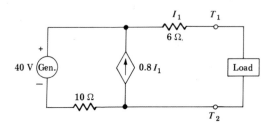

Figure 15-27 Circuit for Prob. 15-13.

15-13 Determine the Thevenin equivalent of the circuit section to the left of terminals $T_1 T_2$ in Fig. 15-27.

15-14 Determine the Thevenin equivalent of the circuit section to the left of terminals $T_1 T_2$ in Fig. 15-28.

15-15 Repeat Prob. 15-14 assuming the controlled voltage source is reversed.

15-16 Determine the Thevenin equivalent of the circuit in Fig. 15-29, as viewed from the $T_1 T_2$ terminals.

15-17 Repeat Prob. 15-16, assuming the controlled voltage source is reversed.

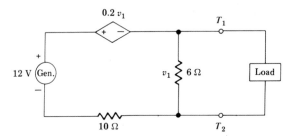

Figure 15-28 Circuit for Probs. 15-14 and 15-15.

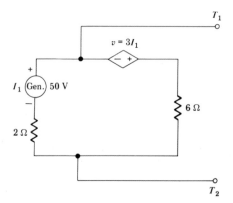

Figure 15-29 Circuit for Probs. 15-16 and 15-17.

Superposition

15-18 Using superposition, determine the current in the 6.0-Ω diagonal resistor shown in Fig. 15-30. The 60-Hz generator voltages are $\mathbf{E}_A = 100\underline{/0°}$ V, $\mathbf{E}_B = 25\underline{/0°}$ V.

15-19 Determine the three ammeter readings in Fig. 15-31 by means of the superposition method. The voltages of the 50-Hz generators are $\mathbf{E}_A = 100\underline{/0°}$ V, $\mathbf{E}_B = 25\underline{/0°}$ V.

15-20 Replace the 100-V and 25-V sinusoidal generators in Prob. 15-19 with 100-V and 25-V batteries, respectively, and determine the ammeter readings. Assume the internal resistance of 25-V battery is 0.20 Ω and that of the 100-V battery is 0.80 Ω.

15-21 Using superposition, determine the ammeter reading in Fig. 15-32. Both generators are operating at 25 Hz.

15-22 Using superposition, determine the steady-state time-domain currents i_1, i_2, and i_3 for the circuit shown in Fig. 15-33.

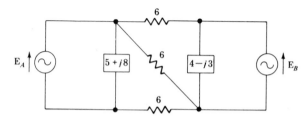

Figure 15-30 Circuit for Prob. 15-18.

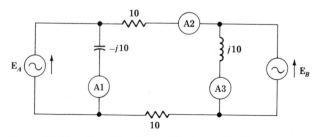

Figure 15-31 Circuit for Probs. 15-19 and 15-20.

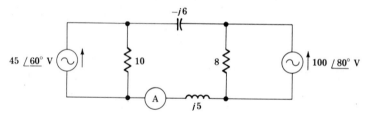

Figure 15-32 Circuit for Prob. 15-21.

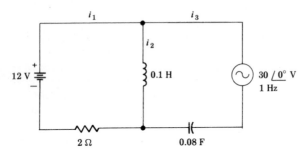

Figure 15-33 Circuit for Prob. 15-22.

15-23 A tachometer generator running at 3600 rpm generates 6.0 V DC with a 0.50-V 7200-Hz ripple voltage; the inductance and resistance of the generator are 1.0 mH and 3.0 Ω, respectively. The generator is connected to a meter whose inductance and resistance are 2.0 mH and 100 Ω, respectively. (*a*) Sketch the circuit; (*b*) using superposition determine the steady-state time-domain current in the circuit.

15-24 Using superposition, determine (*a*) the ammeter reading in Fig. 15-34; (*b*) the active, reactive, and apparent power drawn by the coil.

15-25 Repeat Prob. 15-24 assuming the 12-V and 30-V sinusoidal generators are replaced by 12-V and 30-V batteries, respectively.

Millman

15-26 Determine the Millman equivalent for the following parallel voltage sources whose respective voltages and impedances are 16 V, 1.6 Ω; 14 V, 0.80 Ω; 15 V, 1.2 Ω.

15-27 Determine the Millman equivalent for the following parallel voltage sources whose respective voltages and impedances are $45/\underline{20°}$ V, $2.0/\underline{80°}$ Ω; $50/\underline{0°}$ V, $6/\underline{0°}$ Ω; $60/\underline{-30°}$ V, $4.0/\underline{40°}$ Ω.

Reciprocity

15-28 For the circuit shown in Fig. 15-35*a* determine (*a*) the ammeter reading; (*b*) the transfer impedance between branch *abc* and branch *cf*; (*c*) the current in branch *abc* if the generator is replaced with an ammeter and the ammeter in branch *cf* is replaced with an $80/\underline{20°}$-V, 50-Hz generator, as shown in Fig. 15-35*b*,

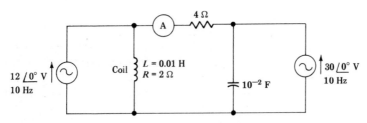

Figure 15-34 Circuit for Probs. 15-24 and 15-25.

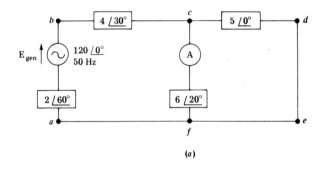

(a)

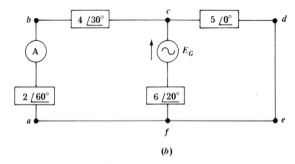

(b)

Figure 15-35 Circuit for Prob. 15-28.

Wye-Delta Conversion

15-29 (a) Convert a balanced delta section of $8.0/25°$ Ω per branch to an equivalent wye section. Sketch both sections. (b) Convert a balanced wye section of $12/40°$ Ω per branch to an equivalent delta section. Sketch both sections.

15-30 For the circuit shown in Fig. 15-36 determine (a) the battery current; (b) the voltage drop across the 4.0-Ω resistor; (c) the heat power drawn by the 4.0-Ω resistor.

15-31 A balanced wye section of $6.0/20°$ Ω per branch is paralleled with a balanced delta section of $3.0/40°$ Ω per branch. Sketch the circuit and determine a single equivalent delta section that can replace the two paralleled sections.

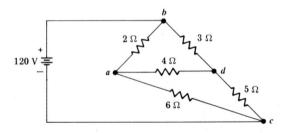

Figure 15-36 Circuit for Prob. 15-30.

CHAPTER 16

RESONANCE

Resonance is a physical condition that exists when the frequency of an external force applied to a structure is equal to the natural frequency of the structure. Some examples of resonance are the vibration of a building by a nearby truck or low-flying airplane, the severe shaking of an automobile under certain road conditions, and the vibration of a tuning fork caused by the striking of a certain piano note.

A structure offers little opposition to motion when excited to its resonance frequency by external forces. If the resonance frequency of the *driver* is sustained and of sufficient amplitude, bridges may collapse, machinery may be damaged, electric circuits may burn out, etc.

A few examples of the many useful applications of resonance in electric circuits are radio, television, lasers, and power-factor improvement.

16-1 RESONANCE IN ELECTRIC CIRCUITS

Resonance in electric circuits is defined as the *steady-state sinusoidal condition that exists in circuits containing capacitance and inductance when the line current is in phase with the driving voltage and the driving frequency is greater than 0 Hz.*

Thus, assuming that a series circuit is at resonance, $\mathbf{V}_T$ and $\mathbf{I}_T$ will have the same phase angle, and the input impedance will be

$$\mathbf{Z}_{in} = \frac{\mathbf{V}_T}{\mathbf{I}_T} = \frac{V_T \underline{/\theta^\circ}}{I_T \underline{/\theta^\circ}} = \frac{V_T}{I_T} \underline{/0^\circ} = Z \underline{/0^\circ}$$

Or, in rectangular form,

$$\mathbf{Z}_{in} = R + j0 \tag{16-1}$$

Similarly, the input admittance of a parallel circuit that is at resonance will be

$$\mathbf{Y}_{in} = \frac{\mathbf{I}_T}{\mathbf{V}_T} = \frac{I_T\underline{/\alpha^\circ}}{V_T\underline{/\alpha^\circ}} = \frac{I_T}{V_T}\underline{/0^\circ} = Y\underline{/0^\circ}$$

Or, in rectangular form,

$$\mathbf{Y}_{in} = G + j0 \tag{16-2}$$

As indicated in Eq. (16-1) for the series circuit, and Eq. (16-2) for the parallel circuit, when a circuit is at resonance, the input impedance or input admittance is the equivalent of an ideal resistance or an ideal conductance, respectively; the circuit draws no reactive power from the source, and its power factor is unity.

16-2   SERIES RESONANCE

In accordance with the definition of resonance, the series circuit shown in Fig. 16-1a will be in resonance if the frequency of the driving voltage is such as to cause the input impedance to become equivalent to the circuit resistance. Since the input impedance for any RLC series circuit is given by

$$\mathbf{Z}_{in} = R + jX_L - jX_C$$

and at resonance

$$\mathbf{Z}_{in} = R$$

Then at the resonance frequency,

$$jX_L - jX_C = 0$$

or

$$X_L = X_C \tag{16-3}$$

Expressing Eq. (16-3) in terms of frequency,

$$2\pi f_r L = \frac{1}{2\pi f_r C}$$

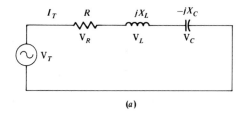

(a)

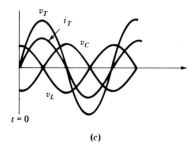

(b)

v_T
i_T
v_C
v_L
$t = 0$

(c)

Figure 16-1 Series circuit at resonance: (a) circuit; (b) phasor diagram; (c) sine waves.

Solving for the frequency,

$$f_r = \frac{1}{2\pi\sqrt{LC}}$$

(16-4)

or

$$\omega_r = \frac{1}{\sqrt{LC}}$$

(16-5)

where f_r = resonance frequency, Hz
    ω_r = resonance frequency, rad/s
    L = inductance, H
    C = capacitance, F

Equations (16-4) and (16-5) determine the driving frequency at which series resonance will occur for a particular combination of inductance and capacitance.

The phasor diagram for a series circuit operating at its resonance frequency is shown in Fig. 16-1b. From Kirchhoff's voltage law,

$$\mathbf{V}_T = \mathbf{V}_R + \mathbf{V}_L + \mathbf{V}_C$$

$$\mathbf{V}_T = \mathbf{I}_T R + \mathbf{I}_T jX_L + \mathbf{I}_T(-jX_C)$$

$$\mathbf{V}_T = \mathbf{I}_T R + \mathbf{I}_T(jX_L - jX_C)$$

At resonance,

$$jX_L - jX_C = 0$$

Hence,

$$\mathbf{V}_L + \mathbf{V}_C = 0$$

$$\mathbf{V}_T = \mathbf{I}_T R = \mathbf{V}_R$$

Thus the voltage across the resistance component of a series circuit, when operating at resonance, is the applied driving voltage.

The sinusoidal waves corresponding to the phasors in Fig. 16-1b are shown in Fig. 16-1c. Note that, at every instant of time, v_C is equal and opposite to v_L.

Graphs of the variations of inductive reactance and capacitive reactance with changes in frequency are shown in Fig. 16-2a. The capacitive reactance is infinite at a frequency of 0 Hz, decreases with increasing frequency, and approaches zero as the frequency approaches infinity. On the other hand, the inductive reactance is zero at a frequency of 0 Hz and increases linearly with increasing frequency. The resonance frequency occurs at the intersection of the two curves. At this frequency, $X_L = X_C$, causing the input impedance to become equal to the circuit resistance.

When a series circuit is at resonance, the effect on the circuit current is the same as though neither inductance nor capacitance is present. The current under such conditions is dependent solely on the resistance of the circuit and the voltage across it. If the circuit resistance is very low, the current at the resonance frequency may be high enough to blow protective fuses, trip circuit breakers, or otherwise cause damage to circuit components.

Figure 16-2b shows how input impedance and rms current of an RLC series circuit vary with frequency. The impedance curve is a graph of

$$Z_{in} = \sqrt{R^2 + (X_L - X_C)^2}$$

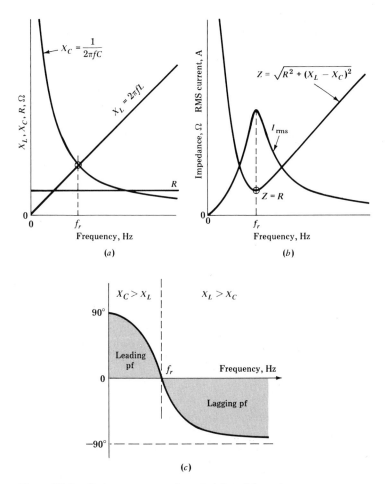

Figure 16-2 Series-resonance characteristics: (*a*) reactance versus frequency; (*b*) current and impedance versus frequency; (*c*) phase angle versus frequency.

The current curve is a graph of

$$I_T = \frac{V_T}{Z_{\text{in}}}$$

To obtain the current curve, the amplitude of the driving voltage is held constant while the frequency is varied. As indicated, the rms current rises from 0 A at 0 Hz to some maximum value at the resonance frequency, and then decreases with further increases in frequency.

A phase-angle plot, called a phase plot, for the RLC series circuit is shown in Fig. 16-2c. At frequencies below resonance $X_C > X_L$; the circuit is predominantly capacitive, causing the current to lead the voltage. At resonance $X_C = X_L$; the circuit is resistive, causing the current to be in phase with the voltage. At frequencies above resonance $X_L > X_C$; the circuit is predominantly inductive, causing the current to lag the voltage.

Example 16-1 A coil whose inductance and resistance are 1.0 mH and 2.0 Ω, respectively, is connected in series with a capacitor and a 120-V 5-kHz supply. Determine (*a*) the value of capacitance that will cause the system to be in resonance; (*b*) the circuit current at the resonance frequency; (*c*) the maximum instantaneous energy stored in the magnetic field of the inductance at the resonance frequency.

Solution
The circuit is similar to that shown in Fig. 16-1.

(*a*) $f_r = \dfrac{1}{2\pi\sqrt{LC}}$

$5000 = \dfrac{1}{2\pi\sqrt{0.001C}}$

Solving for C,

$C = 1.01 \times 10^{-6} = 1.01 \ \mu\text{F}$

(*b*) At series resonance, $Z_{\text{in}} = R$.

$I_T = \dfrac{V_T}{R} = \dfrac{120\underline{/0°}}{2} = 60\underline{/0°} \ \text{A}$

(*c*) $W = \frac{1}{2}LI^2_{\text{max}}$

$I_{\text{max}} = 60\sqrt{2} = 84.85 \ \text{A}$

$W = \frac{1}{2}(0.001)(84.85)^2 = 3.6 \ \text{J}$

16-3 QUALITY FACTOR OF A SERIES RESONANCE CIRCUIT

The quality factor of *any circuit* is defined as 2π times the ratio of the maximum energy stored per cycle to the energy dissipated per cycle *at a given frequency* (generally the resonance frequency). Expressed mathematically,

$$Q = \frac{2\pi \ (\text{maximum energy stored per cycle})}{\text{energy dissipated per cycle}} \qquad (16\text{-}6)$$

The quality factor of a circuit is a *figure of merit* used to measure the ability of the circuit to discriminate between different frequencies.

A comparison of the *frequency-response curves* for two series circuits with the same L and C but different circuit resistance is shown in Fig. 16-3. The circuit *with the lower resistance has a higher Q; it exhibits a sharper resonance peak and a higher value of rms current at the resonance frequency* than does the circuit with the higher resistance. It is also more *selective* in that only a small band of frequencies have high current amplitudes.

The maximum possible energy stored in an RLC circuit at resonance occurs in the inductance when the sinusoidal current through it attains its maximum value, and in the capacitor when the sinusoidal voltage across it attains its maximum value. Furthermore, when the capacitor is releasing energy, the inductor is absorbing it, and vice versa. Thus the maximum amount of energy stored in an RLC circuit at resonance is constant and may be expressed by

$$\text{Maximum energy stored} = \tfrac{1}{2}LI^2_{max} \qquad (16\text{-}7)$$

The energy dissipated per cycle at resonance is equal to the heat-power losses in the resistor times the period of the wave. Thus,

$$\text{Energy dissipated per cycle} = I^2RT_r \qquad (16\text{-}8)$$

where I = rms current at the resonance frequency
T_r = period of the wave at the resonance frequency
I_{max} = amplitude of the current wave at the resonance frequency

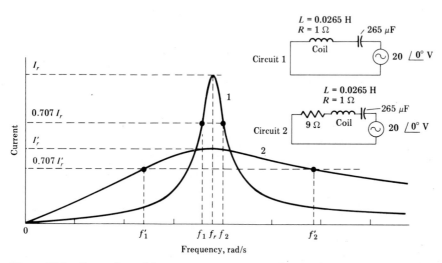

Figure 16-3 Comparison of frequency-response curves for two RLC series circuits with identical values of L and C but different values of resistance.

Substituting Eqs. (16-7) and (16-8) into the defining equation (16-6), and recognizing that the current through all parts of the series circuit is the same,

$$Q_S = \frac{2\pi(\frac{1}{2}LI_{max}^2)}{I^2 R T_r}$$

Since

$$T_r = \frac{1}{f_r}$$

and

$$I_{max} = \sqrt{2}\,I$$

$$Q_S = \frac{2\pi[\frac{1}{2}L(\sqrt{2}\,I)^2]}{I^2 R(1/f_r)} = \frac{I^2 2\pi f_r L}{I^2 R} = \frac{\text{reactive power}}{\text{active power}}$$

$$Q_S = \frac{\omega_r L}{R} \qquad\qquad (16\text{-}9)$$

From Eq. (16-3),

$$\omega_r L = \frac{1}{\omega_r C} \qquad\qquad (16\text{-}10)$$

Hence

$$Q_S = \frac{1}{\omega_r R C} \qquad\qquad (16\text{-}11)$$

From Eq. (16-10),

$$\omega_r = \frac{1}{\sqrt{LC}} \qquad\qquad (16\text{-}12)$$

Substituting Eq. (16-12) into Eq. (16-9) and simplifying,

$$Q_S = \frac{1}{\sqrt{LC}}\frac{L}{R}$$

$$Q_S = \frac{1}{R}\sqrt{\frac{L}{C}} \qquad\qquad (16\text{-}13)$$

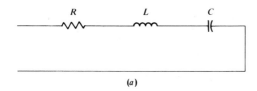

(a)

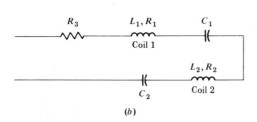

(b)

Figure 16-4   Circuits for Examples 16-2 and 16-3.

Thus, depending on the available data, the quality factor of a series circuit, operating at its resonance frequency, may be determined from any one of the following relationships:

$$Q_S = \frac{\omega_r L}{R} = \frac{1}{\omega_r RC} = \frac{1}{R}\sqrt{\frac{L}{C}} \qquad\qquad (16\text{-}14)$$

Note: Equation set (16-14) applies only to the *ideal RLC* series circuit shown in Fig. 16-4a. Any other circuit configuration, such as that shown in Fig. 16-4b, must be reduced to the ideal form before equation set (16-14) may be used.

Example 16-2   For the circuit shown in Fig. 16-4a, $R = 2.42\ \Omega$, $L = 25.4$ mH, $C = 52.0\ \mu$F. Determine (*a*) the resonance frequency; (*b*) Q_S, using the three relationships in equation set (16-14).

Solution

$$(a)\ f_r = \frac{1}{2\pi\sqrt{LC}} = \frac{1}{2\pi\sqrt{0.0254(52 \times 10^{-6})}} = 138.6\ \text{Hz}$$

$$(b)\ Q_S = \frac{\omega_r L}{R} = \frac{2\pi(138.6)(0.0254)}{2.42} = 9.13$$

$$Q_S = \frac{1}{\omega_r RC} = \frac{1}{2\pi(138.6)(2.42)(52 \times 10^{-6})} = 9.13$$

$$Q_S = \frac{1}{R}\sqrt{\frac{L}{C}} = \frac{1}{2.42}\sqrt{\frac{0.0254}{52 \times 10^{-6}}} = 9.13$$

Example 16-3 For the circuit shown in Fig. 16-4b, R_1, R_2, and R_3 are 0.51 Ω, 1.3 Ω, and 0.24 Ω, respectively; C_1 and C_2 are 25 μF and 62 μF, respectively; L_1 and L_2 are 32 mH and 15 mH, respectively. Determine (a) the resonance frequency; (b) Q_S; (c) $Q_{coil\,1}$; (d) $Q_{coil\,2}$.

Solution
(a) Before the resonance-frequency formula is used, the circuit must be reduced to the ideal form shown in Fig. 16-4a.

$$R_{eq} = R_1 + R_2 + R_3 = 0.51 + 1.3 + 0.24 = 2.05 \ \Omega$$

$$L_{eq} = L_1 + L_2 = 0.032 + 0.015 = 0.047 \ H$$

$$\frac{1}{C_{eq}} = \frac{1}{C_1} + \frac{1}{C_2} = \frac{1}{62 \times 10^{-6}} + \frac{1}{25 \times 10^{-6}}$$

$$C_{eq} = 17.8 \times 10^{-6} = 17.8 \ \mu F$$

$$f_r = \frac{1}{2\pi\sqrt{LC}} = \frac{1}{2\pi\sqrt{0.047 \times 17.8 \times 10^{-6}}} = 174 \ Hz$$

$$(b) \ Q_S = \frac{1}{R}\sqrt{\frac{L}{C}} = \frac{1}{2.05}\sqrt{\frac{0.047}{17.8 \times 10^{-6}}} = 25$$

$$(c) \ Q_{coil\,1} = \frac{\omega_r L_{coil\,1}}{R_{coil\,1}} = \frac{2\pi(174)(0.032)}{0.51} = 68.6$$

$$(d) \ Q_{coil\,2} = \frac{\omega_r L_{coil\,2}}{R_{coil\,2}} = \frac{2\pi(174)(0.015)}{1.3} = 12.6$$

16-4   BANDWIDTH

The general shape of a frequency-response curve that is representative of all resonance-type circuits is shown in Fig. 16-5. The vertical axis represents the applicable variable for the specific circuit.

   The range of frequencies within which the variable does not drop below 70.71 percent of its resonance value is called the passband or bandwidth. Thus, referring to Fig. 16-5, the bandwidth is

$$\boxed{BW = f_2 - f_1}$$

(16-15)

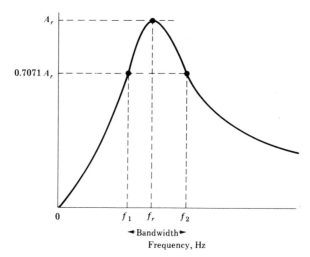

Figure 16-5 General shape of a frequency-response curve.

The frequency extremes of the bandwidth are called the cutoff or corner frequencies. They are also called the half-power frequencies because only half the power is drawn at these frequencies compared with the power drawn at the resonance frequency.

As indicated in Fig. 16-5, the resonance frequency is *not* centrally located with respect to the two half-power frequencies. The relationship between the resonance frequency and the two half-power frequencies is their *geometric mean*:

$$f_r = \sqrt{f_1 f_2} \qquad (16\text{-}16)$$

However, for circuits whose $Q \geq 10$, the resonance frequency is *sufficiently* centered with respect to the two cutoff frequencies that, for all practical purposes,

$$
\begin{aligned}
f_2 &= f_r + \frac{BW}{2} \\[2mm]
&\qquad\qquad Q \geq 10 \\[2mm]
f_1 &= f_r - \frac{BW}{2}
\end{aligned}
\qquad (16\text{-}17)
$$

This is clearly illustrated in Fig. 16-3, where

$$f_r = \frac{1}{2\pi\sqrt{LC}} = \frac{1}{2\pi\sqrt{0.0265(265 \times 10^{-6})}} = 60 \text{ Hz}$$

$$Q_{S1} = \frac{\omega_r L}{R} = \frac{2\pi(60)(0.0265)}{1} = 10$$

$$Q_{S2} = \frac{\omega_r L}{R} = \frac{2\pi(60)(0.0265)}{10} = 1$$

Circuit 1 has a $Q_S = 10$, and the resonance frequency appears to be centrally located with respect to the cutoff frequencies. Circuit 2 has a $Q_S = 1$, and its resonance frequency is not centrally located with respect to its cutoff frequencies.

Example 16-4 A series circuit has a resonance frequency of 150 kHz, a bandwidth of 75 kHz, and $Q_S = 2$. Determine the cutoff frequencies.

Solution
Since $Q_S < 10$, equation set (16-17) cannot be used.

$\text{BW} = f_2 - f_1$	$f_r = \sqrt{f_1 f_2}$
$75,000 = f_2 - f_1$	$150,000 = \sqrt{f_1 f_2}$
$f_2 = (75,000 + f_1)$	$22.5 \times 10^9 = f_1 f_2$

$$22.5 \times 10^9 = f_1(75,000 + f_1)$$

$$f_1^2 + 75,000f_1 - 22.5 \times 10^9 = 0$$

$$f_1 = \frac{-75,000 \pm \sqrt{(75,000)^2 - 4(-22.5 \times 10^9)}}{2}$$

$$f_1 = 117,100, \quad -192,100$$

However, since a negative frequency does not exist,

$$f_1 = 117.1 \text{ kHz}$$

$$\text{BW} = f_2 - f_1$$

$$75,000 = f_2 - 117,100$$

$$f_2 = 192.1 \text{ kHz}$$

Derivation of Equation (16-16)

The rms current at the resonance frequency is

$$I = \frac{V}{Z_r} = \frac{V}{R}$$

At the half-power frequencies, the current is reduced to

$$I_{1/2P} = 0.7071\, I = 0.7071\, \frac{V}{R} = \frac{V}{R/0.7071}$$

Thus the impedance offered by the series circuit at the half-power frequencies is

$$Z_{1/2P} = R/0.7071 = \sqrt{2}\,R \qquad (16\text{-}18)$$

The impedance at any frequency may be determined from

$$Z = \sqrt{R^2 + \left(2\pi f L - \frac{1}{2\pi f C}\right)^2} \qquad (16\text{-}19)$$

By relating Fig. 16-2a to Fig. 16-5, it can be determined that

At the lower half-power frequency (f_1), $X_C > X_L$

At the higher half-power frequency (f_2), $X_L > X_C$

Substituting Eq. (16-18) into Eq. (16-19), and letting $f = f_1$ and f_2, respectively, provides the impedance relationships at the half-power frequencies.

For f=f₁

$$\sqrt{2}\,R = \sqrt{R^2 + \left(\frac{1}{2\pi f_1 C} - 2\pi f_1 L\right)^2}$$

For f=f₂

$$\sqrt{2}\,R = \sqrt{R^2 + \left(2\pi f_2 L - \frac{1}{2\pi f_2 C}\right)^2}$$

Squaring both sides and letting $\omega = 2\pi f$,

$$2R^2 = R^2 + \left(\frac{1}{\omega_1 C} - \omega_1 L\right)^2$$

$$2R^2 = R^2 + \left(\omega_2 L - \frac{1}{\omega_2 C}\right)^2$$

Solving for ω,

$$\omega_1 = \frac{-RC \pm \sqrt{(RC)^2 + 4LC}}{2LC}$$

$$\omega_2 = \frac{RC \pm \sqrt{(RC)^2 + 4LC}}{2LC}$$

Since a negative frequency is meaningless, only the positive root of the frequency equation is significant. Hence,

$$\omega_1 = \frac{-RC + \sqrt{(RC)^2 + 4LC}}{2LC} \qquad \omega_2 = \frac{RC + \sqrt{(RC)^2 + 4LC}}{2LC}$$

(16-20)

Obtaining the product of the two half-power frequencies,

$$\omega_1\omega_2 = \frac{-(RC)^2 + (RC)^2 + 4LC}{4(LC)^2}$$

$$\omega_1\omega_2 = \frac{1}{LC}$$

Substituting $\omega = 2\pi f$,

$$(2\pi f_1)(2\pi f_2) = \frac{1}{LC}$$

$$f_1 f_2 = \frac{1}{4\pi^2 LC}$$

$$\sqrt{f_1 f_2} = \frac{1}{2\pi\sqrt{LC}}$$

From Eq. (16-4), the resonance frequency is

$$f_r = \frac{1}{2\pi\sqrt{LC}}$$

Hence

$$f_r = \sqrt{f_1 f_2}$$

The quality factor of a circuit operating at its resonance frequency may be expressed in terms of the ratio of its resonance frequency to its bandwidth. From Eq. (16-9),

$$Q_S = \frac{2\pi f_r L}{R} = \frac{f_r}{R/2\pi L}$$

(16-21)

From Eq. (16-20),

$$\omega_2 - \omega_1 = \frac{R}{L}$$

or

$$2\pi(f_2 - f_1) = \frac{R}{L}$$

$$2\pi(\text{BW}) = \frac{R}{L}$$

$$\text{BW} = \frac{R}{2\pi L} \tag{16-22}$$

Substituting Eq. (16-22) into Eq. (16-21),

$$\boxed{Q_S = \frac{f_r}{\text{BW}}} \tag{16-23}$$

A circuit with a smaller bandwidth will have a higher Q and, as illustrated in Fig. 16-3, will have a sharper resonance peak and greater *selectivity*.

Example 16-5 Determine the parameters of an RLC series circuit that will resonate at 10,000 Hz, have a bandwidth of 1000 Hz, and draw 15.3 W from a 200-V generator operating at the resonance frequency of the circuit.

Solution
For a series circuit operating at resonance,

$$V_R = V_T = 200 \text{ V}$$

$$P_R = \frac{V_R^2}{R}$$

$$15.3 = \frac{(200)^2}{R}$$

$$R = 2.61 \text{ k}\Omega$$

$$Q_S = \frac{f_r}{\text{BW}}$$

$$Q_S = \frac{10{,}000}{1000} = 10$$

$$Q_S = \frac{2\pi f_r L}{R}$$

$$10 = \frac{2\pi(10{,}000)L}{2610}$$

$$L = 416 \text{ mH}$$

$$f_r = \frac{1}{2\pi\sqrt{LC}}$$

$$10{,}000 = \frac{1}{2\pi\sqrt{0.416C}}$$

$$C = 610 \text{ pF}$$

Example 16-6   An RLC series circuit has a Q_S of 5.1 at a frequency of 100 kHz. Assuming the power dissipation of the circuit is 100 W when drawing a current of 0.80 A, determine (a) the circuit parameters; (b) the bandwidth; (c) the half-power frequencies.

Solution

(a) $P_R = I^2 R$

$$100 = (0.8)^2 R$$

$$R = 156 \ \Omega$$

$$Q_S = \frac{2\pi f_r L}{R}$$

$$5.1 = \frac{2\pi(100{,}000)L}{156}$$

$$L = 1.26 \text{ mH}$$

$$f_r = \frac{1}{2\pi\sqrt{LC}}$$

$$100{,}000 = \frac{1}{2\pi\sqrt{(1.26 \times 10^{-3})C}}$$

$$C = 2.01 \text{ nF}$$

(b) $Q_S = \dfrac{f_r}{BW}$

$5.1 = \dfrac{100{,}000}{BW}$

$BW = 19.6 \text{ kHz}$

(c) $f_r = \sqrt{f_1 f_2}$

$(100{,}000)^2 = f_1 f_2$

$BW = f_2 - f_1$

$19{,}600 = f_2 - f_1$

$(19{,}600 + f_1) = f_2$

Substituting,

$(10)^{10} = f_1(19{,}600 + f_1)$

$f_1^2 + 19{,}600 f_1 - (10)^{10} = 0$

$f_1 = \dfrac{-19{,}600 + \sqrt{(19{,}600)^2 + 4(10)^{10}}}{2}$

$f_1 = 90{,}700 \text{ Hz} = 90.7 \text{ kHz}$

$BW = f_2 - f_1$

$19{,}600 = f_2 - 90{,}700$

$f_2 = 110{,}000 \text{ Hz} = 110 \text{ kHz}$

16-5   RESONANCE RISE IN VOLTAGE

The current in a series circuit operating at its resonance frequency is limited by the resistance alone. Hence decreasing the resistance of the circuit increases the current and thus effects a proportional increase in V_L and V_C. For *any* series RLC circuits,

$$\mathbf{Z} = R + j(X_L - X_C)$$

If the frequency of the source is equal to the resonance frequency of the circuit,

$$X_L = X_C$$

Hence,

$$Z_r = R$$

and

$$I_T = \frac{V_T}{R} \tag{16-24}$$

The magnitudes of the voltage drops across the respective circuit elements are

$$V_R = I_T R$$

$$V_L = I_T X_L \tag{16-25}$$

$$V_C = I_T X_C \tag{16-26}$$

The high rms currents associated with low values of circuit resistance, when a series RLC circuit is at resonance, may produce voltages across the coil and across the capacitor that are many times greater than the driving voltage. Voltages that are significantly higher than the rated voltages of the respective circuit elements may result in destruction of the capacitor and damage to the insulation of the coil. Hence caution must always be exercised when connecting capacitors and coils in series, for if the particular combination of inductance and capacitance is in resonance at the applied frequency, damage to equipment and injury to personnel may occur unless a current-limiting resistor is used.

Example 16-7   A 125-V sinusoidal generator supplies a series circuit consisting of a 20.5-μF capacitor and a coil whose resistance and inductance are 1.06 Ω and 25.4 mH, respectively. The generator frequency is the resonance frequency of the circuit. Determine (a) the resonance frequency; (b) the rms current; (c) the rms voltage across the capacitor; (d) the rms voltage across the coil; (e) the resistance of a resistor that must be connected in series with the circuit to limit the capacitor voltage to 300 V rms.

Solution
The circuit is similar to that shown in Fig. 16-6a.

(a) $f_r = \dfrac{1}{2\pi\sqrt{LC}} = \dfrac{1}{2\pi\sqrt{0.0254(20.5 \times 10^{-6})}} = 220.6$ Hz

(b) At resonance, $\mathbf{Z}_r = R$; hence

$$I = \frac{\mathbf{V}_T}{\mathbf{Z}_r} = \frac{125\underline{/0^\circ}}{1.06} = 117.9\underline{/0^\circ}\ \text{A}$$

(c) $X_C = X_L = 2\pi f_r L = 2\pi(220.6)(0.0254) = 35.21\ \Omega$

$V_C = IX_C = (117.9)(35.21) = 4151\ V$

(d) $V_{coil} = IR + IjX_L = 117.9\underline{/0°}\ (1.06 + j35.21)$

$V_{coil} = (117.9\underline{/0°})(35.23\underline{/88.3°}) = 4154\underline{/88.3°}$

$V_{coil} = 4154\ V$

(e) $V_C = IX_C$

$300 = I(35.21)$

$I = 8.520\ A$

$I = \dfrac{V}{R}$

$8.52 = \dfrac{125}{R}$

$R = 14.67\ \Omega$

Hence the required additional series resistance is

$14.67 = 1.06 + R_X$

$R_X = 13.61\ \Omega$

The voltage across the inductance, and the voltage across the capacitor, at resonance, may be expressed in terms of Q_S by substituting Eq. (16-24) into Eqs. (16-25) and (16-26),

$$V_L = I_T X_L = \frac{V_T}{R}\omega_r L \qquad\qquad V_C = I_T X_C = \frac{V_T}{R}\frac{1}{\omega_r C}$$

$$\boxed{V_L = V_T Q_S} \qquad\qquad \boxed{V_C = V_T Q_S} \qquad\qquad (16\text{-}27)$$

Equation set (16-27) applies to series RLC circuits in ideal form as represented in Fig. 16-6a.

Figure 16-6b shows graphs of rms voltage across R, L, and C, and rms current, as functions of frequency for the series circuit shown in Fig. 16-6a.

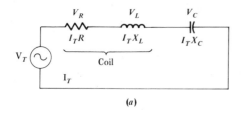

(a)

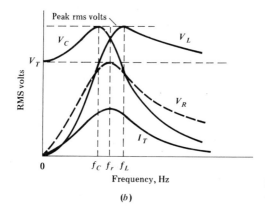

Frequency, Hz

(b)

Figure 16-6 Voltage drop versus frequency for series *RLC* circuit.

At 0 Hz (DC), the capacitor is charged to the source voltage V_T, the current is zero, V_R is zero, and the voltage across the inductance is zero. As the frequency of the source is adjusted upward from 0 Hz, V_R, V_L, V_C, and I_T increase to their respective peak rms values and then decrease. As the driving frequency approaches infinity, V_C, V_R, and I_T approach zero, and V_L approaches V_T.

Although V_L and V_C are equal at resonance, their maximum values do not necessarily occur at the resonance frequency. The frequencies at which these maximums occur may be determined from the following equations:

$$f_C = \frac{1}{2\pi} \sqrt{\frac{1}{LC} - \frac{R^2}{2L^2}}$$ (16-28)

$$f_L = \frac{1}{2\pi} \sqrt{\frac{2}{2LC - R^2C^2}}$$ (16-29)

As the circuit resistance is decreased, f_C and f_L approach f_r, and for values of circuit resistance that cause $Q_S \geq 10$, f_C and f_L will be essentially coincident with f_r.

Derivation of Equations (16-28) and (16-29)

For the circuit in Fig. 16-6*a*,

$$I = \frac{V_T}{Z}$$

$$Z = \sqrt{R^2 + (X_L + X_C)^2}$$

$$V_C = IX_C = \frac{V_T X_C}{Z} \qquad\qquad V_L = IX_L = \frac{V_T X_L}{Z}$$

Squaring both sides,

$$V_C^2 = \frac{V_T^2 X_C^2}{Z^2} \qquad\qquad V_L^2 = \frac{V_T^2 X_L^2}{Z^2}$$

Expressing X_C, X_L, and Z in terms of ω,

$$V_C^2 = \frac{V_T^2}{(\omega C)^2[R^2 + (\omega L - 1/\omega C)^2]} \qquad\qquad V_L^2 = \frac{V_T^2(\omega L)^2}{R^2 + (\omega L - 1/\omega C)^2}$$

Rearranging terms, differentiating with respect to ω, and then setting the resultant equal to zero,

$$V_C^2 = \frac{V_T^2}{(\omega RC)^2 + (\omega^2 LC - 1)^2} \qquad\qquad V_L^2 = \frac{\omega^4 L^2 C^2 V_T^2}{(\omega RC)^2 + (\omega^2 LC - 1)^2}$$

$$\frac{d(V_C^2)}{d\omega} = \frac{-V_T^2[2\omega C^2 R^2 + 2(\omega^2 LC - 1)(2\omega LC)]}{[(\omega RC)^2 + (\omega^2 LC - 1)^2]^2} = 0$$

Solving for ω,

$$\omega_C = \sqrt{\frac{1}{LC} - \frac{R^2}{2L^2}}$$

Following a similar procedure, the value of ω at which V_L has its maximum rms value is

$$\omega_L = \sqrt{\frac{2}{2LC - R^2 C^2}}$$

16-6   PARALLEL RESONANCE

In accordance with the definition of resonance, the parallel circuit shown in Fig. 16-7 will be in resonance if the frequency is such as to cause the input admittance to become equivalent to the circuit conductance.

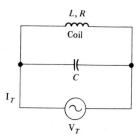

L, R

Coil

C

I_T

V_T

Figure 16-7 Parallel circuit.

The input admittance for the circuit in Fig. 16-7 is

$$\mathbf{Y}_{in} = \frac{1}{R_{coil} + jX_{L\,coil}} + \frac{1}{-jXC} \tag{16-30}$$

Expressing Eq. (16-30) in terms of ω,

$$\mathbf{Y}_{in} + \frac{1}{R_{coil} + j\omega L_{coil}} + \frac{1}{-j(1/\omega C)}$$

Rationalizing the respective denominators, and then separating the real and imaginary terms,

$$\mathbf{Y}_{in} = \frac{R_{coil} - j\omega L_{coil}}{R_{coil}^2 + \omega^2 L_{coil}^2} + j\omega C$$

$$\mathbf{Y}_{in} = \frac{R_{coil}}{R_{coil}^2 + \omega^2 L_{coil}^2} - \frac{j\omega L_{coil}}{R_{coil}^2 + \omega^2 L_{coil}^2} + j\omega C$$

$$\mathbf{Y}_{in} = \frac{R_{coil}}{R_{coil}^2 + \omega^2 L_{coil}^2} + j\left(\frac{-\omega L_{coil}}{R_{coil}^2 + \omega^2 L_{coil}^2} + \omega C \right) \tag{16-31}$$

If the frequency is such as to cause the parallel circuit to be in resonance, the imaginary term will be zero, and

$$\mathbf{Y}_{in} = G + j0 \tag{16-32}$$

Denoting the radian frequency at resonance as ω_r, and equating the real and imaginary terms of Eq. (16-31) with the corresponding terms of Eq. (16-32),

$$G = \frac{R_{coil}}{R_{coil}^2 + \omega_r^2 L_{coil}^2} \tag{16-33}$$

$$j0 = j\left(\frac{-\omega_r L_{coil}}{R_{coil}^2 + \omega_r^2 L_{coil}^2} + \omega_r C \right) \tag{16-34}$$

Solving Eq. (16-34) for ω_r,

$$\omega_r = \sqrt{\frac{1}{L_{coil}C} - \frac{R_{coil}^2}{L_{coil}^2}} \qquad (16\text{-}35)$$

or

$$f_r = \frac{1}{2\pi}\sqrt{\frac{1}{L_{coil}C} - \frac{R_{coil}^2}{L_{coil}^2}} \qquad (16\text{-}36)$$

where f_r = resonance frequency, Hz
ω_r = resonance frequency, rad/s
L_{coil} = inductance of coil, H
C = capacitance, F
R_{coil} = resistance of coil, Ω

Note 1: Equations (16-35) and (16-36) represent a common but special case with no resistance in the capacitive branch.

Note 2: If the parameters of the circuit are such that the radicand in Eqs. (16-35) and (16-36) is negative, that is, if

$$\left(\frac{1}{L_{coil}C} - \frac{R_{coil}^2}{L_{coil}^2}\right) < 0$$

the circuit has no resonance frequency.

Assuming the resistance of the coil is insignificant, that is, $R_{coil} = 0$, Eqs. (16-35) and (16-36) reduce to

$$\begin{array}{cc} \omega_r = \dfrac{1}{\sqrt{L_{coil}C}} \\[4mm] f_r = \dfrac{1}{2\pi\sqrt{L_{coil}C}} \end{array} \qquad R_{coil} = 0 \qquad (16\text{-}37)$$

From Eq. (16-37),

$$\omega_r L_{coil} = \frac{1}{\omega_r C}$$

$$X_L = X_C$$

$$R_{coil} = 0$$

With $R_{coil} = 0$, the conductance of the circuit as expressed in Eq. (16-33) becomes

$$G = 0$$

Thus, for the circuit in Fig. 16-7 if $R_{coil} = 0$, and the driving frequency is the resonance frequency of the circuit, then the admittance as expressed in Eq. (16-32) is

$$Y_{in} = 0 + j0$$

The admittance of the circuit is zero and no current will pass!

$$I = VY_{in} = V(0) = 0$$

Figure 16-8a shows the effect of coil resistance on the resonance frequency of an LC parallel circuit. The graph is a plot of Eq. (16-35), where $C = 7.75\ \mu F$, $L = 0.0020$ H, and the coil resistance is represented by an adjustable resistor.

A graph of admittance *magnitude* versus frequency for the same circuit, assuming a coil resistance of 10 Ω, is shown in Fig. 16-8b. Note that the frequency at which resonance occurs and the frequency at which Y_{in} is a minimum are not the same! Coincidence can occur only for the ideal condition of $R_{coil} = 0$. The equation relating admittance magnitude to radian frequency is obtained from the square root of the sum of the squares of the real and imaginary terms in Eq. (16-31); that is,

$$Y_{in} = \sqrt{\left(\frac{R_{coil}}{R_{coil}^2 + \omega^2 L_{coil}^2}\right)^2 + \left(\frac{-\omega L_{coil}}{R_{coil}^2 + \omega^2 L_{coil}^2} + \omega C\right)^2} \qquad (16\text{-}38)$$

Graphs of input impedance and rms current versus frequency for the circuit in Fig. 16-8b are shown in Fig. 16-8c; the rms value of the sinusoidal driving voltage is held constant at 100 V.

A phase plot for the circuit in Fig. 16-8b is shown in Fig. 16-9. At frequencies below resonance, $X_L < X_C$; this causes $I_L > I_C$, resulting in a lagging current. At resonance, the current is in phase with the voltage. At frequencies above resonance, $X_C < X_L$; this causes $I_C > I_L$, resulting in a leading current.

Example 16-8 A 60-V sinusoidal generator supplies a parallel circuit consisting of a 2.5-μF capacitor and a coil whose resistance and inductance are 260 mH and 15 Ω, respectively. Determine the resonance frequency.

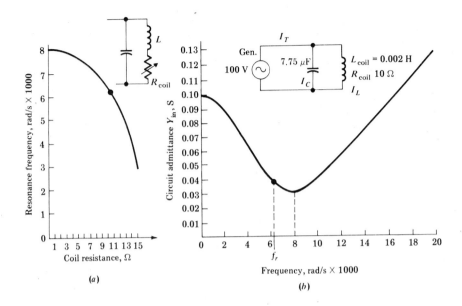

(a)

(b)

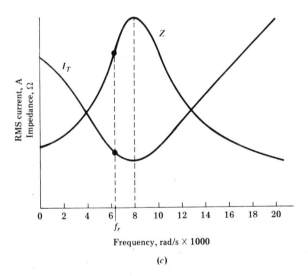

(c)

Figure 16-8 Parallel resonance characteristics: (a) resonance frequency versus coil resistance; (b) circuit admittance versus frequency; (c) circuit current and circuit impedance versus frequency.

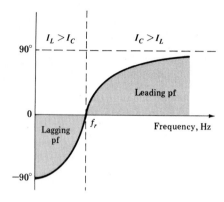

Figure 16-9 Phase angle versus frequency for the parallel circuit shown in Fig. 16-8b.

Solution

The circuit is similar to that shown in Fig. 16-8b.

$$f_r = \frac{1}{2\pi} \sqrt{\frac{1}{L_{\text{coil}}C} - \frac{R^2_{\text{coil}}}{L^2_{\text{coil}}}}$$

$$f_r = \frac{1}{2\pi} \sqrt{\frac{1}{2.5 \times 10^{-6}(0.260)} - \frac{(15)^2}{(0.260)^2}} = 197 \text{ Hz}$$

16-7   **QUALITY FACTOR OF A PARALLEL RESONANCE CIRCUIT**

The quality factor of a parallel resonance circuit is derived using *ideal circuit elements*, as shown in Fig. 16-10, and the defining equation (16-6):

$$Q = \frac{2\pi \text{ (maximum energy stored per cycle)}}{\text{energy dissipated per cycle}} \tag{16-6}$$

At resonance, there is a continuous alternating transfer of energy between the inductor and the capacitor. All the energy released by the inductor is

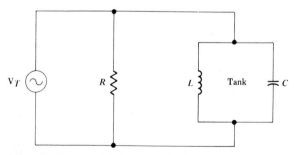

Figure 16-10   Parallel circuit using ideal elements.

absorbed by the capacitor, and when released by the capacitor is absorbed by the inductor. For this reason the parallel LC combination shown in Fig. 16-10 is often called a *tank circuit* or *tank*.

The maximum energy stored in a tank circuit at resonance occurs in the capacitor when the sinusoidal voltage across it attains its maximum value, and in the inductance when the sinusoidal current through it attains its maximum value. Furthermore, when the capacitor is releasing energy, the inductor is absorbing it and vice versa. Thus, the maximum amount of energy stored in the tank circuit at resonance may be expressed by

$$\text{Maximum energy stored} = \tfrac{1}{2}CV_{max}^2 \qquad (16\text{-}39)$$

The energy dissipated in one cycle of the resonance frequency is equal to the heat-power losses in the resistor times the period of the wave.

$$\text{Energy dissipated per cycle} = \frac{V^2}{R}\,T_r \qquad (16\text{-}40)$$

where V_{max} = maximum value of voltage wave
$\quad\quad\quad V$ = rms value of voltage wave
$\quad\quad\quad T_r$ = period of the voltage wave at the resonance frequency

Substituting Eqs. (16-39) and (16-40) into Eq. (16-6),

$$Q_P = \frac{2\pi(1/2\,CV_{max}^2)}{(V^2/R)\,T_r}$$

Substituting,

$$T_r = \frac{1}{f_r} \qquad\qquad\qquad V_{max} = \sqrt{2}\,V$$

$$Q_P = \frac{2\pi[1/2\,C(\sqrt{2}\,V)^2]}{(V^2/R)(1/f_r)} = 2\pi f_r\,CR$$

$$\boxed{Q_P = \omega_r CR = \frac{R}{X_C}} \qquad (16\text{-}41)$$

For the *ideal elements* shown in Fig. 16-10, at resonance

$$\boxed{\omega_r L = \frac{1}{\omega_r C}}$$

Hence Eq. (16-41) may also be written as

$$Q_P = \frac{R}{\omega_r L} = \frac{R}{X_L} \qquad (16\text{-}42)$$

Substituting $\omega_r = 1/\sqrt{LC}$ into Eq. (16-41) provides a third expression for Q_P:

$$Q_P = \frac{CR}{\sqrt{LC}}$$

$$Q_P = R\sqrt{\frac{C}{L}} \qquad (16\text{-}43)$$

Thus, depending on the available data, the quality factor of a parallel resonance circuit, using *ideal elements,* may be determined from any one of the following relationships:

$$Q_P = \omega_R CR = \frac{R}{\omega_R L} = R\sqrt{\frac{C}{L}}$$

$$R_{\text{coil}} = 0 \qquad (16\text{-}44)$$

If the coil is *nonideal,* as shown in Fig. 16-11a, the series-connected elements that make up the coil must be converted to the equivalent parallel elements shown in Fig. 16-11b. This is accomplished by using the series-to-parallel conversion formula developed in Sec. 12-9.

$$R_p = \frac{R_S^2 + X_S^2}{R_S} \qquad \qquad \frac{R_S^2 + X_S^2}{X_S} = X_P$$

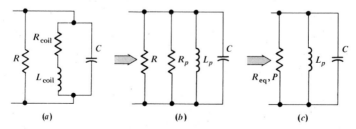

(a)   (b)   (c)

Figure 16-11 Conversion of a nonideal coil to equivalent ideal form.

Dividing through and rearranging terms,

$$R_P = R_S + \frac{X_S^2}{R_S} = R_S + \frac{X_S^2}{R_S}\frac{R_S}{R_S} = R_S + \left(\frac{X_S}{R_S}\right)^2 R_S = R_S(1 + Q_S^2)$$

In terms of coil parameters,

$$\boxed{R_P = R_{\text{coil}}(1 + Q_{\text{coil}}^2)} \tag{16-45}$$

Similarly,

$$X_P = \frac{R_S^2}{X_S} + X_S = \frac{R_S^2}{X_S}\frac{X_S}{X_S} + X_S = \frac{X_S}{Q^2} + X_S$$

$$\boxed{X_P = X_{\text{coil}}\left(1 + \frac{1}{Q_{\text{coil}}^2}\right)} \tag{16-46}$$

Converting the paralleled resistors in Fig. 16-11b to a single equivalent resistor results in the equivalent *ideal circuit* shown in Fig. 16-11c. Relating Eq. (16-44) to Fig. 16-11c,

$$\boxed{Q_P = \omega_r C R_{\text{eq}} = \frac{R_{\text{eq}}}{\omega_r L_P} = R_{\text{eq},P}\sqrt{\frac{C}{L_P}}} \tag{16-47}$$

where

$$R_{\text{eq},P} = \frac{R R_P}{R + R_P}$$

Note: The addition of paralleled resistors does not affect the resonance frequency.

The Q_P of a parallel resonance circuit may also be determined from the ratio of resonance frequency to bandwidth:

$$\boxed{Q_P = \frac{f_r}{\text{BW}}} \tag{16-48}$$

Equation (16-48) is identical to that obtained for a series resonance circuit.

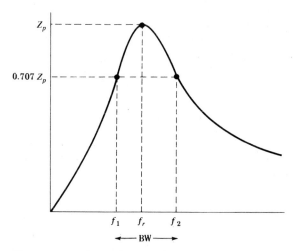

Figure 16-12 Determination of bandwidth for a parallel resonance circuit.

The bandwidth shown in Fig. 16-12 for a parallel resonance circuit is the range of frequencies within which the *impedance* does not drop below 70.71 percent of its value at resonance.

16-8   **SPECIAL CASE** $Q_{coil} \geq 10$

In those applications where $Q_{coil} \geq 10$,

$$
\left.
\begin{aligned}
f_r &= \frac{1}{2\pi\sqrt{LC}} \\[2mm]
X_P &= X_{coil} = X_C \\[2mm]
L_P &= L_{coil}
\end{aligned}
\right\} \quad Q_{coil} \geq 10
$$

(16-49)

(16-50)

(16-51)

Derivation of Equation (16-49)

From Eq. (16-35),

$$
\omega_r^2 = \frac{1}{L_{coil}C} - \frac{R_{coil}^2}{L_{coil}^2}
$$

Rearranging terms,

$$
\omega_r^2 + \frac{R_{coil}^2}{L_{coil}^2} = \frac{1}{L_{coil}C}
$$

(16-52)

However,

$$Q_{\text{coil}} = \frac{\omega_r L_{\text{coil}}}{R_{\text{coil}}} \qquad\qquad (16\text{-}53)$$

Solving Eq. (16-53) for R_{coil}, substituting into Eq. (16-52), and solving for ω_r,

$$\omega_r^2 + \frac{\omega_r^2 L_{\text{coil}}^2}{L_{\text{coil}}^2 Q_{\text{coil}}^2} = \frac{1}{L_{\text{coil}} C}$$

$$\omega_r^2 \left(1 + \frac{1}{Q_{\text{coil}}^2}\right) = \frac{1}{L_{\text{coil}} C}$$

$$\omega_r^2 = \frac{1}{L_{\text{coil}} C} \frac{Q_{\text{coil}}^2}{1 + Q_{\text{coil}}^2}$$

$$\omega_r = \sqrt{\frac{1}{L_{\text{coil}} C} \frac{Q_{\text{coil}}^2}{1 + Q_{\text{coil}}^2}}$$

If $Q \geq 10$,

$$\frac{Q_{\text{coil}}^2}{1 + Q_{\text{coil}}^2} \approx 1$$

and

$$\omega_r = \frac{1}{\sqrt{L_{\text{coil}} C}} \qquad\qquad f_r = \frac{1}{2\pi\sqrt{L_{\text{coil}} C}}$$

Example 16-9 A parallel circuit consisting of a 65-pF capacitor and a coil whose inductance and resistance are 56 μH and 60 Ω, respectively, are connected to the output of a transistor, as shown in Fig. 16-13a. The transistor acting as a current source has a source resistance of 37 kΩ. Determine (a) the resonance frequency; (b) Q_{coil}; (c) Q of the circuit; (d) the bandwidth.

Solution

(a) $f_r = \dfrac{1}{2\pi} \sqrt{\dfrac{1}{L_{\text{coil}} C} - \dfrac{R_{\text{coil}}^2}{L_{\text{coil}}^2}}$

$$f_r = \frac{1}{2\pi} \sqrt{\frac{1}{(56 \times 10^{-6})(65 \times 10^{-12})} - \frac{(60)^2}{(56 \times 10^{-6})^2}} = 2.63 \text{ MHz}$$

(b) $Q_{\text{coil}} = \dfrac{\omega L_{\text{coil}}}{R_{\text{coil}}} = \dfrac{2\pi(2.63 \times 10^6)(56 \times 10^{-6})}{60} = 15.4$

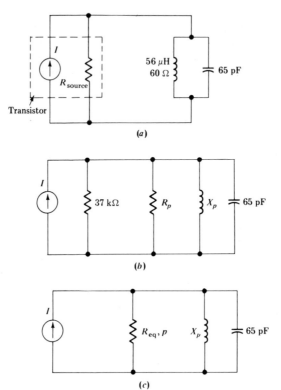

Figure 16-13 Circuits for Example 16-9.

(c) To determine Q_P, it is necessary to convert the coil in Fig. 16-13a to equivalent parallel components as shown in Fig. 16-13b, and then convert the paralleled resistors to a single equivalent resistor as shown in Fig. 16-13c. Figure 16-13c is the equivalent ideal model of the actual circuit in Fig. 16-13a and is the model that should be used when determining Q_P. Thus,

$$R_P = R_{coil}(1 + Q_{coil}^2) = 60[1 + (15.4)^2] = 14.3 \text{ k}\Omega$$

$$R_{eq, P} = \frac{R_G R_P}{R_G + R_P} = \frac{37,000 \times 14,300}{37,000 + 14,300} = 10.3 \text{ k}\Omega$$

$$Q_P = \omega_r C R_{eq, P} = 2\pi(2.63 \times 10^6)(65 \times 10^{-12})(10.3 \times 10^3)$$

$$Q_P = 11$$

(d) $Q_P = \dfrac{f_r}{BW}$

$$11 = \frac{2.63 \times 10^6}{BW}$$

$$BW = 239 \text{ kHz}$$

Example 16-10 Determine (a) the capacitance of a tank circuit that will resonate at 22.3 kHz and have a bandwidth of 4.05 kHz, when using a coil whose resistance and inductance are 56 Ω and 3.2 mH respectively; assume the tank circuit is supplied by a current source whose source resistance is 45.8 kΩ; (b) the amount of additional parallel resistance required to change the bandwidth to 6 kHz.

Solution

(a) the circuit is shown in Fig. 16-14a.

$$Q_{coil} = \frac{\omega_r L_{coil}}{R_{coil}} = \frac{2\pi(22.3 \times 10^3)(0.0032)}{56} = 8$$

$Q_{coil} < 10$; hence special-case formulas cannot be used.

$$R_P = R_{coil}(1 + Q_{coil}^2) = 56[1 + (8)^2] = 3640 \ \Omega$$

$$R_{eq, P} = \frac{R_{source} R_P}{R_{source} + R_P} = \frac{45,800 \times 3640}{45,800 + 3640} = 3372 \ \Omega$$

$$Q_P = \frac{f_r}{BW} = \frac{22.3 \times 10^3}{4.05 \times 10^3} = 5.51$$

$$Q_P = \omega_r C R_{eq, P}$$

$$5.51 = 2\pi(22.3 \times 10^3)(C)(3372)$$

$$C = 11.7 \ nF$$

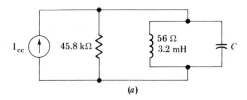

(a)

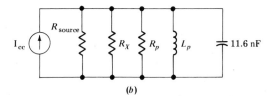

(b)

Figure 16-14 Circuits for Example 16-10.

(b) The equivalent circuit is shown in Fig. 16-14b.

$$Q_P = \frac{f_r}{BW} = \frac{22.3 \times 10^3}{6 \times 10^3} = 3.72$$

$$Q_P = \omega_r C R_{eq,P}$$

$$3.72 = 2\pi(22.3 \times 10^3)(11.7 \times 10^{-9})R_{eq,P}$$

$$R_{eq,P} = 2269 \ \Omega$$

$$\frac{1}{R_{eq,P}} = \frac{1}{R_{source}} + \frac{1}{R_P} + \frac{1}{R_X}$$

$$\frac{1}{2269} = \frac{1}{45.8 \times 10^3} + \frac{1}{3640} + \frac{1}{R_X}$$

$$R_X = 6937 \ \Omega$$

Example 16-11 A tank circuit consisting of a capacitor in parallel with a coil whose inductance and resistance are 1.05 mH and 100 Ω respectively, is driven at its resonance frequency of 600 kHz from a constant-current source. The source consists of a 2.30-mA, 600-kHz, constant-current generator in parallel with a 60-kΩ source resistance. Determine (a) Q_{coil}; (b) capacitance; (c) Q_P; (d) BW; (e) rms voltage across the capacitor; (f) maximum instantaneous energy stored in the capacitor; (g) heat power dissipated.

Solution
The circuit is similar to that shown in Fig. 16-13a.

(a) $Q_{coil} = \dfrac{\omega_r L_{coil}}{R_{coil}} = \dfrac{2\pi(600 \times 10^3)(0.00105)}{100} = 39.58$

(b) Since $Q_{coil} > 10$

$$f_r = \frac{1}{2\pi\sqrt{LC}}$$

$$600 \times 10^3 = \frac{1}{2\pi\sqrt{0.00105C}}$$

$$C = 67 \ pF$$

(c) $R_P = R_{coil}(1 + Q_{coil}^2) = 100[1 + (39.58)^2] = 156.8 \ k\Omega$

$$R_{eq,P} = \frac{R_P R_{source}}{R_P + R_{source}} = \frac{156,800 \times 60,000}{156,800 + 60,000} = 43.39 \ k\Omega$$

$$Q_P = \omega_r C R_{eq,P} = 2\pi 600 \times 10^3(67 \times 10^{-12})(43,390) = 11.0$$

(d) $Q_P = \dfrac{f_r}{BW}$

$11.0 = \dfrac{600 \times 10^3}{BW}$

$BW = 54.5 \text{ kHz}$

(e) The voltage across the capacitor is the voltage across the parallel circuit. Referring to Fig. 16-13c, at resonance, $\mathbf{Z}_{in} = R_{eq,P}$. Therefore,

$V = IR_{eq,P} = 0.0023(43,390) = 99.8 \text{ V}$

(f) $W_C = \frac{1}{2}CV_{max}^2 = \frac{1}{2}(67 \times 10^{-12})(99.8\sqrt{2})^2 = 667 \text{ nJ}$

(g) $P = I^2 R_{eq,P} = (0.0023)^2(43,390) = 230 \text{ mW}$

SUMMARY OF FORMULAS

Series Resonance (see Fig. 16-15)

$$f_r = \dfrac{1}{2\pi\sqrt{LC}} \qquad f_r = \sqrt{f_2 f_1} \qquad \omega_r = 2\pi f_r$$

$$BW = f_2 - f_1 \qquad Q_S = \dfrac{f_r}{BW} \qquad f_2 = f_r + \dfrac{BW}{2}$$

$$Q_S \geq 10$$

$$f_1 = f_r - \dfrac{BW}{2}$$

$$Q_S = \dfrac{\omega_r L_{coil}}{R_{eq,S}} = \dfrac{1}{\omega_r C R_{eq,S}} = \dfrac{1}{R_{eq,S}}\sqrt{\dfrac{L_{coil}}{C}}$$

$$R_{eq,S} = R_{coil} + R_{source} + R_X$$

$$V_L = V_C = V_T Q_S \qquad\qquad \mathbf{Z}_{in} = R_{eq,S}$$

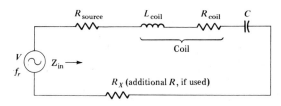

Figure 16-15 Circuit for series resonance formulas.

Parallel Resonance   (see Fig. 16-16)

$$f_r = \frac{1}{2\pi}\sqrt{\frac{1}{L_{coil}C} - \frac{R_{coil}^2}{L_{coil}^2}} \qquad f_r = \sqrt{f_2 f_1} \qquad \omega_r = 2\pi f_r$$

$$\text{BW} = f_2 - f_1 \qquad\qquad Q_P = \frac{f_r}{\text{BW}} \qquad \begin{matrix} f_2 = f_r + \dfrac{\text{BW}}{2} \\[2mm] f_1 = f_r - \dfrac{\text{BW}}{2} \end{matrix} \quad Q_P \geq 10$$

$$R_P = R_{coil}(1 + Q_{coil}^2) \qquad X_P = X_{coil}\left(1 + \frac{1}{Q_{coil}^2}\right)$$

$$Q_P = \omega_r C R_{eq,P} = \frac{R_{eq,P}}{\omega_r L_P} = R_{eq,P}\sqrt{\frac{C}{L_P}}$$

$$Q_{coil} = \frac{\omega_r L_{coil}}{R_{coil}} \qquad\qquad Z_{in} = R_{eq,P} = R_{in}$$

$$\frac{1}{R_{eq,P}} = \frac{1}{R_{source}} + \frac{1}{R_P} + \frac{1}{R_X}$$

R_X = additional resistance, if used

$$\left.\begin{matrix} f_r = \dfrac{1}{2\pi\sqrt{L_{coil}C}} \\[4mm] X_P = X_{coil} = X_C \\[3mm] L_P = L_{coil} \end{matrix}\right\} \quad Q_{coil} \geq 10$$

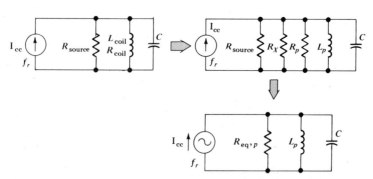

Figure 16-16   Circuit for parallel resonance formulas.

ROBLEMS

Series Resonance

16-1 A coil whose resistance and inductance are 40 Ω and 50 mH, respectively, is connected in series with a 450-pF capacitor and a generator. Determine (*a*) the resonance frequency; (*b*) the circuit impedance at the resonance frequency; (*c*) the circuit current if the generator is operating at the resonance frequency and has an rms voltage of 60 V and a series-connected source resistance of 10 Ω; (*d*) the voltage drop across the capacitor.

16-2 A 25-μF capacitor is connected in series with a coil whose inductance is 0.50 mH. Determine (*a*) the resonance frequency; (*b*) the resistance of the coil if a 40-V generator operating at the resonance frequency causes a circuit current of 3.6 mA; (*c*) the *Q* of the coil.

16-3 A coil whose inductance and resistance are 3.0 mH and 20 Ω, respectively, is connected in series with a capacitor and a 12-V 5.0 kHz source. Determine (*a*) the value of capacitance that will cause the system to be in resonance; (*b*) the circuit current at the resonance frequency; (*c*) the maximum instantaneous energy stored in the capacitor at the resonance frequency; (*d*) the *Q* of the circuit.

16-4 A 60-Hz sinusoidal generator supplies energy to a series circuit consisting of a 4.0-Ω capacitive reactance, and a coil whose resistance and inductive reactance are 2.0 Ω and 3.0 Ω, respectively. The reactances were calculated at 60 Hz. The driving voltage in polar form is $100/30°$ V. Sketch the circuit and determine (*a*) the circuit impedance; (*b*) the circuit current in polar form; (*c*) sketch the impedance diagram and the phasor diagram; (*d*) write the equations for the driving voltage and the current as functions of time; (*e*) determine the current if the driving voltage remains at $100/30°$ V but the frequency is adjusted to the resonance value; (*f*) the voltage drop across the capacitor at resonance; (*g*) the voltage drop across the coil at resonance.

16-5 A 120-V 20-Hz generator supplies a series circuit consisting of a 5.0-Ω capacitive reactance, a 1.6-Ω resistor, and a coil whose resistance and inductive reactance are 3.0 Ω and 1.2 Ω, respectively. Sketch the circuit and determine (*a*) the input impedance; (*b*) the circuit current; (*c*) the voltage drop across coil; (*d*) the resonance frequency.

16-6 Determine the *Q* of a 200-turn coil operating at 2000 Hz, whose resistance and inductance are 10 Ω and 0.040 H, respectively.

16-7 A coil whose resistance and inductance are 5.0 Ω and 32 mH, respectively, is connected in series with a 796-pF capacitor. Sketch the circuit and determine (*a*) the resonance frequency of the circuit; (*b*) the quality factor; (*c*) the bandwidth; (*d*) the rms current if the circuit is driven at its resonance frequency by a 120-V sinusoidal generator; (*e*) the voltage drop across the capacitor for the conditions in part *d*; (*f*) the steady-state current if the generator is replaced by a 120-V battery.

16-8 A 400-V 200-Hz sinusoidal generator is connected in series with a capacitor and a coil whose resistance and inductance are 0.020 Ω and 0.0060 H, respectively. If the circuit is in resonance at 200 Hz, determine (*a*) the capacitance of the capacitor; (*b*) the circuit current; (*c*) the voltage drop across the capacitor; (*d*) the maximum instantaneous energy stored in the magnetic field of the coil; (*e*) the *Q* of the circuit at 200 Hz; (*f*) the bandwidth; (*g*) the cutoff frequencies; (*h*) the steady-state current if the generator is replaced by a 100-V battery.

16-9 A series resonance circuit has a resistance of 1000 Ω and cutoff frequencies of 20,000 Hz and 100,000 Hz. Determine (*a*) the bandwidth; (*b*) the resonance frequency; (*c*) the quality factor at the resonance frequency; (*d*) the inductance; (*e*) the capacitance.

16-10 Determine the resonance frequency, corner frequencies, bandwidth, and quality factor for the frequency-response curve shown in Fig. 16-17.

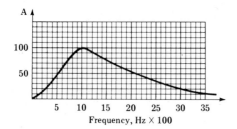

Figure 16-17 Frequency-response curve for Prob. 16-10.

Parallel Resonance

16-11 Figure 16-18 shows three different arrangements of ideal circuit elements in parallel with a 120-V 60-Hz generator. The resistance of the resistor is 10 Ω and the inductive reactance and capacitive reactance are each 4.0 Ω at the 60-Hz frequency. Determine the reading of each ammeter for the three circuit configurations.

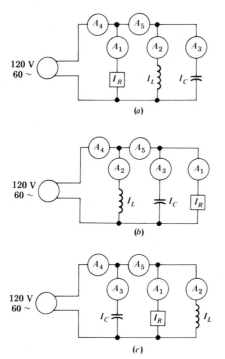

Figure 16-18 Circuits for Prob. 16-11.

16-12 A 240-V sinusoidal generator supplies current to a parallel circuit containing the following ideal circuit elements: (a) 0.50-Ω resistor, a 10-mH inductance, and a 2500-μF capacitor. Sketch the circuit and determine (a) the feeder current if the generator is operating at the resonance frequency; (b) the resonance frequency; (c) the maximum instantaneous voltage across the inductance.

16-13 A 0.50-μF capacitor is connected in parallel with a coil whose resistance and inductance are 1.0 Ω and 2.0 H, respectively. The parallel circuit is supplied by a 100-V sinusoidal generator that is operating at the resonance frequency of the circuit. Sketch the circuit and determine (a) the resonance frequency; (b) the input impedance; (c) the current; (d) the steady-state current if the sinusoidal generator is replaced by a 100-V battery.

16-14 A parallel section consisting of a 5.4-μF capacitor in parallel with a coil whose resistance and inductance are 18.8 Ω and 8.0 mH, respectively, is driven at its resonance frequency by a 240-V sinusoidal generator. Sketch the circuit and determine the resonance frequency.

16-15 A 450-V 60-Hz source supplies energy to a parallel circuit consisting of a $25/\underline{30°}$-Ω branch and a $12/\underline{-40°}$-Ω branch. Determine the resistance, capacitance, or inductance of a pure circuit element that, if connected in series with the source, will cause the system to be in resonance.

16-16 A series-parallel circuit consisting of a $(5 - j3)$-Ω impedance in series with a parallel section consisting of $(4 + j2)$-Ω and $(2 + j3)$-Ω impedances is supplied by a 490-V 30-Hz generator. Sketch the circuit and determine (a) the input impedance; (b) the input current; (c) the current in the $(4 + j2)$-Ω impedance; (d) is the series-parallel circuit operating at its resonance frequency? Explain.

16-17 A tank circuit is supplied by a current source whose source resistance is 56 kΩ. The tank circuit is composed of a 56-nF capacitor in parallel with a coil whose inductance and resistance are 35 mH and 80 Ω, respectively. Determine (a) f_r; (b) Q_{coil}; (c) R_{in}; (d) Q_p; (e) BW, (f) cutoff frequencies.

16-18 (a) Determine the capacitance required for a tank circuit that uses a 35-μH 60-Ω coil and resonates at 1.65 MHz. The tank circuit is to be fed from a 1.65-MHz 2.0-mA constant-current generator whose source resistance is 60 kΩ. (b) Determine the rms voltage across the capacitor and the capacitor current.

16-19 Repeat Prob. 16-18 assuming the existing coil is replaced with a 350-μH 40-Ω coil.

16-20 A tank circuit consisting of a coil whose resistance and inductance are 40.6 Ω and 21.5 mH, respectively, is connected in parallel with a capacitor and supplied by a 1000-Hz 125-V generator of negligible impedance. Assuming the circuit is at resonance determine (a) Q_{coil}; (b) the capacitance of the capacitor; (c) Q_p; (d) BW; (e) the feeder current.

16-21 A current source consisting of a sinusoidal 2.6-mA constant-current generator in parallel with a 60-kΩ resistor supplies a tank circuit whose parameters are $C = 105$ nF, $L_{coil} = 10.5$ mH, $R_{coil} = 106$ Ω. Determine (a) f_r; (b) Q_{coil}; (c) Q_p; (d) BW. If the current source is operating at the resonance frequency of the tank circuit, determine (e) the voltage across the capacitor; (f) the current in the coil.

CHAPTER 17

FILTER CIRCUITS

Filter circuits, called filters, are two-port networks used to block or pass a specific range of frequencies. *Low-pass* filters allow the passage of low frequencies but block higher frequencies; *high-pass* filters pass high frequencies but block low frequencies; *bandpass* filters pass a specific range of frequencies but block higher and lower frequencies; *bandstop* filters, also known as *band-elimination, band-rejection,* or *band-suppression* filters, or *wave traps,* block a specific range of frequencies but pass all higher and lower frequencies.

Depending on the desired characteristics, filters may be designed with *RL, RC,* and *RLC* circuits in various combinations. Adjustable filters are used in radio and TV sets to enable the listener or viewer to tune the set to a desired station. Tuning is accomplished by adjusting the resonance frequency of an *RLC* circuit so that it will pass the desired frequency and reject all others.

The circuits discussed in this chapter are limited to the most elementary types of filters and serve mainly as an introduction to filter theory. A detailed analysis of the large variety of filters used in electronic circuits is available in more advanced texts.

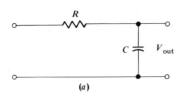

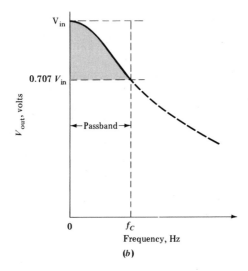

Figure 17-1 Simple *RC* low-pass filter circuit and frequency-response curve.

17-1 LOW-PASS FILTERS

Figure 17-1a shows a simple *RC* circuit used as a low-pass filter, and Fig. 17-1b shows the corresponding passband. The bandwidth of the passband is

$$BW = f_2 - f_1$$

$$BW = f_c - 0 = f_c$$

Frequency f_c is the *cutoff frequency*; it is the frequency above which the output voltage drops below 70.7 percent of the input voltage.

The relationship between R, C, and f_c may be determined by applying the voltage-divider equation to Fig. 17-1a and solving for the value of f_c that will cause V_{out} to equal 0.707 V_{in}. Thus,

$$V_{out} = V_{in} \frac{X_C}{\sqrt{R^2 + X_C^2}} \qquad (17\text{-}1)$$

At the cutoff frequency,

$$V_{out} = 0.707 \ V_{in} \qquad\qquad (17\text{-}2)$$

Substituting Eq. (17-2) into Eq. (17-1), expressing the resultant equation in terms of the cutoff frequency, and then solving for f_c,

$$0.707 \ V_{in} = V_{in} \ \frac{X_C}{\sqrt{R^2 + X_C^2}}$$

$$(0.707)^2 = \frac{X_C^2}{R^2 + X_C^2}$$

$$0.5R^2 + 0.5X_C^2 = X_C^2$$

$$R^2 = X_C^2$$

$$R = X_C$$

$$R = \frac{1}{2\pi f_c C}$$

$$\boxed{f_c = \frac{1}{2\pi RC}} \qquad\qquad (17\text{-}3)$$

In designing a low-pass filter, a choice of R or C is made, and the other parameter is determined from Eq. (17-3).

Example 17-1   Design an RC low-pass filter that will have a cutoff frequency of 300 Hz.

Solution
Assume $R = 1 \ k\Omega$

$$f_c = \frac{1}{2\pi RC}$$

$$300 = \frac{1}{2\pi(1000)C}$$

$$C = 531 \ nF$$

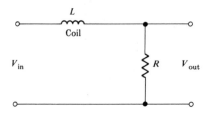

Figure 17-2 Simple *RL* low-pass filter circuit.

Figure 17-2 shows a simple *RL* circuit used as a low-pass filter. Following the same procedure used in developing the *RC* filter, and assuming R_{coil} is negligible,

$$V_{out} = V_{in} \frac{R}{\sqrt{R^2 + X_L^2}} \tag{17-4}$$

$$0.707 V_{in} = V_{in} \frac{R}{\sqrt{R^2 + X_L^2}}$$

$$(0.707)^2 = \frac{R^2}{R^2 + X_L^2}$$

$$X_L = R$$

$$2\pi f_c L = R$$

$$\boxed{f_c = \frac{R}{2\pi L}} \tag{17-5}$$

Example 17-2 (*a*) Design an *RL* low-pass filter that will have a cutoff frequency of 1200 Hz; (*b*) if the rms input voltage to the filter is 10 V, determine the output voltage at 0 Hz, 200 Hz, 600 Hz, 1200 Hz, 5000 Hz, and 10,000 Hz.

Solution
(*a*) Selecting $R = 2\ k\Omega$

$$f_c = \frac{R}{2\pi L}$$

$$1200 = \frac{2000}{2\pi L}$$

$$L = 265\ mH$$

(b) Substituting the design parameters into Eq. (17-4),

$$V_{out} = V_{in} \frac{R}{\sqrt{R^2 + X_L^2}} = 10 \frac{2000}{\sqrt{(2000)^2 + [2\pi f(0.265)]^2}}$$

Substituting the specified frequencies into the output equation, calculating, and tabulating results in

f	V_{out}
0	10.00
200	9.86
600	8.95
1,200	7.07
5,000	2.34
10,000	1.19

7-2 HIGH-PASS FILTERS

Figure 17-3a shows a simple RC circuit used as a high-pass filter, and Fig. 17-3b shows the corresponding passband. Following the same procedure as used before,

$$V_{out} = V_{in} \frac{R}{\sqrt{R^2 + X_C^2}} \qquad (17\text{-}6)$$

$$0.707\, V_{in} = V_{in} \frac{R}{\sqrt{R^2 + X_C^2}}$$

$$(0.707)^2 = \frac{R^2}{R^2 + X_C^2}$$

$$X_C = R$$

$$\frac{1}{2\pi f_c C} = R$$

$$\boxed{f_c = \frac{1}{2\pi RC}} \qquad (17\text{-}7)$$

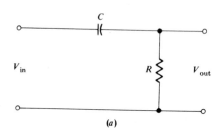

(a)

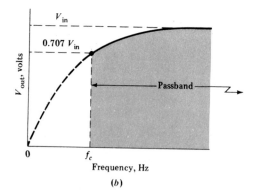

Frequency, Hz

(b)

Figure 17-3 (a) Simple RC high-pass filter circuit; (b) frequency-response curve.

Note: The equation for the cutoff frequency of a high-pass RC filter is identical to that for the low-pass filter. However, the respective equations for the output voltages [Eqs. (17-1) and (17-6)] are different.

Figure 17-4 shows a simple RL circuit used as a high-pass filter. Applying the voltage-divider equation and solving for the cutoff frequency,

$$V_{out} = V_{in} \frac{X_L}{\sqrt{R^2 + X_L^2}} \tag{17-8}$$

$$\boxed{f_c = \frac{R}{2\pi L}} \tag{17-9}$$

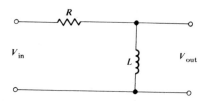

Figure 17-4 Simple RL high-pass filter circuit.

Example 17-3 (*a*) Design an *RL* high-pass filter that will have a cutoff frequency of 1200 Hz; (*b*) if the rms input voltage is 10 V, determine the output voltage at 0 Hz, 200 Hz, 600 Hz, 5000 Hz, and 10,000 Hz.

Solution
(*a*) Selecting $R = 2 \text{ k}\Omega$,

$$f_c = \frac{R}{2\pi L}$$

$$1200 = \frac{2000}{2\pi L}$$

$$L = 265 \text{ mH}$$

(*b*) Substituting the design parameters into Eq. (17-8),

$$V_{\text{out}} = V_{\text{in}} = \frac{X_L}{\sqrt{R^2 + X_L^2}} = 10 \, \frac{2\pi f(0.265)}{\sqrt{(2000)^2 + [2\pi f(0.265)]^2}}$$

Substituting the specified frequencies into the output equation, calculating, and tabulating results in

f	V_{out}
0	0
200	1.64
600	4.47
1,200	7.07
5,000	9.72
10,000	9.93

17-3   **BANDPASS FILTERS**

A bandpass filter that uses a series *RLC* circuit whose resonance frequency and bandwidth provide the desired passband is shown in Fig. 17-5*a*. Combining the paralleled resistors R_0 and R_{load} results in the equivalent circuit shown in Fig. 17-5*b*, where

$$R_T = \frac{R_0 R_{\text{load}}}{R_0 + R_{\text{load}}}$$

$$f_r = \frac{1}{2\pi\sqrt{LC}}$$

$$\text{BW} = \frac{f_r}{Q_S} \qquad Q_S = \frac{\omega_r L}{R_T}$$

$$V_{out} = V_{in} \frac{R_T}{\sqrt{R_T^2 + (X_L - X_C)^2}}$$

$$V_{out} = V_{in} \frac{R_T}{\sqrt{R_T^2 + [2\pi f L - 1/(2\pi f C)]^2}}$$

A typical frequency-response curve for a bandpass filter is shown in Fig. 17-5c.

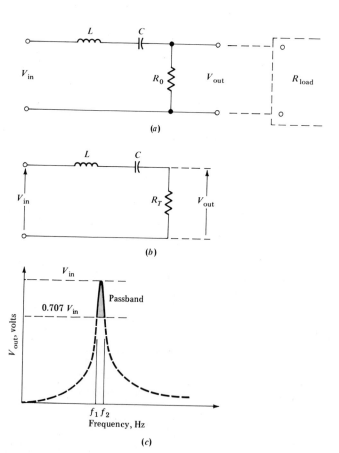

(a)

(b)

(c)

Figure 17-5 (a) and (b) Bandpass filter circuits using a series resonance circuit; (c) frequency-response curve.

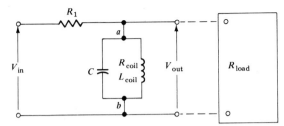

Figure 17-6 Bandpass filter circuit using a parallel resonance circuit.

A parallel resonance circuit that would result in similar bandpass characteristics is shown in Fig. 17-6. In this circuit, assuming the current drawn by the load is insignificant,

$$V_{out} = V_{in} \frac{Z_{ab}}{R_1 + Z_{ab}}$$

where

$$Z_{ab} = \frac{(R_{coil} + jX_{coil})(-jX_C)}{(R_{coil} + jX_{coil}) + (-jX_C)}$$

Example 17-4 Assuming the circuit parameters in Fig. 17-5a are $L_{coil} = 50$ mH, $C = 127$ nF, $R_{out} = 63$ Ω, $R_{load} = 600$ Ω, determine (a) the resonance frequency; (b) the bandwidth; (c) the cutoff frequencies; (d) the voltage output at the resonance frequency, the corner frequencies, and at $f = 10f_r$, assuming V_{in} is 30 V.

Solution

(a) $f_r = \dfrac{1}{2\pi\sqrt{LC}} = \dfrac{1}{2\pi\sqrt{(0.05)(127 \times 10^{-9})}} = 2$ kHz

(b) $R_T = \dfrac{R_0 R_{load}}{R_0 + R_{load}} = \dfrac{63(600)}{63 + 600} = 57$ Ω

$Q_S = \dfrac{\omega_r L}{R_T} = \dfrac{2\pi(2000)(0.05)}{57} = 11$

$Q_S = \dfrac{f_r}{BW}$

$11 = \dfrac{2000}{BW}$

$BW = 182$

(c) Since $Q_S > 10$

$$f_2 = f_r + \frac{BW}{2} = 2000 + \frac{182}{2} = 2091 \text{ Hz}$$

$$f_1 = f_r - \frac{BW}{2} = 2000 - \frac{182}{2} = 1909 \text{ Hz}$$

(d) $V_{out} = V_{in} \dfrac{R_T}{\sqrt{R_T^2 + (X_L - X_C)^2}}$

At the resonance frequency ($X_L = X_C$). Hence,

$$V_{out} = V_{in} \frac{R_T}{R_T} = 30 \frac{57}{57} = 30 \text{ V}$$

At $f_1 = 1909$ Hz,

$$X_L = 2\pi f L = 2\pi(1909)(0.05) = 600 \text{ } \Omega$$

$$X_C = \frac{1}{2\pi f C} = \frac{1}{2\pi(1909)(127 \times 10^{-9})} = 656 \text{ } \Omega$$

$$V_{out} = 30 \frac{57}{\sqrt{(57)^2 + (600 - 656)^2}} = 21.4 \text{ V}$$

At $f_2 = 2091$,

$$X_L = 2\pi(2091)(0.05) = 656 \text{ } \Omega$$

$$X_C = \frac{1}{2\pi(2091)(127 \times 10^{-9})} = 600 \text{ } \Omega$$

$$V_{out} = 30 \frac{57}{\sqrt{(57)^2 + (656 - 600)^2}} = 21.4 \text{ V}$$

At $f = 10 f_r = 10 \, (2000) = 20{,}000$

$$X_L = 2\pi(20{,}000)(0.05) = 6.28 \times 10^3 \text{ } \Omega$$

$$X_C = \frac{1}{2\pi(20{,}000)(127 \times 10^{-9})} = 62.7 \text{ } \Omega$$

$$V_{out} = 30 \frac{57}{\sqrt{(57)^2 + (6280 - 62.7)^2}} = 0.28 \text{ V}$$

Example 17-5 Design a series-resonance-type bandpass filter that has cutoff frequencies of 25 kHz and 23 kHz. The load resistance is 50 kΩ, and the only available coil has an inductance of 45 mH and negligible resistance.

Solution
The circuit is similar to that shown in Fig. 17-5a.

$$BW = f_2 - f_1 = 25{,}000 - 23{,}000 = 2 \text{ kHz}$$

$$f_2 = f_r + \frac{BW}{2}$$

$$25{,}000 = f_r + \frac{2000}{2}$$

$$f_r = 24{,}000 \text{ Hz}$$

$$f_r = \frac{1}{2\pi\sqrt{LC}}$$

$$24{,}000 = \frac{1}{2\pi\sqrt{0.045C}}$$

$$C = 978 \text{ pF}$$

$$Q_S = \frac{f_r}{BW} = \frac{24{,}000}{2000} = 12$$

$$Q_S = \frac{\omega_r L}{R_T}$$

$$12 = \frac{2\pi(24{,}000)(0.045)}{R_T}$$

$$R_T = 565 \ \Omega$$

$$\frac{1}{R_T} = \frac{1}{R_0} + \frac{1}{R_{\text{load}}}$$

$$\frac{1}{565} = \frac{1}{R_0} + \frac{1}{50{,}000}$$

$$R_0 = 571 \ \Omega$$

Thus the design parameters for the bandpass filter are

$C = 978 \text{ pF}$

$R_0 = 571 \,\Omega$

$L = 45 \text{ mH}$

17-4 BANDSTOP FILTERS

The bandstop filter shown in Fig. 17-7a uses a series RLC circuit whose bandwidth and resonance frequency determine the frequency range of the stop band.

A typical frequency-response curve for a bandstop filter is shown in Fig. 17-7b. For input frequencies within the stop-band region, $V_{out} < 0.707 \, V_{in}$.

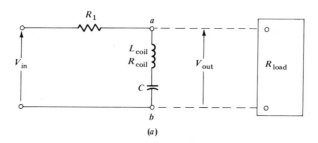

(a)

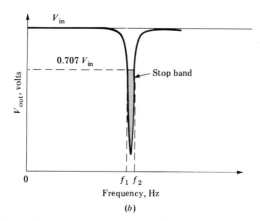

(b)

Figure 17-7 (a) Bandstop filter circuit using a series resonance circuit; (b) frequency-response curve.

The resonance frequency and bandwidth of the bandstop filter shown in Fig. 17-7a may be determined from

$$f_r = \frac{1}{2\pi\sqrt{LC}}$$

$$Q_S = \frac{f_r}{\mathrm{BW}} = \frac{\omega_r L_{\mathrm{coil}}}{R_1 + R_{\mathrm{coil}}}$$

Assuming the current drawn by the load is insignificant,

$$V_{\mathrm{out}} = V_{\mathrm{in}} \frac{Z_{ab}}{Z_{\mathrm{circ}}}$$

$$\mathbf{Z}_{ab} = R_{\mathrm{coil}} + j(X_L - X_C)$$

$$\mathbf{Z}_{\mathrm{circ}} = R_1 + R_{\mathrm{coil}} + j(X_L - X_C)$$

$$Z_{ab} = \sqrt{R_{\mathrm{coil}}^2 + (X_L - X_C)^2}$$

$$Z_{\mathrm{circ}} = \sqrt{(R_{\mathrm{coil}} + R_1)^2 + (X_L - X_C)^2}$$

$$V_{\mathrm{out}} = V_{\mathrm{in}} \frac{\sqrt{R_{\mathrm{coil}}^2 + (X_L + X_C)^2}}{\sqrt{(R_{\mathrm{coil}} + R_1)^2 + (X_L - X_C)^2}}$$

A *parallel resonance* circuit that would result in similar bandstop characteristics is shown in Fig. 17-8. In this circuit, assuming the current drawn by the load is insignificant,

$$\mathbf{V}_{\mathrm{out}} = \mathbf{V}_{\mathrm{in}} \frac{R_0}{R_0 + \mathbf{Z}_{ab}}$$

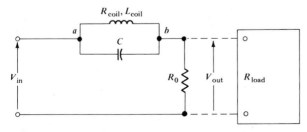

Figure 17-8 Bandstop filter circuit using a parallel resonance circuit.

where

$$Z_{ab} = \frac{(R_{\text{coil}} + jX_{L\,\text{coil}})(-jX_C)}{(R_{\text{coil}} + jX_{L\,\text{coil}}) + (-jX_C)}$$

Example 17-6 Assume the circuit parameters in Fig. 17-7a are $R_1 = 1000\ \Omega$, $L_{\text{coil}} = 160$ mH, $R_{\text{coil}} = 2.0\ \Omega$, and $C = 396$ pF. Determine $(a) f_r$; (b) the BW; (c) the cutoff frequencies; (d) the output voltage at the resonance frequency, the cutoff frequencies, and 100 Hz and 30 kHz.

Solution

$(a)\ f_r = \dfrac{1}{2\pi\sqrt{LC}} = \dfrac{1}{2\pi\sqrt{0.160(396 \times 10^{-12})}} = 20$ kHz

$(b)\ Q_S = \dfrac{\omega_r L}{R_S} = \dfrac{2\pi(20{,}000)(0.160)}{2 + 1000} = 20$

$\qquad Q = \dfrac{f_r}{\text{BW}}$

$\qquad 20 = \dfrac{20{,}000}{\text{BW}}$

$\qquad \text{BW} = 1000$ Hz

(c) Since $Q > 10$,

$f_2 = f_r + \dfrac{\text{BW}}{2} = 20{,}000 + \dfrac{1000}{2} = 20{,}500$ Hz

$f_1 = f_r - \dfrac{\text{BW}}{2} = 20{,}000 - \dfrac{1000}{2} = 19{,}500$ Hz

(d) At resonance, $X_L = X_C$; hence

$V_{\text{out}} = V_{\text{in}}\dfrac{\sqrt{(R_{\text{coil}})^2}}{\sqrt{(R_{\text{coil}} + R_1)^2}} = V_{\text{in}}\dfrac{2}{1002} = 0.002\ V_{\text{in}}$

At $f_1 = 19{,}500$ Hz,

$X_L = 2\pi(19{,}500)(0.160) = 19.6$ kΩ

$X_C = \dfrac{1}{2\pi(19{,}500)(396 \times 10^{-12})} = 20.6$ kΩ

$$X_L - X_C = -1000$$

$$V_{out} = V_{in} \frac{\sqrt{(2)^2 + (-1000)^2}}{\sqrt{(2 + 1000)^2 + (-1000)^2}} = 0.706 \; V_{in}$$

At $f_2 = 20{,}500$ Hz,

$$X_L = 2\pi(20{,}500)(0.160) = 20.6 \text{ k}\Omega$$

$$X_C = \frac{1}{2\pi(20{,}500)(396 \times 10^{-12})} = 19.6 \text{ k}\Omega$$

$$X_L - X_C = 1000 \; \Omega$$

$$V_{out} = V_{in} \frac{\sqrt{(2)^2 + (1000)^2}}{\sqrt{(1002)^2 + (1000)^2}} = 0.706 \; V_{in}$$

At 100 Hz,

$$X_L = 2\pi(100)(0.160) = 100.5 \; \Omega$$

$$X_C = \frac{1}{2\pi(100)(396 \times 10^{-12})} = 4 \text{ M}\Omega$$

$$X_L - X_C = 100.5 - 4 \times 10^6 = -4 \times 10^6$$

$$V_{out} = V_{in} \frac{\sqrt{(2)^2 + (-4 \times 10^6)^2}}{\sqrt{(1002)^2 + (-4 \times 10^6)^2}} = 0.999 \; V_{in}$$

At 30 kHz,

$$X_L = 2\pi(30{,}000)(0.160) = 30.1 \text{ k}\Omega$$

$$X_C = \frac{1}{2\pi(30{,}000)(396 \times 10^{-12})} = 13.4 \text{ k}\Omega$$

$$X_L - X_C = 30{,}100 - 13{,}400 = 16.7 \text{ k}\Omega$$

$$V_{out} = V_{in} \frac{\sqrt{(2)^2 + (16{,}700)^2}}{\sqrt{(1002)^2 + (16{,}700)^2}} = 0.998 \; V_{in}$$

Example 17-7 (*a*) Determine the capacitance required for a series resonance bandstop filter that will block 85 kHz. The inductance and resistance of the coil are 60 mH and 15 Ω, respectively. $R_1 = 2000 \; \Omega$, and $R_{load} = 1.4$ MΩ. (*b*) Determine the bandwidth.

Solution

(a) Referring to Fig. 17-7a,

$$f_r = \frac{1}{2\pi\sqrt{LC}}$$

$$85,000 = \frac{1}{2\pi\sqrt{0.06C}}$$

$$C = 58 \text{ pF}$$

(b) $Q_S = \dfrac{\omega_r L}{R} = \dfrac{2\pi(85,000)(0.06)}{15 + 2000} = 15.9$

$$Q_S = \frac{f_r}{BW}$$

$$15.9 = \frac{85,000}{BW}$$

$$BW = 5.34 \text{ kHz}$$

17-5   DOUBLE-RESONANT FILTER

A double-resonant filter such as those shown in Fig. 17-9a and b has two resonance frequencies. There is one frequency at which parallel resonance occurs and another frequency at which series resonance occurs. The parallel resonance frequency of the tank circuit determines the rejected frequency, and the series resonance frequency determines the accepted frequency.

The use of two capacitors and one coil, as shown in Fig. 17-9a and b, will result in

$$f_{\text{reject}} > f_{\text{pass}}$$

Replacing C_2 with L_2, shown with broken lines, will result in

$$f_{\text{reject}} < f_{\text{pass}}$$

Example 17-8 Assume C_1 in Fig. 17-9a has a capacitance of 3.5 nF. Determine the remaining parameters required in order that the filter will reject a 100-kHz signal but accept 50 kHz.

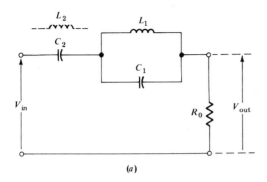

(a)

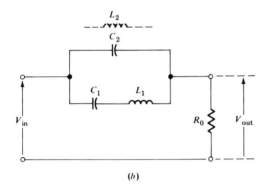

(b)

Figure 17-9 Double-resonant filter circuits.

Solution

The resonance frequency of the stop band is determined by the tank circuit. Assuming $Q_{coil} \geq 10$,

$$f_r = \frac{1}{2\pi\sqrt{LC}}$$

$$100{,}000 = \frac{1}{2\pi\sqrt{L_1(3.5 \times 10^{-9})}}$$

$$L_1 = 724 \ \mu H$$

The impedance of the tank circuit at 50 kHz is

$$X_{L1} = 2\pi(50{,}000)(724 \times 10^{-6}) = 227 \ \Omega$$

$$X_{C1} = \frac{1}{2\pi(50{,}000)(3.5 \times 10^{-9})} = 910 \ \Omega$$

$$\mathbf{Z}_{tank} = \frac{(jX_{L1})(-jX_{C1})}{(jX_{L1}) + (-jX_{C1})} = \frac{X_{L1}X_{C1}}{j(X_{L1} - X_{C1})}$$

$$\mathbf{Z}_{tank} = \frac{(227)(910)}{j(227 - 910)} = j302$$

Thus, at 50 kHz the tank circuit behaves as a pure inductive reactance of 302 Ω. To cause series resonance at 50 kHz, capacitor C_2 must have a capacitive reactance of 302 Ω.

$$X_C = \frac{1}{2\pi f C}$$

$$302 = \frac{1}{2\pi(50,000)C}$$

$$C = 10.5 \text{ nF}$$

Example 17-9 Assuming L_1 in Fig. 17-9b is 2.5 mH, determine the remaining parameters so that the filter will reject 150 kHz and accept 200 kHz.

Solution

The resonance frequency of the passband is determined by L_1 and C_1.

$$f_r = \frac{1}{2\pi\sqrt{L_1 C_1}}$$

$$200,000 = \frac{1}{2\pi\sqrt{0.025 C_1}}$$

$$C_1 = 25 \text{ pF}$$

At 150 kHz, the impedance of the series LC branch is

$$\mathbf{Z}_{ser} = jX_{L1} - jX_{C1}$$

$$X_{L1} = 2\pi(150,000)(0.025) = 23.6 \text{ k}\Omega$$

$$X_{C1} = \frac{1}{2\pi(150,000)(25 \times 10^{-12})} = 42.5 \text{ k}\Omega$$

$$\mathbf{Z}_{ser} = j23,600 - j42,500 = -j18.900 \ \Omega$$

Thus the series branch is in effect a capacitive reactance of 18,900 Ω at 150 kHz. For tank resonance to occur the parallel branch must have an inductive reactance of 18,900 Ω.

$$X_{L2} = 2\pi f L_2$$

$$18,900 = 2\pi(150,000)L_2$$

$$L_2 = 20 \text{ mH}$$

SUMMARY OF FORMULAS

Low-Pass and High-Pass Filters

RC filter

$$f_c = \frac{1}{2\pi RC}$$

LR filter

$$f_c = \frac{R}{2\pi L}$$

Bandpass and Bandstop Filters ($Q \geq 10$)

$$f_r = \frac{1}{2\pi\sqrt{LC}} \qquad f_2 = f_r + \frac{BW}{2}$$

$$Q = \frac{f_r}{BW} \qquad f_1 = f_r - \frac{BW}{2}$$

$$Q_S = \frac{\omega_r L}{R_T} \qquad BW = f_2 - f_1$$

PROBLEMS

17-1 Design an RC low-pass filter that will have a cutoff frequency of 800 Hz. Assume $R = 2$ kΩ.

17-2 Design an RC low-pass filter that will have a cutoff frequency of 2000 Hz. Assume $C = 80$ nF.

17-3 Design an RL low-pass filter that uses a 1500-Ω resistor and has a cutoff frequency of 1600 Hz.

17-4 Design an RL low-pass filter that uses a 25-mH coil and has a cutoff frequency of 4000 Hz.

17-5 Design a high-pass RC filter that will have a cutoff frequency of 1600 Hz. Assume $R = 1.25$ kΩ.

17-6 Design a high-pass RC filter that uses a 650-pF capacitor and has a cutoff frequency of 9000 Hz.

17-7 Design a high-pass LR filter that will have a cutoff frequency of 10 kHz and uses a 1600-Ω resistor.

17-8 Design a series-resonance-type bandpass filter that has cutoff frequencies of 15 kHz and 35 kHz. The load resistance is 60 kΩ, and the coil has an inductance of 50 mH and negligible resistance.

17-9 Assume the circuit parameters for the series-resonance bandpass filter in Fig. 17-5a are $C = 1.8$ pF, $L_{coil} = 25$ mH, $R_{out} = 52$ Ω, and $R_{load} = 9000$ Ω. Determine (a) the resonance frequency; (b) the bandwidth; (c) the cutoff frequencies; (d) the output voltage at the resonance frequency, cutoff frequencies, and 10 f_r if the input voltage is 60 V.

17-10 Assume the circuit parameters for the bandstop filter shown in Fig. 17-7a are $R_1 = 1500$ Ω, $L_{coil} = 140$ mH, $R_{coil} = 1.5$ Ω, and $C = 300$ pF. Determine (a) f_r; (b) the bandwidth; (c) the cutoff frequencies; (d) the output voltage at f_r, 800 Hz, and 900 Hz and 50 kHz if the input voltage is 30 V. Assume R_{load} has a negligible effect on the filter.

17-11   Determine the required capacitance for a series-resonance bandstop filter that will block 65 kHz. The load resistance is 50 kΩ, R_1 is 3000 Ω, and the coil inductance and resistance are 55 mH and 10 Ω, respectively. Determine the voltage across the load at the resonance frequency if V_{in} = 80 V.

17-12   Assume capacitance C_1 for the double-resonant filter shown in Fig. 17-9b is 2.1 nF. Determine the remaining parameters that will block 90 kHz and accept 100 kHz.

17-13   Assume capacitor C_1 in Fig. 17-9a has a capacitance of 6.5 nF. Determine the remaining parameters that will block 75 kHz and accept 20 kHz.

CHAPTER 18

TRANSFORMERS

When two or more coils are arranged with respect to one another, so that all or part of the flux caused by the current in one coil passes through the window of the other coil or coils, the coils are said to be *magnetically coupled*. This is illustrated in Fig. 18-1a, where current in one coil causes flux to pass through the window of another coil. When coils are specifically arranged for the purpose of providing magnetic coupling, the arrangement is called a *transformer*.

In transformer applications using ferromagnetic cores, such as that shown in Fig. 18-1b, *it will be assumed that the normal operating flux levels in the core will always be below the knee of the magnetization curve.* Hence, unless otherwise specified, the ferromagnetic cores referred to throughout this chapter are assumed to be fashioned from homogeneous material of constant permeability.

18-1   COEFFICIENT OF COUPLING AND MUTUAL INDUCTANCE

Figure 18-2a shows two coils wound around a ferromagnetic core. Coil 1 is connected in series with a battery and a switch, and coil 2 is *open-circuited*. An open-circuited coil is one that has nothing connected to its terminals. Coil 1 is called the primary coil (or primary) because it is connected to the driver, and coil 2 is called the secondary coil (or secondary).

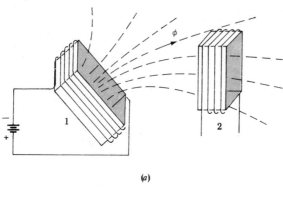

(a)

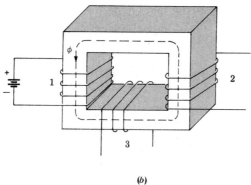

(b)

Figure 18-1 (*a*) Magnetic coupling through air;
(*b*) magnetic coupling using a ferromagnetic core.

When the switch in Fig. 18-2*a* is closed, the rise in current through the primary coil causes a buildup of flux through the primary and secondary windows. However, because of *leakage around the primary coil*, as shown in Fig. 18-2*a*, not all the flux produced by the primary current passes through the secondary window. The component of primary flux that passes through the secondary window is equal to the total primary flux minus the leakage flux. Thus,

$$\boxed{\phi_{12} = \phi_1 - \phi_{11}}$$

(18-1)

where ϕ_1 = total flux through the window of coil 1 *due to the current in coil 1*

ϕ_{11} = leakage flux around coil 1 *due to the current in coil 1*

ϕ_{12} = flux through the window of coil 2 *due to the current in coil 1*

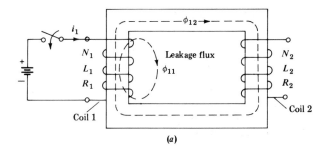

(a)

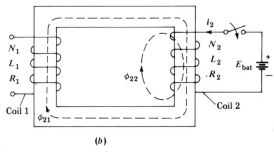

(b)

Figure 18-2 Transformers exhibiting leakage flux.

The ratio of the flux through the secondary window to the total primary flux is called the *coefficient of coupling k.*

$$
k = \frac{\phi_{12}}{\phi_1}
\tag{18-2}
$$

From Faraday's law, the voltages generated in coils 1 and 2 by the rate of change of flux through their respective windows are

$$
v_1 = N_1 \frac{d\phi_1}{dt}
\tag{18-3}
$$

$$
v_2 = N_2 \frac{d\phi_{12}}{dt}
\tag{18-4}
$$

From Eq. (18-2),

$$
\phi_{12} = k\phi_1
\tag{18-5}
$$

Substituting Eq. (18-5) into Eq. (18-4),

$$v_2 = N_2 k \frac{d\phi_1}{dt} \tag{18-6}$$

The ratio of secondary to primary coil voltages may be obtained by dividing Eq. (18-6) by Eq. (18-3). Thus,

$$\frac{v_2}{v_1} = \frac{N_2 k (d\phi_1/dt)}{N_1 (d\phi_1/dt)}$$

$$\boxed{\frac{v_2}{v_1} = k \frac{N_2}{N_1}} \tag{18-7}$$

Equation (18-7) shows that the ratio of voltages, *secondary to primary*, is equal to the *secondary to primary* turns ratio multiplied by the coefficient of coupling.

The voltage induced in the primary coil may be expressed in terms of its self-inductance and the rate of change of current through it. Thus, *for the primary coil* in Fig. 18-2a,

$$v_1 = L_1 \frac{di_1}{dt} \tag{18-8}$$

The voltage induced in the secondary coil cannot be expressed in terms of a changing current in the secondary, because there is no secondary current; the secondary is not connected to anything! *The voltage induced in the secondary is caused by a changing current in the primary. Hence, the secondary voltage may be expressed in terms of the rate of change of primary current, and the mutual inductance of the two coupled coils.*

A mathematical relationship that expresses the secondary voltage as a function of the rate of change of primary current and a mutual inductance coefficient is obtained by substituting Eqs. (8-10) and (18-8) into Eq. (18-7) and simplifying. Thus, from Eq. (8-10), Chap. 8,

$$L_1 = \frac{N_1^2}{\mathcal{R}} \qquad L_2 = \frac{N_2^2}{\mathcal{R}} \tag{18-9}$$

Rearranging terms,

$$N_1 = \sqrt{L_1 \mathcal{R}} \tag{18-10}$$

$$N_2 = \sqrt{L_2 \mathcal{R}} \tag{18-11}$$

Substituting Eqs. (18-8), (18-10), and (18-11) into Eq. (18-7) and simplifying

$$\frac{v_2}{L_1(di_1/dt)} = k\frac{\sqrt{L_2\mathscr{R}}}{\sqrt{L_1\mathscr{R}}} \tag{18-12}$$

$$v_2 = k\sqrt{L_1L_2}\frac{di_1}{dt} \tag{18-13}$$

Defining,

$$\boxed{M = k\sqrt{L_1L_2}} \tag{18-14}$$

where M = mutual inductance coefficient of two coupled coils, H
       k = coefficient of coupling
      L_1 = self-inductance of coil 1, H
      L_2 = self-inductance of coil 2, H

The mutual inductance of two coupled coils is a measure of the ability of a changing current in one coil to induce a voltage in another coil. Substituting Eq. (18-14) into Eq. (18-13),

$$\boxed{v_2 = M\frac{di_1}{dt}} \tag{18-15}$$

As indicated in Eq. (18-15), the voltage induced in the secondary (coil 2) is a function of a changing current in the primary (coil 1).

The mutual inductance of two coupled coils may also be expressed in terms of the ratio of the change in flux through the secondary window to the change in primary current that caused it. This very useful relationship is obtained from Eqs. (18-4) and (18-15). Equating the two and solving for M,

$$M\frac{di_1}{dt} = N_2\frac{d\phi_{12}}{dt} \tag{18-16}$$

Solving for M,

$$\boxed{M = N_2\frac{d\phi_{12}}{di_1}} \tag{18-17}$$

If the coils are interchanged, that is, if the driver is connected to coil 2 and coil 1 is open-circuited as shown in Fig. 18-2b, then Eq. (18-16) becomes

$$M \frac{di_2}{dt} = N_1 \frac{d\phi_{21}}{dt}$$

$$\boxed{M = N_1 \frac{d\phi_{21}}{di_2}} \tag{18-18}$$

where ϕ_{21} = flux through the window of coil 1 *due to the current in coil 2.*

Note: If operation is in the linear region of the magnetization curve,

$$\frac{d\phi_1}{di_1} = \frac{\Delta\phi_1}{\Delta I_1} \qquad \frac{d\phi_{12}}{di_1} = \frac{\Delta\phi_{12}}{\Delta I_1} \qquad \frac{d\phi_{21}}{di_2} = \frac{\Delta\phi_{21}}{\Delta I_2}$$

and the values of M obtained from Eqs. (18-17) and (18-18) are identical.

Example 18-1 A 24-V battery is connected to coil 1 of a two-winding transformer, as shown in Fig. 18-3a, and the secondary coil is open-circuited. The coil parameters are

$$R_1 = 10\ \Omega \qquad N_1 = 100 \text{ turns}$$
$$R_2 = 30\ \Omega \qquad N_2 = 160 \text{ turns}$$

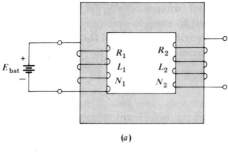

(a)

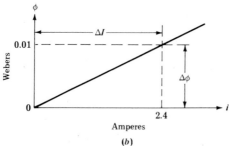

Amperes

(b)

Figure 18-3 Circuit diagram and magnetization curve for Example 18-1.

If the flux through the window of coil 2 is 0.008 Wb and that in coil 1 is 0.01 Wb, determine (a) the self-inductance of coil 1; (b) the mutual inductance; (c) the coefficient of coupling; (d) the self-inductance of coil 2.

Solution

(a) $I_1 = \dfrac{E_{bat}}{R_1} = \dfrac{24}{10} = 2.4$ A

Assume operation of the core flux is in the linear region (normal inductance) as shown in Fig. 18-3b. From the defining equation of inductance,

$$L_i = N_1 \frac{d\phi_1}{di_1} = N_1 \frac{\Delta\phi_1}{\Delta I_1} = 100 \frac{0.01}{2.4} = 0.417 \text{ H}$$

(b) $M = N_2 \dfrac{d\phi_{12}}{di_1} = N_2 \dfrac{\Delta\phi_{12}}{\Delta I_1} = 160 \dfrac{0.008}{2.4} = 0.533$ H

(c) $k = \dfrac{\phi_{12}}{\phi_1} = \dfrac{0.008}{0.01} = 0.8$

(d) $M = k\sqrt{L_1 L_2}$

$0.533 = 0.8\sqrt{0.417\, L_2}$

$L_2 = 1.064$H

Example 18-2 Figure 18-4 shows a transformer connected through a switch to a 120-V battery. The diode connected across the primary coil is used to discharge the energy stored within the magnetic field when the switch is opened. Assuming the switch is closed and the current in the primary coil has reached its steady-state value, determine (a) the steady-state primary current; (b) the steady-state primary flux; (c) that part of the primary flux that passes through the window of the secondary; (d) the self-inductances of the primary and secondary coils; (e) the mutual inductance; (f) the instantaneous voltage induced in the 250-turn coil, if at the instant the switch is opened, the current in coil 1 starts to fall at the rate of 11.42 A/s; (g) the instantaneous voltage induced in the 300-turn coil for the conditions in part f.

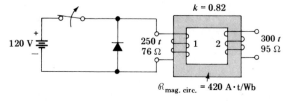

Figure 18-4 Circuit diagram for Example 18-2.

Solution

(a) $I_1 = \dfrac{E_{bat}}{R} = \dfrac{120}{76} = 1.579$ A

(b) $\phi_1 = \dfrac{N_1 I_1}{\mathcal{R}} = \dfrac{250(1.579)}{420} = 0.940$ Wb

(c) $\phi_{12} = k\phi_1 = 0.82(0.940) = 0.771$ Wb

(d) $L_1 = \dfrac{N_1^2}{\mathcal{R}} = \dfrac{(250)^2}{420} = 148.8$ H

or, using another method,

$$L_1 = N_1 \dfrac{\Delta\phi_1}{\Delta i_1} = \dfrac{(250)(0.940)}{1.579} = 148.8 \text{ H}$$

$$L_2 = \dfrac{N_2^2}{\mathcal{R}} = \dfrac{(300)^2}{420} = 214.3 \text{ H}$$

(e) $M = k\sqrt{L_1 L_2} = 0.82\sqrt{(148.8)(214.3)} = 146.4$ H

or using another method,

$$M = N_2 \dfrac{\Delta\phi_{12}}{\Delta i_1} = \dfrac{300(0.771)}{1.579} = 146.4 \text{ H}$$

(f) $V_1 = L_1 \dfrac{di_1}{dt} = 148.8(11.42) = 1699.3$ V

(g) $v_2 = M \dfrac{di_1}{dt} = 146.4(11.42) = 1671.9$ V

18-2   TRANSFORMER WITH A SINUSOIDAL DRIVING VOLTAGE

The application of a sinusoidal driving voltage to one coil of a transformer will cause a sinusoidally varying flux through the windows of the coupled coils. The voltage generated in each coil, as determined from Faraday's law, is

$$v = N \dfrac{d\phi}{dt} \tag{18-19}$$

where $\dfrac{d\phi}{dt}$ = rate of change of flux through the window of the respective coil.

The sinusoidal variation of flux through the coil windows, expressed as a function of time, is

$$\phi = \Phi_{max} \sin 2\pi f t \qquad (18\text{-}20)$$

The rms value of the sinusoidal voltage generated in each coil by the rate of change of sinusoidal flux through its respective window is

$$\boxed{V = 4.44 N f\, \Phi_{max}} \qquad (18\text{-}21)$$

where V = generated voltage, V rms
   N = number of series-connected turns of wire in the coil
   f = frequency, Hz
   Φ_{max} = maximum instantaneous value of the sinusoidal flux through the window of the coil, Wb

Derivation of Equation (18-21)

Substituting Eq. (18-20) into Eq. (18-19) and performing the indicated operations,

$$v = N\frac{d(\Phi_{max} \sin 2\pi f t)}{dt}$$

$$v = 2\pi f N\Phi_{max} \cos 2\pi f t \qquad (18\text{-}22)$$

Since $\cos 2\pi f t$ ranges between $+1$ and -1, the maximum value of the voltage is

$$V_{max} = 2\pi f N\Phi_{max} \qquad (18\text{-}23)$$

The rms value is obtained by dividing the maximum value by $\sqrt{2}$. Thus

$$V = \frac{2\pi}{\sqrt{2}} f N\Phi_{max}$$

$$V = 4.44 N f\Phi_{max}$$

Example 18-3   Determine the amplitude of a 20-Hz sinusoidal flux wave that will generate 50 V rms in a 30-turn coil.

Solution

$$V = 4.44 N f\, \Phi_{max}$$

$$50 = 4.44\,(30)(20)\Phi_{max}$$

$$\Phi_{max} = 0.01877\ \text{Wb}$$

18-3 TRANSFORMER WITH NO CURRENT IN SECONDARY

Figure 18-5a shows the primary coil of a transformer connected in series with a battery and a switch. The secondary coil is open-circuited. When the switch is closed, the buildup of flux through the primary window induces an opposing voltage in the primary coil. *In accordance with Lenz's law, the direction of this countervoltage (also called counterelectromotive force or cemf) will be such as to oppose the changing flux that caused it.*

As discussed in Sec. 8-1 of Chap. 8, the driving voltage has to force the current against the opposing voltage produced by the changing flux. In accomplishing this, part of the driving voltage is "used up" is overcoming the induced countervoltage. This "loss" of voltage, *called a voltage drop*, is equal but opposite to the induced voltage.

An equivalent circuit model of a transformer primary with a DC driver and no current in the secondary is shown in Fig. 18-5b. Applying Kirchhoff's voltage law to the primary circuit,

$$E_{\text{bat}} = v_1 + i_1 R_1 \qquad (18\text{-}24)$$

where

$$v_1 = N_1 \frac{d\phi_1}{dt} \qquad (18\text{-}25)$$

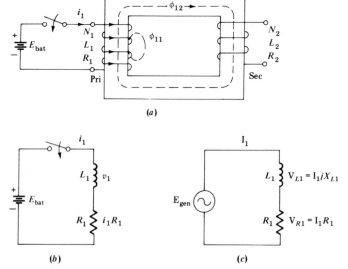

(a)

(b)   (c)

Figure 18-5 (a) Transformer with a DC driver; (b) equivalent circuit for (a); (c) equivalent circuit for (a) assuming the battery is replaced by a sinusoidal driver.

Substituting Eq. (18-25) into Eq. (18-24),

$$E_{bat} = N_1 \frac{d\phi_1}{dt} + i_1 R_1 \qquad (18\text{-}26)$$

Or, in terms of inductance and the rate of change of current,

$$E_{bat} = L_1 \frac{di_1}{dt} + i_1 R_1 \qquad (18\text{-}27)$$

The driver in Fig. 18-5a is a battery. Hence the current will have attained its steady-state value in approximately five time constants. At steady state, neither the primary current nor the flux through the primary window will be changing. Thus, at steady state,

$$\frac{d\phi_1}{dt} = 0$$

$$\frac{di_1}{dt} = 0$$

and Eqs. (18-24), (18-26), and (18-27) reduce to

$$E_{bat} = i_1 R_1 \qquad (18\text{-}28)$$

If the battery in Fig. 18-5a is replaced by a sinusoidal generator, the equivalent circuit model will be that shown in Fig. 18-5c. Voltage v_1 and current i_1 will be sinusoidal, and Kirchhoff's voltage equation, expressed in phasor form, becomes

$$\mathbf{E}_{gen} = \mathbf{I}_1 j X_{L1} + \mathbf{I}_1 R_1 \qquad (18\text{-}29)$$

Equation (18-29) is identical to the Kirchhoff's voltage law equation for a sinusoidal driving voltage applied to an impedance coil. *The primary of a transformer acts as an impedance coil if there is no current in the secondary.*

Example 18-4 The transformer shown in Fig. 18-6a has a coefficient of coupling of 0.70, and primary and secondary turns of 270 and 540, respectively. The primary has a resistance of 10 Ω and draws a current of 2 A when connected across a 120-V 60-Hz supply. Determine (a) the rms voltage induced in the secondary under no-load conditions; (b) the primary current if the primary is inadvertently connected to a 120-V DC generator.

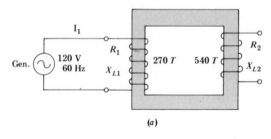

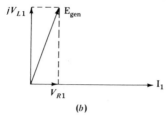

(b)

Figure 18-6 Circuit and phasor diagram for Example 18-4.

Solution

(a) Applying Kirchhoff's voltage law to the primary,

$$\mathbf{E}_{gen} = \mathbf{I}_1 jX_{L1} + \mathbf{I}_1 R_1 \tag{18-30}$$

Defining,

$$V_{L1} = I_1 X_{L1} = \text{rms voltage drop due to the primary inductance}$$

$$V_{R1} = I_1 R_1 = \text{rms voltage drop due to the primary resistance}$$

Equation (18-30) may be rewritten as

$$\mathbf{E}_{gen} = V_{R1} + jV_{L1} \tag{18-31}$$

The phasor diagram representing Eq. (18-31) is shown in Fig. 18-6b. The magnitude of $\mathbf{E}_{gen}$, as obtained from the phasor diagram, is

$$E_{gen} = \sqrt{V_{R1}^2 + V_{L1}^2} \tag{18-32}$$

The rms value for E_{gen} is given as 120 V. The rms value for V_{R1} is

$$I_1 R_1 = 2(10) = 20 \text{ V}$$

Substituting the known voltages in Eq. (18-32), and solving for V_{L1},

$$120 = \sqrt{20^2 + V_{L1}^2}$$

$$V_{L1} = 118.32 \text{ V}$$

From the voltage-ratio formula,

$$\frac{V_2}{V_1} = \frac{N_2}{N_1} \times k$$

$$\frac{V_2}{118.32} = \frac{540}{270} \, 0.7$$

$$V_2 = 165.65 \text{ V}$$

(b) $X_{L1} = 2\pi(0)L = 0$

$$\mathbf{Z} = R + jX_{L1} = (10 + j0) = 10 \ \Omega$$

$$I_1 = \frac{E_{\text{gen}}}{Z_1} = \frac{120}{10} = 12 \text{ A}$$

18-4 TRANSFORMER WITH INDUCED CURRENT IN SECONDARY

Figure 18-7 shows a transformer with a resistor (called a load resistor or load) connected across the secondary terminals. When the switch in the primary circuit is closed, the buildup of primary current causes a buildup of flux ϕ_{12} through the window of the secondary coil. The changing flux through the secondary window generates a secondary voltage v_2 that causes current i_2.

In accordance with Lenz's law, the secondary current i_2 will set up a flux of its own, called ϕ_2, that will be in a direction to oppose the changing of flux ϕ_{12} through its window. One component of flux ϕ_2, called ϕ_{22}, takes the leakage path around the secondary coil, as shown in Fig. 18-7; the other component, called ϕ_{21}, passes through the window of the primary coil. The relationship between the secondary flux and its components is

$$\phi_2 = \phi_{22} + \phi_{21} \tag{18-33}$$

where ϕ_2 = total flux through window of coil 2 *due to the current in coil 2*
ϕ_{22} = leakage flux around coil 2 *due to the current in coil 2*
ϕ_{21} = flux through the window of coil 1 *due to the current in coil 2*

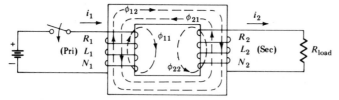

Figure 18-7 Transformer with induced current in the secondary.

Referring to Fig. 18-7, the net flux through the window of coil 1 is the difference between the opposing fluxes. Thus,

$$\phi_{p,\text{win}} = \phi_{11} + \phi_{12} - \phi_{21}$$

From Eq. 18-1,

$$\phi_1 = \phi_{11} + \phi_{12}$$

Hence

$$\phi_{p,\text{win}} = \phi_1 - \phi_{21} \tag{18-34}$$

Thus the countervoltage induced in the primary coil by the net rate of change of flux through its window is

$$v_1 = N_1 \frac{d\phi_{p,\text{win}}}{dt}$$

or

$$v_1 = N_1 \frac{d(\phi_1 - \phi_{21})}{dt} \tag{18-35}$$

Expanding Eq. (18-35),

$$v_1 = N_1 \frac{d\phi_1}{dt} - N_1 \frac{d\phi_{21}}{dt} \tag{18-36}$$

Comparing Eq. (18-36) with Eq. (18-25) will show that there is less countervoltage in the primary when a resistor load is connected to the secondary than when the secondary is open-circuited. Since a circuit with less countervoltage will have a higher primary current, it may be concluded that *connecting a resistor load to the secondary of a transformer will cause an increase in the primary current.*

Applying Kirchhoff's voltage law to the primary circuit of Fig. 18-7, the driving voltage E_{bat} is equal to the IR drop plus the countervoltage V_1. That is,

$$E_{\text{bat}} = i_1 R_1 + v_1 \tag{18-37}$$

Substituting Eq. (18-36) into Eq. (18-37),

$$E_{\text{bat}} = i_1 R_1 + N_1 \frac{d\phi_1}{dt} - N_1 \frac{d\phi_{21}}{dt} \tag{18-38}$$

Expressing Eq. (18-38) in terms of self- and mutual inductances, and the rates of change of primary and secondary currents, respectively,

$$E_{bat} = i_1 R_1 + L_1 \frac{di_1}{dt} - M \frac{di_2}{dt} \qquad (18\text{-}39)$$

Note: If the load is disconnected from the secondary, i_2 is zero, and Eq. (18-39) becomes the same as Eq. (18-27). Furthermore, if the driver is a battery, the current i_1 will attain its steady-state value in approximately 5 time constants; di_1/dt and di_2/dt will be zero, and Eq. (18.39) reduces to

$$E_{bat} = i_1 R_1$$

The voltage v_2 generated in the secondary acts as a driver, causing secondary current i_2. Applying Kirchhoff's voltage law to the secondary circuit of Fig. 18-7,

$$v_2 = i_2 R_{load} + i_2 R_2 \qquad (18\text{-}40)$$

Voltage v_2 is caused by the *net* rate of change of flux through the secondary window. From Faraday's law,

$$v_2 = N_2 \frac{d\phi_{s,\,win}}{dt} \qquad (18\text{-}41)$$

Referring to Fig. 18-7, the net flux through the window of the secondary is

$$\phi_{s,\,win} = \phi_{12} - \phi_{21} - \phi_{22} \qquad (18\text{-}42)$$

From Eq. (18-33),

$$-\phi_2 = -\phi_{22} - \phi_{21} \qquad (18\text{-}43)$$

Substituting Eq. (18-43) into Eq. (18-42),

$$\phi_{s,\,win} = \phi_{12} - \phi_2$$

Thus, from Eq. (18-41),

$$V_2 = N_2 \frac{d(\phi_{12} - \phi_2)}{dt}$$

or

$$V_2 = N_2 \frac{d\phi_{12}}{dt} - N_2 \frac{d\phi_2}{dt} \qquad (18\text{-}44)$$

Substituting Eq. (18-44) into Eq. (18-40),

$$N_2 \frac{d\phi_{12}}{dt} - N_2 \frac{d\phi_2}{dt} = i_2 R_{\text{load}} + i_2 R_2 \tag{18-45}$$

Expressing Eq. (18-45) in terms of self-inductance, mutual inductance, and the corresponding primary and secondary currents,

$$M \frac{di_1}{dt} - L_2 \frac{di_2}{dt} = i_2 R_{\text{load}} + i_2 R_2 \tag{18-46}$$

Rearranging the terms,

$$0 = i_2 R_{\text{load}} + i_2 R_2 + L_2 \frac{di_2}{dt} - M \frac{di_1}{dt} \tag{18-47}$$

Although derived for a circuit containing a DC driver, Eqs. (18-39) and (18-47) hold true for any type of driver.

If the driver in Fig. 18-7 is a sinusoidal generator instead of a battery, Eqs. (18-39) and (18-47) may be written in phasor form, and $\mathbf{Z}_{\text{load}}$ may be substituted for R_{load}. Thus,

$$\mathbf{E}_{\text{gen}} = \mathbf{I}_1 R_1 + \mathbf{I}_1 j X_{L1} - \mathbf{I}_2 j X_M \tag{18-48}$$

$$0 = \mathbf{I}_2 R_{\text{load}} + \mathbf{I}_2 R_2 + \mathbf{I}_2 j X_{L2} - \mathbf{I}_1 j X_M \tag{18-49}$$

where $X_{L1} = 2\pi f L_1$
$X_{L2} = 2\pi f L_2$
$X_M = 2\pi f M$

The *mutual reactance* X_M can be determined from the coefficient of coupling and the respective inductive reactances of the coupled coils:

$$X_M = k\sqrt{X_{L1}X_{L2}} \tag{18-50}$$

Equation (18-50) is obtained by multiplying both sides of Eq. (18-14) by $2\pi f$. Thus,

$$(2\pi f)M = (2\pi f)k\sqrt{L_1 L_2}$$

$$2\pi f M = k\sqrt{(2\pi f L_1)(2\pi f L_2)}$$

$$X_M = k\sqrt{X_{L1}X_{L2}}$$

The effect that the magnitude and phase angle of the secondary current has on the primary current may be demonstrated by solving Eq. (18-48) for I_1. Thus

$$\boxed{I_1 = \frac{E_{gen} + I_2 jX_M}{R_1 + jX_{L1}}} \tag{18-51}$$

If no load is connected to the secondary, $I_2 = 0$, and Eq. (18-51) becomes

$$I_1 = \frac{E_{gen}}{R_1 + jX_{L1}}$$

The magnitude and phase angle of primary current I_1, in Eq. (18-51), depend on the magnitude and phase angle of secondary current I_2. Thus the impedance and power factor of the load have a significant effect on the primary current.

Example 18-5 A 60-Hz 240-V generator connected to the primary of a transformer causes $20\underline{/10°}$ A in a load connected to the secondary. The coefficient of coupling is 0.75, and the parameters of the transformer (measured at 60 Hz) are $R_1 = 2.3 \, \Omega$, $X_{L1} = 80 \, \Omega$, $R_2 = 4.8 \, \Omega$, $X_{L2} = 100 \, \Omega$. Determine (a) the magnitude and phase angle of the primary current; (b) the primary current if the load is removed; (c) the primary current if a 240-V DC source is inadvertently connected to the primary.

Solution

(a) $X_M = k\sqrt{X_{L1}X_{L2}} = 0.75\sqrt{(80)(100)} = 67.1 \, \Omega$

$$I_1 = \frac{E_{gen} + I_2 jX_M}{R_1 + jX_{L1}} = \frac{(240\underline{/0°}) + (20\underline{/10°})(j67.1)}{2.3 + j80} = 16.5\underline{/1.3°} \text{ A}$$

(b) $I_1 = \dfrac{E_{gen}}{R_1 + jX_{L1}} = \dfrac{240\underline{/0°}}{2.3 + j80} = 3\underline{/-88.4°} \text{ A}$

(c) $I_1 = \dfrac{E_{bat}}{R_1 + jX_{L1}}$

At 0 Hz,

$R_1 = 2.3$

$X_{L1} = 2\pi f L_1 = 2\pi(0)L_1 = 0$

$I_1 = \dfrac{240}{2.3 + j0} = 104 \text{ A}$

18-5   IDEAL TRANSFORMER

The ideal transformer shown in Fig. 18-8a is a hypothetical transformer that has no leakage flux; its windings have no resistance, the permeability of the iron is infinite, it requires no current to maintain the flux, and there are no hysteresis and no eddy-current losses in the iron.

Although the ideal transformer cannot be physically realized, its mathematical relationships provide a very useful tool in the analysis of iron-core transformers. Specific applications are in the analysis of power and distribution transformers, and impedance-matching transformers.

The coefficient of coupling of the ideal transformer is 1. Hence its voltage ratio is

$$\boxed{\frac{E_2}{E_1} = \frac{N_2}{N_1}} \qquad\qquad (18\text{-}52)$$

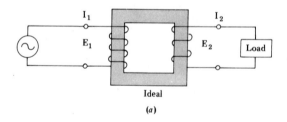

Ideal

(a)

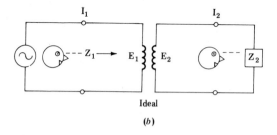

Ideal

(b)

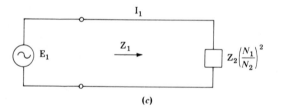

(c)

Figure 18-8   (a) Ideal transformer; (b) and (c) equivalent circuits.

Since the ideal transformer has no losses, the volt-ampere input to the primary is equal to the volt-ampere output of the secondary. That is,

$$\boxed{E_1 I_1 = E_2 I_2}$$ (18-53)

Although products $E_1 I_1$ and $E_2 I_2$ are complex numbers, they do not represent phasor power. The respective phasor powers are $E_1 I_1^*$ and $E_2 I_2^*$.

Expressing Eq. (18-53) as a current ratio,

$$\frac{I_1}{I_2} = \frac{E_2}{E_1}$$ (18-54)

Substituting the turns ratio from Eq. (18-52) into Eq. (18-54),

$$\boxed{\frac{I_1}{I_2} = \frac{N_2}{N_1}}$$ (18-55)

The input impedance of an ideal transformer and its connected load may be obtained by applying Ohm's law to the primary and secondary windings, and then manipulating the equations. Referring to the elementary diagram of the ideal transformer shown in Fig. 18-8b, and applying Ohm's law to the primary,

$$I_1 = \frac{E_1}{Z_1}$$ (18-56)

Impedance Z_1 is called the input impedance of the ideal transformer; it is the impedance "seen" when looking into the primary terminals.

Applying Ohm's law to the secondary,

$$I_2 = \frac{E_2}{Z_2}$$ (18-57)

Impedance Z_2 is the impedance looking out from the secondary terminals.

Dividing Eq. (18-57) by Eq. (18-56),

$$\frac{I_2}{I_1} = \frac{E_2/Z_2}{E_1/Z_1}$$

$$\frac{I_2}{I_1} = \frac{E_2 Z_1}{E_1 Z_2}$$

Substituting the appropriate turns ratios from Eqs. (18-52) and (18-55), and simplifying

$$\frac{N_1}{N_2} = \frac{N_2 Z_1}{N_1 Z_2}$$

$$\boxed{Z_1 = Z_2 \left(\frac{N_1}{N_2}\right)^2} \tag{18-58}$$

Thus, as viewed from the primary side, an ideal transformer multiplies the impedance of the secondary circuit by the square of the turns ratio. Impendance Z_1 is called the *reflected impedance*; it is the impedance of the secondary circuit *referred to the primary side*. Figure 18-8c represents the equivalent series circuit of an ideal transformer and its reflected load.

18-6   POWER AND DISTRIBUTION TRANSFORMERS

Power and distribution transformers are iron-core transformers designed for use on constant-frequency sinusoidal systems to transform energy from one voltage level to another. Transformers rated 500 kVA or smaller are generally classified as distribution transformers, and those rated in excess of 500 kVA are classified as power transformers. Applications range from small doorbell transformers to very large units for long-distance transmission of large amounts of energy.

No-Load Conditions

The no-load or *exciting current* of a power or distribution transformer is the current drawn by the primary when the secondary circuit is open. Under such conditions, as shown in Fig. 18-9a, the primary behaves as an impedance coil, and its current for a given driving voltage and frequency is limited by its cemf and effective resistance. A detailed analysis of the no-load condition is given in Sec. 18-3. Designating the exciting current as I_0, the primary current at no load will be

$$I_{p,\,\text{no load}} = I_0$$

Load Conditions

Connecting a resistor load to the secondary terminals, as shown in Fig. 18-9b, causes an increase in the primary current. (This phenomenon was discussed in Sec. 18-4.) The additional current is called the load component of primary current $I_{p,\,\text{load}}$. Thus, under load conditions, the primary current becomes

$$I_p = I_0 + I_{p,\,\text{load}}$$

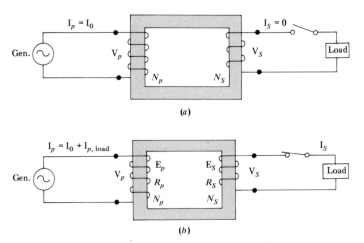

Figure 18-9   Distribution transformer: (a) no load; (b) loaded.

In a well-designed power or distribution transformer, operating at rated or near-rated load, the exciting component is very small when compared with the load component. Under such conditions, very little error is introduced if I_0 is assumed to be negligible. Hence, for practical considerations, when analyzing *loaded* transformers, it is acceptable to assume

$$\mathbf{I}_p = \mathbf{I}_{p,\,\text{load}}$$

In the steady-state analysis of loaded power and distribution transformers, a simple *equivalent-circuit model* that uses an ideal transformer with series-connected resistors and reactors is generally used in place of the actual transformer. The parameters for the ideal transformer and its series-connected reactors are derived from the magnetically induced voltages of the actual transformer. Thus, applying Kirchhoff's voltage law to the primary circuit of Fig. 18-9b, and neglecting the exciting current I_0,

$$\mathbf{V}_p = \mathbf{I}_{p,\,\text{load}} R_p + \mathbf{E}_p \qquad (18\text{-}59)$$

where $\mathbf{V}_p$ = driving voltage applied to primary circuit
   $\mathbf{E}_p$ = voltage drop generated in primary coil (cemf)
   R_p = resistance of primary winding

Expressing $\mathbf{E}_p$ as a function of flux and frequency, and noting that the modulus of ϕ in all related equations is Φ_{max}, Eq. (18-59) becomes

$$\mathbf{V}_p = \mathbf{I}_{p,\,\text{load}} R_p + 4.44\, N_p f \phi_{p,\,\text{win}} \qquad (18\text{-}60)$$

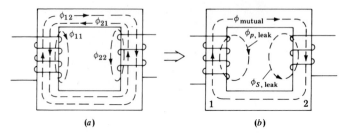

Figure 18-10 Component fluxes in a transformer core.

As previously seen in Fig. 18-7, and resketched in Fig. 18-10a, the flux through the primary window is

$$\phi_{p,\,win} = (\phi_{12} - \phi_{21}) + \phi_{11}$$

Defining,

$$\phi_{11} = \phi_{p,\,leak}, \text{ primary leakage flux}$$

$$(\phi_{12} - \phi_{21}) = \phi_{mutual}, \text{flux common to both primary and secondary windows}$$

$$\phi_{p,\,win} = \phi_{mutual} + \phi_{p,\,leak} \tag{18-61}$$

This is shown in Fig. 18-10b. Substituting Eq. (18-61) into Eq. (18-60) and expanding,

$$V_p = I_{p,\,load} R_p + 4.44\, N_p f \phi_{mutual} + 4.44\, N_p f \phi_{p,\,leak} \tag{18-62}$$

Expressed in simplified form,

$$V_p = I_{p,\,load} R_p + E'_p + E_{p,\,leak} \tag{18-63}$$

where

$$E'_p = 4.44\, N_p f \phi_{mutual} \tag{18-64}$$

$$E_{p,\,leak} = 4.44\, N_p f \phi_{p,\,leak}$$

A comparison of Eq. (18-63) with Eq. (18-59) shows that the cemf E_p generated in the primary of the actual transformer has been divided into two components; component $E_{p,\,leak}$ accounts for the voltage drop generated by the leakage flux in the primary, and component E'_p accounts for the voltage drop generated by the mutual flux in the primary.

Similarly, applying Kirchhoff's voltage law to the secondary of Fig. 18-9*b*,

$$\mathbf{E}_s = \mathbf{I}_s R_s + \mathbf{V}_s \tag{18-65}$$

where $\mathbf{E}_s$ = driving voltage induced in the secondary winding
$\mathbf{V}_s$ = voltage drop across the load
R_s = resistance of the secondary winding

Expressing $\mathbf{E}_s$ in terms of flux and frequency,

$$4.44\, N_s f \phi_{s,\,\text{win}} = \mathbf{I}_s R_s + \mathbf{V}_s \tag{18-66}$$

From Fig. 18-10*a*,

$$\phi_{s,\,\text{win}} = (\phi_{12} - \phi_{21}) - \phi_{22}$$

$$\phi_{s,\,\text{win}} = \phi_{\text{mutual}} - \phi_{s,\,\text{leak}}$$

Substituting into Eq. (18-66) and expanding,

$$4.44\, N_s f \phi_{\text{mutual}} - 4.44\, N_s f \phi_{s,\,\text{leak}} = \mathbf{I}_s R_s + \mathbf{V}_s \tag{18-67}$$

Expressed in simplified form,

$$\mathbf{E}'_s - \mathbf{E}_{s,\,\text{leak}} = \mathbf{I}_s R_s + \mathbf{V}_s \tag{18-68}$$

Rearranging terms,

$$\mathbf{E}'_s = \mathbf{E}_{s,\,\text{leak}} + \mathbf{I}_s R_s + \mathbf{V}_s \tag{18-69}$$

where

$$\mathbf{E}'_s = 4.44\, N_s f \phi_{\text{mutual}} \tag{18-70}$$

$$\mathbf{E}_{s,\,\text{leak}} = 4.44\, N_s f \phi_{s,\,\text{leak}}$$

A comparison of Eq. (18-68) with Eq. (18-65) shows that the driving voltage $\mathbf{E}_s$ induced in the secondary of the actual transformer has been divided into two components; component $\mathbf{E}'_s$ accounts for the voltage generated by the mutual flux in the secondary, and component $\mathbf{E}_{s,\,\text{leak}}$ accounts for the voltage drop generated by the leakage flux in the secondary.

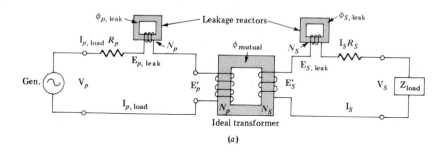

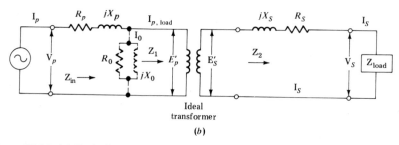

Figure 18-11 (a) Equivalent-circuit model of a loaded transformer; (b) elementary diagram of (a).

Figure 18-11a represents the equivalent-circuit model of the loaded transformer. The model satisfies Kirchhoff's voltage laws as expressed by Eqs. (18-63) and (18-69). The ideal transformer has the same turns ratio N_s/N_p as the actual transformer it replaces, and the series-connected reactors (X_p, X_s), called *leakage reactors*, generate voltage drops in the model that are equal to the respective voltage drops caused by leakage flux in the actual transformer. The series-connected resistors (R_p, R_s) generate voltage drops in the model that are equal to the respective voltage drops caused by the resistance of the actual transformer coils.

Figure 18-11b is an elementary diagram of the circuit shown in Fig. 18-11a. The R_0 and X_0 in the dotted insert account for the exciting current $\mathbf{I}_0$ of the actual transformer. The current drawn by R_0 accounts for the hysteresis and eddy current losses in the iron, and the current drawn by X_0 accounts for $\mathbf{\Phi}_{\text{mutual}}$.

Inductive reactances X_p and X_s are called the primary and secondary leakage reactances, respectively, and are expressed in ohms. Expressing the leakage drops in terms of the leakage reactances,

$$\mathbf{E}_{p,\text{leak}} = \mathbf{I}_{p,\text{load}} jX_p$$

$$\mathbf{E}_{s,\text{leak}} = \mathbf{I}_s jX_s$$

Note: Leakage reactances X_p and X_s must not be confused with X_{L1} and X_{L2} developed in Sec. 18-3. Reactances X_p and X_s are based on *leakage flux* through the respective windows, whereas X_{L1} and X_{L2} are based on *total flux* through the respective windows.

Applying Kirchhoff's voltage law to the primary and secondary circuits of Fig. 18-11b, and ignoring the exciting current,

$$\mathbf{V}_p = \mathbf{I}_{p,\,load} R_p + \mathbf{I}_{p,\,load} jX_p + \mathbf{E}'_p \qquad (18\text{-}71)$$

$$\mathbf{E}'_s = \mathbf{I}_s jX_s + \mathbf{I}_s R_s + \mathbf{V}_s \qquad (18\text{-}72)$$

A phasor diagram representing the voltage drops in the secondary circuit, assuming a 0.8-pf (36.87°) lagging load, is shown in Fig. 18-12a, and a tip-to-tail summation is shown in Fig. 18-12b. The tip-to-tail summation, representative of the phasor summation in Eq. (18-72), results in voltage phasor $\mathbf{E}'_s$.

A phasor diagram of primary voltage drops and a tip-to-tail summation that satisfies Eq. (18-71) is shown in Fig. 18-12c. Note that $\mathbf{E}'_p$ is in phase with $\mathbf{E}'_s$, and $\mathbf{I}_{p,\,load}$ is in phase with $\mathbf{I}_s$. This must be so because the two windings

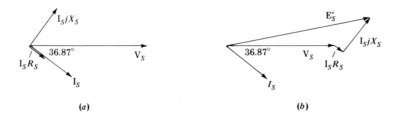

(a)   (b)

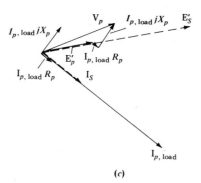

(c)

Figure 18-12 Phasor diagrams showing (a) component voltage drops in the secondary, assuming a 0.8-pf lagging load; (b) tip-to-tail summation of the voltage phasors in (a); (c) primary voltage drops and a tip-to-tail summation.

of the ideal transformer are linked by the same flux, ϕ_{mutual}. Assuming a turns ratio of $N_s/N_p = 3/1$,

$$\frac{\mathbf{E}'_s}{\mathbf{E}'_p} = \frac{N_s}{N_p} = \frac{3}{1} \qquad \frac{\mathbf{I}_s}{\mathbf{I}_{p,\text{load}}} = \frac{N_p}{N_s} = \frac{1}{3}$$

$$\mathbf{E}'_s = 3\mathbf{E}'_p \qquad \mathbf{I}_s = \tfrac{1}{3}\mathbf{I}_{p,\text{load}}$$

18-7 INPUT IMPEDANCE OF A LOADED TRANSFORMER

The equivalent-circuit model shown in Fig. 18-11b is very useful for analyzing the individual effects of resistance and leakage flux in the primary and secondary windings. However, in order to simplify transformer calculations, the actual transformer and its connected load are often replaced by the equivalent series circuit shown in Fig. 18-13.

Development of the equivalent series circuit is accomplished by referring the impedance of the secondary circuit in Fig. 18-11b to the primary side. Thus, using Eq. (18-58) for the ideal transformer,

$$\mathbf{Z}_1 = \mathbf{Z}_2 \left(\frac{N_p}{N_s}\right)^2 \tag{18-73}$$

where

$$\mathbf{Z}_2 = R_s + jX_s + \mathbf{Z}_{\text{load}} \tag{18-74}$$

Neglecting the exciting component, shown with dotted lines in Fig. 18-11b, the overall input impedance of the primary circuit is

$$\mathbf{Z}_{\text{in}} = R_p + jX_p + \mathbf{Z}_1 \tag{18-75}$$

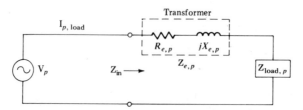

Figure 18-13 Equivalent series circuit model of a transformer.

Substituting Eqs. (18-73) and (18-74) into Eq. (18-75) and rearranging terms,

$$\mathbf{Z}_{\text{in}} = R_p + jX_p + (R_s + jX_s + \mathbf{Z}_{\text{load}})\left(\frac{N_p}{N_s}\right)^2$$

$$\mathbf{Z}_{\text{in}} = \left[R_p + R_s\left(\frac{N_p}{N_s}\right)^2\right] + j\left[X_p + X_s\left(\frac{N_p}{N_s}\right)^2\right] + \mathbf{Z}_{\text{load}}\left(\frac{N_p}{N_s}\right)^2$$

Defining,

$$R_{e,p} = \left[R_p + R_s\left(\frac{N_p}{N_s}\right)^2\right]$$

$$X_{e,p} = \left[X_p + X_s\left(\frac{N_p}{N_s}\right)^2\right]$$

$$\mathbf{Z}_{e,p} = R_{e,p} + jX_{e,p}$$

$$\mathbf{Z}_{\text{load},p} = \mathbf{Z}_{\text{load}}\left(\frac{N_p}{N_s}\right)^2$$

then,

$$\mathbf{Z}_{\text{in}} = (R_{e,p} + jX_{e,p}) + \mathbf{Z}_{\text{load},p} = \mathbf{Z}_{e,p} + \mathbf{Z}_{\text{load},p} \qquad (18\text{-}76)$$

Thus, as indicated in Eq. (18-76), the transformer and its connected load may be replaced by the equivalent series circuit shown in Fig. 18-13. Applying Kirchhoff's voltage law to this circuit,

$$\mathbf{V}_p = \mathbf{I}_{p,\text{load}}(R_{e,p} + jX_{e,p} + \mathbf{Z}_{\text{load},p})$$

$$\mathbf{V}_p = \mathbf{I}_{p,\text{load}}(\mathbf{Z}_{e,p} + \mathbf{Z}_{\text{load},p})$$

$$\mathbf{V}_p = \mathbf{I}_{p,\text{load}}\mathbf{Z}_{\text{in}}$$

Parameters $R_{e,p}$, $X_{e,p}$, and $\mathbf{Z}_{e,p}$ are called the equivalent resistance, equivalent reactance, and equivalent impedance, respectively, of the transformer as viewed from the primary side, and impedance $\mathbf{Z}_{\text{load},p}$ is the load impedance referred to the primary side.

Since transformers may be used to step up or step down voltage, either side may be the primary. Hence, to avoid confusion, the parameters provided by the manufacturer are specified as referred to the high-voltage side (high side) or referred to the low-voltage side (low side); the terms primary and secondary are avoided when specifying parameters. Conversion from low-side (LS) values to high-side (HS) values and vice versa may be accomplished by using the following extension of Eq. (18-73):

$$\mathbf{Z}_{e,\,HS} = \mathbf{Z}_{e,\,LS}\left(\frac{N_{HS}}{N_{LS}}\right)^2 \tag{18-77}$$

The turns ratio of a transformer is defined as the ratio of the number of turns in the high side to the number of turns in the low side, and is numerically equal to the ratio of induced voltages. That is,

$$\frac{N_{HS}}{N_{LS}} = \frac{E'_{HS}}{E'_{LS}}$$

However, if information regarding the turns ratio is not available, and an appropriate test cannot be made, the nameplate voltage ratio may be used as a rough approximation. That is,

$$\frac{N_{HS}}{N_{LS}} \approx \frac{V_{HS}}{V_{LS}}$$

The operating limits or range of a transformer are stamped on its nameplate. Thus, a nameplate rating of 2300/240 V, 60 Hz, 100 kVA indicates the following: (1) If used for step-down operation, the high-voltage winding should not be connected to a driving voltage that is appreciably higher than 2300 V. (2) If used for step-up operation, the low-voltage winding should not be connected to a driving voltage that is appreciably higher than 240 V; (3) the system frequency should be approximately 60 Hz; (4) the ampere loading of the secondary should not exceed the value obtained by dividing rated secondary voltage into rated volt-amperes.

Example 18-6   The equivalent resistance and equivalent reactance (high side) for a certain 25 kVA 2400/600-V, 60-Hz transformer are 2.8 Ω and 6.0 Ω, respectively. If a load impedance of $10\underline{/20°}$ Ω is connected to the low side, determine (a) the equivalent input impedance of the combined transformer and load; (b) the primary current if 2400 V is applied to the high side.

Solution
(a) Using the equivalent circuit shown in Fig. 18-13, referring the load impedance to the high side, and using $N_{HS}/N_{LS} \approx V_{HS}/V_{LS}$,

$$\mathbf{Z}_{in} = (R_{e,\,HS} + jX_{e,\,HS}) + \mathbf{Z}_{load,\,HS}$$

$$\mathbf{Z}_{in} = (2.8 + j6.0) + 10\underline{/20°}\left(\frac{2400}{600}\right)^2$$

$\mathbf{Z}_{in} = (2.8 + j6.0) + (150.35 + j54.72)$

$\mathbf{Z}_{in} = (153.15 + j60.72) = 164.75\underline{/21.63°}\ \Omega$

(b) From Ohm's law,

$$\mathbf{I}_{p,\,\mathrm{HS}} = \frac{\mathbf{V}_{\mathrm{HS}}}{\mathbf{Z}_{in,\,\mathrm{HS}}} = \frac{2400\underline{/0°}}{164.75\underline{/21.63°}} = 14.57\underline{/-21.63°}\ \mathrm{A}$$

Example 18-7 A 50-kVA, 4160/600-V, 60-Hz transformer has an equivalent impedance referred to the high side of $9\underline{/50°}\ \Omega$. Determine (a) the input impedance of the combined transformer and load if the secondary is supplying rated kVA at 600 V to an 0.80-pf lagging load; (b) the input voltage to the primary that results in the given output.

Solution
(a) The power-factor angle is

$\theta = \cos^{-1} 0.80 = 36.86°$ lagging

The phasor power supplied by the secondary is

$\mathbf{S}_s = 50{,}000\underline{/36.86°}\ \mathrm{VA}$

$\mathbf{S}_s = \mathbf{V}_s\mathbf{I}_s^*$

$50{,}000\underline{/36.86} = (600\underline{/0°})\mathbf{I}_s^*$

$\mathbf{I}_s^* = 83.33\underline{/36.86°}$

$\mathbf{I}_s = 83.33\underline{/-36.86°}\ \mathrm{A}$

From Ohm's law,

$$\mathbf{Z}_{\mathrm{load}} = \frac{\mathbf{V}_s}{\mathbf{I}_s} = \frac{600\underline{/0°}}{83.33\underline{/-36.86°}} = 7.20\underline{/36.86°}\ \Omega$$

$\mathbf{Z}_{\mathrm{load,\,HS}} = 7.20\underline{/36.86°}\left(\dfrac{4160}{600}\right)^2 = 346.11\underline{/36.86°}\ \Omega$

$\mathbf{Z}_{in,\,\mathrm{HS}} = R_{e,\,\mathrm{HS}} + jX_{e,\,\mathrm{HS}} + \mathbf{Z}_{\mathrm{load,\,HS}}$

$\mathbf{Z}_{in,\,\mathrm{HS}} = 9.50\underline{/50°} + 346.11\underline{/36.86°}$

$\mathbf{Z}_{in,\,\mathrm{HS}} = 355.37\underline{/37.21°}\ \Omega$

(b) From the equivalent-circuit diagram of Fig. 18-13,

$$\mathbf{V}_p = \mathbf{I}_{p,\,load}\,\mathbf{Z}_{in}$$

Using the voltage ratio as an approximation of the turns ratio,

$$\frac{\mathbf{I}_s}{\mathbf{I}_{p,\,load}} = \frac{N_p}{N_s} \approx \frac{V_p}{V_s}$$

$$\mathbf{I}_{p,\,load} = \frac{600}{4160}\,83.33\underline{/-36.86^\circ} = 12.02\underline{/-36.86^\circ}\ \text{A}$$

Thus,

$$\mathbf{V}_p = (12.02\underline{/-36.86^\circ})(355.37\underline{/37.21^\circ}) = 4271.55\underline{/0.35^\circ}\ \text{V}$$

18-8   AUTOTRANSFORMER

The autotransformer is a transformer that accomplishes the desired transformer action within one coil as compared with two or more coils of a standard transformer. The autotransformer requires less copper and iron, hence occupies less space than its two-winding counterpart. It is used extensively for reduced-voltage starting of AC motors, balance coils for three-wire DC generators, voltage step-up or step-down for transmission lines, small-toy operation such as model electric trains, and many other uses.

Figure 18-14a illustrates the construction details of an autotransformer. The entire coil is the high-voltage side and the tapped section is the low-voltage side. The position of the tap determines the turns ratio, and hence the voltage ratio of the autotransformer. The number of turns embraced by the low side is generally expressed as a percentage of the total. Thus, assuming a 20 percent tap, the low-side voltage will be 20 percent of the high-side voltage. The number of turns embraced by the driver is called the primary turns N_p, and the number of turns embraced by the load is designated the secondary turns N_s. Connections for step-down operation and step-up operation are shown in Fig. 18-14b and c, respectively.

The single-winding construction of the autotransformer results in less winding resistance and less leakage flux than that in an equivalent two-winding transformer. Hence the voltage drops due to the resistance and leakage effects are generally small enough to be neglected, permitting the relationships developed for the ideal transformer to be applied. Thus, for the autotransformer,

$$\frac{\mathbf{V}_s}{\mathbf{V}_p} = \frac{N_s}{N_p} \qquad \frac{\mathbf{I}_s}{\mathbf{I}_{p,\,load}} = \frac{N_p}{N_s}$$

$$\mathbf{V}_p\mathbf{I}_{p,\,load} = \mathbf{V}_s\mathbf{I}_s$$

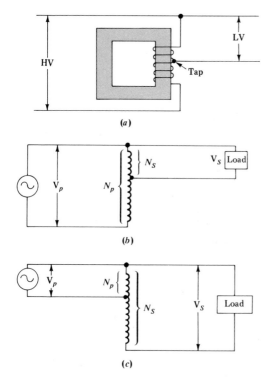

Figure 18-14 Autotransformer:
(*a*) construction details;
(*b*) elementary diagram showing step-down operation;
(*c*) elementary diagram showing step-up operation.

The current and voltage relationships in a two-winding transformer and in an equivalent autotransformer are shown in Fig. 18-15*a* and *b*, respectively. The autotransformer's 33 1/3 percent tap provides the same turns ratio as does the two-winding transformer. Each transformer supplies 240 V to a 2-Ω resistor, and for purposes of comparison, it will be assumed that both transformers are ideal.

The secondary current in each transformer, as determined by Ohm's law, is

$$\mathbf{I}_s = \frac{\mathbf{V}_s}{\mathbf{Z}_{\text{load}}} = \frac{240\underline{/0^\circ}}{2\underline{/0^\circ}} = 120\underline{/0^\circ} \text{ A}$$

The corresponding primary currents are

$$\frac{\mathbf{I}_s}{\mathbf{I}_{p,\text{load}}} = \frac{N_p}{N_s}$$

$$\frac{120\underline{/0^\circ}}{\mathbf{I}_{p,\text{load}}} = \frac{300}{100}$$

$$\mathbf{I}_{p,\text{load}} = 40\underline{/0^\circ} \text{ A}$$

With the primary and secondary currents known, the application of Kirchhoff's current law to node "a" in Fig. 18-15b indicates that current I_x must be $80\underline{/0°}$ A directed toward node "a";

$$40\underline{/0°} + I_x = 120\underline{/0°}$$

$$I_x = 80\underline{/0°}\ A$$

It is interesting to note that the 80 A in the 100 t section of the autotransformer is equivalent to the superposition of the 40 A down and 120 A up in the two-winding transformer.

The theory of load transfer in the autotransformer may be explained with the aid of Fig. 18-15c, which illustrates somewhat differently the electrical connections of Fig. 18-15b. For the load conditions shown, the 40 A supplied by the generator feed the load directly but in series with the lower 200 turns of the autotransformer. The upper 100 turns act as the "secondary" of a 2/1 ratio transformer whose "primary" is the lower 200 turns. The voltage across the "secondary" is 240 V, and the voltage across the "primary" is 480 V. The current in the "secondary," as determined from the turns ratio, is

$$\frac{\text{"}I_s\text{"}}{\text{"}I_{p,\,load}\text{"}} = \frac{\text{"}N_p\text{"}}{\text{"}N_s\text{"}}$$

$$\frac{\text{"}I_s\text{"}}{40} = \frac{200}{100}$$

$$\text{"}I_s\text{"} = 80\ A$$

In accordance with Lenz's law, the instantaneous direction of current in the "secondary" is always in a direction to oppose the change in flux that produced it. Hence, if the instantaneous direction of current in the "primary" is downward, that which is in the "secondary" must be upward. The 80 A of the "secondary" add to the 40 A of the generator to supply 120 A to the load.

From the standpoint of power transfer, part of the kVA supplied to the load is by direct conduction from the generator and part by the medium of transformer action. Thus, referring to Fig. 18-15b,

$$\text{kVA conducted} = \frac{I_p V_s}{1000} = \frac{40 \times 240}{1000} = 9.6\ \text{kVA}$$

$$\text{kVA transformed} = \frac{(I_s - I_p)V_s}{1000} = \frac{(120-40)(240)}{1000} = 19.2\ \text{kVA}$$

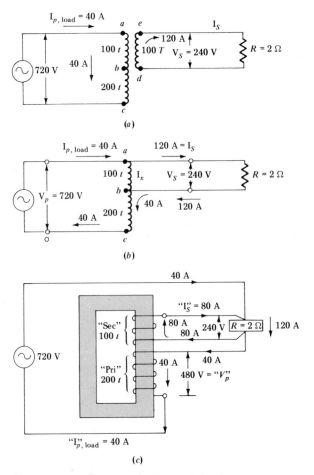

Figure 18-15 Current and voltage relationships in (a) a two-winding transformer; (b) an equivalent autotransformer; (c) rearranged connections of (b).

The kVA transformed may also be obtained by subtracting the kVA conducted from the total kVA. Thus,

$$\text{kVA transformed} = \frac{I_s V_s - I_p V_s}{1000} = \frac{120 \times 240 - 40 \times 240}{1000} = 19.2 \text{ kVA}$$

Example 18-8 A 10-kVA, 4160/450-V, 25-Hz autotransformer is used for step-down operation and is delivering rated kVA at rated voltage and 0.85 pf lagging. The maximum instantaneous core flux at rated high-side voltage is 0.052 Wb. Neglecting losses and leakage, determine (a) the current to the load; (b) the current supplied by the generator;

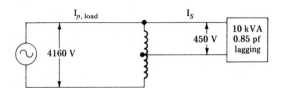

Figure 18-16 Circuit for Example 18-8.

(c) the number of turns of wire in the transformer coil; (d) the number of turns of wire embraced by the secondary; (e) the kVA conducted to the load, and the kVA transformed.

Solution
The circuit diagram is shown in Fig. 18-16.

(a) $\theta = \cos^{-1} 0.85 = 31.79°$

$\mathbf{S}_s = 10{,}000\underline{/31.79°}$ VA

$\mathbf{S}_s = \mathbf{V}_s \mathbf{I}_s^*$

$10{,}000\underline{/31.79°} = (450\underline{/0°})\mathbf{I}_s^*$

$\mathbf{I}_s^* = 22.22\underline{/31.79°}$

$\mathbf{I}_s = 22.22\underline{/-31.79°}$ A

(b) $\dfrac{\mathbf{I}_s}{\mathbf{I}_{p,\text{load}}} = \dfrac{N_p}{N_s}$ where $\dfrac{N_p}{N_s} = \dfrac{4160}{450}$

$\dfrac{22.22\underline{/-31.79°}}{\mathbf{I}_{p,\text{load}}} = \dfrac{4160}{450}$

$\mathbf{I}_{p,\text{load}} = 2.4\underline{/-31.79°}$ A

(c) $V_p = 4.44 \, N_p f\phi_{\text{max}}$

$4160 = 4.44 \, N_P(25)(0.052)$

$N_p = 720.7$ turns

(d) $\dfrac{V_s}{V_P} = \dfrac{N_s}{N_p}$

$\dfrac{450}{4160} = \dfrac{N_s}{720.7}$

$N_s = 78$ turns

(e) kVA conducted $= \dfrac{2.4 \times 450}{1000} = 1.08 \text{ kVA}$

kVA transformed $= 10 - 1.08 = 8.92 \text{ kVA}$

18-9   **SLIDE-WIRE AUTOTRANSFORMER**

A slide-wire autotransformer, shown in Fig. 18-17a, is a very efficient voltage divider and for currents greater than 1 A is far superior to the slide-wire rheostat shown in Fig. 18-17b.

If the slide in Fig. 18-17a is adjusted to supply 100 A at 200 V to a resistor load, the apparent power input to the autotransformer will be

$$V_p I_p = V_s I_s = 200(100) = 20 \text{ kVA}$$

The slide-wire rheostat does not use transformer action; it relies on IR drops to obtain the desired output voltage and output current. The slide position that provides the desired output may be determined by applying loop analysis. Thus,

For loop *abcda*,

$$0 = 100(2) - I_1 R_1 \qquad\qquad (18\text{-}78)$$

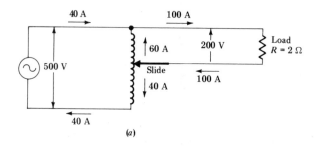

(a)

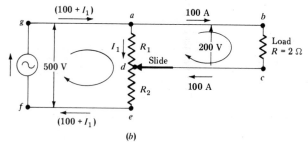

(b)

Figure 18-17   (a) Slide-wire autotransformer; (b) slide-wire rheostat.

For loop *gadefg*,

$$500 = I_1 R_1 + (100 + I_1)R_2 \tag{18-79}$$

Assuming an arbitrary value of 6 Ω for the total resistance of the rheostat,

$$R_1 + R_2 = 6\,\Omega \tag{18-80}$$

Solving Eq. (18-80) for R_2 and Eq. (18-78) for I_1, substituting the respective values into Eq. (18-79), and simplifying,

$$R_1^2 - R_1 - 12 = 0$$

Solving,

$$R_1 = 4\,\Omega$$

Since

$$R_1 + R_2 = 6$$

$$R_2 = 2\,\Omega$$

Substituting into Eq. (18-78),

$$0 = 100(2) - I_1(4)$$

$$I_1 = 50\,\text{A}$$

Hence the generator current is

$$I_{gen} = 100 + 50 = 150\,\text{A}$$

Thus the apparent power input to the rheostat is

$$S = VI = 500(150) = 75\,\text{kVA}$$

The saving in input current and power when using a slide-wire autotransformer instead of a slide-wire rheostat is 110 A and 55 kVA.

18-10   TRANSFORMERS FOR IMPEDANCE MATCHING

The impedance-modifying characteristics of a transformer have applications in those circuits where it is desirable to match the resistance of a given load to the resistance of a generator for the purpose of maximizing the power

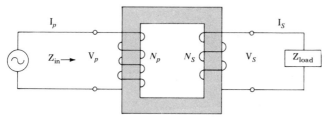

Figure 18-18   Ideal transformer for impedance matching.

delivered to the load. In such applications, the generator, or source, is usually the output of an amplifier.

Transformers designed for impedance-matching purposes closely approach the ideal. Hence the current, voltage, and impedance relationships previously developed for the ideal transformer are applicable. Assuming the transformer in Fig. 18-18 is designed for impedance matching, I_0 is negligible, and $I_p = I_{p,\text{load}}$, then

$$\frac{V_s}{V_p} = \frac{N_s}{N_p} \qquad \frac{I_s}{I_p} = \frac{N_p}{N_s}$$

$$V_p I_p = V_s I_s$$

$$Z_{\text{in}} = Z_{\text{load}}\left(\frac{N_p}{N_s}\right)^2$$

It should be noted that only the magnitude of Z_{load} is affected by the transformer; *the phase angle is not altered.*

Example 18-9   Determine the required turns ratio for an impedance-matching transformer that will provide maximum power transfer to a 6-Ω loudspeaker from an amplifier whose output impedance is 50,000 Ω resistive. The circuit is shown in Fig. 18-19a.

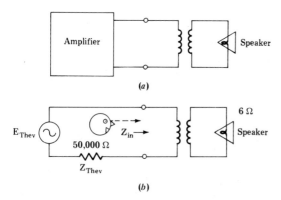

Figure 18-19   Circuits for Example 18-9.

Solution
The output impedance of an amplifier is its Thevenin impedance. Thus the amplifier acts as a generator whose Thevenin impedance is 50,000 Ω, as shown in Fig. 18-19b. For maximum power transfer, the amplifier must "see" a load whose resistance is equal to its own resistance. Placing a transformer between the load and the amplifier causes the amplifier to "see" an impedance equal to

$$\mathbf{Z}_{in} = \left(\frac{N_p}{N_s}\right)^2 \mathbf{Z}_{load}$$

Thus, for maximum power transfer, $\mathbf{Z}_{in}$ must equal 50,000 Ω.

$$50,000 = \left(\frac{N_p}{N_s}\right)^2 \quad (6)$$

$$\frac{N_p}{N_s} = 91.3$$

Example 18-10 An impedance-matching transformer is used to couple a 20-Ω resistive load to an amplifier whose Thevenin equivalent is

$$\mathbf{Z}_{Thev} = 1000\underline{/0°}\ \Omega$$

$$\mathbf{E}_{Thev} = 25\underline{/0°}\ V$$

Determine (a) the turns ratio required for maximum power transfer; (b) the primary current; (c) the secondary current; (d) the secondary voltage; (e) the active power delivered to the 20-Ω load; (f) the active power delivered to the 20-Ω load if the transformer is not used; (g) the percentage increase in delivered power when the impedance-matching transformer is used.

Solution
The circuit is shown in Fig. 18-20a. (a) For a perfect match the $\mathbf{Z}_{in}$ of the combined transformer and load must equal $\mathbf{Z}_{Thev}^*$. Thus,

$$\mathbf{Z}_{Thev}^* = \mathbf{Z}_{in} = \left(\frac{N_p}{N_s}\right)^2 \mathbf{Z}_{load}$$

$$1000 = \left(\frac{N_p}{N_s}\right)^2 20$$

$$\frac{N_p}{N_s} = 7.07$$

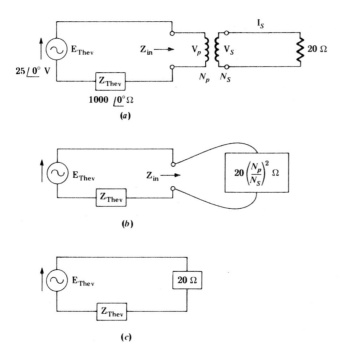

Figure 18-20 Circuits for Example 18-10.

(b) The equivalent circuit for perfect matching is shown in Fig. 18-20b. Applying Ohm's law,

$$I_p = \frac{E_{Thev}}{Z_{Thev} + Z_{in}} = \frac{25\underline{/0°}}{1000 + 1000} = 0.0125\underline{/0°} \text{ A}$$

(c) $\dfrac{I_s}{I_p} = \dfrac{N_p}{N_s}$

$$\frac{I_s}{0.0125\underline{/0°}} = \frac{7.07}{1}$$

$I_s = 0.0884\underline{/0°} \text{ A}$

(d) Referring to Fig. 18-20a, $V_s = I_s Z_{load}$

$V_s = 20\,I_s = 20(0.0884)\underline{/0°} = 1.768 \text{ V}$

(e) $P = I_{load}^2 R_{load}$

$P = (0.0884)^2(20) = 156.3 \text{ mW}$

(*f*) The circuit for part *f* is shown in Fig. 18-20*c*.

$$I = \frac{\mathbf{E}_{\text{Thev}}}{\mathbf{Z}_{\text{Thev}} + 20} = \frac{25\underline{/0°}}{1000 + 20} = 0.0245 \text{ A}$$

$$P = I_{\text{load}}^2 R_{\text{load}}$$

$$P = (0.0245)^2(20) = 12 \text{ mW}$$

(*g*) $\dfrac{0.1563 - 0.0120}{0.0120} \, 100 = 1202.5$ percent

The addition of an impedance-matching transformer resulted in a 1202.5 percent increase in power delivered to the load, as compared with the power delivered without the transformer.

SUMMARY OF FORMULAS

General

$$\phi_1 = \phi_{11} + \phi_{12} \qquad \phi_2 = \phi_{22} + \phi_{21} \qquad \frac{L_2}{L_1} = \left(\frac{N_2}{N_1}\right)^2$$

$$k = \frac{\phi_{12}}{\phi_1} \qquad\qquad \frac{v_2}{v_1} = k\frac{N_2}{N_1}$$

$$M = N_2 \frac{d\phi_{12}}{di_1} = N_1 \frac{d\phi_{21}}{di_2} = k\sqrt{L_1 L_2}$$

Sinusoidal Driver

$$V = 4.44 \, Nf\phi_{\max} \qquad X_M = k\sqrt{X_{L1} X_{L2}}$$

$$\mathbf{E}_{\text{gen}} = \mathbf{I}_1 R_1 + \mathbf{I}_1 jX_{L1} - \mathbf{I}_2 jX_M$$

$$0 = \mathbf{I}_2 \mathbf{Z}_{\text{load}} + \mathbf{I}_2 jX_{L2} - \mathbf{I}_1 jX_M \qquad \mathbf{I}_1 = \frac{\mathbf{E}_{\text{gen}} + \mathbf{I}_2 jX_M}{R_1 + jX_{L1}}$$

Power and Distribution Transformers

$$\frac{N_{\text{HS}}}{N_{\text{LS}}} = \frac{\mathbf{E}'_{\text{HS}}}{\mathbf{E}'_{\text{LS}}} \approx \frac{\mathbf{V}_{\text{HS}}}{\mathbf{V}_{\text{LS}}} \qquad \frac{\mathbf{I}_{\text{HS}}}{\mathbf{I}_{\text{LS}}} = \frac{N_{\text{LS}}}{N_{\text{HS}}}$$

$$\mathbf{Z}_{e,\text{HS}} = \mathbf{Z}_{e,\text{LS}}\left(\frac{N_{\text{HS}}}{N_{\text{LS}}}\right)^2 \qquad \mathbf{Z}_{\text{in, HS}} = \mathbf{Z}_{e,\text{HS}} + \mathbf{Z}_{\text{load, HS}}$$

$$\mathbf{Z}_{\text{load, HS}} = \mathbf{Z}_{\text{load, LS}}\left(\frac{N_{\text{HS}}}{N_{\text{LS}}}\right)^2$$

Impedance-Matching and Ideal Transformers

$$\frac{V_s}{V_p} = \frac{N_s}{N_p} \qquad \frac{I_s}{I_p} = \frac{N_p}{N_s}$$

$$V_p I_p = V_s I_s \qquad Z_{in} = Z_{load}\left(\frac{N_p}{N_s}\right)^2$$

PROBLEMS

18-1 A DC current of 5 A through a 20-turn coil causes a flux of 0.080 Wb in a 100-turn coil. Determine the mutual inductance.

18-2 Two coils have a mutual inductance of 10 H and a coefficient of coupling of 0.90. The primary and secondary coils have turns of 100 and 300, respectively. The resistances of the primary and secondary windings are 2.0 Ω and 8.0 Ω, respectively. Sketch the circuit and determine (a) the primary current required to produce a flux of 0.80 Wb in the secondary coil; (b) the self-inductance of the primary; (c) the self-inductance of the secondary.

18-3 Two coils have a mutual inductance of 15 H and a coefficient of coupling of 0.80. The primary and secondary coils have turns of 100 and 250, respectively. The resistances of the primary and secondary windings are 2.0 Ω and 6.0 Ω, respectively. Sketch the circuit and determine (a) the primary current required to produce a flux of 1.0 Wb in the secondary coil; (b) the self-inductance of the primary; (c) the self-inductance of the secondary.

18-4 The application of 120 V from a DC generator to the primary of a transformer results in a primary flux of 0.50 Wb. The primary and secondary windings have 200 and 100 turns, respectively. The resistance of the primary winding is 4.0 Ω and that of the secondary is 3.0 Ω. Assume the coefficient of coupling to be 0.9. Sketch the circuit and determine (a) the mutual inductance; (b) the self-inductance of the secondary.

18-5 The application of 240 V from a DC generator to the primary of a transformer results in a primary flux of 0.30 Wb. The primary and secondary windings have 400 and 100 turns, respectively. The resistance of the primary winding is 8.0 Ω and that of the secondary is 3.0 Ω. Assume the coefficient of coupling is 0.95. Sketch the circuit and determine (a) the mutual inductance; (b) the self-inductance of the primary and secondary coils.

18-6 A 250-turn coil and a 300-turn coil have resistances of 76 Ω and 95 Ω, respectively. The coefficient of coupling is 0.82, and the reluctance of the magnetic circuit is 420 A · t/Wb. Assume a 120-V 0-Hz driver is connected to the 250-turn coil. Sketch the circuit and determine (a) the flux linking the secondary; (b) the mutual inductance; (c) the instantaneous value of secondary voltage if at the instant the switch is opened the current starts to fall at the rate of 0.8064 A/s (assume a diode is properly connected across the primary coil to provide a path for the released energy); (d) the instantaneous value of primary cemf for the conditions in part c.

18-7 A 100-turn coil and a 200-turn coil have resistances of 50 Ω and 80 Ω, respectively. The coefficient of coupling is 0.75, and the reluctance of the magnetic circuit is 3000 A · t/Wb. The application of 150-V 0 Hz to the 100-turn coil produces a primary flux of 0.10 Wb. Sketch the circuit and determine (a) the secondary flux; (b) the mutual inductance; (c) the instantaneous value of secondary voltage if at the instant the switch is opened the current starts to fall at the rate of 45.0 A/s (assume a diode is properly connected across the primary to provide a path for the released energy); (d) the instantaneous value of primary cemf.

18-8 A 40-turn coil and a 20-turn coil have resistances of 10 Ω and 5.0 Ω, respectively. They are both linked by a doughnut-shaped ferromagnetic core whose reluctance is 2400 A · t/Wb. The coefficient of coupling is 0.80. The 40-turn coil is connected to a 60-V battery. Sketch the circuit and determine (a) steady-state flux through the window of the 40-turn coil; (b) flux through the secondary window; (c) self-inductance of the 40-turn coil; (d) self-inductance of the 20-turn coil; (e) mutual inductance.

18-9 Determine the peak value of the sinusoidal flux in the iron core of a transformer that has a primary coil of 200 turns and is connected to a 240-V, 60-Hz 50-kVA generator.

18-10 A certain 60-Hz transformer has two windings of 1000 and 250 turns, respectively. What is the operating voltage of a 60-Hz alternator that causes a sinusoidal flux of 0.495 Wb (maximum) when connected to the 1000-turn coil?

18-11 (a) Calculate the amplitude of the flux wave required for a 100-kVA, 25-Hz, 440/880-V transformer (assume that the 440-V coil has 170 turns); (b) write the equation of the flux wave.

18-12 A 60-Hz 120-V generator connected to the primary of a transformer causes $15/-60°$ A in a load connected to the secondary. If the coefficient of coupling is 0.86 and the parameters of the transformer are $R_1 = 1.8\ \Omega$, $X_{L1} = 57\ \Omega$, $R_2 = 4.6\ \Omega$, $X_{L2} = 150\ \Omega$, determine (a) the magnitude and phase angle of the primary current; (b) repeat part a assuming the load current is $15/+60°$ A; (c) the primary current if the load is removed; (d) the primary and secondary current if a 120-V battery is inadvertently connected to the primary.

18-13 A 60-V 25-Hz generator supplies a current of $4.0/25°$ A to the primary of a transformer whose secondary is connected to some unknown impedance. The parameters of the transformer measured at 25 Hz are $R_1 = 3.9\ \Omega$, $X_{L1} = 65\ \Omega$, $R_2 = 1.4\ \Omega$, $X_{L2} = 42\ \Omega$, $k = 0.92$. Determine (a) the magnitude and phase angle of the secondary current; (b) the primary current if the load is removed.

18-14 A 4160/208-V 25-Hz transformer has its low side connected to a load impedance of $0.010/30°\ \Omega$. The equivalent impedance of the transformer referred to the high side is $2.0/70°\ \Omega$. Sketch the circuit and determine (a) the input impedance of the combined transformer and load; (b) the primary current if the actual driving voltage is 4000 V.

18-15 A 2300/120-V 60-Hz transformer has an equivalent impedance, referred to the high side, of $4/70°\ \Omega$. The load connected to the low-side terminals is $0.04/10°\ \Omega$. Sketch the circuit and determine (a) the input impedance of the combined transformer and load; (b) the primary current if the rated voltage is applied to the high side.

18-16 A 25-kVA, 2200/600-V, 60-Hz transformer has the following parameters: $R_{HS} = 1.40\ \Omega$, $X_{HS} = 3.2\ \Omega$, $R_{LS} = 0.11\ \Omega$, $X_{LS} = 0.25\ \Omega$. (a) Find the input voltage required to obtain an output of 25 kVA at 600 V and 0.80 pf lagging; (b) assuming the high side is inadvertently connected to a 2200-V DC driver, determine the steady-state high-side and low-side currents.

18-17 An autotransformer has a total of 600 turns with a tap embracing 200 turns. A driver of 100 V at 60 Hz is impressed across the 200 turns, and a 30-Ω resistor load is connected across the 600 turns. Sketch the circuit diagram and, neglecting leakage and losses, determine (a) the voltage across the load; (b) the current in the load; (c) the input current to the transformer; (d) on the diagram, indicate the magnitude and relative directions to and from the generator, in the load, in the 200-turn section, and in the 400-turn section.

18-18 A 280-turn autotransformer has a secondary tapped at 14 turns. The input voltage to the high side is 240 V at 60 Hz. The secondary is connected to a 0.0060-Ω resistor. Neglect losses and leakage effects. Sketch the circuit and determine (a) the voltage at the load; (b) the secondary current; (c) the primary current; (d) the magnitude of current in section of transformer winding embraced by the secondary load.

18-19 The input voltage and current to an autotransformer are 400 V and 50 A, respectively. The transformer is used for step-down operation and the section of the winding embraced by the load has a current of 30 A. The losses and leakage effects may be neglected. Sketch the circuit and calculate (a) the current to the load; (b) the total volt-amperes supplied to the load; (c) the voltage at the load; (d) the kVA conducted to the load; (e) the kVA transformed.

18-20 An 801-turn autotransformer is connected to a 450-V 25-Hz supply, and a 0.20-Ω resistor load is connected to a 50-V tap. Sketch the circuit and determine (a) the number of turns embraced by the tap; (b) the current to the load; (c) the input current to the transformer; (d) the kVA conducted to the load; (e) the kVA transformed; (f) the total kVA supplied to the load.

18-21 An autotransformer with a 25 percent tap is used for step-down operation from a 600-V 60-Hz supply. The load connected to the secondary terminals is 30 kVA at unity power factor. Sketch the circuit and determine (a) the voltage at the secondary terminals; (b) the secondary current; (c) the primary current.

18-22 An ideal transformer is required to match an 8.0-Ω speaker coil to an amplifier whose output impedance is 10,000 Ω. Determine the turns ratio of the transformer.

18-23 An ideal transformer has a 50/20 turns ratio. Determine the input impedance at the primary terminals if a 6.0-Ω load is connected to the 20-turn secondary.

18-24 An ideal transformer is used to couple a 50-Ω resistor load to a generator whose internal impedance is $400/0°$ Ω. If the coil connected to the generator has 100 turns, determine the required number of secondary turns so that maximum power will be transferred to the load.

18-25 The primary and secondary turns of an impedance-matching transformer are 200 and 600, respectively. Determine (a) the input impedance at the primary terminals if a $20/30°$-Ω load is connected to the 600-turn secondary; (b) the rms primary current if the driving voltage is 50 sin (300t) V; (c) the secondary current (rms); (d) the secondary voltage (rms); (e) the active power drawn by the load.

18-26 A 100-Hz generator supplies a 2.0-Ω resistor through an ideal transformer whose primary to secondary turns ratio is 200/600. The rms voltage across the load is 120 V. Sketch the circuit and determine (a) the secondary current; (b) the primary voltage; (c) the primary current; (d) the input impedance.

CHAPTER 19

MAGNETICALLY COUPLED NETWORKS

This chapter extends network theory to magnetically coupled circuits. It combines network theory developed in Chap. 14 with transformer theory developed in Chap. 18. Magnetically coupled networks include two or more coupled coils and may have voltage or current sources connected to one or more of the coupled coils.

19-1 TRANSFORMERS WITH DRIVERS IN SERIES WITH EACH COIL

The flux contributions caused by the respective coil currents in Fig. 19-1a and b are defined as follows:

ϕ_{11} = leakage flux around coil 1 *due to the current in coil 1*
ϕ_{12} = flux through the window of coil 2 *due to the current in coil 1*
ϕ_1 = total flux through the window of coil 1 *due to the current in coil 1*

Thus,

$$\phi_1 = \phi_{11} + \phi_{12} \tag{19-1}$$

ϕ_{22} = leakage flux around coil 2 *due to the current in coil 2*
ϕ_{21} = flux through the window of coil 1 *due to the current in coil 2*
ϕ_2 = total flux through the window of coil 2 *due to the current in coil 2*

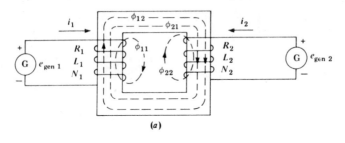

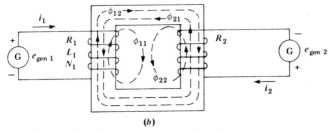

Figure 19-1 Transformers with drivers in series with each coil.

Thus

$$\boxed{\phi_2 = \phi_{22} + \phi_{21}} \tag{19-2}$$

For the transformer shown in Fig. 19-1a, the drivers are connected in such a manner as to cause the fluxes set up by the respective currents to be in the same direction within the transformer core. For this condition, the net flux through the window of coil 1 is

$$\phi_{\text{win coil 1}} = \phi_{12} + \phi_{11} + \phi_{21} = \phi_1 + \phi_{21}$$

The net flux through the window of coil 2 is

$$\phi_{\text{win coil 2}} = \phi_{21} + \phi_{22} + \phi_{12} = \phi_2 + \phi_{12}$$

The voltage drops in the transformer coils, due to the rate of change of flux through the respective windows, are

$$v_1 = N_1 \frac{d(\phi_1 + \phi_{21})}{dt} = N_1 \frac{d\phi_1}{dt} + N_1 \frac{d\phi_{21}}{dt} \tag{19-3}$$

$$v_2 = N_2 \frac{d(\phi_2 + \phi_{12})}{dt} = N_2 \frac{d\phi_2}{dt} + N_2 \frac{d\phi_{12}}{dt} \tag{19-4}$$

Applying Kirchhoff's voltage law to the respective circuits of coil 1 and coil 2 in Fig. 19-1a

$$e_{\text{gen 1}} = i_1 R_1 + v_1 \tag{19-5}$$

$$e_{\text{gen 2}} = i_2 R_2 + v_2 \tag{19-6}$$

Substituting Eqs. (19-3) and (19-4) into Eqs. (19-5) and (19-6), respectively:

$$e_{\text{gen 1}} = i_1 R_1 + N_1 \frac{d\phi_1}{dt} + N_1 \frac{d\phi_{21}}{dt} \tag{19-7}$$

$$e_{\text{gen 2}} = i_2 R_2 + N_2 \frac{d\phi_2}{dt} + N_2 \frac{d\phi_{12}}{dt} \tag{19-8}$$

Expressing Eqs. (19-7) and (19-8) in terms of self-inductance, mutual inductance, and the rates of change of the respective primary and secondary currents,

$$e_{\text{gen 1}} = i_1 R_1 + L_1 \frac{di_1}{dt} + M \frac{di_2}{dt}$$

$$e_{\text{gen 2}} = i_2 R_2 + L_2 \frac{di_2}{dt} + M \frac{di_1}{dt} \tag{19-9}$$

If the drivers are sinusoidal, equation set (19-9) may be written in phasor form. Thus, with fluxes additive, as in Fig. 19-1a,

$$\mathbf{E}_{\text{gen 1}} = \mathbf{I}_1 R_1 + \mathbf{I}_1 j X_{L1} + \mathbf{I}_2 j X_M$$

$$\mathbf{E}_{\text{gen 2}} = \mathbf{I}_2 R_2 + \mathbf{I}_2 j X_{L2} + \mathbf{I}_1 j X_M \tag{19-10}$$

For the transformer shown in Fig. 19-1b, the drivers are connected in such a manner as to cause the fluxes set up by the respective currents to be in opposite directions within the transformer core. For this condition, the net flux through the window of coil 1 is

$$\phi_{\text{win coil 1}} = \phi_{12} + \phi_{11} - \phi_{21} = \phi_1 - \phi_{21}$$

The net flux through the window of coil 2 is

$$\phi_{\text{win coil 2}} = \phi_{22} + \phi_{21} - \phi_{12} = \phi_2 - \phi_{12}$$

The voltage drops in the transformer coils, due to the rate of change of flux through the respective windows, are

$$v_1 = N_1 \frac{d(\phi_1 - \phi_{21})}{dt} = N_1 \frac{d\phi_1}{dt} - N_1 \frac{d\phi_{21}}{dt} \tag{19-11}$$

$$v_2 = N_2 \frac{d(\phi_2 - \phi_{12})}{dt} = N_2 \frac{d\phi_2}{dt} - N_2 \frac{d\phi_{12}}{dt} \tag{19-12}$$

Applying Kirchhoff's voltage law to the respective circuits of coil 1 and coil 2 in Fig. 19-1b,

$$e_{\text{gen 1}} = i_1 R_1 + v_1 \tag{19-13}$$

$$e_{\text{gen 2}} = i_2 R_2 + v_2 \tag{19-14}$$

Substituting Eqs. (19-11) and (19-12) into Eqs. (19-13) and (19-14), respectively,

$$e_{\text{gen 1}} = i_1 R_1 + N_1 \frac{d\phi_1}{dt} - N_1 \frac{d\phi_{21}}{dt}$$

$$e_{\text{gen 2}} = i_2 R_2 + N_2 \frac{d\phi_2}{dt} - N_2 \frac{d\phi_{12}}{dt} \tag{19-15}$$

Expressing equation set (19-15) in terms of self-inductance, mutual inductance, and the respective rate of change of current,

$$e_{\text{gen 1}} = i_1 R_1 + L_1 \frac{di_1}{dt} - M \frac{di_2}{dt}$$

$$e_{\text{gen 2}} = i_2 R_2 + L_2 \frac{di_2}{dt} - M \frac{di_1}{dt} \tag{19-16}$$

If the drivers are sinusoidal, equation set (19-16) may be written in phasor form. Thus, with fluxes subtractive as in Fig. 19-1b,

$$\mathbf{E}_{\text{gen 1}} = \mathbf{I}_1 R_1 + \mathbf{I}_1 j X_{L1} - \mathbf{I}_2 j X_M$$

$$\mathbf{E}_{\text{gen 2}} = \mathbf{I}_2 R_2 + \mathbf{I}_2 j X_{L2} - \mathbf{I}_1 j X_M \tag{19-17}$$

In summary, for any two-winding transformer, with any driver,

$$e_{\text{gen }1} = i_1 R_1 + L_1 \frac{di_1}{dt} + M \frac{di_2}{dt}$$

$$e_{\text{gen }2} = i_2 R_2 + L_2 \frac{di_2}{dt} + M \frac{di_1}{dt}$$

(19-18)

With a sinusoidal driver,

$$\mathbf{E}_{\text{gen }1} = \mathbf{I}_1 R_1 + \mathbf{I}_1 j X_{L1} + \mathbf{I}_2 j (X_M)$$

$$\mathbf{E}_{\text{gen }2} = \mathbf{I}_2 R_2 + \mathbf{I}_2 j X_{L2} + \mathbf{I}_1 j (X_M)$$

(19-19)

where M and X_M are positive if ϕ_{12} and ϕ_{21} are in the same direction
M and X_M are negative if ϕ_{12} and ϕ_{21} are in opposite directions
M and X_M are negative if there is no generator in the secondary circuit

A positive M or X_M indicates that the voltage drop induced in one coil by a changing current in another coil is a positive voltage drop. A negative M or X_M indicates that the voltage drop induced in one coil by a changing current in another coil is a negative voltage drop.

Example 19-1   For the circuit shown in Fig. 19-2,

$\mathbf{E}_{G1} = 100\underline{/0^\circ}\ \text{V}$ $\mathbf{E}_{G2} = 20\underline{/90^\circ}\ \text{V}$

$X_{L1} = 10\ \Omega$ $X_{L2} = 30\ \Omega$

$R_1 = 2\ \Omega$ $R_2 = 5\ \Omega$

Assuming the coefficient of coupling is 0.85, determine $\mathbf{I}_1$ and $\mathbf{I}_2$.

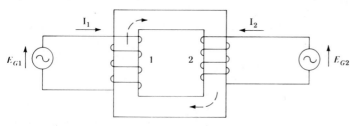

Figure 19-2   Circuit for Example 19-1.

Solution

$$X_M = k\sqrt{X_{L1} X_{L2}} = 0.85\sqrt{(10)(30)} = 14.7 \ \Omega$$

As indicated by the given directions of current and associated fluxes, the sign of X_M is positive. Substituting into equation set (19-19), and rearranging terms,

$$100\underline{/0°} = I_1(2 + j10) + I_2(j14.7)$$

$$20\underline{/90°} = I_2(5 + j30) + I_1(j14.7)$$

Solving,

$$I_1 = \frac{\Delta_{Z1}}{\Delta_Z} = \frac{\begin{vmatrix} 100\underline{/0°} & j14.7 \\ 20\underline{/90°} & j14.7 \end{vmatrix}}{\begin{vmatrix} (2 + j10) & j14.7 \\ (5 + j30) & j14.7 \end{vmatrix}} = 5.04\underline{/87.2°}A$$

$$I_2 = \frac{\Delta_{Z2}}{\Delta_Z} = \frac{\begin{vmatrix} (2 + j10) & 100\underline{/0°} \\ (5 + j30) & 20\underline{/90°} \end{vmatrix}}{\begin{vmatrix} (2 + j10) & j14.7 \\ (5 + j30) & j14.7 \end{vmatrix}} = 10.23\underline{/-94.5°}A$$

19-2   POLARITY MARKS OF COUPLED COILS

The relative polarity of coupled coils must be considered when making circuit calculations involving coupled coils, connecting transformers in parallel, connecting transformers to synchroscopes, wattmeters, or power-factor meters, and in other applications where the relative direction of the output voltage of the transformer is significant. If not properly connected, transformers, instruments, and associated equipment may be destroyed and personnel seriously injured or killed.†

Polarity marks, consisting of ± markings, paint marks such as ●, ▲, ■, etc., or appropriate letter designations, are usually placed on one terminal of each coupled coil to indicate the *direction or sense of the coil windings* with respect to one another. The directions of the coil currents, relative to the respective polarity marked terminals, may be used to determine the sign (+ or −) of M and X_M.

If, at the same instant of time, the currents in the two coupled coils are both entering their respective polarity terminals or both leaving their respective polarity terminals, as shown in Fig. 19-3a and b, the sign of M and X_M is positive (+). This is so because the flux contributions from both coils are additive.

† C. I. Hubert, *Preventive Maintenance of Electrical Equipment*, McGraw-Hill Book Company, New York, 1969.

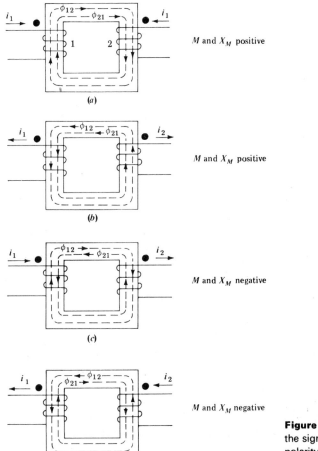

M and X_M positive

(a)

M and X_M positive

(b)

M and X_M negative

(c)

M and X_M negative

Figure 19-3 Determining the signs of M and X_M from polarity marks and the directions of coil currents.

(d)

If the current in one coil enters its polarity terminal at the same time that the current in the other coil leaves its polarity terminal, as shown in Fig. 19-3c *and d, the sign of M and X$_M$ is negative* $(-)$. In this case the flux contributions from both coils are subtractive.

It should be noted that a negative M and a negative X$_M$ sign is also the normal condition for a transformer that has a driving voltage connected to only one coil and a load connected to the other coil. This was discussed in Sect. 18-4 of Chap. 18.

If the coil windings are available for inspection, as shown in Fig. 19-4, the polarity terminals may be determined by applying the right-hand-flux rule to the *coil wrappings*. When the switch in the primary circuit is closed, flux ϕ_{12} will be clockwise as determined by the right-hand-flux rule. As

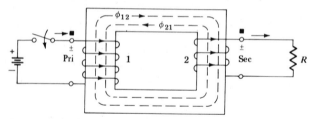

Figure 19-4 Determination of polarity marks from coil
wrappings in a two-winding transformer.

determined from Lenz's law, the direction of flux ϕ_{21} produced by the induced current in coil 2 will be in a direction to oppose the buildup of ϕ_{12}.
Thus the direction of ϕ_{21} must be counterclockwise. With the direction of
ϕ_{21} known, application of the right-hand-flux rule to the secondary coil will
determine the direction of the secondary current. *The primary terminal that
has the current from the external circuit going into it, and the secondary terminal
that has the current coming out of it, are assigned polarity marks.* The polarity
terminals are indicated in Fig. 19-4 by ± marks or ■ marks.

In the case of three or more mutually coupled coils, such as that shown
in Fig. 19-5, different polarity marks must be designated for each pair of
coupled coils. The designated polarity marks for the coils in Fig. 19-5 are

Coils 1 and 2 ▲
Coils 2 and 3 ■
Coils 3 and 1 ●

*The polarity terminals of coupled coils must be determined one pair at a time
while disregarding the other coils.* Although voltage sources and loads are not
shown in Fig. 19-5, the procedure is identical to that outlined for the two
coupled coils in Fig. 19-4.

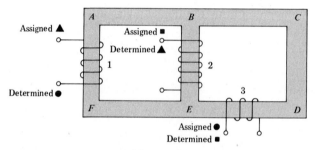

Figure 19-5 Determination of polarity marks from coil
wrappings in a three-winding transformer.

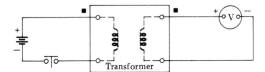

Figure 19-6 Determination of polarity marks by a voltmeter test.

For coils 1 and 2, arbitrarily assign the designated ▲ mark to one terminal of coil 1, as shown in Fig. 19-5, and send direct current *into* that terminal. The resultant flux will pass through the windows of coils 1 and 2 in the *FEBAF* direction. The current induced in coil 2 generates an opposing flux in direction *BEFAB*. Using the right-hand rule, determine the direction of current in coil 2 that generates this opposing flux, and scribe a ▲ mark on the terminal of coil 2 where the current exits.

For coils 2 and 3, arbitrarily assign the designated ■ mark to one terminal of coil 2 and send direct current into that terminal. The resultant flux will pass through the windows of coils 2 and 3 in the *BCDEB* direction. The current induced in coil 3 generates an opposing flux in direction *DCBED*. Using the right-hand rule, determine the direction of current in coil 3 that generates this opposing flux, and scribe a ■ mark on the terminal of coil 3 where the current exits.

Similarly, for coils 3 and 1, arbitrarily assign the designated ● mark to one terminal of coil 3 and send direct current into that terminal. The resultant flux will pass through the windows of coils 3 and 1 in the *EFABCDE* direction. The current induced in coil 1 generates an opposing flux in direction *FEDCBAF*. Using the right-hand rule, determine the direction of current in coil 1, and scribe a ● mark on the terminal of coil 1 where the current exits.

If the transformer coils are not available for inspection, the polarity marks may be determined by using the test procedure illustrated in Fig. 19-6. Connect a battery and a push button in series with one coil, and a DC voltmeter across the other coil. Make a momentary contact by depressing and then quickly releasing the push button. If the voltmeter deflects upscale when the button is depressed and deflects downscale when the button is released, *the terminal that connects to the plus (+) of the battery and the terminal that connects to the plus (+) of the voltmeter are the polarity terminals.* If the voltmeter does not deflect upscale when the button is depressed, interchange the voltmeter connections and repeat the test.

9-3   COUPLED COILS IN SERIES

The equivalent inductance of two series-connected *uncoupled coils*, shown in Fig. 19-7a, is

$$L_{eq} = L_1 + L_2$$

(19-20)

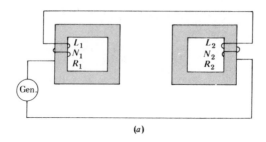

(a)

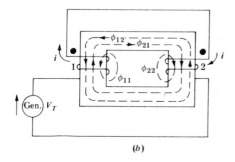

(b)

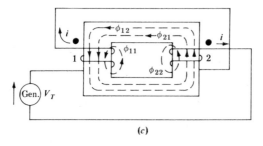

(c)

Figure 19-7 (a) Uncoupled coils in series; (b) coupled coils in series (fluxes subtracting); (c) coupled coils in series (fluxes adding).

However, if the same two coils are physically arranged so that their respective magnetic fluxes pass through each other's windows, as shown in Fig. 19-7b and c, the equivalent inductance of the series combination is

$$L_{eq} = L_1 + L_2 \pm 2M \qquad (19\text{-}21)$$

The positive sign is used if both coil currents are entering their respective polarity terminals or both coil currents are leaving their respective polarity terminals as shown in Fig. 19-7c. The negative sign is used if the current in one coil enters its polarity terminal at the same time that the current in the other coil leaves its polarity terminal, as shown in Fig. 19-7b.

If a sinusoidal driver is used, the equivalent inductive reactance is

$$(2\pi f)L_{eq} = (2\pi f)L_1 + (2\pi f)L_2 \pm (2\pi f)2M$$

or

$$\boxed{X_{L,\,eq} = X_{L1} + X_{L2} \pm 2X_M} \qquad (19\text{-}22)$$

The $\pm 2M$ in Eq. (19-21) can be justified by considering the net flux through the windows of each coil. In the case of Fig. 19-7b, the flux contributions of the two coils are in opposite directions. Hence,

$$\phi_{\text{win coil 1}} = \phi_{11} + \phi_{12} - \phi_{21} = \phi_1 - \phi_{21}$$

$$\phi_{\text{win coil 2}} = \phi_{22} + \phi_{21} - \phi_{12} = \phi_2 - \phi_{12}$$

The induced voltage in the coils, caused by the rate of change of flux through the respective windows, is

$$v_1 = N_1 \frac{d(\phi_1 - \phi_{21})}{dt} = N_1 \frac{d\phi_1}{dt} - N_1 \frac{d\phi_{21}}{dt}$$

$$v_2 = N_2 \frac{d(\phi_2 - \phi_{12})}{dt} = N_2 \frac{d\phi_2}{dt} - N_2 \frac{d\phi_{12}}{dt}$$

In terms of inductance and the rate of change of current,

$$v_1 = L_1 \frac{di_1}{dt} - M \frac{di_2}{dt} \qquad (19\text{-}23)$$

$$v_2 = L_2 \frac{di_2}{dt} - M \frac{di_1}{dt} \qquad (19\text{-}24)$$

Applying Kirchhoff's voltage law to the circuit in Fig. 19-7b,

$$v_T = i_1 R_1 + v_1 + i_2 R_2 + v_2 \qquad (19\text{-}25)$$

Substituting Eqs. (19-23) and (19-24) into Eq. (19-25),

$$v_T = i_1 R_1 + \left(L_1 \frac{di_1}{dt} - M \frac{di_2}{dt} \right) + i_2 R_2 + \left(L_2 \frac{di_2}{dt} - M \frac{di_1}{dt} \right) \qquad (19\text{-}26)$$

Since the two coils are in series,

$$i_1 = i_2 = i$$

Substituting i for i_1 and i_2 in Eq. (19-26), and simplifying,

$$v_T = iR_1 + L_1 \frac{di}{dt} - M \frac{di}{dt} + iR_2 + L_2 \frac{di}{dt} - M \frac{di}{dt}$$

$$v_T = (R_1 + R_2)i + (L_1 + L_2 - 2M) \frac{di}{dt} \tag{19-27}$$

where $L_1 + L_2 - 2M$ = equivalent inductance of two series-connected mutually coupled coils whose fluxes are in opposition

In the case of Fig. 19-7c, the flux contributions of the two coils are in the same direction. Hence,

$$\phi_{\text{win coil 1}} = \phi_{11} + \phi_{12} + \phi_{21} = \phi_1 + \phi_{21}$$

$$\phi_{\text{win coil 2}} = \phi_{22} + \phi_{21} + \phi_{12} = \phi_2 + \phi_{12}$$

Following the development that was outlined for the preceding case, Kirchhoff's voltage-law equation works out to be

$$v_T = (R_1 + R_2)i + (L_1 + L_2 + 2M) \frac{di}{dt} \tag{19-28}$$

where $L_1 + L_2 + 2M$ = equivalent inductance of two series-connected mutually coupled coils whose fluxes are additive

If the driver in Fig. 19-7b and c is sinusoidal, Eqs. (19.27) and (19-28) become, respectively,

$$\mathbf{V}_T = \mathbf{I}[(R_1 + R_2) + j(X_{L1} + X_{L2} - 2X_M)] \tag{19-29}$$

$$\mathbf{V}_T = \mathbf{I}[(R_1 + R_2) + j(X_{L1} + X_{L2} + 2X_M)] \tag{19-30}$$

Example 19-2   A 120-V 50-Hz driver is connected to the coupled circuit shown in Fig. 19-7b. The coil parameters are

$L_1 = 0.2$ H $R_1 = 70 \, \Omega$

$L_2 = 0.8$ H $R_2 = 80 \, \Omega$

$k = 0.75$

Determine (a) the steady-state sinusoidal current; (b) the steady-state current if the connections to coil 2 are interchanged; (c) the steady-state current if the supply voltage is 120 V DC.

Solution

(a) $X_{L1} = 2\pi(50)(0.2) = 62.83 \ \Omega$

$X_{L2} = 2\pi50(0.8) = 251.33 \ \Omega$

$X_M = k\sqrt{X_{L1}X_{L2}} = 0.75\sqrt{(62.83)(251.33)} = 94.25 \ \Omega$

Since M is negative for Fig. 19-7b,

$\mathbf{V}_T = \mathbf{I}[(R_1 + R_2) + j(X_{L1} + X_{L2} - 2X_M)]$

$120\underline{/0°} = \mathbf{I}\{(70 + 80) + j[62.83 + 251.33 - 2(94.25)]\}$

$$\mathbf{I} = \frac{120\underline{/0°}}{195.68\underline{/39.95°}} = 0.61\underline{/-39.95°} \ \text{A}$$

(b) If the connections to coil 2 are interchanged, the mutual inductance M will be positive. Hence,

$\mathbf{V}_T = \mathbf{I}[(R_1 + R_2) + j(X_{L1} + X_{L2} + 2X_M)]$

$120\underline{/0°} = \mathbf{I}\{(70 + 80) + j[62.83 + 251.33 + 2(94.25)]\}$

$120\underline{/0°} = \mathbf{I}(524.56\underline{/73.38°})$

$\mathbf{I} = 0.23\underline{/-73.38°} \ \text{A}$

(c) With a DC driving voltage,

$X_{L1} = 2\pi(0)L_1 = 0$

$X_{L2} = 2\pi(0)L_2 = 0$

$X_M = k\sqrt{X_{L1}X_{L2}} = 0$

Hence,

$V = I(R_1 + R_2)$

$120 = I(70 + 80)$

$I = 0.80 \ \text{A}$

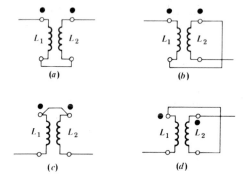

(a)   (b)   (c)   (d)

Figure 19-8 Circuits for Example 19-3.

Example 19-3 For each sketch in Fig. 19-8, determine the equivalent inductance of the series-connected coupled coils. The self-inductances of L_1 and L_2 are 2 H and 8 H, respectively, and $k = 1$. Note: The sign ($+$ or $-$) of M may be determined from the polarity marks of the coupled coils, and an assumed direction of current through the series connection.

Solution

$$M = k\sqrt{L_1 L_2} = 1\sqrt{(2)(8)} = 4\,H$$

For Fig. 19-8a and c, an assumed direction of current through the series connection will show current entering one dot and leaving the other dot. Hence M is negative, and

$$L_{eq} = L_1 + L_2 - 2M = 2 + 8 - 2(4) = 2\,H$$

For Fig. 19-8b and d, an assumed direction of current through the series connection will show the current entering both dots or leaving both dots. Hence M is positive, and

$$L_{eq} = L_1 + L_2 + 2M = 2 + 8 + 2(4) = 18\,H$$

Example 19-4 For the coupled coils shown in Fig. 19-9, determine (a) X_M; (b) R_{eq}; (c) $X_{L,\,eq}$; (d) Z_{eq}.

Solution

(a) $X_M = k\sqrt{X_{L1} X_{L2}} = 0.7\sqrt{(2)(4)} = 1.98\,\Omega$

(b) $R_{eq} = 3 + 5 = 8\,\Omega$

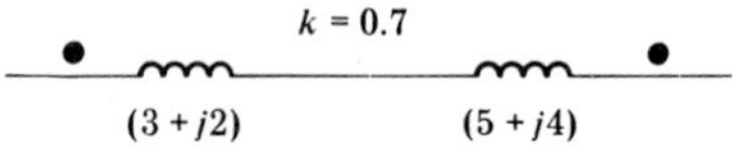

$k = 0.7$

(3 + j2)   (5 + j4)

Figure 19-9 Circuit for Example 19-4.

(c) An assumed direction of current through the series connection of coupled coils will show current entering one dot and leaving the other dot. Hence X_M is negative, and

$$X_{L, \text{eq}} = X_{L1} + X_{L2} - 2X_M$$

$$X_{L, \text{eq}} = 2 + 4 - 2(1.98) = 2.04 \ \Omega$$

(d) $\mathbf{Z}_{\text{eq}} = R_{\text{eq}} + jX_{\text{eq}}$

$$\mathbf{Z}_{\text{eq}} = (8 + j2.04) = 8.26\underline{/14.31°} \ \Omega$$

19-4 **PROCEDURE FOR LOOP-CURRENT ANALYSIS OF NETWORKS CONTAINING COUPLED COILS**

The guidelines for the efficient writing of loop equations, as outlined in Sec. 14-5 of Chap. 14, are applicable to networks containing coupled coils. The only modification is the addition of mutual induction sections shown in Table 19-1. The inner parentheses are reserved for the mutual-inductance parameters, called *coupling parameters*.

The step-by-step procedure for entering data in Table 19-1 is explained using the circuit shown in Fig. 19-10a. Figure 19-10b is an elementary diagram of the circuit, with both loop currents assumed in the clockwise direction. The mutual reactance is

$$X_M = k\sqrt{X_{L1}X_{L2}} = 0.8\sqrt{(8)(2)} = 3.2 \ \Omega$$

Procedure

1. Prepare a blank format for the two loops, using Table 19-1 as a guide.
2. Leaving the inner parentheses blank, enter all driving voltages and branch impedances in the table, using the procedure outlined in Sec. 14-5 for uncoupled circuits. This is shown in Table 19-2 for the example problem.
3. Coupling parameters, loop 1:
    (a) *If there are series-connected coils coupled in loop 1*, enter $\pm 2jX_M$ in the inner parentheses associated with $\mathbf{I}_1$; the sign is determined by the direction of $\mathbf{I}_1$ relative to the dotted terminals of the two coupled coils. If there are no series-connected

Table 19-1 Loop-current format for coupled circuits

Loop	Driving voltage	Voltage drops					
1	[ ] = + [+ ( )	]$\mathbf{I}_1$ + [+ ( )	]$\mathbf{I}_2$ + ··· + [+ ( )	]$\mathbf{I}_n$
2	[ ] = + [+ ( )	]$\mathbf{I}_1$ + [+ ( )	]$\mathbf{I}_2$ + ··· + [+ ( )	]$\mathbf{I}_n$
⋮	⋮ ⋮	⋮	⋮	⋮	⋮	⋮
n	[ ] = + [+ ( )	]$\mathbf{I}_1$ + [+ ( )	]$\mathbf{I}_2$ + ··· + [+ ( )	]$\mathbf{I}_n$

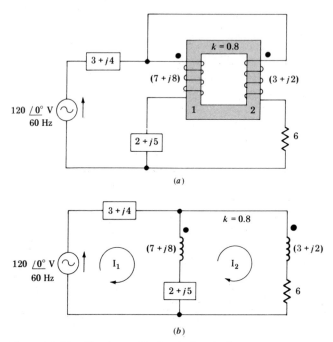

Figure 19-10 Circuit used in developing the loop-current format for coupled circuits: (*a*) connection diagram; (*b*) elementary diagram.

coupled coils in loop 1, enter zero in the inner parentheses. **The coupling parameter is (0) for this component in the example problem,** and is so entered in parentheses in Table 19-3.

(*b*) If current I_2 is in a coil that is mutually coupled to a coil in loop 1, enter $\pm jX_M$ in the inner parentheses associated with current I_2; the sign is determined by the polarity marks of the coupled coils, and the respective direction of the associated currents I_1 and I_2. **The coupling parameter is $+j3.2$ for this component of the example problem,** and is so entered in parentheses in Table 19-3.

Table 19-2 Coupling not included

Loop	Driving voltage	Voltage drops
1	$[120\underline{/0°}]$	$= [(3 + j4) + (7 + j8) + (2 + j5) + (\quad)]I_1$
		$+ [-(2 + j5) - (7 + j8) + (\quad)]I_2$
2	$[0]$	$= [-(2 + j5) - (7 + j8) + (\quad)]I_1$
		$+ [(2 + j5) + (7 + j8) + (3 + j2) + 6 + (\quad)]I_2$

Table 19-3 Coupling included

Loop	Driving voltage	Voltage drops
1	$[120\underline{/0^\circ}] = [(12 + j17) + (0)]\mathbf{I}_1 + [-\ (9 + j13) + (+j3.2)]\mathbf{I}_2$	
2	$[0] = [-(9 + j13) + (+j3.2)]\mathbf{I}_1 + [(18 + j15) + (-2j3.2)]\mathbf{I}_2$	

4. Coupling parameters, loop 2:

   (a) *If there are series-connected coils coupled in loop 2, enter* $\pm 2jX_M$ *in the inner parentheses associated with* $\mathbf{I}_2$; *the sign is determined by the direction of* $\mathbf{I}_2$ *relative to the dotted terminals of the two coupled coils. If there are no coupled coils in series in loop 2, enter zero in the inner parentheses.* **The coupling parameter is** $-2(j3.2)$ **for this component of the example problem,** and is so entered in parentheses in Table 19-3.

   (b) *If current* $\mathbf{I}_1$ *is in a coil that is mutually coupled to a coil in loop 2, enter* $\pm jX_M$ *in the inner parentheses associated with current* $\mathbf{I}_1$; *the sign is determined by the polarity marks of the coupled coils, and the respective directions of the associated currents* $\mathbf{I}_1$ *and* $\mathbf{I}_2$. **The coupling parameter is** $+j3.2$ **for this component of the example problem,** and is so entered in parentheses in Table 19-3.

Simplifying the equations in Table 19-3,

$$120\underline{/0^\circ} = (12 + j17)\mathbf{I}_1 + (-9 - j9.8)\mathbf{I}_2$$

$$0 = (-9 - j9.8)\mathbf{I}_1 + (18 + j8.6)\mathbf{I}_2$$

Example 19-5 (a) Determine the reading of the three ammeters in Fig. 19-11a; (b) repeat part a, assuming the circuit was inadvertently connected to a 100-V battery instead of a 100-V sinusoidal generator.

Solution
An elementary diagram of Fig. 19-11a, with the assumed loop currents, is shown in Fig. 19-11b. The mutual reactances of the coupled coils are

$$X_{M,AB} = k\sqrt{X_{LA}X_{LB}} = 1\sqrt{(1)(9)} = 3\ \Omega$$

$$X_{M,CD} = k\sqrt{X_{LC}X_{LD}} = 1\sqrt{(2)(8)} = 4\ \Omega$$

(a) Using Table 19-1 as a guide, the three loop equations are

Loop 1

$$[0] = [(5 + j1) + (0)]\mathbf{I}_1 + [0 + (-j3)]\mathbf{I}_2 + [0 + (0)]\mathbf{I}_3$$

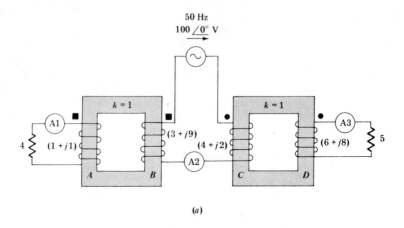

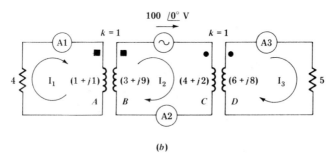

Figure 19-11 Circuit for Example 19-5: (*a*) connection diagram; (*b*) elementary diagram.

Loop 2

$$[100\underline{/0^\circ}] = [0 + (-j3)]\mathbf{I}_1 + [(7 + j11) + (0)]\mathbf{I}_2 + [0 + (-j4)]\mathbf{I}_3$$

Loop 3

$$[0] = [0 + (0)]\mathbf{I}_1 + [0 + (-j4)]\mathbf{I}_2 + [(11 + j8) + (0)]\mathbf{I}_3$$

Simplifying,

$$0 = (5 + j1)\mathbf{I}_1 - j3\mathbf{I}_2 + 0\mathbf{I}_3$$

$$100 = -j3\mathbf{I}_1 + (7 + j11)\mathbf{I}_2 - j4\mathbf{I}_3$$

$$0 = 0\mathbf{I}_1 - j4\mathbf{I}_2 + (11 + j8)\mathbf{I}_3$$

Solving for the three currents,

$$
\Delta_{\mathbf{z}} = \begin{vmatrix} +(5+j1) & (-j3) & 0 \\ -(-j3) & (7+j11) & (-j4) \\ +(0) & (-j4) & (11+j8) \end{vmatrix}
$$

$$
\Delta_{\mathbf{z}} = (5+j1)[(7+j11)(11+j8) - (-j4)(-j4)] - (-j3)[(-j3)(11+j8) - (0)(-j4)]
$$
$$
+ 0[(-j3)(-j4) - (7+j11)(0)]
$$

$$
\Delta_{\mathbf{z}} = 963.6\underline{/93.16°}
$$

$$
\Delta_{\mathbf{z}1} = \begin{vmatrix} +(0) & (-j3) & 0 \\ -(100) & (7+j11) & (-j4) \\ +(0) & (-j4) & (11+j8) \end{vmatrix}
$$

$$
\Delta_{\mathbf{z}1} = 0 - 100[(-j3)(11+j8) - (-j4)(0)] + 0 = 4080.4\underline{/126.03°}
$$

$$
\Delta_{\mathbf{z}2} = \begin{vmatrix} +(5+j1) & 0 & 0 \\ -(-j3) & (100) & (-j4) \\ +(0) & 0 & (11+j8) \end{vmatrix}
$$

$$
\Delta_{\mathbf{z}2} = (5+j1)[(100)(11+j8) - 0] - (-j3)[0 - 0] + 0[0 - 0] = 6936.8\underline{/47.34°}.
$$

$$
\Delta_{\mathbf{z}3} = \begin{vmatrix} +(5+j1) & (-j3) & 0 \\ -(-j3) & (7+j11) & (100) \\ +(0) & (-j4) & 0 \end{vmatrix}
$$

$$
\Delta_{\mathbf{z}3} = (5+j1)[(7+j11)(0) - (-j4)(100)] - (-j3)[(-j3)(0) - (-j4)(0)]
$$
$$
+ (0)[(-j3)(100) - (7+j11)(0)]
$$

$$
\Delta_{\mathbf{z}3} = (5.10\underline{/11.31°})(400\underline{/90°}) = 2040\underline{/101.31°}
$$

$$
\mathbf{I}_1 = \frac{\Delta_{\mathbf{z}1}}{\Delta_{\mathbf{z}}} = \frac{4080.44\underline{/126.03°}}{963.63\underline{/93.16°}} = 4.23\underline{/32.87°}\ \text{A} \qquad A_1 \text{ reads } 4.23\ \text{A}
$$

$$
\mathbf{I}_2 = \frac{\Delta_{\mathbf{z}2}}{\Delta_{\mathbf{z}}} = \frac{6936.75\underline{/47.34°}}{963.63\underline{/93.16°}} = 7.20\underline{/-45.82°}\ \text{A} \qquad A_2 \text{ reads } 7.20\ \text{A}
$$

$$
\mathbf{I}_3 = \frac{\Delta_{\mathbf{z}3}}{\Delta_{\mathbf{z}}} = \frac{2040\underline{/101.31°}}{963.63\underline{/93.16°}} = 2.12\underline{/8.15°}\ \text{A} \qquad A_3 \text{ reads } 2.12\ \text{A}
$$

(b) Steady-state currents I_1 and I_3 are zero, and steady-state current I_2 is

$$I_2 = \frac{100}{3+4} = 14.29 \text{ A}$$

Example 19-6 (a) Determine the ammeter readings for the circuit shown in Fig. 19-12a. Assume the coefficient of coupling between coils is 1.0. (b) Repeat part a, assuming the three coils have separate iron cores as shown in Fig. 19-12c. (c) Determine the active and reactive power supplied by each generator in Fig. 19-12c.

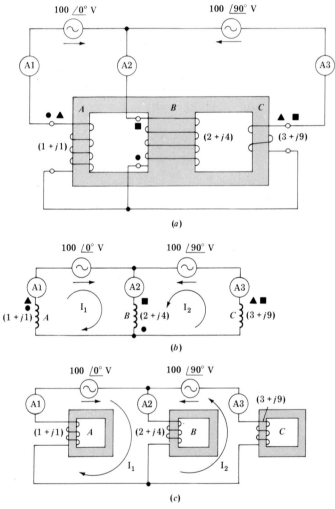

(a)

(b)

(c)

Figure 19-12 Circuit diagram for Example 19-6: (a) connection diagram; (b) elementary diagram; (c) circuit diagram assuming separate iron cores for each coil.

Solution
An elementary diagram for the circuit is shown in Fig. 19-12b. The mutual reactances are

$$X_{M,AB} = 1\sqrt{(1)(4)} = 2 \,\Omega$$

$$X_{M,BC} = 1\sqrt{(4)(9)} = 6 \,\Omega$$

$$X_{M,CA} = 1\sqrt{(9)(1)} = 3 \,\Omega$$

(a) Using Table 19-1 as a guide, the two loop equations are

Loop	Drivers	Voltage drops
1	$100/\underline{0^\circ} = [(1 + j1) + (2 + j4) + \boxed{(2j2)}]\mathbf{I}_1 + [(2 + j4) + \boxed{(j2 + j3 - j6)}]\mathbf{I}_2$	(19-31)
2	$100/\underline{90^\circ} = [(2 + j4) + \boxed{(j2 + j3 - j6)}]\mathbf{I}_1 + [(2 + j4) + (3 + j9) + \boxed{(-2j6)}]\mathbf{I}_2$	(19-32)

The circled portions of the equations represent the coupling parameters. Simplifying,

$$100/\underline{0^\circ} = (3 + j9)\mathbf{I}_1 + (2 + j3)\mathbf{I}_2$$

$$100/\underline{90^\circ} = (2 + j3)\mathbf{I}_1 + (5 + j1)\mathbf{I}_2$$

$$\Delta_\mathbf{z} = \begin{vmatrix} (3 + j9) & (2 + j3) \\ (2 + j3) & (5 + j1) \end{vmatrix} = (3 + j9)(5 + j1) - (2 + j3)(2 + j3)$$

$$\Delta_\mathbf{z} = (11 + j36) = 37.67/\underline{73.02^\circ}$$

$$\Delta_{\mathbf{z}1} = \begin{vmatrix} 100/\underline{0^\circ} & (2 + j3) \\ 100/\underline{90^\circ} & (5 + j1) \end{vmatrix} = (100/\underline{0^\circ})(5 + j1) - (100/\underline{90^\circ})(2 + j3)$$

$$\Delta_{\mathbf{z}1} = (100/\underline{0^\circ})(5.10/\underline{11.31^\circ}) - (100/\underline{90^\circ})(3.61/\underline{56.31^\circ}) = 806.72/\underline{-7.14^\circ}$$

$$\Delta_{\mathbf{z}2} = \begin{vmatrix} (3 + j9) & 100/\underline{0^\circ} \\ (2 + j3) & 100/\underline{90^\circ} \end{vmatrix} = (3 + j9)(100/\underline{90^\circ}) - (2 + j3)(100/\underline{0^\circ})$$

$$\Delta_{\mathbf{z}2} = 1100.57/\underline{-179.98^\circ}$$

$$\mathbf{I}_1 = \frac{\Delta_{\mathbf{z}1}}{\Delta_\mathbf{z}} = \frac{806.72/\underline{-7.14^\circ}}{37.67/\underline{73.02^\circ}} = 21.42/\underline{-80.16^\circ}\,\text{A}$$

$$\mathbf{I}_2 = \frac{\Delta_{\mathbf{z}2}}{\Delta_\mathbf{z}} = \frac{1100.57/\underline{-179.98^\circ}}{37.67/\underline{73.02^\circ}} = 29.22/\underline{-253.00^\circ}\,\text{A}$$

Thus, ammeter A_1 indicates 21.42 A, and ammeter A_3 indicates 29.2 A. Ammeter A_2 indicates a value equal to $\mathbf{I}_1 + \mathbf{I}_2 = 21.42\underline{/-80.16°} + 29.22\underline{/-253.00°} = 8.40\underline{/125.53°}$ A. Ammeter A_2 reads 8.40 A.

(b) The elementary diagram for the circuit shown in Fig. 19-12c corresponds to that shown in Fig. 19-12b, except that the coils are not magnetically coupled. Thus Eqs. (19-31) and (19-32), with the coupling parameters removed, become the equations for the circuit in Fig. 19-12c. The modified equations are

Loop 1

$$100\underline{/0°} = [(1 + j1) + (2 + j4)]\mathbf{I}_1 + [(2 + j4)]\mathbf{I}_2$$

Loop 2

$$100\underline{/90°} = [(2 + j4)]\mathbf{I}_1 + [(2 + j4) + (3 + j9)]\mathbf{I}_2$$

Simplifying,

$$100\underline{/0°} = (3 + j5)\mathbf{I}_1 + (2 + j4)\mathbf{I}_2$$

$$100\underline{/90°} = (2 + j4)\mathbf{I}_1 + (5 + j13)\mathbf{I}_2$$

$$\Delta_Z = \begin{vmatrix} (3 + j5) & (2 + j4) \\ (2 + j4) & (5 + j13) \end{vmatrix} = (3 + j5)(5 + j13) - (2 + j4)(2 + j4) = 61.22\underline{/128.37°}$$

$$\Delta_{Z1} = \begin{vmatrix} 100\underline{/0°} & (2 + j4) \\ 100\underline{/90°} & (5 + j13) \end{vmatrix} = (100\underline{/0°})(5 + j13) - (100\underline{/90°})(2 + j4)$$

$$\Delta_{Z1} = 1421.34\underline{/50.72°}$$

$$\Delta_{Z2} = \begin{vmatrix} (3 + j5) & 100\underline{/0°} \\ (2 + j4) & 100\underline{/90°} \end{vmatrix} = (3 + j5)(100\underline{/90°}) - (2 + j4)(100\underline{/0°})$$

$$\Delta_{Z2} = 706.97\underline{/-171.88°}$$

$$\mathbf{I}_1 = \frac{\Delta_{Z1}}{\Delta_Z} = \frac{1421.34\underline{/50.72°}}{61.22\underline{/128.37°}} = 23.22\underline{/-77.65°} \text{ A}$$

$$\mathbf{I}_2 = \frac{\Delta_{Z2}}{\Delta_Z} = \frac{706.97\underline{/-171.88°}}{61.22\underline{/128.37°}} = 11.55\underline{/-300.25°} \text{ A}$$

$$\mathbf{I}_1 + \mathbf{I}_2 = (23.22\underline{/-77.65°}) + (11.55\underline{/-300.25°}) = 16.67\underline{/-49.67°} \text{ A}$$

Thus the ammeter indications for the uncoupled circuit are

$$A_1 = 23.22 \text{ A}$$

$$A_2 = 16.67 \text{ A}$$

$$A_3 = 11.55 \text{ A}$$

(c) For the generator in loop 1,

$$\mathbf{S} = \mathbf{VI}^* = (100\underline{/0°})(23.22\underline{/+77.65°})$$

$$\mathbf{S} = 2322\underline{/77.65°} = (496.64 + j2268.27) \text{ VA}$$

$$P = 496.6 \text{ W}$$

$$Q = 2268.3 \text{ var}$$

For the generator in loop 2,

$$\mathbf{S} = \mathbf{VI}^* = (100\underline{/90°})(11.55\underline{/+300.25°})$$

$$\mathbf{S} = 1155\underline{/390.25°} = (997.73 + j581.86) \text{ VA}$$

$$P = 997.7 \text{ W}$$

$$Q = 581.9 \text{ var}$$

9-5   EQUIVALENT T-SECTION OF A TRANSFORMER

When making circuit calculations involving transformers, it is often helpful if each transformer is replaced by an equivalent T-section. Figure 19-13b shows three T-connected impedances $\mathbf{Z}_A$, $\mathbf{Z}_B$, and $\mathbf{Z}_C$ whose parameters are such as to result in the same load current as is produced by the magnetically coupled circuit in Fig. 19-13a. The equivalent T-section is sometimes referred to as an equivalent Y-section.

When transformers are replaced by equivalent T-sections, the resultant *conductively coupled circuit* can be readily converted to a Thevenin equivalent or Norton equivalent, as shown in Fig. 19-13c and d, respectively.

Determination of T-Section Parameters

1. Write the loop equations for the magnetically coupled circuit.
2. Write the loop equations for the conductively coupled equivalent circuit.
3. Equate the coefficients of the respective currents and solve for the T-parameters $\mathbf{Z}_A$, $\mathbf{Z}_B$, and $\mathbf{Z}_C$.

Example 19-7   Determine the equivalent T-section parameters for the transformer in Fig. 19-13a.

Solution

$$X_M = k\sqrt{X_{L1}X_{L2}} = 1\sqrt{(4)(2)} = 2.83 \text{ } \Omega$$

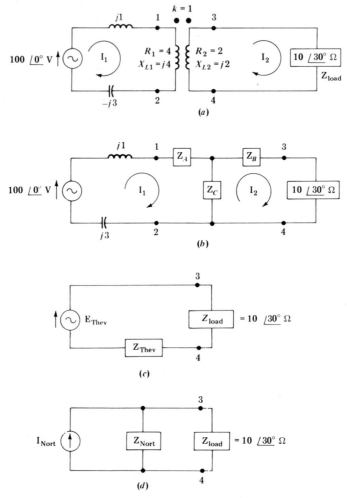

Figure 19-13 Development of equivalent T-sections of a transformer: (a) transformer circuit; (b) equivalent T-section; (c) Thevenin equivalent; (d) Norton equivalent.

For the magnetically coupled circuit,

Loop 1

$$[100\underline{/0°}] = [j1 + 4 + j4 - j3 + (0)]\mathbf{I}_1 + [0 - (j2.83)]\mathbf{I}_2$$

Loop 2

$$[0] = [0 - (j2.83)]\mathbf{I}_1 + [2 + j2 + 10\underline{/30°} + (0)]\mathbf{I}_2$$

Simplifying,

$$100 = (4 + j2)\mathbf{I}_1 + (-j2.83)\mathbf{I}_2 \tag{19-33}$$

$$0 = (-j2.83)\mathbf{I}_1 + (2 + j2 + 10\underline{/30°})\mathbf{I}_2 \tag{19-34}$$

For the conductively coupled circuit,

Loop 1

$$[100\underline{/0°}] = [j1 + \mathbf{Z}_A + \mathbf{Z}_C - j3]\mathbf{I}_1 + [-\mathbf{Z}_C]\mathbf{I}_2 \tag{19-35}$$

Loop 2

$$[0] = [-\mathbf{Z}_C]\mathbf{I}_1 + [\mathbf{Z}_C + \mathbf{Z}_B + 10\underline{/30°}]\mathbf{I}_2 \tag{19-36}$$

For the two circuits to be equivalent, the coefficients of the currents in loop equations (19-35) and (19-36) must be equal to the respective coefficients of the currents in Eqs. (19-33) and (19-34). Thus,

$$4 + j2 = \mathbf{Z}_A + \mathbf{Z}_C - j2$$

$$-j2.83 = -\mathbf{Z}_C$$

$$2 + j2 + 10\underline{/30°} = \mathbf{Z}_C + \mathbf{Z}_B + 10\underline{/30°}$$

Solving the three simultaneous equations,

$$\mathbf{Z}_A = 4 + j1.17 \ \Omega$$

$$\mathbf{Z}_B = 2 - j0.83 \ \Omega$$

$$\mathbf{Z}_C = +j2.83 \ \Omega$$

Example 19-8 Replace the transformers in Fig. 19-14a by equivalent T-sections.

Solution
Figure 19-14b represents the circuit with the equivalent T-sections whose parameters are to be determined. The loop equations for the magnetically coupled circuit of Fig. 19-14a was previously determined in Example 19-5 to be

Loop 1

$$0 = (5 + j1)\mathbf{I}_1 + (-j3)\mathbf{I}_2 + (0)\mathbf{I}_3$$

Loop 2

$$100 = (-j3)\mathbf{I}_1 + (7 + j11)\mathbf{I}_2 + (-j4)\mathbf{I}_3$$

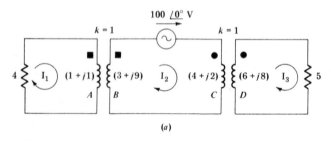

(a)

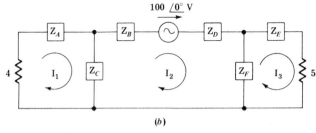

(b)

Figure 19-14   Coupled circuits and equivalent T-sections for Example 19-8.

Loop 3

$$0 = (0)\mathbf{I}_1 + (-j4)\mathbf{I}_2 + (11 + j8)\mathbf{I}_3$$

The loop equations for the conductively coupled equivalent circuit in Fig. 19-14b are

Loop 1

$$0 = [4 + \mathbf{Z}_A + \mathbf{Z}_C]\mathbf{I}_1 + [-\mathbf{Z}_C]\mathbf{I}_2 + [0]\mathbf{I}_3$$

Loop 2

$$100 = [-\mathbf{Z}_C]\mathbf{I}_1 + [\mathbf{Z}_C + \mathbf{Z}_B + \mathbf{Z}_D + \mathbf{Z}_F]\mathbf{I}_2 + [-\mathbf{Z}_F]\mathbf{I}_3$$

Loop 3

$$0 = [0]\mathbf{I}_1 + [-\mathbf{Z}_F]\mathbf{I}_2 + [\mathbf{Z}_F + \mathbf{Z}_E + 5]\mathbf{I}_3$$

Equating the respective coefficients,

$$5 + j1 = 4 + \mathbf{Z}_A + \mathbf{Z}_C$$

$$-j3 = -\mathbf{Z}_C$$

$$7 + j11 = \mathbf{Z}_C + \mathbf{Z}_B + \mathbf{Z}_D + \mathbf{Z}_F$$

$$-j4 = -\mathbf{Z}_F$$

$$11 + j8 = \mathbf{Z}_F + \mathbf{Z}_E + 5$$

Solving the five simultaneous equations,

$\mathbf{Z}_A = 1 - j2$ $\mathbf{Z}_E = 6 + j4$

$\mathbf{Z}_C = j3$ $\mathbf{Z}_F = j4$

$\mathbf{Z}_B + \mathbf{Z}_D = 7 + j4$

Referring to Fig. 19-14b, $\mathbf{Z}_B$ is in series with $\mathbf{Z}_D$ and is therefore represented as one impedance $(\mathbf{Z}_B + \mathbf{Z}_D)$.

UMMARY OF FORMULAS

$$L_{eq} = L_1 + L_2 + 2M$$

$$X_{L,\,eq} = X_{L1} + X_{L2} \pm 2X_M$$

ROBLEMS

19-1 Using the proper hand rule, assign correct polarity marks to the coupled coils in Fig. 19-15a.

19-2 (a) In Fig. 19-15b show the directions of flux produced by all four coils when the switch is closed and the current is building up; (b) assign correct polarity marks to all four coils.

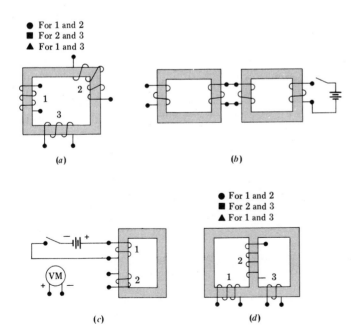

Figure 19-15 Circuits for Probs. 19-1, 19-2, 19-3, and 19-4.

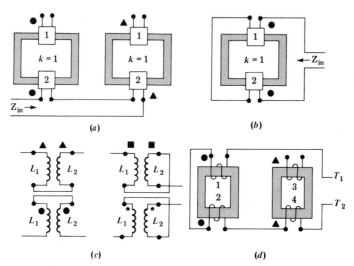

Figure 19-16   Circuit for Probs. 19-5, 19-6, 19-7, and 19-8.

19-3   Referring to Fig. 19-15c, properly connect coil 2 to the voltmeter, so that the momentary deflection will be upscale when the switch is closed.

19-4   Assign correct polarity marks to the coupled coils in Fig. 19-15d.

19-5   Determine the input impedance for the circuit in Fig. 19-16a. The impedances of the respective coils are $\mathbf{Z}_1 = (5 + j9)\,\Omega$, $\mathbf{Z}_2 = (3 + j4)\,\Omega$.

19-6   Determine the input impedance for the circuit shown in Fig. 19-16b. The impedances of the respective coils are $\mathbf{Z}_1 = (5 + j9)\,\Omega$, $\mathbf{Z}_2 = (3 + j4)\,\Omega$.

19-7   For each connection shown in Fig. 19-16c, determine the overall inductance of the coupled coils in series. $L_1 = 3\,\text{H}$, $L_2 = 5\,\text{H}$, $k = 0.60$.

19-8   The coil impedances in Fig. 19-16d are $\mathbf{Z}_1 = (3 + j8)$, $\mathbf{Z}_2 = (1 + j2)$, $\mathbf{Z}_3 = (5 + j1)$, $\mathbf{Z}_4 = (6 + j3)$. If $k = 1.0$, determine the impedance measured at terminals $T_1 T_2$.

19-9   For the circuit in Fig. 19-17 calculate (a) the mutual reactance; (b) the mutual inductance; (c) write the loop equations; (d) determine the ammeter reading; (e) the voltage drop across the $(2 + j5)$-Ω impedance.

19-10   Write the loop equations for the circuit in Fig. 19-18 and determine (a) the ammeter readings; (b) the voltage drop across the $(5 + j7)$-Ω impedance; (c) the apparent power, active power, and reactive power drawn by the $(5 + j7)$-Ω impedance.

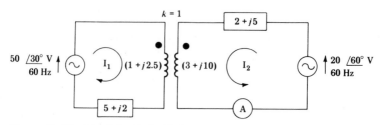

Figure 19-17   Circuit for Prob. 19-9.

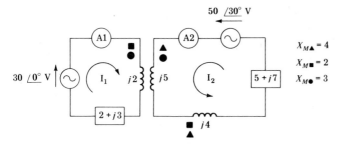

Figure 19-18 Circuit for Probs. 19-10 and 19-11.

19-11 Repeat Prob. 19-10 assuming the 30-V and 50-V sinusoidal generators in Fig. 19-18 are replaced by 30-V and 50-V batteries, respectively.

19-12 Write the loop equations for the circuit in Fig. 19-19 and determine (*a*) the ammeter readings; (*b*) the active and reactive power supplied by each generator; (*c*) repeat parts *a* and *b* with the I_2 arrow reversed.

19-13 (*a*) Write the loop equations for the circuit in Fig. 19-20 and determine the ammeter reading; (*b*) determine the coefficient of coupling between coils *j*5 and *j*6; (*c*) if the *j*5 coil has 300 turns, how many turns are there in the *j*6 coil?

19-14 If the 100-V sinusoidal generator in Fig. 19-20 is replaced by a 100-V battery, what would the ammeter read?

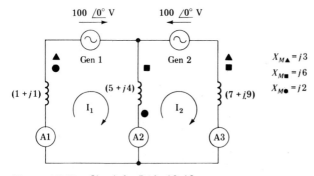

Figure 19-19 Circuit for Prob. 19-12.

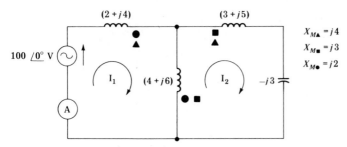

Figure 19-20 Circuit for Probs. 19-13 and 19-14.

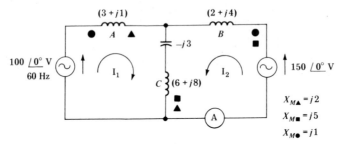

Figure 19-21 Circuit for Prob. 19-15.

19-15 Write the loop equations for the circuit shown in Fig. 19-21 and determine (a) the ammeter reading; (b) the voltage drop across the capacitor; (c) the resistance and inductance of coil B.

19-16 (a) Determine the polarity marks for the coupled coils in Fig. 19-22; (b) sketch the corresponding elementary circuit diagrams; (c) write the loop equations and determine the ammeter reading (assume $k = 1$).

19-17 (a) Determine the polarity marks for the coupled coils in Fig. 19-23; (b) sketch the corresponding elementary diagram; (c) write the loop equations and determine the ammeter readings (assume $k = 1$).

19-18 Write the loop equations for the circuit in Fig. 19-24 and determine (a) the ammeter reading (assume $k = 1$); (b) the voltage drop across the capacitor; (c) the kvars drawn by the capacitor.

19-19 Write the loop equations for the circuit in Fig. 19-25 and determine (a) the ammeter reading (assume $k = 1$); (b) the active and reactive power supplied by the generator.

19-20 (a) If the 100-V sinusoidal generator in Fig. 19-25 is replaced by a 100-V battery, what will the ammeter read? Determine (b) the voltage drop across the capacitor; (c) the energy stored at steady state.

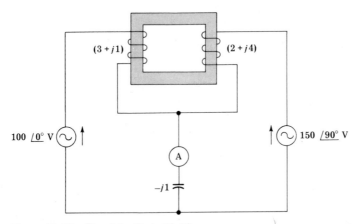

Figure 19-22 Circuit for Prob. 19-16.

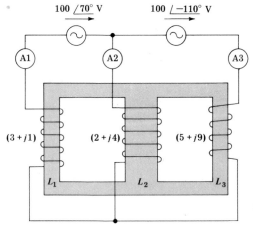

Figure 19-23 Circuit for Prob. 19-17.

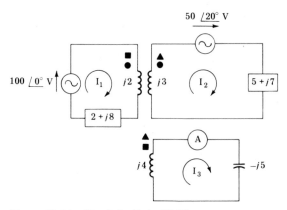

Figure 19-24 Circuit for Prob. 19-18.

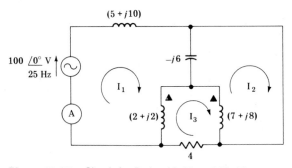

Figure 19-25 Circuit for Probs. 19-19 and 19-20.

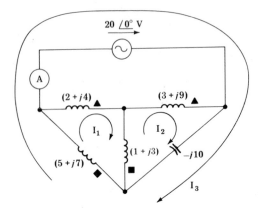

Figure 19-26 Circuit for Probs. 19-21 and 19-22.

19-21 Write the loop equations for the circuit in Fig. 19-26 and determine (*a*) the ammeter reading (assume $k = 1$); (*b*) the apparent power, active power, and reactive power supplied by the generator.

19-22 If the 20-V sinusoidal driver in Fig. 19-26 is replaced by a 20-V battery, what will the ammeter read? (*b*) Determine the heat power drawn by the circuit.

19-23 Determine the parameters of an equivalent T-section that can be used to replace the transformer in Fig. 19-27.

19-24 Determine the parameters of an equivalent T-section that can be used to replace the transformer in Fig. 19-28.

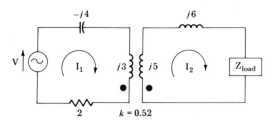

Figure 19-27 Circuit for Prob. 19-23.

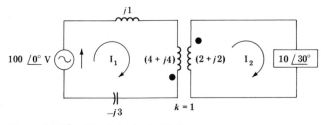

Figure 19-28 Circuit for Prob. 19-24.

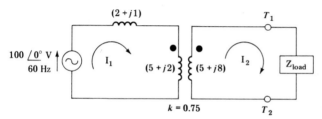

Figure 19-29 Circuit for Prob. 19-25.

19-25 (*a*) Determine the parameters of an equivalent T-section that can be used to replace the transformer in Fig. 19-29; (*b*) the Thevenin equivalent to the left of terminals $T_1 T_2$; (*c*) $\mathbf{Z}_{\text{load}}$ for maximum power transfer; (*d*) the active and reactive power delivered to the load.

19-26 (*a*) Determine the parameters of an equivalent T-section that can be used to replace the transformer in Fig. 19-30; (*b*) the Thevenin equivalent to the left of the $(8 - j4)$ impedance; (*c*) the current in the $(8 - j4)$ impedance.

19-27 Determine (*a*) the parameters of an equivalent T-section that can be used to replace the transformer in Fig. 19-31; (*b*) the Thevenin equivalent to the left of the capacitor; (*c*) the current to the capacitor; (*d*) the maximum instantaneous charge in the capacitor.

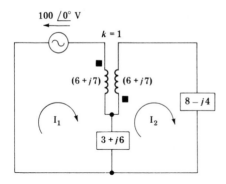

Figure 19-30 Circuit for Prob. 19-26.

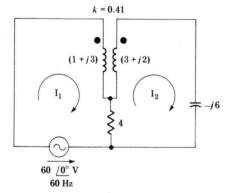

Figure 19-31 Circuit for Prob. 19-27.

CHAPTER 20
TWO-PORT NETWORKS

A port is defined as a pair of terminals at which a signal may enter or leave a network. A one-port network has one pair of terminals, a two-port network has two pairs of terminals, a three-port network has three pairs of terminals, etc.

The one-port network shown in Fig. 20-1a is typical of the types of circuits discussed in Chaps. 4 and 11. The two-port network may be a filter, as shown in Fig. 20-1b, an impedance-matching circuit, or some other circuit that operates on or modifies the signal as it passes through the two-port. The three-port, shown as a block diagram in Fig. 20-1c, may be a three-winding transformer or a mixer that combines signals from two different sources, etc. An example of a mixer is the circuit that combines the incoming frequency signals in a radio set with that of an oscillator.

Although there are many practical networks that have more than two or three ports, this chapter is devoted to the most common type, the two-port network.

20-1 TWO-PORT NETWORKS

A two-port network, also called a four-terminal or two-terminal-pair network, is shown in Fig. 20-2 as a "black box" with two input terminals and two output terminals. The contents of the box may be a transformer, filter, amplifier, transmission line, transistor, etc.

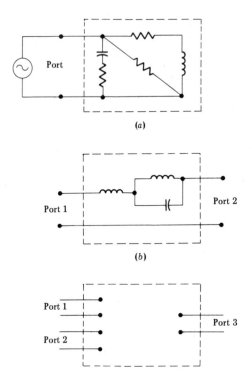

(a)

(b)

(c)

Figure 20-1 (a) One-port network; (b) two-port network; (c) three-port network.

Two-port network theory was developed for use in those situations where the relationship between the input and output signals (current and voltage) of a two-port network is of greater interest than is its internal circuitry. It can be used to identify the input-output characteristics of a network whose contents are unknown or whose circuitry is of such complexity that the methods of analysis previously discussed are too cumbersome if not impossible to use.

In applying two-port theory, the actual two-port is transformed into an *equivalent two-port model, with specially defined parameters* that have the same input-output relationship as does the actual two-port. The most commonly used two-port parameters are impedance parameters (z_{11}, z_{12},

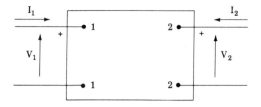

Figure 20-2 "Black-box" representation of a two-port network.

z_{21}, z_{22}), admittance parameters (y_{11}, y_{12}, y_{21}, y_{22}), and hybrid parameters (h_{11}, h_{12}, h_{21}, h_{22}).

Impedance and admittance parameters are generally used when the elements of the two-port consist of R, L, and C. Hybrid parameters are generally used in transistor applications because they are the easiest to obtain experimentally, and provide the most useful information.

Two-port parameters derived from networks containing inductance and capacitance are frequency-dependent.

The equivalent two-port parameters of a network may be calculated from current and voltage measurements made at the input and output terminals without actually knowing the specific elements that make up the network. *When making the measurements (or using other network techniques to determine the voltage and currents), it is essential that the assumed directions of current entering the ports, and the polarities of the terminal voltages, be the same as those indicated in Fig.* 20-2. Once determined, two-port parameters provide the means for identifying the characteristics of the two-port, and aid in problem solving.

20-2 z-PARAMETERS

The simple two-port network shown in Fig. 20-3a will be used as the vehicle to define the equivalent **z**-network and its associated **z**-parameters.

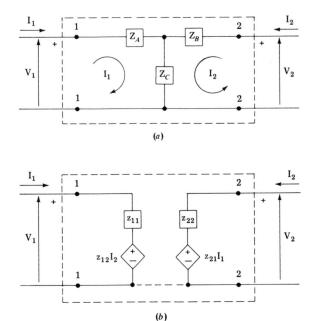

(a)

(b)

Figure 20-3 (a) Two-port network; (b) **z**-parameter model.

The loop equations for Fig. 20-3a are

$$V_1 = (Z_A + Z_C)I_1 + Z_C I_2$$

$$V_2 = Z_C I_1 + (Z_B + Z_C)I_2$$

(20-1)

Defining the coefficients of the currents in equation set (20-1) to be the z-parameters,

$$z_{11} = Z_A + Z_C$$

$$z_{12} = Z_C$$

(20-2)

$$z_{21} = Z_C$$

$$z_{22} = Z_B + Z_C$$

Equation set (20-1) becomes

$$\boxed{\begin{aligned} V_1 &= z_{11}I_1 + z_{12}I_2 \\ V_2 &= z_{21}I_1 + z_{22}I_2 \end{aligned}}$$

(20-3)

The two-port z-parameter model described by equation set (20-3) is shown in Fig. 20-3b. Note that the z-parameters are expressed in ohms and the *z-parameter model includes two controlled voltage sources.* The dotted line joining the ports is included when a common connection is used.

The z-parameters may be determined experimentally from phasor current and phasor voltage measurements made during open-circuit tests of the "unknown" two-port represented by Fig. 20-2. Each port in turn is open-circuited, as shown in Fig. 20-4a and b, and the parameters are determined in the following manner: Opening the output port (Fig. 20-4a) causes I_2 to equal 0, and equation set (20-3) reduces to

$$V_1 = z_{11}I_1 \bigg|$$

$$V_2 = z_{21}I_1 \bigg|_{I_2 = 0}$$

(20-4)

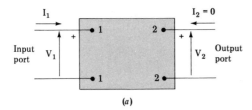

(a)

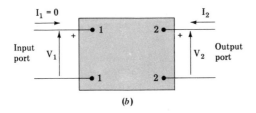

(b)

Figure 20-4   Open-circuit tests for determining the **z**-parameters of an unknown two-port.

Opening the input port (Fig. 20-4*b*) causes I_1 to equal 0, and equation set (20-3) reduces to

$$\left. \begin{aligned} V_1 &= z_{12} I_2 \\ V_2 &= z_{22} I_2 \end{aligned} \right|_{I_1 = 0} \tag{20-5}$$

Solving equation sets (20-4) and (20-5) for the respective **z**-parameters,

$$\left. \begin{array}{cc} z_{11} = \dfrac{V_1}{I_1} & z_{12} = \dfrac{V_1}{I_2} \\[2ex] z_{21} = \dfrac{V_2}{I_1}\Big|_{I_2 = 0} & z_{22} = \dfrac{V_2}{I_2}\Big|_{I_1 = 0} \end{array} \right. \tag{20-6}$$

The following descriptive symbols and descriptive names are commonly associated with the **z**-parameters:

$z_{11} = z_i$ open-circuit input impedance, Ω

$z_{21} = z_f$ open-circuit forward transfer impedance, Ω

$z_{12} = z_r$ open-circuit reverse transfer impedance, Ω

$z_{22} = z_0$ open-circuit output impedance, Ω

Example 20-1 Determine the z-parameters for the circuit shown in Fig. 20-5a. The output port includes a controlled voltage source.

Solution

Since the actual parameters of the circuit are known, and the circuit is relatively simple, the z-parameters may be determined by writing the two loop equations

$$\mathbf{V}_1 = [3 + (6 + j4)]\mathbf{I}_1 + [6 + j4]\mathbf{I}_2$$

$$\mathbf{V}_2 - 2\mathbf{I}_1 = [6 + j4]\mathbf{I}_1 + [6 + j4]\mathbf{I}_2$$

Simplifying,

$$\mathbf{V}_1 = [9 + j4]\mathbf{I}_1 + [6 + j4]\mathbf{I}_2$$

$$\mathbf{V}_2 = [8 + j4]\mathbf{I}_1 + [6 + j4]\mathbf{I}_2$$

Thus the z-parameters are

$$\mathbf{z}_{11} = (9 + j4)\,\Omega \qquad \mathbf{z}_{12} = (6 + j4)\,\Omega$$

$$\mathbf{z}_{21} = (8 + j4)\,\Omega \qquad \mathbf{z}_{22} = (6 + j4)\,\Omega$$

The corresponding z-parameter model is shown in Fig. 20-5b.

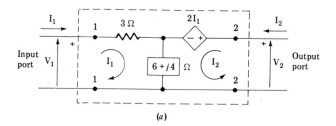

(a)

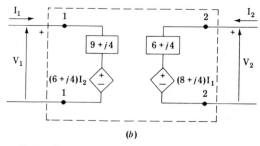

(b)

Figure 20-5 Circuits for Example 20-1: (a) original circuit; (b) equivalent z-parameter model.

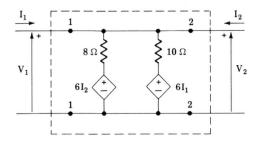

Figure 20-6 z-parameter model for Example 20-2.

Example 20-2 The following open-circuit currents and voltages were determined experimentally for an unknown two-port:

$$\mathbf{V}_1 = 100\underline{/0^\circ}\ \text{V} \qquad \mathbf{V}_1 = 30\underline{/0^\circ}\ \text{V}$$

$$\mathbf{V}_2 = 75\underline{/0^\circ}\ \text{V} \qquad \mathbf{V}_2 = 50\underline{/0^\circ}\ \text{V}$$

$$\mathbf{I}_1 = 12.5\underline{/0^\circ}\ \text{A} \Big|_{I_2 = 0} \qquad \mathbf{I}_2 = 5\underline{/0^\circ}\ \text{A} \Big|_{I_1 = 0}$$

Determine the **z**-parameters and sketch the equivalent **z**-parameter model.

Solution

$$\mathbf{z}_{11} = \frac{\mathbf{V}_1}{\mathbf{I}_1} \Big| = \frac{100}{12.5} = 8\ \Omega \qquad \mathbf{z}_{12} = \frac{\mathbf{V}_1}{\mathbf{I}_2} = \frac{30}{5} = 6\ \Omega$$

$$\mathbf{z}_{21} = \frac{\mathbf{V}_2}{\mathbf{I}_1} \Big| = \frac{75}{12.5} = 6\ \Omega \qquad \mathbf{z}_{22} = \frac{\mathbf{V}_2}{\mathbf{I}_2} \Big| = \frac{50}{5} = 10\ \Omega$$

$$\Big|_{I_2 = 0} \qquad\qquad\qquad\qquad \Big|_{I_1 = 0}$$

The **z**-parameter model is shown in Fig. 20-6.

20-3   y-PARAMETERS

The simple two-port network shown in Fig. 20-7*a* will be the vehicle to define the equivalent **y**-network and its associated **y**-parameters. The node equations for Fig. 20-7*a* are

$$\mathbf{I}_1 = (\mathbf{Y}_A + \mathbf{Y}_B)\mathbf{V}_1 - \mathbf{Y}_B\mathbf{V}_2$$

$$\mathbf{I}_2 = -\mathbf{Y}_B\mathbf{V}_1 + (\mathbf{Y}_C + \mathbf{Y}_B)\mathbf{V}_2$$

(20-7)

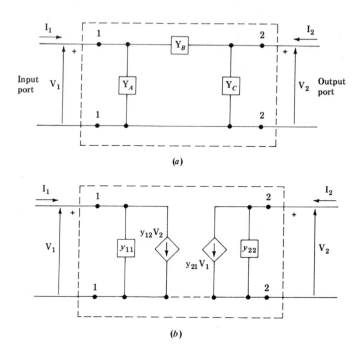

Figure 20-7 (a) Two-port network; (b) y-parameter model.

Defining the coefficients of the voltages in equation set (20-7) to be the y-parameters,

$$y_{11} = Y_A + Y_B$$

$$y_{12} = -Y_B$$

$$y_{21} = -Y_B \qquad\qquad (20\text{-}8)$$

$$y_{22} = Y_C + Y_B$$

Equation-set (20-7) becomes

$$\boxed{\begin{aligned} I_1 &= y_{11}V_1 + y_{12}V_2 \\ I_2 &= y_{21}V_1 + y_{22}V_2 \end{aligned}} \qquad (20\text{-}9)$$

The two-port y-parameter model described by equation set (20-9) is shown in Fig. 20-7b. Note that the y-parameters are expressed in siemens, and the y-parameter model includes two controlled current sources. The dotted line joining the ports is used when a common connection is used.

The **y**-parameters may be determined experimentally from phasor-current and phasor-voltage measurements made during *short-circuit tests* of the "unknown" two-port represented by Fig. 20-1. Each port in turn is short-circuited, as shown in Fig. 20-8*a* and *b*, and the parameters are determined in the following manner: Shorting the output port (Fig. 20-8*a*) causes $\mathbf{V}_2$ to equal 0, and equation set (20-9) reduces to

$$\left. \begin{array}{l} \mathbf{I}_1 = \mathbf{y}_{11}\mathbf{V}_1 \\[2mm] \mathbf{I}_2 = \mathbf{y}_{21}\mathbf{V}_1 \end{array} \right|_{\mathbf{V}_2 = 0} \tag{20-10}$$

Shorting the input port (Fig. 20-8*b*) causes $\mathbf{V}_1$ to equal 0, and equation set (20-9) reduces to

$$\left. \begin{array}{l} \mathbf{I}_1 = \mathbf{y}_{12}\mathbf{V}_2 \\[2mm] \mathbf{I}_2 = \mathbf{y}_{22}\mathbf{V}_2 \end{array} \right|_{\mathbf{V}_1 = 0} \tag{20-11}$$

Solving equation sets (20-10) and (20-11) for the respective **y**-parameters,

$$\left. \begin{array}{ll} \mathbf{y}_{11} = \dfrac{\mathbf{I}_1}{\mathbf{V}_1} & \mathbf{y}_{12} = \dfrac{\mathbf{I}_1}{\mathbf{V}_2} \\[4mm] \mathbf{y}_{21} = \dfrac{\mathbf{I}_2}{\mathbf{V}_1} \Big|_{\mathbf{V}_2 = 0} & \mathbf{y}_{22} = \dfrac{\mathbf{I}_2}{\mathbf{V}_2} \Big|_{\mathbf{V}_1 = 0} \end{array} \right. \tag{20-12}$$

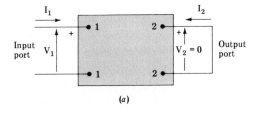

(a)

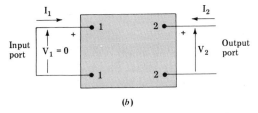

(b)

Figure 20-8 Short-circuit tests for determining the **y**-parameters of an unknown two-port.

The following descriptive symbols and descriptive names are commonly associated with the **y-parameters**:

$y_{11} = y_i$ short-circuit input admittance, S

$y_{21} = y_f$ short-circuit forward transfer admittance, S

$y_{12} = y_r$ short-circuit reverse transfer admittance, S

$y_{22} = y_0$ short-circuit output admittance, S

Example 20-3 Determine the y-parameters for the circuit shown in Fig. 20-9a.

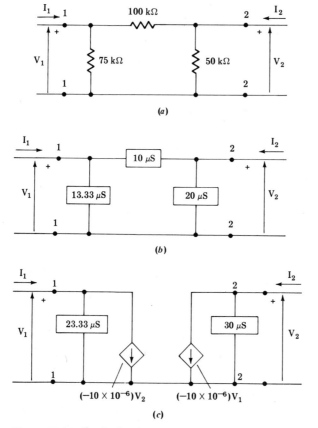

Figure 20-9 Circuits for Example 20-3: (*a*) original circuit; (*b*) circuit expressed in admittance form; (*c*) equivalent y-parameter model.

Solution

Since the actual parameters of the circuit are known, and the circuit is relatively simple, the **y**-parameters may be determined by writing the two node equations. Converting the impedances to admittances,

$$100 \text{ k}\Omega \Rightarrow 10 \text{ } \mu\text{S}$$

$$75 \text{ k}\Omega \Rightarrow 13.33 \text{ } \mu\text{S}$$

$$50 \text{ k}\Omega \Rightarrow 20 \text{ } \mu\text{S}$$

Figure 20-9b shows the circuit expressed in terms of admittance.

$$\mathbf{I}_1 = [13.33 + 10]10^{-6}\mathbf{V}_1 - [10]10^{-6} \mathbf{V}_2$$

$$\mathbf{I}_2 = -[10]10^{-6} \mathbf{V}_1 + [10 + 20]10^{-6} \mathbf{V}_2$$

Thus, the **y**-parameters are

$$\mathbf{y}_{11} = 23.33 \text{ } \mu\text{S} \qquad \mathbf{y}_{12} = -10 \text{ } \mu\text{S}$$

$$\mathbf{y}_{21} = -10 \text{ } \mu\text{S} \qquad \mathbf{y}_{22} = 30 \text{ } \mu\text{S}$$

The corresponding **y**-parameter model is shown in Fig. 20-9c.

Example 20-4 The following short-circuit currents and voltages were determined experimentally for an unknown two-port:

$\mathbf{I}_1 = 3 \text{ mA}$	$\mathbf{I}_1 = -1 \text{ mA}$
$\mathbf{I}_2 = -0.6 \text{ mA}$	$\mathbf{I}_2 = 12 \text{ mA}$
$\mathbf{V}_1 = 24 \text{ V}$	$\mathbf{V}_2 = 40 \text{ V}$
$\mathbf{V}_2 = 0$	$\mathbf{V}_1 = 0$

(a) Determine the **y**-parameters and sketch the equivalent **y**-parameter model.
(b) If a 100-V DC source is connected to the input port and a 5000-Ω resistor load is connected to the output port, determine the current and power drawn by the load.
(c) Determine the current input from the 100-V source.

Solution

$$(a) \text{ } \mathbf{y}_{11} = \frac{\mathbf{I}_1}{\mathbf{V}_1}\bigg|_{\mathbf{V}_2 = 0} = \frac{0.003}{24} = 125 \text{ } \mu\text{S} \qquad \mathbf{y}_{12} = \frac{\mathbf{I}_1}{\mathbf{V}_2}\bigg|_{\mathbf{V}_1 = 0} = \frac{-0.001}{40} = -25 \text{ } \mu\text{S}$$

$$\mathbf{y}_{21} = \frac{\mathbf{I}_2}{\mathbf{V}_1}\bigg|_{\mathbf{V}_2 = 0} = \frac{-0.0006}{24} = -25 \text{ } \mu\text{S} \qquad \mathbf{y}_{22} = \frac{\mathbf{I}_2}{\mathbf{V}_2}\bigg|_{\mathbf{V}_1 = 0} = \frac{0.012}{40} = 300 \text{ } \mu\text{S}$$

The **y**-parameter model is shown in Fig. 20-10a.

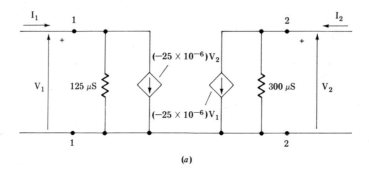

(a)

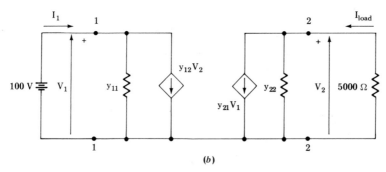

(b)

Figure 20-10 y-parameter models for Example 20-4.

(b) The circuit for part b is shown in Fig. 20-10b. The admittance of the load is $1/5000 = 200\ \mu S$. The node equation for node 2 is

$$-y_{21}V_1 = [y_{22} + 200 \times 10^{-6}]V_2$$

$$-(-25 \times 10^{-6})V_1 = [300 \times 10^{-6} + 200 \times 10^{-6}]V_2$$

Substituting 100 for V_1, and solving for V_2,

$$25(100) = 500V_2$$

$$V_2 = 5\ V$$

Applying Ohm's law to the 5000-Ω load,

$$I_{load} = \frac{5}{5000} = 1\ mA$$

The power drawn by the load is

$$P_{load} = I_{load}^2 R_{load} = (0.001)^2(5000) = 5\ mW$$

(c) The node equation for node 1 is

$$[\mathbf{I}_1 - \mathbf{y}_{12}\mathbf{V}_2] = [\mathbf{y}_{11}]\mathbf{V}_1$$

$$\mathbf{I}_1 - (-25 \times 10^{-6})\mathbf{V}_2 = (125 \times 10^{-6})\mathbf{V}_1$$

Substituting the known values of $\mathbf{V}_1$ and $\mathbf{V}_2$, and solving,

$$\mathbf{I}_1 - (-25 \times 10^{-6})5 = (125 \times 10^{-6})100$$

$$\mathbf{I}_1 = 12.4 \text{ mA}$$

20-4 h-PARAMETERS

Hybrid parameters are determined from both open-circuit and short-circuit measurements and as such include impedance and admittance parameters. The h-parameter model includes a controlled voltage source and a controlled current source as shown in Fig. 20-11. The loop equation for the input port is

$$\mathbf{V}_1 - \mathbf{h}_{12}\mathbf{V}_2 = \mathbf{h}_{11}\mathbf{I}_1$$

The node equation for the output port is

$$\mathbf{I}_2 - \mathbf{h}_{21}\mathbf{I}_1 = \mathbf{h}_{22}\mathbf{V}_2$$

Rearranging terms,

$$\boxed{\begin{aligned} \mathbf{V}_1 &= \mathbf{h}_{11}\mathbf{I}_1 + \mathbf{h}_{12}\mathbf{V}_2 \\ \mathbf{I}_2 &= \mathbf{h}_{21}\mathbf{I}_1 + \mathbf{h}_{22}\mathbf{V}_2 \end{aligned}}$$

(20-13)

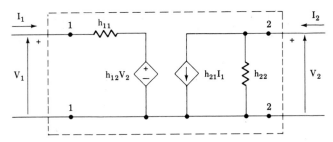

Figure 20-11 h-parameter model of a two-port network.

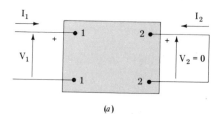

(a)

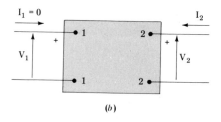

(b)

Figure 20-12 Test circuits for determining h-parameters: (a) short-circuit test; (b) open-circuit test.

The **h**-parameters may be determined experimentally from phasor-current and phasor-voltage measurements made during the following short-circuit and open-circuit tests: With the output port shorted (Fig. 20-12a), $V_2 = 0$, and equation set (20-13) reduces to

$$V_1 = h_{11} I_1 \Big|$$
$$I_2 = h_{21} I_1 \Big|_{V_2 = 0} \tag{20-14}$$

With the input terminals open-circuited (Fig. 20-12b), $I_1 = 0$, and equation set (20-13) reduces to

$$V_1 = h_{12} V_2 \Big|$$
$$I_2 = h_{22} V_2 \Big|_{I_1 = 0} \tag{20-15}$$

Solving equation sets (20-14) and (20-15) for the respective **h**-parameters,

$$\left. h_{11} = \frac{V_1}{I_1} \quad h_{12} = \frac{V_1}{V_2} \right.$$
$$\left. h_{21} = \frac{I_2}{I_1} \right|_{V_2 = 0} \quad h_{22} = \frac{I_2}{V_2} \right|_{I_1 = 0} \tag{20-16}$$

The following descriptive symbols and descriptive names are commonly associated with the **h-parameters**:

$\mathbf{h_{11}} = \mathbf{h_i}$ = short-circuit input impedance, Ω

$\mathbf{h_{21}} = \mathbf{h_f}$ = short-circuit forward current ratio

$\mathbf{h_{12}} = \mathbf{h_r}$ = open-circuit reverse voltage ratio

$\mathbf{h_{22}} = \mathbf{h_o}$ = open-circuit output admittance, S

Example 20-5 A two-port network containing a common-emitter transistor amplifier is connected to a 2-mV signal source and a 10-kΩ resistor load as shown in Fig. 20-13a. The **h-parameters** for the two-port are

$h_i = h_{11} = 1.4\ k\Omega$ $\qquad$ $h_r = h_{12} = 3.4 \times 10^{-4}$

$h_f = h_{21} = 44$ $\qquad$ $h_o = h_{22} = 27\ \mu S$

Determine (a) the voltage across the load; (b) the current to the load; (c) the voltage gain V_2/V_1 of the circuit.

Solution
(a) The equivalent **h-parameter** circuit is shown in Fig. 20-13b. The loop equation for the input port is

$$\mathbf{V_1} - \mathbf{h_r V_2} = \mathbf{I_1 h_i}$$

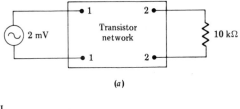

(a)

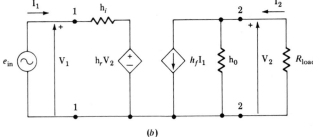

(b)

Figure 20-13 Circuits for Example 20-5: (a) original circuit; (b) equivalent **h**-parameter model.

The node equation for the output port is

$$-h_f I_1 = \left[h_o + \frac{1}{R_{load}} \right] V_2$$

Substituting the given parameters,

$$0.002 - 3.4 \times 10^{-4} V_2 = 1400 I_1 \tag{20-17}$$

$$-44 I_1 = [27 \times 10^{-6} + 100 \times 10^{-6}] V_2 \tag{20-18}$$

Solving Eq. (20-18) for I_1, and then substituting into Eq. (20-17),

$$I_1 = -2.886 \times 10^{-6} V_2$$

$$0.002 - 3.4 \times 10^{-4} V_2 = 1400(-2.886 \times 10^{-6}) V_2$$

$$V_2 = -0.54 \text{ V}$$

(b) $\mathbf{I}_{load} = \dfrac{\mathbf{V}_{load}}{R_{load}} = \dfrac{-0.54}{10,000} = -54 \ \mu A$

(c) The voltage gain for the circuit is

$$A_v = \frac{\mathbf{V}_2}{\mathbf{V}_1} = \frac{-0.54}{0.002} = -270$$

The minus sign indicates a $180°$ phase shift between $\mathbf{V}_2$ and $\mathbf{V}_1$.

20-5   INPUT IMPEDANCE OF A TWO-PORT

The input impedance of any two-port may be determined by applying Ohm's law to the input terminals, or may be determined from the load impedance and the equivalent two-port parameters.

To obtain $\mathbf{Z}_{in}$ from Ohm's law, substitute the measured or calculated values into

$$\mathbf{Z}_{in} = \frac{\mathbf{V}_1}{\mathbf{I}_1} \tag{20-19}$$

The expression for the input impedance, in terms of two-port parameters, may be obtained by writing the corresponding network equations for each port, manipulating the equations, and then substituting into the defining equation $\mathbf{Z}_{in} = \mathbf{V}_1/\mathbf{I}_1$.

Referring to Fig. 20-14a, the loop equation for the input port is

$$\mathbf{V}_1 - \mathbf{h}_r \mathbf{V}_2 = \mathbf{h}_i \mathbf{I}_1 \tag{20-20}$$

The node equation for the output port is

$$-\mathbf{h}_f \mathbf{I}_1 = \mathbf{V}_2 \left[\mathbf{h}_o + \frac{1}{\mathbf{Z}_{\text{load}}} \right] \tag{20-21}$$

Solving Eq. (20-21) for $\mathbf{V}_2$ and substituting into Eq. (20-20),

$$\mathbf{V}_2 = \frac{-\mathbf{h}_f \mathbf{I}_1 \mathbf{Z}_{\text{load}}}{\mathbf{h}_o \mathbf{Z}_{\text{load}} + 1} \tag{20-22}$$

$$\mathbf{V}_1 - \mathbf{h}_r \frac{-\mathbf{h}_f \mathbf{I}_1 \mathbf{Z}_{\text{load}}}{\mathbf{h}_o \mathbf{Z}_{\text{load}} + 1} = \mathbf{h}_i \mathbf{I}_1 \qquad .$$

Solving for $\mathbf{V}_1$,

$$\mathbf{V}_1 = \mathbf{I}_1 \left[\mathbf{h}_i - \frac{\mathbf{h}_r \mathbf{h}_f \mathbf{Z}_{\text{load}}}{\mathbf{h}_o \mathbf{Z}_{\text{load}} + 1} \right] \tag{20-23}$$

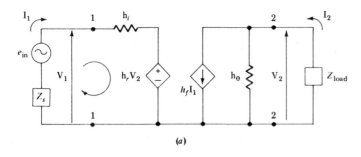

(a)

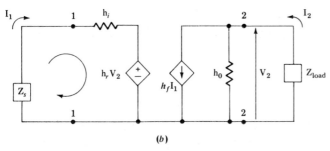

(b)

Figure 20-14   Circuits for determining $\mathbf{Z}_{\text{in}}$, $\mathbf{Z}_{\text{out}}$, A_v, and A_i.

Substituting Eq. (20-23) into the defining equation (20-19),

$$Z_{in} = \frac{V_1}{I_1} = h_i - \frac{h_r h_f Z_{load}}{h_o Z_{load} + 1} \qquad (20\text{-}24)$$

20-6 OUTPUT IMPEDANCE OF A TWO-PORT

The output impedance of any two-port may be determined from the ratio of output voltage to output current, with the *source voltage set to zero.* That is,

$$Z_{out} = \left.\frac{V_2}{I_2}\right|_{e_{in}=0} \qquad (20\text{-}25)$$

The expression for the output impedance, in terms of two-port parameters, may be obtained by writing the corresponding network equations for each port, manipulating the equations, and then substituting into the defining equation $Z_{out} = V_2/I_2$.

Referring to Fig. 20-14b, where the source voltage is set to zero but the source impedance is maintained, the loop equation for the input port is

$$-h_r V_2 = (h_i + Z_s)I_1 \qquad (20\text{-}26)$$

The node equation for node 2 at the output port is

$$I_2 - h_f I_1 = h_o V_2 \qquad (20\text{-}27)$$

Solving Eq. (20-26) for I_1, substituting into Eq. (20-27), and then rearranging the terms,

$$I_1 = \frac{-h_r V_2}{h_i + Z_s}$$

$$-h_f \frac{-h_r V_2}{h_i + Z_s} = h_o V_2 - I_2$$

$$V_2\left[h_o - \frac{h_r h_f}{h_i + Z_s}\right] = I_2 \qquad (20\text{-}28)$$

$$\frac{V_2}{I_2} = \frac{1}{h_o - \dfrac{h_r h_f}{h_i + Z_s}} \qquad (20\text{-}29)$$

Thus,

$$\boxed{Z_{out} = \frac{V_2}{I_2} = \cfrac{1}{h_o - \cfrac{h_r h_f}{h_i + Z_s}}}$$ (20-30)

Although the above developments for input and output impedances were derived in terms of **h**-parameters, similar developments in terms of **x**-parameters and **y**-parameters can also be accomplished using the circuits in Figs. 20-3*b* and 20-7*b*, respectively.

Example 20-6 Given the following two-port parameters:

$h_i = 1.4 \text{ k}\Omega$	$h_r = 3.0 \times 10^{-4}$
$h_f = 40$	$h_o = 25 \ \mu\text{S}$
$Z_{gen} = 1.2 \text{ k}\Omega$	$Z_{load} = 15 \text{ k}\Omega$

Determine the input and output impedances.

Solution

$$Z_{in} = h_i - \frac{h_r h_f Z_{load}}{h_o Z_{load} + 1}$$

$$Z_{in} = 1400 - \frac{(3.0 \times 10^{-4})(40)(15,000)}{(25 \times 10^{-6})(15,000) + 1} = 1269 \ \Omega$$

$$Z_{out} = \cfrac{1}{h_o - \cfrac{h_r h_f}{h_i + Z_s}} = \cfrac{1}{25 \times 10^{-6} - \cfrac{(3.0 \times 10^{-4})40}{1400 + 1200}}$$

$$Z_{out} = 49 \text{ k}\Omega$$

20-7   DETERMINATION OF VOLTAGE GAIN AND   CURRENT GAIN FROM TWO-PORT PARAMETERS

The voltage gain A_v is defined as the ratio V_2/V_1. Thus, from Eqs. (20-22) and (20-23),

$$A_v = \frac{V_2}{V_1} = \cfrac{\cfrac{-h_f I_1 Z_{load}}{h_o Z_{load} + 1}}{I_1 \left[h_i - \cfrac{h_r h_f Z_{load}}{h_o Z_{load} + 1} \right]}$$

$$\boxed{A_v = \frac{-h_f Z_{load}}{h_i (h_o Z_{load} + 1) - h_r h_f Z_{load}}}$$ (20-31)

The current gain A_i is defined as the ratio I_2/I_1. Thus, from Eqs. (20-21) and from Ohm's law, respectively,

$$I_1 = \frac{(h_o + 1/Z_{\text{load}})V_2}{-h_f} \qquad I_2 = -\frac{V_2}{Z_{\text{load}}}$$

$$A_i = \frac{I_2}{I_1} = \frac{-V_2/Z_{\text{load}}}{[(h_o + 1/Z_{\text{load}})V_2]/-h_f}$$

$$\boxed{A_i = \frac{h_f}{h_o Z_{\text{load}} + 1}} \qquad\qquad (20\text{-}32)$$

Although the above developments for voltage gain and current gain were derived in terms of **h**-parameters, similar derivations may be accomplished for **x**-parameters and **y**-parameters.

Example 20-7 Determine the voltage and current gains for the two-port described in Example 20-6.

Solution

$$A_v = \frac{-h_f Z_{\text{load}}}{h_i(h_o Z_{\text{load}} + 1) - h_r h_f Z_{\text{load}}}$$

$$A_v = \frac{-40(15{,}000)}{1400[(25 \times 10^{-6})(15{,}000) + 1] - (3.0 \times 10^{-4})(40)(15{,}000)}$$

$$A_v = -343.8$$

The minus sign indicates a 180° phase shift between V_2 and V_1.

$$A_i = \frac{h_f}{h_o Z_{\text{load}} + 1} = \frac{40}{(25 \times 10^{-6})(15{,}000) + 1} = 29$$

20-8   PARAMETER CONVERSION

Conversion from one set of parameters to another may be accomplished through the use of Table 20-1, where $\Delta \mathbf{y}$ represents the determinant of the **y** matrix.

$$\Delta\mathbf{y} = \begin{vmatrix} y_{11} & y_{12} \\ y_{21} & y_{22} \end{vmatrix} = y_{11}y_{22} - y_{21}y_{12}$$

Table 20-1 Parameter conversion table

From → To ↓	h		y		z	
h_{11} h_{12}			$\dfrac{1}{y_{11}}$	$\dfrac{-y_{12}}{y_{11}}$	$\dfrac{\Delta z}{z_{22}}$	$\dfrac{z_{12}}{z_{22}}$
h_{21} h_{22}			$\dfrac{y_{21}}{y_{11}}$	$\dfrac{\Delta y}{y_{11}}$	$\dfrac{-z_{21}}{z_{22}}$	$\dfrac{1}{z_{22}}$
y_{11} y_{12}	$\dfrac{1}{h_{11}}$	$\dfrac{-h_{12}}{h_{11}}$			$\dfrac{z_{22}}{\Delta z}$	$\dfrac{-z_{12}}{\Delta z}$
y_{21} y_{22}	$\dfrac{h_{21}}{h_{11}}$	$\dfrac{\Delta h}{h_{11}}$			$\dfrac{-z_{21}}{\Delta z}$	$\dfrac{z_{11}}{\Delta z}$
z_{11} z_{12}	$\dfrac{\Delta h}{h_{22}}$	$\dfrac{h_{12}}{h_{22}}$	$\dfrac{y_{22}}{\Delta y}$	$\dfrac{-y_{12}}{\Delta y}$		
z_{21} z_{22}	$\dfrac{-h_{21}}{h_{22}}$	$\dfrac{1}{h_{22}}$	$\dfrac{-y_{21}}{\Delta y}$	$\dfrac{y_{11}}{\Delta y}$		

Similarly,

$$\Delta z = \begin{vmatrix} z_{11} & z_{12} \\ z_{21} & z_{22} \end{vmatrix} = z_{11}z_{22} - z_{21}z_{12}$$

$$\Delta h = \begin{vmatrix} h_{11} & h_{12} \\ h_{21} & h_{22} \end{vmatrix} = h_{11}h_{22} - h_{21}h_{12}$$

SUMMARY OF FORMULAS

$$z_{11} = \left.\frac{V_1}{I_1}\right| \qquad z_{12} = \left.\frac{V_1}{I_2}\right|$$

$$z_{21} = \left.\frac{V_2}{I_1}\right|_{I_2 = 0} \qquad z_{22} = \left.\frac{V_2}{I_2}\right|_{I_1 = 0}$$

$$y_{11} = \left.\frac{I_1}{V_1}\right|_{V_2=0} \qquad y_{12} = \left.\frac{I_1}{V_2}\right|$$

$$y_{21} = \left.\frac{I_2}{V_1}\right|_{V_2=0} \qquad y_{22} = \left.\frac{I_2}{V_2}\right|_{V_1=0}$$

$$h_{11} = \left.\frac{V_1}{I_1}\right| \qquad h_{12} = \left.\frac{V_1}{V_2}\right|$$

$$h_{21} = \left.\frac{I_2}{I_1}\right|_{V_2=0} \qquad h_{22} = \left.\frac{I_2}{V_2}\right|_{I_1=0}$$

$$Z_{in} = \frac{V_1}{I_1} = h_i - \frac{h_r h_f Z_{load}}{h_o Z_{load} + 1}$$

$$Z_{out} = \frac{V_2}{I_2} = \cfrac{1}{h_o - \cfrac{h_r h_f}{h_i + Z_s}}$$

$$A_v = \frac{V_2}{V_1} = \frac{-h_f Z_{load}}{h_i(h_o Z_{load} + 1) - h_r h_f Z_{load}}$$

$$A_i = \frac{I_2}{I_1} = \frac{h_f}{h_o Z_{load} + 1}$$

PROBLEMS

20-1 Determine the equivalent z-parameters for the two-port network in Fig. 20-15.

20-2 Determine the equivalent z-parameters for the two-port network in Fig. 20-16.

20-3 The following open-circuit currents and voltages were determined experimentally for an unknown two-port. Measurements were made at 300 Hz.

$$\left.\begin{array}{l} V_1 = 208.1\underline{/54.8°} \\ V_2 = 133.1\underline{/-133°} \\ I_1 = 10\underline{/0°} \end{array}\right|_{I_2=0} \qquad \left.\begin{array}{l} V_1 = 53.24\underline{/-133°} \\ V_2 = 79.8\underline{/25.54°} \\ I_2 = 4\underline{/0°} \end{array}\right|_{I_1=0}$$

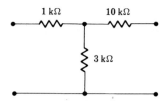

Figure 20-15 Circuit for Prob. 20-1.

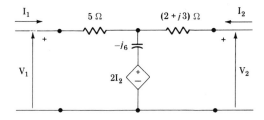

Figure 20-16 Circuit for Prob. 20-2.

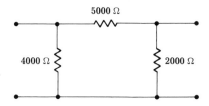

Figure 20-17 Circuit for Prob. 20-4.

Determine (*a*) the equivalent **z**-parameters for the two-port; (*b*) the output current if a 20-V 300-Hz generator is connected to the input port and a 10-Ω resistor is connected to the output port.

20-4 Determine the equivalent y-parameters for the two-port in Fig. 20-17.

20-5 The y-parameters of a certain two-port network are $y_{11} = 14$ mS, $y_{12} = -10$ mS, $y_{21} = -10$ mS, $y_{22} = 12$ mS. Assuming a 50-V DC source is connected to the input port and a 100-Ω resistor load is connected to the output port, determine (*a*) the current and power drawn by the load; (*b*) the battery current.

20-6 A 0.020-V 100-kHz source and a 2000-Ω resistor load are connected to the respective input and output ports of a certain common-emitter transistor amplifier. The parameters of the equivalent two-port are $h_{11} = 1.2$ kΩ, $h_{12} = 0.001$, $h_{21} = 45$, $h_{22} = 100$ μS. Determine (*a*) the voltage across the load; (*b*) the current to the load; (*c*) the voltage gain V_2/V_1 of the circuit.

20-7 The equivalent two-port network of a common-base transistor amplifier has the following parameters: $h_i = 35$ Ω, $h_f = -0.98$, $h_r = 260 \times 10^{-6}$, $h_o = 0.3$ μS. Assuming a sinusoidal input voltage of 100 mV and a load resistance of 10.0 kΩ, determine (*a*) the voltage across the load; (*b*) the voltage gain V_2/V_1 of the circuit.

20-8 Using the **h**-parameters in Prob. 20-6 and assuming the source resistance is 1.0 kΩ, determine (*a*) Z_{in}; (*b*) Z_{out}; (*c*) A_v; (*d*) A_i.

20-9 Using the **h**-parameters in Prob. 20-7 and assuming the source resistance is 800 Ω, determine (*a*) Z_{in}; (*b*) Z_{out}; (*c*) A_v; (*d*) A_i.

CHAPTER 21

POLYPHASE SYSTEM

A polyphase system is an arrangement of two or more generators, usually in a single frame, each supplying pulsating power at different intervals of time to a common load. It is similar to a multicylinder gasoline engine in which each cylinder supplies pulsating power at different intervals to a common crankshaft. One outstanding advantage of a polyphase system is that it serves to provide a more uniform flow of energy. The power waves of each generator are timed with respect to one another so that they do not reach zero value at the same time. Thus the resultant power at any instant will never be zero. Another advantage of a polyphase system is the greater output obtained from a given volume and weight of machine.

The three-phase system is the most common in current usage for the transmission and distribution of large quantities of electrical power. The two-phase system has applications in feedback control systems, where two-phase servomotors are used for position control. Circuits using 6 phase, 12 phase, and higher are used in some electronic rectification circuits to supply a relatively low-ripple direct current to a load.

21-1 THREE-PHASE GENERATION

A three-phase generator has three separate but identical armature windings, called *phases*, that are acted on by one system of magnets. This is shown in very elementary form in Fig. 21-1a, where the flux of the rotating magnets

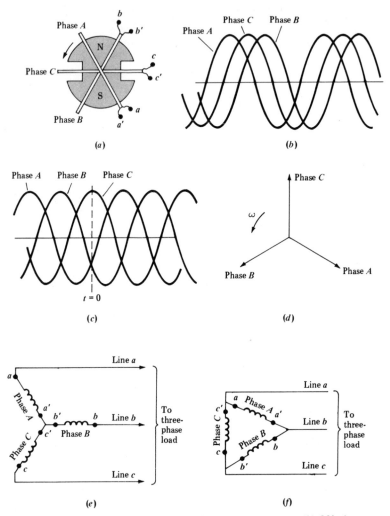

Figure 21-1 Three-phase system: (*a*) elementary generator; (*b*) 60° phase relationship; (*c*) 120° phase relationship; (*d*) phasor diagram; (*e*) wye-connected phases; (*f*) delta-connected phases.

passes, in turn, through the windows of the three armature windings. Although the sinusoidal voltages generated in each of the three phases have the same amplitude, the physical spacing is such that the three voltages reach their respective maximums 60° apart, as shown in Fig. 21-1*b*. However, when connecting the three phases in a wye or delta arrangement for power transmission and distribution, one of the phases (phase *C*) is deliberately reversed to obtain a symmetrical 120° relationship as shown in Fig. 21-1*c*. The corresponding phasor diagram is shown in Fig. 21-1*d*, with the phasors "frozen" at $t = 0$. The wye and delta connections, shown in Fig. 21-1e and f, respectively, are used to reduce the number of cables required for power transmission and distribution, and in motor, generator, and transformer applications.

1-2   WYE CONNECTION

In the wye connection shown in Fig. 21-2a the primed terminals a', b', and c' are connected together to form a *common junction* (or *common*), and the unprimed terminals are connected to the output lines. *The voltage between any line and the common is called the phase voltage, and the voltage between any two lines is called the line-to-line voltage or simply the line voltage.*

The phasor diagram, showing the magnitudes and angles of the phase voltages, is shown in Fig. 21-2b. The subscripts indicate the direction of voltage measurement. Thus,

$$\mathbf{E}_{a'a} = -\mathbf{E}_{aa'}$$

$$\mathbf{E}_{b'b} = -\mathbf{E}_{bb'}$$

$$\mathbf{E}_{c'c} = -\mathbf{E}_{cc'}$$

The voltage between any two lines may be determined by traversing the circuit from one line to the other, adding the voltages en route (phasor sum). Referring to Fig. 21-2a, the voltage measured *from* line a *to* line b is the voltage from a to a' of phase A added to the voltage from b' to b of phase B. Using subscript notations,

$$\mathbf{E}_{a\,to\,b} = \mathbf{E}_{a\,to\,a'} + \mathbf{E}_{b'\,to\,b}$$

Or in simplified form, with the word "to" implied,

$$\mathbf{E}_{ab} = \mathbf{E}_{aa'} + \mathbf{E}_{b'b} \tag{21-1}$$

Similarly,

$$\mathbf{E}_{bc} = \mathbf{E}_{bb'} + \mathbf{E}_{c'c} \tag{21-2}$$

$$\mathbf{E}_{ca} = \mathbf{E}_{cc'} + \mathbf{E}_{a'a} \tag{21-3}$$

Figure 21-2c shows the graphical construction of the line voltages represented by Eqs. (21-1), (21-2), and (21-3). The three line voltages are equal in magnitude but displaced from each other by 120°, and the set of line-voltage

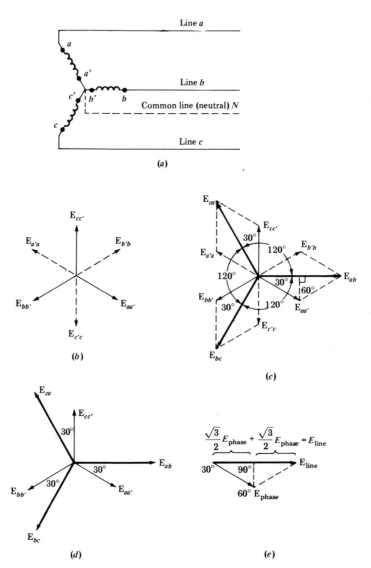

Figure 21-2 Wye connection: (a) line and neutral connections; (b) phase voltages; (c) graphical determination of line voltages; (d) phase and line voltages in standard form; (e) geometry of line and phase voltages.

phasors are displaced from the set of phase-voltage phasors by 30°. Since all phasors are rotating at the same angular velocity, for purposes of standardization in problem solving and analysis, the phasors are "frozen" with line voltage phasor $\mathbf{E}_{ab}$ at 0°. Figure 21-2d shows the three line voltages and the three phase-voltages of the wye connection in *standard form*.

The magnitude of the line voltage, in terms of the magnitude of the phase voltage, may be determined from the geometry of the phasor diagram in Fig. 21-2e:

$$E_{ab} = \sqrt{3}\,E_{aa'} = \sqrt{3}\,E_{b'b}$$

$$E_{bc} = \sqrt{3}\,E_{bb'} = \sqrt{3}\,E_{c'c}$$

$$E_{ca} = \sqrt{3}\,E_{cc'} = \sqrt{3}\,E_{a'a}$$

Thus, for the wye connection,

$$E_{\text{line}} = \sqrt{3}\,E_{\text{phase}}$$

(21-4)

where E_{line} = magnitude of any line-to-line voltage
E_{phase} = magnitude of any line to common voltage

A fourth cable, called the *common line* or *neutral line*, is often brought out from the common point of the wye-connected generator, as shown in Fig. 21-2a. The availability of the neutral line makes it possible to use all three phase-voltage as well as all three line voltages. Thus the line-to-neutral voltages are

$$\mathbf{E}_{aN} = \mathbf{E}_{aa'}$$

$$\mathbf{E}_{bN} = \mathbf{E}_{bb'}$$

(21-5)

$$\mathbf{E}_{cN} = \mathbf{E}_{cc'}$$

21-3   DELTA CONNECTION

In the delta connection, shown in Fig. 21-3a, the three generator phases are connected together to form a closed loop. The connections are a' to b, b' to c, and c' to a. The nodes, or junction points, of the delta form the output terminals of the generator. There is no common connection for the three phases; hence the delta cannot have a neutral line.

As indicated in Fig. 21-3a, the line voltage and its corresponding phase voltage are the same. Thus, for the delta connection, the line voltages are

$$\mathbf{E}_{ab} = \mathbf{E}_{aa'}$$

$$\mathbf{E}_{bc} = \mathbf{E}_{bb'}$$

(21-6)

$$\mathbf{E}_{ca} = \mathbf{E}_{cc'}$$

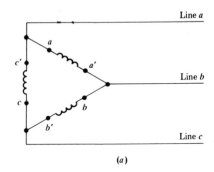

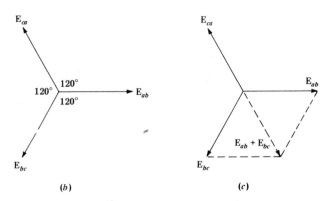

Figure 21-3   Delta connection: (a) line connections; (b) line voltages in standard form; (c) determination of sum of voltages around the delta loop.

The corresponding phasor diagram, with line-voltage phasor $\mathbf{E}_{ab}$ "frozen" at 0°, is the *standard form* used for analysis and problem solving in delta-connected systems. This is shown in Fig. 21-3b.

Although the three phases of the delta-connected generator form a closed loop, the phasor sum of the three driving voltages around the loop equals zero. This can be proved by adding the phasor voltages around the closed loop formed by the delta in Fig. 21-3a.

$$\mathbf{E}_{\text{resultant}} = \mathbf{E}_{aa'} + \mathbf{E}_{bb'} + \mathbf{E}_{cc'} \tag{21-7}$$

Substituting equation set (21-6) into Eq. (21-7),

$$\mathbf{E}_{\text{resultant}} = \mathbf{E}_{ab} + \mathbf{E}_{bc} + \mathbf{E}_{ca}$$

The graphical addition of these phasors, shown in Fig. 21-3c, indicates that the sum of the voltages around the delta loop equals zero.

$$\mathbf{E}_{ab} + \mathbf{E}_{bc} = -\mathbf{E}_{ca}$$

thus

$$\mathbf{E}_{ab} + \mathbf{E}_{bc} + \mathbf{E}_{ca} = 0$$

Since the sum of the driving voltages around the closed loop formed by the delta is zero, there can be no current circulating around the delta loop. The driving voltage generated in each phase supplies current to its own connected load; it does not supply current for circulation around the delta.

21-4   CALCULATING LINE CURRENTS TO ANY THREE-PHASE CIRCUIT

Depending on the complexity of the circuit, the current supplied by each line of a three-phase source may be determined by Ohm's law, Kirchhoff's current law, or loop analysis. Furthermore, if there is no neutral line connected to the load, the load does not know, nor does it care, whether the generator is wye or delta; the only significant factors are the magnitude and phase angles of the supply voltages.

The voltage of the source and the voltage printed on the nameplates of three-phase motors and other three-phase apparatus *are always assumed to be the rms line values,* unless otherwise specified.

When solving for the current in each of the three lines, it is convenient (but not critical) to assume the direction of the phasor current in each line to be from the source to the load. However, once assigned, the assumed directions must not be changed.

Example 21-1 (a) Determine the readings of the three ammeters in Fig. 21-4a. The circuit is supplied by a three-phase, 240-V, 60-Hz source. (b) Write the three time-domain equations for the currents in part a.

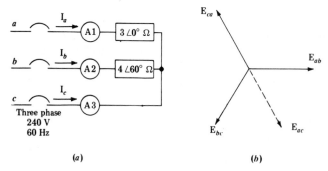

(a)   (b)

Figure 21-4   Circuit and phasor diagram for Example 21-1.

Solution

(a) As indicated in Fig. 21-4a, the assigned directions of the three currents are from the generator to the load. The $3/0°$ impedance is connected directly to terminals a and c of the generator, and the $4/60°$ impedance is connected directly to terminals b and c of the generator. Hence, in accordance with Ohm's law,

$$\mathbf{I}_a = \frac{\mathbf{E}_{ac}}{3/0°}$$

The phasor voltage is considered from a to c because the assumed direction of phasor current is *from the source to the load*. The phase angle of the voltage phasor is determined from the phasor diagram in Fig. 21-4b, which represents the line-to-line voltages of the source. From Fig. 21-4b,

$$\mathbf{E}_{ca} = 240/120°\text{ V}$$

However,

$$\mathbf{E}_{ac} = -\mathbf{E}_{ca} = -240/120°\text{ V}$$

Hence,

$$\mathbf{I}_a = \frac{-240/120°}{3/0°} = -80/120°\text{ A.}$$

Ammeter 1 will read 80 A.
Similarly,

$$\mathbf{I}_b = \frac{\mathbf{E}_{bc}}{4/60°} = \frac{240/-120°}{4/60°} = 60/-180°\text{ A}$$

Ammeter 2 will read 60 A.

Current $\mathbf{I}_c$ can be determined by applying Kirchhoff's current law to node 1 in Fig. 21-4a, The phasor sum of the currents to a node is equal to the phasor sum of the current leaving the node. Thus,

$$\mathbf{I}_a + \mathbf{I}_b + \mathbf{I}_c = 0$$

$$-80/120° + 60/-180° + \mathbf{I}_c = 0$$

$$\mathbf{I}_c = -72.11/-106.1°\text{ A}$$

Ammeter 3 will read 72.11 A.

(b) $\omega = 2\pi f = 2\pi(60) = 377$ rad/s

$$i_a = -80\sqrt{2}\sin(377t + 120°)$$
$$i_b = 60\sqrt{2}\sin(377t - 180°)$$
$$i_c = -72.11\sqrt{2}\sin(377t - 106.1°)$$

Example 21-2 Determine the three ammeter readings for the circuit shown in Fig. 21-5a. The circuit is supplied by a three-phase, 450-V, 60-Hz source. $\mathbf{Z}_1 = 5\underline{/10°}$, $\mathbf{Z}_2 = 9\underline{/30°}$, $\mathbf{Z}_3 = 10\underline{/80°}$.

Solution

Line current $\mathbf{I}_a$ has two components, one component in branch $\mathbf{Z}_1$ and the other in branch $\mathbf{Z}_2$. The branch currents may be determined by Ohm's law, and the line current by applying Kirchhoff's current law to node 1. Thus,

$$\mathbf{I}_a = \frac{\mathbf{E}_{ab}}{\mathbf{Z}_1} + \frac{\mathbf{E}_{ac}}{\mathbf{Z}_2}$$

From the phasor diagram in Fig. 21-5b,

$$\mathbf{E}_{ab} = 450\underline{/0°}\ \mathrm{V}$$

$$\mathbf{E}_{ac} = -\mathbf{E}_{ca} = -(450\underline{/120°})\ \mathrm{V}$$

$$\mathbf{I}_a = \frac{450\underline{/0°}}{5\underline{/10°}} + \frac{-(450\underline{/120°})}{9\underline{/30°}} = 110.29\underline{/-36.52°}\ \mathrm{A}$$

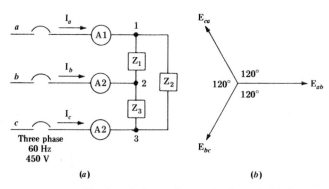

(a)   (b)

Figure 21-5 Circuit and phasor diagram for Example 21-2.

Ammeter 1 will read 110.29 A.
For node 2,

$$\mathbf{I}_b = \frac{\mathbf{E}_{ba}}{\mathbf{Z}_1} + \frac{\mathbf{E}_{bc}}{\mathbf{Z}_3}$$

From the phasor diagram

$$\mathbf{E}_{ba} = -\mathbf{E}_{ab} = -(450\underline{/0^\circ})\ \text{V}$$

$$\mathbf{E}_{bc} = 450\underline{/-120^\circ}\ \text{V}$$

$$\mathbf{I}_b = \frac{-(450\underline{/0^\circ})}{5\underline{/10^\circ}} + \frac{450\underline{/-120^\circ}}{10\underline{/80^\circ}} = 134.54\underline{/166.67^\circ}\ \text{A}$$

Ammeter 2 reads 134.54 A.
Similarly,

$$\mathbf{I}_c = \frac{\mathbf{E}_{ca}}{\mathbf{Z}_2} + \frac{\mathbf{E}_{cb}}{\mathbf{Z}_3}$$

$$\mathbf{E}_{ca} = 450\underline{/120^\circ}\ \text{V} \qquad \mathbf{E}_{cb} = -\mathbf{E}_{bc} = -(450\underline{/-120^\circ})\ \text{V}$$

$$\mathbf{I}_c = \frac{450\underline{/120^\circ}}{9\underline{/30^\circ}} + \frac{-450\underline{/-120^\circ}}{10\underline{/80^\circ}} = 54.64\underline{/39.30^\circ}\ \text{A}$$

Ammeter 3 reads 54.64 A.

Example 21-3 Determine the four ammeter readings in Fig. 21-6a. The circuit is supplied from a 240-V, three-phase, 60-Hz source.

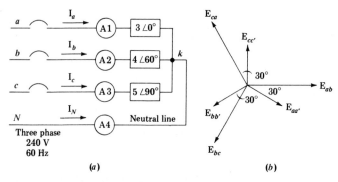

Figure 21-6 Circuit and phasor diagram for Example 21-3.

Solution

The voltage drops across the respective impedances may be determined from equation set (21-5) and the phasor diagram in Fig. 21-6b.

$$\mathbf{E}_{aN} = \mathbf{E}_{aa'} = \frac{240}{\sqrt{3}}\underline{/-30^\circ} = 138.56\underline{/-30^\circ}\text{ V}$$

$$\mathbf{E}_{bN} = \mathbf{E}_{bb'} = \frac{240}{\sqrt{3}}\underline{/-150^\circ} = 138.56\underline{/-150^\circ}\text{ V}$$

$$\mathbf{E}_{cN} = \mathbf{E}_{cc'} = \frac{240}{\sqrt{3}}\underline{/90^\circ} = 138.56\underline{/90^\circ}\text{ V}$$

Applying Ohm's law to the three impedances,

$$\mathbf{I}_a = \frac{\mathbf{E}_{aa'}}{3\underline{/0^\circ}} = \frac{138.56\underline{/-30^\circ}}{3\underline{/0^\circ}} = 46.19\underline{/-30^\circ}\text{ A}$$

$$\mathbf{I}_b = \frac{\mathbf{E}_{bb'}}{4\underline{/60^\circ}} = \frac{138.56\underline{/-150^\circ}}{4\underline{/60^\circ}} = 34.64\underline{/-210^\circ}\text{ A}$$

$$\mathbf{I}_c = \frac{\mathbf{E}_{cc'}}{5\underline{/90^\circ}} = \frac{138.56\underline{/90^\circ}}{5\underline{/90^\circ}} = 27.71\underline{/0^\circ}\text{ A}$$

The neutral current $\mathbf{I}_N$ may be obtained by applying Kirchhoff's current law to node k:

$$\mathbf{I}_a + \mathbf{I}_b + \mathbf{I}_c + \mathbf{I}_N = 0$$

$$46.19\underline{/-30^\circ} + 34.64\underline{/-210^\circ} + 27.71\underline{/0^\circ} + \mathbf{I}_N = 0$$

$$(40.00 - j23.10) + (-30 + j17.32) + (27.71 + j0) + \mathbf{I}_N = 0$$

$$\mathbf{I}_N = -38.15\underline{/-8.71^\circ}\text{ A}$$

Hence,

Ammeter A_1 reads 46.2 A

Ammeter A_2 reads 34.6 A

Ammeter A_3 reads 27.7 A

Ammeter A_4 reads 38.1 A

Example 21-4 Determine the three ammeter readings for the circuit shown in Fig. 21-7a. The circuit is supplied by a three-phase, 240-V, 60-Hz source.

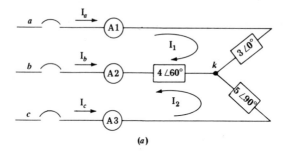

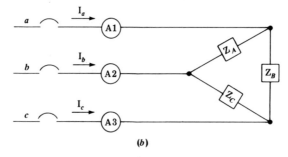

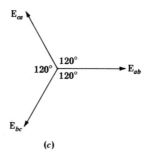

Figure 21-7 (a) Circuit for loop analysis of Example 21-4; (b) circuit using the wye-delta transformation; (c) standard phasor diagram.

Solution

Voltages $\mathbf{E}_{ak}$, $\mathbf{E}_{bk}$, and $\mathbf{E}_{ck}$ are not known (there is no neutral connection to the generator). Hence, Ohm's law cannot be used to determine the currents in the three impedances. However, the problem can be solved by using loop analysis or by converting the wye to an equivalent delta. The phase angles of the three line voltages, as obtained from the standard phasor diagram in Fig. 21-7c, are

$$\mathbf{E}_{ab} = 240\underline{/0°}\text{ V} \qquad \mathbf{E}_{bc} = 240\underline{/-120°}\text{ V} \qquad \mathbf{E}_{ca} = 240\underline{/120°}\text{ V}$$

Using loop analysis and the loop currents shown in Fig. 21-7a,

$$[\mathbf{E}_{ab}] = [3\underline{/0^\circ} + 4\underline{/60^\circ}]\mathbf{I}_1 + [4\underline{/60^\circ}]\mathbf{I}_2$$

$$[\mathbf{E}_{cb}] = [4\underline{/60^\circ}]\mathbf{I}_1 + [5\underline{/90^\circ} + 4\underline{/60^\circ}]\mathbf{I}_2$$

Simplifying,

$$240\underline{/0^\circ} = 6.08\underline{/34.72^\circ}\,\mathbf{I}_1 + 4\underline{/60^\circ}\,\mathbf{I}_2$$

$$-240\underline{/-120^\circ} = 4\underline{/60^\circ}\,\mathbf{I}_1 + 8.70\underline{/76.71^\circ}\,\mathbf{I}_2$$

$$\Delta_\mathbf{z} = \begin{vmatrix} 6.08\underline{/34.72^\circ} & 4\underline{/60^\circ} \\ 4\underline{/60^\circ} & 8.70\underline{/76.71^\circ} \end{vmatrix} = 37.14\underline{/107.75^\circ}$$

$$\Delta_{\mathbf{z}1} = \begin{vmatrix} 240\underline{/0^\circ} & 4\underline{/60^\circ} \\ -240\underline{/-120^\circ} & 8.70\underline{/76.71^\circ} \end{vmatrix} = 1537.29\underline{/51.36^\circ}$$

$$\Delta_{\mathbf{z}2} = \begin{vmatrix} 6.08\underline{/34.72^\circ} & 240\underline{/0^\circ} \\ 4\underline{/60^\circ} & -240\underline{/-120^\circ} \end{vmatrix} = 864.90\underline{/133.93^\circ}$$

$$\mathbf{I}_1 = \frac{\Delta_{\mathbf{z}1}}{\Delta_\mathbf{z}} = \frac{1537.29\underline{/51.36^\circ}}{37.14\underline{/107.75^\circ}} = 41.39\underline{/-56.39^\circ}\ \mathrm{A}$$

$$\mathbf{I}_2 = \frac{\Delta_{\mathbf{z}2}}{\Delta_\mathbf{z}} = \frac{864.90\underline{/133.93^\circ}}{37.14\underline{/107.75^\circ}} = 23.29\underline{/26.18^\circ}\ \mathrm{A}$$

Referring to Fig. 21-7a,

$$\mathbf{I}_a = \mathbf{I}_1 = 41.4\underline{/-56.4^\circ}\ \mathrm{A}$$

$$\mathbf{I}_c = \mathbf{I}_2 = 23.3\underline{/26.2^\circ}\ \mathrm{A}$$

$$\mathbf{I}_b = -(\mathbf{I}_1 + \mathbf{I}_2) = -(41.4\underline{/-56.4^\circ} + 23.3\underline{/26.2^\circ}) = -50\underline{/-28.91^\circ}\ \mathrm{A}$$

Therefore,

Ammeter A_1 reads 41.4 A

Ammeter A_2 reads 50.0 A

Ammeter A_3 reads 23.3 A

Using the wye-delta transformation technique outlined in Sec. 15-7, the impedances for the equivalent delta, shown in Fig. 21-7b, are

$$Z_A = \frac{(3\underline{/0°})(5\underline{/90°}) + (5\underline{/90°})(4\underline{/60°}) + (4\underline{/60°})(3\underline{/0°})}{5\underline{/90°}}$$

$$Z_A = \frac{37.16\underline{/107.74°}}{5\underline{/90°}} = 7.43\underline{/17.74°} \ \Omega$$

$$Z_B = \frac{37.16\underline{/107.74°}}{4\underline{/60°}} = 9.29\underline{/47.74°} \ \Omega$$

$$Z_C = \frac{37.16\underline{/107.74°}}{3\underline{/0°}} = 12.39\underline{/107.74°} \ \Omega$$

The three line currents to the equivalent delta are

$$I_a = \frac{E_{ab}}{Z_A} + \frac{E_{ac}}{Z_B} = \frac{240\underline{/0°}}{7.43\underline{/17.74°}} + \frac{-(240\underline{/120°})}{9.29\underline{/47.74°}}$$

$$I_a = 41.4\underline{/-56.4°} \ A$$

$$I_b = \frac{E_{ba}}{Z_A} + \frac{E_{bc}}{Z_C} = \frac{-(240\underline{/0°})}{7.43\underline{/17.74°}} + \frac{240\underline{/-120°}}{12.39\underline{/107.74°}}$$

$$I_b = 50.0\underline{/151.1°} = -50\underline{/-28.91°} \ A$$

$$I_c = \frac{E_{ca}}{Z_B} + \frac{E_{cb}}{Z_C} = \frac{240\underline{/120°}}{9.29\underline{/47.74°}} + \frac{-(240\underline{/-120°})}{12.39\underline{/107.74°}}$$

$$I_c = 23.3\underline{/26.2°} \ A$$

As expected, the three line currents obtained through the application of the wye-delta transformation are identical to those obtained by using loop analysis.

21-5 BALANCED THREE-PHASE LOADS—A SPECIAL CASE

A three-phase load is considered to be balanced, if all three branches of the load (wye or delta) have identical impedances. A three-phase motor is one example of a balanced load.

The currents in each of the three phases of a balanced three-phase load are equal in magnitude and displaced from each other by 120°; as a consequence, the three line currents are also equal in magnitude and displaced from each other by 120°.

Example 21-5 Calculate the phase currents and the line currents for the balanced delta load shown in Fig. 21-8a.

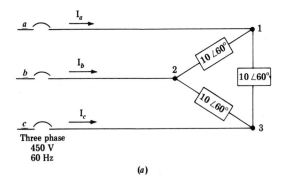

(a)

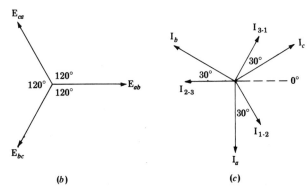

(b)   (c)

Figure 21-8 (a) Circuit for Example 21-5; (b) standard phasor diagram; (c) phasor diagram showing all phase and line currents.

Solution
The three phase-currents are

$$I_{1-2} = \frac{E_{ab}}{10\underline{/60°}}$$

$$I_{2-3} = \frac{E_{bc}}{10\underline{/60°}}$$

$$I_{3-1} = \frac{E_{ca}}{10\underline{/60°}}$$

From the phasor diagram in Fig. 21-8b,

$$\mathbf{E}_{ab} = 450\underline{/0°} \text{ V} \qquad \mathbf{E}_{bc} = 450\underline{/-120°} \text{ V} \qquad \mathbf{E}_{ca} = 450\underline{/120°} \text{ V}$$

Hence the three phase-currents are

$$\mathbf{I}_{1-2} = \frac{450\underline{/0°}}{10\underline{/60°}} = 45\underline{/-60°} \text{ A}$$

$$\mathbf{I}_{2-3} = \frac{450\underline{/-120°}}{10\underline{/60°}} = 45\underline{/-180°} \text{ A}$$

$$\mathbf{I}_{3-1} = \frac{450\underline{/120°}}{10\underline{/60°}} = 45\underline{/60°} \text{ A}$$

The three line currents are

$$\mathbf{I}_a = \frac{\mathbf{E}_{ab}}{10\underline{/60°}} + \frac{\mathbf{E}_{ac}}{10\underline{/60°}} = \frac{450\underline{/0°}}{10\underline{/60°}} + \frac{-(450\underline{/120°})}{10\underline{/60°}} = 77.9\underline{/-90°} \text{ A}$$

$$\mathbf{I}_b = \frac{\mathbf{E}_{ba}}{10\underline{/60°}} + \frac{\mathbf{E}_{bc}}{10\underline{/60°}} = \frac{-(450\underline{/0°})}{10\underline{/60°}} + \frac{450\underline{/-120°}}{10\underline{/60°}} = 77.9\underline{/150°} \text{ A}$$

$$\mathbf{I}_c = \frac{\mathbf{E}_{ca}}{10\underline{/60°}} + \frac{\mathbf{E}_{cb}}{10\underline{/60°}} = \frac{450\underline{/120°}}{10\underline{/60°}} + \frac{-(450\underline{/-120°})}{10\underline{/60°}} = 77.9\underline{/30°} \text{ A}$$

The phasor diagram in Fig. 21-8c shows the phase currents and line currents for the balanced delta load.

The ratio of line-current magnitude to phase-current magnitude for the *balanced delta load* in Fig. 21-8a is

$$\frac{I_{\text{line}}}{I_{\text{phase}}} = \frac{77.94}{45} = 1.7320 = \sqrt{3}$$

This relationship holds true for any balanced delta-connected load. Thus

> For a balanced delta load
>
> $$I_{\text{line}} = \sqrt{3}\, I_{\text{phase}}$$

(21-8)

Example 21-6 Calculate the phase currents and the line currents for the balanced wye load in Fig. 21-9a.

Solution

$$\mathbf{I}_a = \frac{\mathbf{E}_{aN}}{5\underline{/20°}} \qquad \mathbf{I}_b = \frac{\mathbf{E}_{bN}}{5\underline{/20°}} \qquad \mathbf{I}_c = \frac{\mathbf{E}_{cN}}{5\underline{/20°}}$$

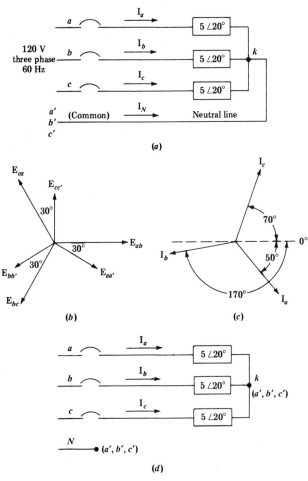

Figure 21-9 (*a*) Circuit diagram for Example 21-6; (*b*) standard phasor diagram; (*c*) phasor diagram of line currents; (*d*) circuit diagram showing neutral line disconnected.

From the phasor diagram in Fig. 21-9*b*,

$$\mathbf{E}_{aN} = \mathbf{E}_{aa'} = \frac{120}{\sqrt{3}}\underline{/-30°} = 69.28\underline{/-30°}\ \text{V}$$

$$\mathbf{E}_{bN} = \mathbf{E}_{bb'} = \frac{120}{\sqrt{3}}\underline{/-150°} = 69.28\underline{/-150°}\ \text{V}$$

$$\mathbf{E}_{cN} = \mathbf{E}_{cc'} = \frac{120}{\sqrt{3}}\underline{/90°} = 69.28\underline{/90°}\ \text{V}$$

Hence the phase currents, which are equal to their corresponding line currents, are

$$\mathbf{I}_a = \frac{69.28 / -30°}{5 / 20°} = 13.86 / -50° \text{ A}$$

$$\mathbf{I}_b = \frac{69.28 / -150°}{5 / 20°} = 13.86 / -170° \text{ A}$$

$$\mathbf{I}_c = \frac{69.28 / 90°}{5 / 20°} = 13.86 / 70° \text{ A}$$

Figure 21-9c shows the phasor diagram for the three line currents in a balanced wye load.

The current in the neutral line, as determined by applying Kirchhhoff's current law to node k in Fig. 21-9a, is

$$\mathbf{I}_a + \mathbf{I}_b + \mathbf{I}_c + \mathbf{I}_N = 0$$

$$13.86 / -50° + 13.86 / -170° + 13.86 / 70° + \mathbf{I}_N = 0$$

$$(8.91 - j10.62) + (-13.65 - j2.41) + (4.74 + j13.02) + \mathbf{I}_N = 0$$

$$(0 + j0) + \mathbf{I}_N = 0$$

$$\mathbf{I}_N = 0$$

As demonstrated in Example 21-6, *there is no current in the neutral line of a balanced wye system.* Hence the neutral line is not needed and may be omitted without altering the magnitude and angle of the phase currents. This is shown in Fig. 21-9d. Furthermore, because of the balanced load, the potential at the wye junction (node k) will be the same without the neutral line as with it; node k in Fig. 21-9d has the same common potential a', b', c' as node k in Fig. 21-9a.

Hence, when calculating the current in any branch of a balanced wye load, with or without a neutral line, the voltage across each branch will be the respective phase voltage $\mathbf{E}_{aa'}$, $\mathbf{E}_{bb'}$, $\mathbf{E}_{cc'}$.

21-6 PARALLEL-CONNECTED LOADS IN A THREE-PHASE SYSTEM

The calculation of currents in a three-phase system of parallel-connected loads may be accomplished by using Ohm's law, Kirchhoff's current law, and the standardized phasor diagram of line and phase voltages.

Example 21-7 Determine the ammeter indications for the circuit shown in Fig. 21-10a.

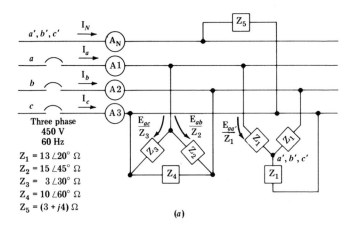

a', b', c'

a

b

c

Three phase
450 V
60 Hz

$Z_1 = 13 \angle 20° \; \Omega$
$Z_2 = 15 \angle 45° \; \Omega$
$Z_3 = \; 3 \angle 30° \; \Omega$
$Z_4 = 10 \angle 60° \; \Omega$
$Z_5 = (3 + j4) \; \Omega$

(a)

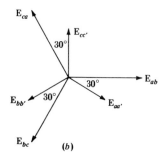

(b)

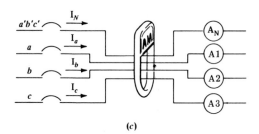

(c)

Figure 21-10 (a) Circuit diagram for Example 21-7;
(b) standard phasor diagram; (c) physical concept of the accuracy
check.

Solution

The source voltages, determined with the aid of the standardized phasor diagram in Fig. 21-10*b*, are

$$\mathbf{E}_{ab} = 450\underline{/0°} \text{ V} \qquad \mathbf{E}_{aa'} = \frac{450}{\sqrt{3}}\underline{/-30°} = 259.81\underline{/-30°} \text{ V}$$

$$\mathbf{E}_{bc} = 450\underline{/-120°} \text{ V} \qquad \mathbf{E}_{bb'} = \frac{450}{\sqrt{3}}\underline{/-150°} = 259.81\underline{/-150°} \text{ V}$$

$$\mathbf{E}_{ca} = 450\underline{/120°} \text{ V} \qquad \mathbf{E}_{cc'} = \frac{450}{\sqrt{3}}\underline{/90°} = 259.81\underline{/90°} \text{ V}$$

Since the wye load is *balanced*, the wye junction is at the same potential as the neutral line (*a'*, *b'*, *c'*). The current in each line is determined by the phasor summation of branch currents fed by that line, and the current in each branch is determined by Ohm's law and the known voltage drop across the branch impedance. The heavy arrows in Fig. 21-10*a* show the branch currents that make up line current I_a. Thus,

The current in line a is

$$\mathbf{I}_a = \frac{\mathbf{E}_{ab}}{\mathbf{Z}_2} + \frac{\mathbf{E}_{ac}}{\mathbf{Z}_3} + \frac{\mathbf{E}_{aa'}}{\mathbf{Z}_1}$$

$$\mathbf{I}_a = \frac{450\underline{/0°}}{15\underline{/45°}} + \frac{-(450\underline{/120°})}{3\underline{/30°}} + \frac{259.81\underline{/-30°}}{13\underline{/20°}} = 189.61\underline{/-79.65°} \text{ A}$$

Ammeter A_1 reads 190 A.

The current in line b is

$$\mathbf{I}_b = \frac{\mathbf{E}_{ba}}{\mathbf{Z}_2} + \frac{\mathbf{E}_{bc}}{\mathbf{Z}_4} + \frac{\mathbf{E}_{bb'}}{\mathbf{Z}_1}$$

$$\mathbf{I}_b = \frac{-(450\underline{/0°})}{15\underline{/45°}} + \frac{450\underline{/-120°}}{10\underline{/60°}} + \frac{259.81\underline{/-150°}}{13\underline{/20°}} = 87.71\underline{/168.33°} \text{ A}$$

Ammeter A_2 reads 87.7 A.

The current in line c is

$$\mathbf{I}_c = \frac{\mathbf{E}_{ca}}{\mathbf{Z}_3} + \frac{\mathbf{E}_{cb}}{\mathbf{Z}_4} + \frac{\mathbf{E}_{cc'}}{\mathbf{Z}_5} + \frac{\mathbf{E}_{cc'}}{\mathbf{Z}_1}$$

$$\mathbf{I}_c = \frac{450\underline{/120°}}{3\underline{/30°}} + \frac{-(450\underline{/-120°})}{10\underline{/60°}} + \frac{259.81\underline{/90°}}{(3 + j4)} + \frac{259.81\underline{/90°}}{13\underline{/20°}}$$

$$\mathbf{I}_c = 220.70\underline{/64.96°}$$

Ammeter A_3 reads 221 A.

The current in the neutral line is

$$\mathbf{I}_N = \frac{\mathbf{E}_{c'c}}{\mathbf{Z}_5} = \frac{-(259.81\underline{/90°})}{(3 + j4)} = -51.96\underline{/36.87°} \text{ A}$$

Ammeter A_N reads 52.0 A.

Accuracy Check

If *all* lines coming from a three-phase source (or for that matter, any AC source) are passed through the window of a hook-on ammeter, as shown in Fig. 21-10c, the ammeter will read 0 A. In mathematical terms, the phasor sum of *all* currents through the window of the hook-on ammeter in Fig. 21-10c is zero;

$$\mathbf{I}_a + \mathbf{I}_b + \mathbf{I}_c + \mathbf{I}_N = 0$$

The hook-on ammeter in Fig. 21-10c verifies Kirchhoff's current law.

Thus an accuracy check of line-current calculations may be made by obtaining the phasor summation of the currents in *all* feeder lines. If all calculations are accurate, the phasor summation should be zero.

Applying the accuracy check to Example 21-7, and *considering all current phasors to be directed from the source to the load,*

$$\sum \mathbf{I}\text{'s} = \mathbf{I}_a + \mathbf{I}_b + \mathbf{I}_c + \mathbf{I}_N$$

$$\sum \mathbf{I}\text{'s} = 189.61\underline{/-79.65°} + 87.71\underline{/168.33°} + 220.70\underline{/64.96°} - 51.96\underline{/36.87°}$$

$$\sum \mathbf{I}\text{'s} = 0.01\underline{/9.46°} \text{ A}$$

As indicated, there is an error of 0.01 A at an angle of 9.46° in the summation. Considering the magnitude of the currents in each of the four lines, an error of 0.01 A represents a relatively insignificant error and is attributable to round-off errors in the individual line-current calculations.

21-7   **PHASE SEQUENCE**

The phase sequence, or *phase rotation* of a polyphase source, is the *order or sequence in which the voltage waves reach their maximum positive values.* The phase sequence of the source can have a profound effect on the performance of the load. For example, reversing the phase sequence of the driving voltage to a three-phase motor will reverse its direction of rotation; reversing the phase sequence of the driving voltage applied to an unbalanced three-phase load could cause major changes in the magnitudes and phase angles of the line currents; reversing the phase sequence of a three-phase generator that is to be paralleled with another three-phase generator can cause extensive damage to both machines.

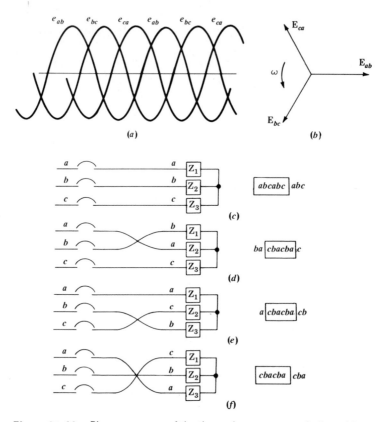

Figure 21-11 Phase sequence of the three-phase system as indicated by (*a*) voltage waves; (*b*) phasors. (*c*), (*d*), (*e*), (*f*) changing the phase sequence by interchanging any two lines.

The phase sequence indicated by the voltage waves in Fig. 21-11*a*, for a three-phase source, is

$$E_{ab}, E_{bc}, E_{ca}, E_{ab}, E_{bc}, E_{ca}, \ldots$$

For convenience, this may be simplified by denoting only the first subscript or only the second subscript in the voltage sequence. Thus, using the first subscripts, the sequence is written as

$\boxed{abcabc}abc \cdots$ or simply *abc*

Using the second subscripts, the sequence is written as

$bc\boxed{abcabc}a \cdots$ which is also *abc*

The phase sequence, represented by the corresponding phasor diagram shown in Fig. 21-11b, is *abc*

When analyzing circuit diagrams involving three-phase loads, it is convenient to define the phase sequence *at the load*, reading from top to bottom or left to right as applicable. For example, in Fig. 21-11c, reading repetitively from top to bottom, the phase sequence at the load is *abcabcabc* $\cdots$, or simply *abc*.

Interchanging any two of the three cables causes the *sequence at the load* to be reversed. This is shown in Fig. 21-11d, e, and f and is indicated in the following tabulation:

Cables interchanged	Phase sequence
a and b	ba$\boxed{cbacba}$c $\cdots$ or cba
b and c	a$\boxed{cbacba}$cb $\cdots$ or cba
c and a	$\boxed{cbacba}$cba $\cdots$ or cba

It should be noted that the three-phase system has only two possible sequences: *abc* or *cba*.

Example 21-8 (*a*) Determine the three ammeter readings for the circuit shown in Fig. 21-12a; (*b*) reverse the phase sequence by interchanging line "*a*" and line "*b*," as shown in Fig. 21-12b, and determine the ammeter readings.

Solution

(*a*) Using the standardized phasor diagram in Fig. 21-12c, the three line currents for Fig. 21-12a are

Line "a"

$$\mathbf{I}_a = \frac{\mathbf{E}_{ab}}{\mathbf{Z}_1} + \frac{\mathbf{E}_{ac}}{\mathbf{Z}_3} = \frac{300\underline{/0°}}{10\underline{/25°}} + \frac{-(300\underline{/120°})}{15\underline{/0°}} = 47.78\underline{/-38.89°} \text{ A}$$

Ammeter A_1 reads 47.8 A.

Line "b"

$$\mathbf{I}_b = \frac{\mathbf{E}_{ba}}{\mathbf{Z}_1} + \frac{\mathbf{E}_{bc}}{\mathbf{Z}_2} = \frac{-(300\underline{/0°})}{10\underline{/25°}} + \frac{300\underline{/-120°}}{20\underline{/60°}} = 44.05\underline{/163.27°} \text{ A}$$

Ammeter A_2 reads 44.1 A.

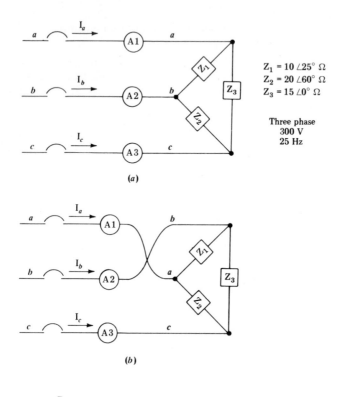

$Z_1 = 10 \angle 25° \ \Omega$
$Z_2 = 20 \angle 60° \ \Omega$
$Z_3 = 15 \angle 0° \ \Omega$

Three phase
300 V
25 Hz

(a)

(b)

(c)

Figure 21-12 (a) Three-phase circuit for Example 21-8; (b) three-phase circuit with reversed phase sequence; (c) standard phasor diagram.

Line "c"

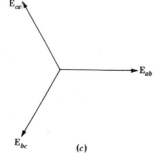

$$\mathbf{I}_c = \frac{\mathbf{E}_{ca}}{\mathbf{Z}_3} + \frac{\mathbf{E}_{cb}}{\mathbf{Z}_2} = \frac{300\underline{/120°}}{15\underline{/0°}} + \frac{-(300\underline{/-120°})}{20\underline{/60°}} = 18.03\underline{/73.90°} \ \text{A}$$

Ammeter A_3 reads 18.0 A.

(*b*) Interchanging lines "*a*" and "*b*," as shown in Fig. 21-12*b*, causes the phase sequence at the load to be *bacbacbac* or simply *cba*.

Line "a"

$$\mathbf{I}_a = \frac{\mathbf{E}_{ab}}{\mathbf{Z}_1} + \frac{\mathbf{E}_{ac}}{\mathbf{Z}_2} = \frac{300/0°}{10/25°} + \frac{-(300/120°)}{20/60°} = 32.35/-52.51° \text{ A}$$

Ammeter A_1 reads 32.4 A.

Line "b"

$$\mathbf{I}_b = \frac{\mathbf{E}_{ba}}{\mathbf{Z}_1} + \frac{\mathbf{E}_{bc}}{\mathbf{Z}_3} = \frac{-300/0°}{10/25°} + \frac{300/-120°}{15/0°} = 37.48/-172.89° \text{ A}$$

Ammeter A_2 reads 37.5 A.

Line "c"

$$\mathbf{I}_c = \frac{\mathbf{E}_{cb}}{\mathbf{Z}_3} + \frac{\mathbf{E}_{ca}}{\mathbf{Z}_2} = \frac{-(300/-120°)}{15/0°} + \frac{300/120°}{20/60°} = 35.00/60° \text{ A}$$

Ammeter A_3 reads 35.0 A.

The following tabulation emphasizes the effect of phase sequence on the line currents to an unbalanced three-phase load.

	AMPERES	
Line	Sequence abc	Sequence cba
a	47.8	32.4
b	44.1	37.5
c	18.0	35.0

1-8 THREE-PHASE TO TWO-PHASE TRANSFORMATION

A two-phase system consists of two sinusoidal voltages equal in magnitude but displaced from each other by 90°. The two-phase system has applications in control systems that utilize two-phase motors. One example of such a system is the control-rod drive for nuclear power plants.

A Scott connection of transformers is the most common method for transforming a three-phase system of voltages to a two-phase system and vice versa. The Scott connection requires two transformers, one with a primary tap at $0.50N_p$ and the other with a primary tap at $0.866N_p$, as shown in Fig. 21-13*a*.

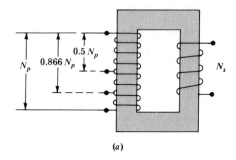

(a)

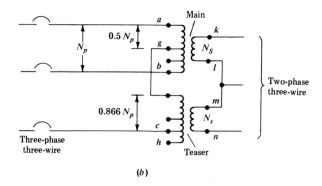

(b)

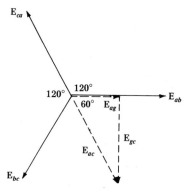

Figure 21-13 Three-phase to two-phase transformation; (a) transformer construction; (b) Scott connection; (c) phasor diagram of the Scott connection.

The circuit for the Scott connection is shown in Fig. 21-13b. The primary of one transformer (called the main transformer) is connected across two of the incoming three-phase lines, and its center tap is connected to the second transformer (called the teaser transformer). The teaser utilizes only 86.6 percent of its primary turns; the extra turns between c and h have no effect on the operation of the transformer and are not used. Even though some taps are not used, it is more economical to bring out the same taps on both transformers than to stock transformers with different tap arrangements. The two-phase output may be either two-phase four-wire or the more common two-phase three-wire as shown in Fig. 21-13b.

The theory of the Scott connection is developed by applying Kirchhoff's voltage law to the primary side and, with the aid of a phasor diagram, proving that a 90° relationship exists between the two secondary voltages. Thus, from Fig. 21-13b,

$$\mathbf{E}_{ac} = \mathbf{E}_{ag} + \mathbf{E}_{gc} \tag{21-9}$$

$$\mathbf{E}_{ag} = \tfrac{1}{2}\mathbf{E}_{ab}$$

$$\mathbf{E}_{ac} = -\mathbf{E}_{ca}$$

The standard phasor diagram for the three-phase system is drawn with solid lines in Fig. 21-13c. Phasors $\mathbf{E}_{ac}$ and $\mathbf{E}_{ag}$, shown with broken lines, represent two components of Eq. (21-9). A line joining the tips of phasors $\mathbf{E}_{ag}$ and $\mathbf{E}_{ac}$ provides the third phasor $\mathbf{E}_{gc}$; this must be so because it satisfies the tip-to-tail addition indicated by Kirchhoff's voltage equation (21-9).

Examination of the triangle formed by these three phasors shows that the length of the side $\mathbf{E}_{ag}$ is equal to one-half the length of side $\mathbf{E}_{ac}$ and has an included angle of 60°, indicating a 30°, 60°, 90° triangle. From the geometry of the phasor diagram, primary voltage $\mathbf{E}_{ab}$ of the main transformer leads primary voltage $\mathbf{E}_{gc}$ of the teaser by 90°. It therefore follows, by transformer action, that secondary voltage $\mathbf{E}_{kl}$ must lead secondary voltage $\mathbf{E}_{mn}$ by 90°.

To prove that the two secondary voltages are equal in magnitude, it is merely necessary to show that the volts/turn ratio of each of the two primaries is the same:

$$\text{Volts/turn of main} = \frac{E_{ab}}{N_p}$$

$$\text{Volts/turn of teaser} = \frac{E_{gc}}{0.866\, N_p} = \frac{E_{ac}\sin 60°}{0.866\, N_p} = \frac{E_{ac}}{N_p}$$

However,

$$E_{ab} = E_{ac}$$

Hence

$$\text{Volts/turn of teaser} = \frac{E_{ab}}{N_p}$$

Since the volts/turn of the two transformers are the same, and they have the same number of secondary turns, the two secondary voltages are equal. Thus the two secondary voltages are equal in magnitude but displaced in phase by 90°, which is the requirement for a two-phase system.

SUMMARY OF FORMULAS

Wye (*bal*)

$$V_{\text{line}} = \sqrt{3}\,E_{\text{br}}$$

$$I_{\text{line}} = I_{\text{br}}$$

Delta (*bal*)

$$V_{\text{line}} = E_{\text{br}}$$

$$I_{\text{line}} = \sqrt{3}\,I_{\text{br}}$$

PROBLEMS

21-1   A 240-V, three-phase, 60-Hz supply is connected to an unbalanced delta load. The impedances are connected in the following manner: $3\underline{/20°}$ Ω between lines a and b, $4\underline{/10°}$ Ω between lines a and c, $5\underline{/35°}$ Ω between lines b and c. (*a*) Sketch the circuit and include ammeters in the three lines; (*b*) sketch the standardized phasor diagram and calculate the ammeter readings.

21-2   An unbalanced delta load and a single-phase load is connected to a three-phase, 100-V 60-Hz system. The delta-connected impedances $2\underline{/-30°}$ Ω, $4\underline{/0°}$ Ω, and $5\underline{/60°}$ Ω are connected, respectively, to lines b and c, b and a, c and a. The single-phase impedance of $2\underline{/0°}$ Ω is connected between lines a and b. (*a*) Sketch the circuit and include ammeters in the three lines; (*b*) sketch the standardized phasor diagram and determine the ammeter readings.

21-3   A three-wire, three-phase, 240-V, 60-Hz system supplies power to the following loads: three $(5 + j10)$-Ω impedances connected in wye, and a $(2 + j6)$-Ω impedance connected between line a and line b. (*a*) Sketch the circuit and include ammeters in the three lines; (*b*) sketch the standardized phasor diagram and determine the ammeter readings.

21-4   A four-wire, three-phase, 450-V, 60-Hz system supplies power to the following loads: three impedances $(13\underline{/20°}$ each) connected in wye; an unbalanced delta load consisting of $15\underline{/45°}$ Ω between lines a and b, $3\underline{/30°}$ Ω between lines a and c, and $10\underline{/60°}$ Ω between lines b and c; a single-phase load of $(3 + j4)$ Ω between line c and the neutral line. (*a*) Sketch the circuit and include ammeters in all four lines; (*b*) sketch the standardized phasor diagram and determine the ammeter readings.

21-5   A wye-connected load and a delta-connected load are supplied by a four-wire, 400-V, three-phase, 50-Hz system. The connections for the wye load are $2\underline{/20°}$ Ω to line a, $30\underline{/50°}$ Ω to line c, and $6\underline{/75°}$ Ω to line b. The connections for the delta load are $50\underline{/30°}$ Ω between lines a and b, $25\underline{/-60°}$ Ω between lines b and c, and $17.3\underline{/90°}$ Ω between lines a and c. (*a*) Sketch the circuit and include an ammeter in line b; (*b*) sketch the standardized phasor diagram and determine the ammeter reading.

21-6 A three-phase, three-wire, 450-V, 60-Hz generator supplies power to the following loads: a balanced delta load whose impedance is $20/30°$ Ω per leg, a balanced wye-connected load whose impedance is $40/0°$ Ω per leg, and a single-phase load of $38/0°$ Ω connected between lines a and b. (a) Sketch the circuit and include ammeters to measure the line current to the delta load, the line current to the wye load, the line current to the 38-Ω load, and the total line current supplied by the generator in line a; (b) sketch the standardized phasor diagram and determine the ammeter readings.

21-7 A 208-V, three-phase, 60-Hz, four-wire system supplies power to a balanced wye-connected load of $6/50°$ Ω per leg, a balanced delta-connected load of $4/25°$ Ω per leg, and a single-phase load of $2/10°$ Ω connected between line b and the neutral line. (a) Sketch the circuit. (b) What should a clip-on ammeter indicate in each of the incoming lines? (c) Repeat part b, assuming an open occurred in the delta leg between lines b and c.

21-8 The impedances for the circuit in Fig. 21-14 are $Z_1 = 10/20°$, $Z_2 = 10/50°$, $Z_3 = 10/80°$, $Z_4 = 10/-10°$, $Z_5 = 10/-40°$. Sketch the standardized phasor diagram and determine the ammeter readings.

21-9 Determine the ammeter readings for the circuit shown in Fig. 21-15.

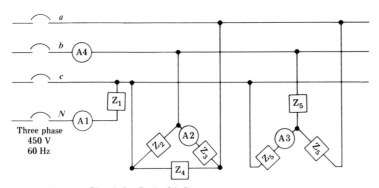

Figure 21-14 Circuit for Prob. 21-8.

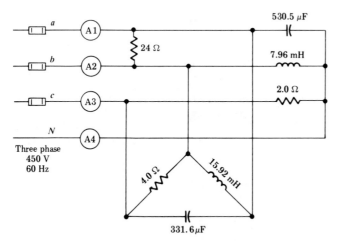

Figure 21-15 Circuit for Probs. 21-9 and 21-10.

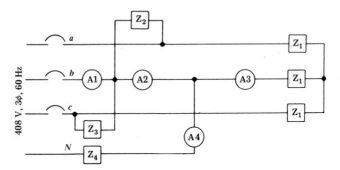

Figure 21-16   Circuits for Prob. 21-11.

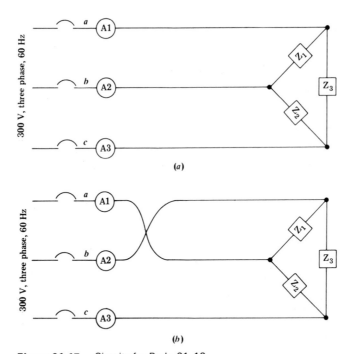

Figure 21-17   Circuits for Prob. 21-12.

21-10   Determine the ammeter readings for the circuit shown in Fig. 21-15, assuming the fuse in line b is blown.

21-11   Determine the ammeter readings for the circuit shown in Fig. 21-16. The impedances are $\mathbf{Z}_1 = (2 + j3.5)$, $\mathbf{Z}_2 = (3 - j4)$, $\mathbf{Z}_3 = (4 - j4)$, $\mathbf{Z}_4 = (2 + j0)$.

21-12   The impedances for the circuit shown in Fig. 21-17a are $\mathbf{Z}_1 = 8\underline{/20°}$ $\mathbf{Z}_2 = 15\underline{/65°}$, and $\mathbf{Z}_3 = 10\underline{/0°}$. (a) Determine the ammeter readings; (b) determine the ammeter readings if the phase sequence to the load is reversed as shown in Fig. 21-17b.

CHAPTER 22

THREE-PHASE POWER

The following discussions of three-phase power are based on the single-phase power relationships developed in Chap. 13.

ACTIVE POWER, REACTIVE POWER, APPARENT POWER, AND POWER FACTOR IN THE THREE-PHASE SYSTEM (BALANCED OR UNBALANCED)

The total active power delivered to any three-phase circuit (balanced or unbalanced) is equal to the sum of the active power components drawn by the individual branches. That is,

$$P_{3\phi} = \sum P_{br}$$

where

$$P_{br} = V_{br} I_{br} \cos \angle \frac{\mathbf{V}_{br}}{\mathbf{I}_{br}}$$

(22-1)

Similarly, the total reactive power delivered to any three-phase circuit (balanced or unbalanced) is equal to the sum of the reactive power components drawn by the individual branches. That is,

$$Q_{3\phi} = \sum Q_{br}$$

where

$$Q_{br} = V_{br} I_{br} \sin \qquad (22\text{-}2)$$

Note: In Eqs. (22-1) and (22-2), the phase angle

is measured *from* the current phasor *to* the voltage phasor. A counterclockwise measurement is a positive angle, and a clockwise measurement is a negative angle. Failure to consider the direction of angular measurement may introduce serious errors in reactive-power calculations.

The total apparent power delivered to any three-phase system (balanced or unbalanced) is equal to the right-angle summation of the total active and total reactive components. That is,

$$S_{3\phi} = (P_{3\phi} + jQ_{3\phi}) = S_{3\phi}\underline{/\theta}$$

where

$$S_{3\phi} = \sqrt{P_{3\phi}^2 + Q_{3\phi}^2}$$

Angle θ has significance only in a balanced system, where it represents the phase angle between the phase current and the corresponding phase voltage and is identifiable as such on the associated phasor diagram. However, *if the system is unbalanced, angle θ is not identifiable on the phasor diagram, and the angle has no significance.*

The power factor of a three-phase system (balanced or unbalanced) is defined as the ratio of total system active power to total system apparent power.

$$pf = \frac{P_{3\phi}}{S_{3\phi}}$$

If the system is balanced, the power factor of the system is identical to the power factor of each phase.

Example 22-1 Determine the active power, reactive power, apparent power, and power factor of the system shown in Fig. 22-1a.

Solution

Using Ohm's law and the standard phasor diagram for the three-phase driving voltages,

$$\mathbf{I}_a = \frac{\mathbf{E}_{aa'}}{\mathbf{Z}_1} = \frac{(480/\sqrt{3})/-30°}{6/20°} = 46.19/-50° \text{ A}$$

$$\mathbf{I}_b = \frac{\mathbf{E}_{bb'}}{\mathbf{Z}_2} = \frac{(480/\sqrt{3})/-150°}{8/40°} = 34.64/-190° \text{ A}$$

$$\mathbf{I}_c = \frac{\mathbf{E}_{cc'}}{\mathbf{Z}_3} = \frac{(480/\sqrt{3})/90°}{10/0°} = 27.71/90° \text{ A}$$

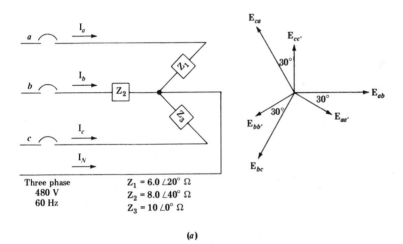

Three phase
480 V
60 Hz

$Z_1 = 6.0 \angle 20° \ \Omega$
$Z_2 = 8.0 \angle 40° \ \Omega$
$Z_3 = 10 \angle 0° \ \Omega$

(a)

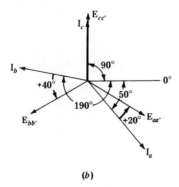

(b)

Figure 22-1 (a) Circuit and standard phasor diagram for Example 22-1; (b) measuring the angle *from* a current phasor *to* a voltage phasor.

Power drawn by Z_1:

The branch current and branch voltage for the Z_1 branch are

$$I_a = 46.19\underline{/-50°}\ A \qquad E_{aa'} = \frac{480}{\sqrt{3}}\underline{/-30°} = 277.13\underline{/-30°}\ V$$

As indicated in Fig. 22-1b, the angle measured *from* the branch-current phasor *to* the branch-voltage phasor is

$= +20°$

Substituting the branch current, branch voltage, and included angle into Eqs. (22-1) and (22-2),

$P_{Z1} = (277.13)(46.19)\cos 20°$ $Q_{Z1} = (277.13)(46.19)\sin 20°$

$P_{Z1} = 12{,}028.66$ W $Q_{Z1} = 4378.07$ var

Power drawn by Z_2:

The branch current and branch voltage for the Z_2 branch are

$$I_b = 34.64\underline{/-190°}\ A \qquad E_{bb'} = 277.13\underline{/-150°}\ V$$

As indicated in Fig. 22-1b, the angle measured *from* the branch-current phasor *to* the branch-voltage phasor is

$= +40°$

Thus,

$P_{Z2} = (277.13)(34.64)\cos 40°$ $Q_{Z2} = (277.13)(34.64)\sin 40°$

$P_{Z2} = 7353.86$ W $Q_{Z2} = 6170.62$ var

Power drawn by Z_3:

The branch current and branch voltage for the Z_3 branch are

$$I_c = 27.71\underline{/90°}\ A \qquad E_{cc'} = 277.13\underline{/90°}\ V$$

The angle measured *from* the branch-current phasor *to* the branch-voltage phasor is

$= 0°$

Thus,

$P_{Z3} = (277.13)(27.71)\cos 0°$ $Q_{Z3} = (277.13)(27.71)\sin 0°$

$P_{Z3} = 7679.27 \text{ W}$ $Q_{Z3} = 0$

The total three-phase active power supplied by the source is

$P_{3\phi} = 12{,}028.66 + 7353.86 + 7679.27 = 27{,}061.79 \text{ W}$

The total three-phase reactive power supplied by the source is

$Q_{3\phi} = 4378.07 + 6170.62 + 0 = 10{,}548.69 \text{ var}$

The total three-phase apparent power supplied by the source is

$S_{3\phi} = P_{3\phi} + jQ_{3\phi} = 27{,}061.79 + j10{,}548.69$

$S_{3\phi} = 29{,}045.06\underline{/21.30°}$

$S_{3\phi} = 29.045 \text{ kVA}$

The power factor of the system is

$$\text{pf}_{3\phi} = \frac{P_{3\phi}}{S_{3\phi}} = \frac{27{,}061.79}{29{,}045.06} = 0.93$$

A similar procedure may be used to calculate the power supplied to a delta load or any combination of parallel-connected three-phase and single-phase loads.

2-2   ACTIVE POWER IN A BALANCED THREE-PHASE SYSTEM

If a three-phase load is *balanced*, the total active power supplied by the three-phase source is equal to three times the active power drawn by one branch.

$$P_{3\phi,\text{bal}} = 3P_{\text{br}} \tag{22-3}$$

Substituting Eq. (22-1) into Eq. (22-3),

$$P_{3\phi,\text{bal}} = 3V_{\text{br}}I_{\text{br}}\cos \sphericalangle \frac{\mathbf{V}_{\text{br}}}{\mathbf{I}_{\text{br}}} \tag{22-4}$$

where V_{br} = rms voltage across one branch
     I_{br} = rms current through the same branch

$\sphericalangle \dfrac{\mathbf{V}_{\text{br}}}{\mathbf{I}_{\text{br}}}$ = angle measured *from* a branch-current phasor *to* the respective branch-voltage phasor, also called the *branch power-factor angle*

Equation (22-4) may be expressed in terms of line current and line voltage by making the following substitutions:

For the balanced delta load

$$V_{br} = V_{line}$$

$$I_{br} = I_{line}/\sqrt{3}$$

For the balanced wye load

$$V_{br} = V_{line}/\sqrt{3}$$

$$I_{br} = I_{line}$$

Substituting the preceding relationships into Eq. (22-4),

$$P_{3\phi,\,bal} = 3V_{line}\,\frac{I_{line}}{\sqrt{3}}\,\cos\overset{\textstyle \mathbf{V}_{br}}{\underset{\textstyle \mathbf{I}_{br}}{\big<}} \qquad\qquad P_{3\phi,\,bal} = \frac{3V_{line}}{\sqrt{3}}\,I_{line}\,\cos\overset{\textstyle \mathbf{V}_{br}}{\underset{\textstyle \mathbf{I}_{br}}{\big<}}$$

Thus, whether we have a wye- or delta-connected load, if the load impedances are balanced,

$$P_{3\phi,\,bal} = \sqrt{3}\,V_{line}\,I_{line}\,\cos\overset{\textstyle \mathbf{V}_{br}}{\underset{\textstyle \mathbf{I}_{br}}{\big<}} \tag{22-5}$$

Note: The power factor of a *balanced* three-phase load is equal to the power factor of the branches and may be determined from

$$pf_{br} = \cos\overset{\textstyle \mathbf{V}_{br}}{\underset{\textstyle \mathbf{I}_{br}}{\big<}}$$

Expressing Eq. (22-5) in terms of power factor,

$$P_{3\phi,\,bal} = \sqrt{3}\,V_{line}\,I_{line}\,pf_{br}$$

Example 22-2 A three-phase, 450-V, 60-Hz source supplies power to a *balanced* delta-connected resistor load. If the line current is 100 A, determine the active power consumed by the load.

Solution
The power factor of a resistor load is 1.0.

$$P_{3\phi} = \sqrt{3}\,V_{line}\,I_{line}\,pf$$

$$P_{3\phi} = \sqrt{3}\,(450)(100)(1) = 77.94\ kW$$

Example 22-3 A three-phase, 600-V, 25-Hz source supplies power to a balanced three-phase motor load. If the line current is 40 A and the power factor of the motor is 0.80, determine the active power drawn by the motor.

Solution

$$P_{3\phi} = \sqrt{3}\, V_{\text{line}} I_{\text{line}}\, \text{pf}$$

$$P_{3\phi} = \sqrt{3}(600)(40)(0.8) = 33.26 \text{ kW}$$

2-3   REACTIVE POWER IN A BALANCED THREE-PHASE SYSTEM

If a three-phase load is balanced, the total reactive power supplied by the three-phase source is equal to three times the reactive power drawn by one branch.

$$Q_{3\phi,\,\text{bal}} = 3Q_{\text{br}} \tag{22-6}$$

From Eq. (22-2),

$$Q_{\text{br}} = V_{\text{br}} I_{\text{br}} \sin \angle\left(\mathbf{V}_{\text{br}}, \mathbf{I}_{\text{br}}\right) \tag{22-7}$$

Substituting Eq. (22-7) into Eq. (22-6),

$$Q_{3\phi,\,\text{bal}} = 3V_{\text{br}} I_{\text{br}} \sin \angle\left(\mathbf{V}_{\text{br}}, \mathbf{I}_{\text{br}}\right) \tag{22-8}$$

Expressing Eq. (22-8) in terms of line voltage and line current,

$$\boxed{Q_{3\phi,\,\text{bal}} = \sqrt{3}\, V_{\text{line}} I_{\text{line}} \sin \angle\left(\mathbf{V}_{\text{br}}, \mathbf{I}_{\text{br}}\right)} \tag{22-9}$$

2-4   APPARENT POWER IN A BALANCED THREE-PHASE SYSTEM

The apparent power delivered to a *balanced* three-phase load is equal to three times the apparent power drawn by one branch.

$$S_{3\phi,\,\text{bal}} = 3 S_{\text{br}} \tag{22-10}$$

where

$$S_{\text{br}} = V_{\text{br}} I_{\text{br}} \tag{22-11}$$

Substituting Eq. (22-11) into Eq. (22-10), and expressing the result in terms of line voltage and line current,

$$S_{3\phi, \text{bal}} = 3 \, V_{\text{br}} I_{\text{br}}$$

$$\boxed{S_{3\phi, \text{bal}} = \sqrt{3} \, V_{\text{line}} I_{\text{line}}} \tag{22-12}$$

Examination of Eqs. (22-5) and (22-9) indicates that they may be represented as the two legs of a right triangle whose hypotenuse is Eq. (22-12). This is shown in Fig. 22-2a for a lagging pf load, and is called a power diagram. From the power diagram,

$$S_{3\phi, \text{bal}} = \sqrt{P_{3\phi, \text{bal}}^2 + Q_{3\phi, \text{bal}}^2} \tag{22-13}$$

$$\theta = \cos^{-1} \frac{P_{3\phi, \text{bal}}}{S_{3\phi, \text{bal}}} = \text{pf angle} \tag{22-14}$$

$$\text{pf} = \cos \theta = \frac{P_{3\phi, \text{bal}}}{S_{3\phi, \text{bal}}} \tag{22-15}$$

$$P_{3\phi, \text{bal}} = S_{3\phi, \text{bal}} \cos \theta = S_{3\phi, \text{bal}} \, \text{pf} \tag{22-16}$$

$$Q_{3\phi, \text{bal}} = S_{3\phi, \text{bal}} \sin \theta \tag{22-17}$$

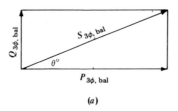

(a)

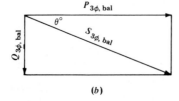

(b)

Figure 22-2 Power diagram: (a) lagging pf; (b) leading pf.

Expressed in complex numbers,

$$S_{3\phi,\,bal} = P_{3\phi,\,bal} + jQ_{3\phi,\,bal} \qquad (22\text{-}18)$$

$$S_{3\phi,\,bal} = S_{3\phi,\,bal}\underline{/\theta^\circ} \qquad (22\text{-}19)$$

Note: If the system vars are leading, the imaginary term in Eq. (22-18) and angle θ in Eq. (22-19) are negative as shown in Fig. 22-2b.

Example 22-4 A 25-hp induction motor is operating at rated load from a three-phase, 450-V, 60-Hz system. The efficiency and power factor of the motor are 87 percent and 90 percent, respectively. Determine (*a*) the active power in kW; (*b*) the apparent power in kVA; (*c*) the reactive power in kvar; (*d*) the line current in amperes.

Solution
A three-phase motor is a balanced load, and the phase angle for the given motor load is $\theta = \cos^{-1} 0.90 = 25.84^\circ$.

(*a*) $P_{3\phi} = \dfrac{hp(746)}{\text{eff.}} = \dfrac{25(746)}{0.87} = 21{,}436.78$ W

$\qquad P_{3\phi} = 21.44\ kW$

(*b*) $pf = \dfrac{P_{3\phi}}{S_{3\phi}}$

$\qquad 0.9 = \dfrac{21.44}{S_{3\phi}}$

$\qquad S_{3\phi,\,bal} = 23.82\ kVA$

(*c*) $Q_{3\phi,\,bal} = S_{3\phi,\,bal}\ \sin\theta$

$\qquad Q_{3\phi,\,bal} = 23.82 \sin 25.84^\circ = 10.38\ kvar$

(*d*) $P_{3\phi,\,bal} = \sqrt{3}\ V_{line} I_{line}\ pf$

$\qquad 21{,}436.8 = \sqrt{3}\,(450)\,I_{line}\,(0.90)$

$\qquad I_{line} = 30.56\ A$

Example 22-5 A 460-V, three-phase, 60-Hz source supplies energy to the following three-phase balanced loads: a 200-hp induction-motor load operating at 94 percent efficiency and 0.88 pf lagging, a 50-kW resistance-heating load, and a combination of miscellaneous loads totaling 40 kW at a 0.70 lagging power factor. Determine (*a*) the total kW supplied; (*b*) the total kvar supplied; (*c*) the total apparent power; (*d*) the system power factor; (*e*) the line current supplied by the source.

Solution

The three-phase system and its one-line counterpart are shown in Fig. 22-3a and b, respectively.

(a) Active power:

$$P_M = \frac{hp(746)}{eff.}$$ $$P_H = 50 \text{ kW}$$ $$P_{misc} = 40 \text{ kW}$$

$$P_M = \frac{200(746)}{0.94(1000)}$$

$$P_M = 158.72 \text{ kW}$$

$$P_T = 158.72 + 50 + 40 = 248.72 \text{ kW}$$

(b) Reactive power:

$$\theta_{mot} = \cos^{-1} 0.88$$ $$\theta_{heat} = 0°$$ $$\theta_{misc} = \cos^{-1} 0.70$$

$$\theta_{mot} = 28.36°$$ $$\theta_{misc} = 45.57°$$

From the power diagrams in Fig. 22-3c,

$$\tan\theta = \frac{Q_{mot}}{P_{mot}}$$ $$Q_{heat} = 0 \text{ kvar}$$ $$\tan\theta = \frac{Q_{misc}}{P_{misc}}$$

$$\tan 28.36° = \frac{Q_{mot}}{158.72}$$ $$\tan 45.57° = \frac{Q_{misc}}{40}$$

$$Q_{mot} = 85.68 \text{ kvar}$$ $$Q_{misc} = 40.81 \text{ kvar}$$

$$Q_T = Q_{mot} + Q_{heat} + Q_{misc}$$

$$Q_T = 85.68 + 0 + 40.81 = 126.49 \text{ kvar}$$

(c) Apparent power:

$$S_{3\phi, \text{bal}} = P_{3\phi, \text{bal}} + jQ_{3\phi, \text{bal}}$$

$$S_{3\phi, \text{bal}} = 248.72 + j126.49 = 279\underline{/27.0°} \text{ kVA}$$

(d) Power factor:

$$pf = \cos 26.96° = 0.89 \text{ lagging}$$

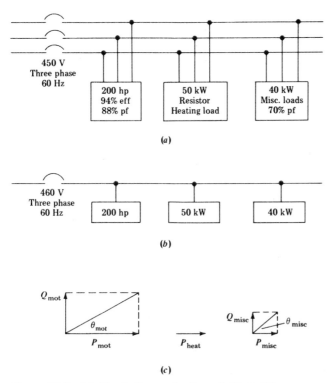

Figure 22-3 (a) Circuit diagram for Example 22-5; (b) one-line diagram; (c) power diagrams.

(e) Line current:

$$S_{3\phi} = \sqrt{3}\, V_{\text{line}} I_{\text{line}}$$

$$279{,}040 = \sqrt{3}\,(460)\, I_{\text{line}}$$

$$I_{\text{line}} = 350.2\ \text{A}$$

Example 22-6 A balanced wye-connected load and a balanced delta-connected load are supplied by a three-phase, 480-V 50-Hz generator, as shown in Fig. 22-4. The branch impedances of the wye and delta loads are $10\underline{/30°}\ \Omega$ and $20\underline{/-50°}\ \Omega$, respectively. Determine (a) the active and reactive power drawn by each three-phase load; (b) the system kW, kvar, kVA, and power factor.

Solution
(a) Determine the phasor voltage and phasor current for *any* one branch of each three-phase load, and then substitute into the power equation for balanced three-phase loads.

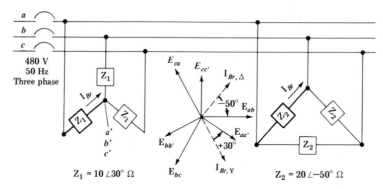

Figure 22-4 Circuit and phasor diagram for Example 22-6.

The branches selected for this example are drawn with heavy lines in Fig. 22-4. The branch currents are determined by using the standardized phasor diagram and Ohm's law. Thus,

<table>
<tr>
<td>

For the selected branch of the wye bank

$$\mathbf{I}_{br, Y} = \frac{\mathbf{E}_{aa'}}{\mathbf{Z}_1} = \frac{480/\sqrt{3}\,\underline{/-30°}}{10/30°}$$

$$\mathbf{I}_{br, Y} = 27.71\underline{/-60°}\ A$$

$$\mathbf{V}_{br, Y} = \mathbf{E}_{aa'} = 277.13\underline{/-30°}\ V$$

</td>
<td>

For the selected branch of the delta bank

$$\mathbf{I}_{br, \Delta} = \frac{\mathbf{E}_{ab}}{\mathbf{Z}_2} = \frac{480/0°}{20/-50°}$$

$$\mathbf{I}_{br, \Delta} = 24\underline{/50°}\ A$$

$$\mathbf{V}_{br, \Delta} = \mathbf{E}_{ab} = 480/0°\ V$$

</td>
</tr>
</table>

$$P_{3\phi,\,bal} = 3\,V_{br}I_{br}\cos\!\underset{\mathbf{I}_{br}}{\overset{\mathbf{V}_{br}}{\diagup}}$$

Using the phasor diagram, measure the angle *from* the branch-current phasor *to* the respective branch-voltage phasor. The direction is shown with heavy arrows in the phasor diagram.

$$P_{Y,\,bal} = 3(277.13)(27.71)\cos(+30°) \qquad P_{\Delta,\,bal} = 3(480)(24)\cos(-50°)$$

$$P_{Y,\,bal} = 19{,}951.2\ W \qquad\qquad P_{\Delta,\,bal} = 22{,}214.7\ W$$

$$Q_{3\phi,\,bal} = 3\,V_{br}I_{br}\sin\!\underset{\mathbf{I}_{br}}{\overset{\mathbf{V}_{br}}{\diagup}}$$

$$Q_{Y, bal} = 3(277.13)(27.71)\sin(+30°)$$

$$Q_{Y, bal} = 11,518.9 \text{ var}$$

$$Q_{\Delta, bal} = 3(480)(24)\sin(-50°)$$

$$Q_{\Delta, bal} = -26,474.5 \text{ var}$$

(b) $P_T = P_Y + P_\Delta = 19,951.2 + 22,214.7 = 42,165.9 = 42.2 \text{ kW}$

$$Q_T = Q_Y + Q_\Delta = 11,518.9 - 26,474.5 = -14,955.6 \text{ var} = -15.0 \text{ kvar}$$

The negative sign indicates leading reactive power.

$$S_T = P_T + jQ_T$$

$$S_T = 42,165.9 - j14,955.6 = 44,739.6 \underline{/-19.53°} \text{ VA}$$

$$S_T = 44.74 \text{ kVA}$$

$$pf = \cos(-19.53°) = 0.94 \text{ leading}$$

22-5   MEASUREMENT OF ACTIVE POWER IN THE THREE-PHASE THREE-WIRE SYSTEM (BALANCED OR UNBALANCED)

The active power drawn by any three-phase three-wire system can be measured with two wattmeters as shown in Fig. 22-5a or by a polyphase wattmeter whose connections are shown in Fig. 22-5b. If the polyphase wattmeter is of the electromechanical type, it will have two wattmeter elements in a single housing, and the common pointer will indicate the total three-phase power. If the polyphase wattmeter is of the digital type, the digital readout will be the total three-phase power.

In those applications where the line current is greater than the current rating of the wattmeter, current transformers (called C-Ts) are used to reduce the current to the meter. This is shown in Fig. 22-5c. The short heavy line represents the primary, and the double loop represents the secondary. When C-Ts are used, each wattmeter reading must be multiplied by its associated C-T ratio.

In order to obtain correct readings, the two ± terminals of each wattmeter must have the same instantaneous polarity. This is accomplished by connecting the ± current terminal and the associated ± potential terminal to the same line, as shown in Fig. 22-5. The ± terminals are called the polarity terminals.

Each wattmeter in Fig. 22-5a indicates the product of the rms line voltage across its potential coil, the rms line current through its current terminals, and the cosine of the angle between the respective line-voltage and line-current

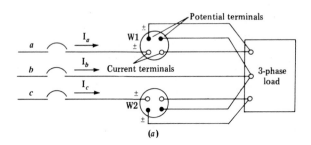

(a)

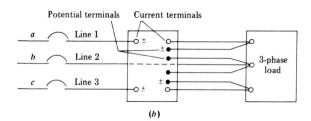

(b)

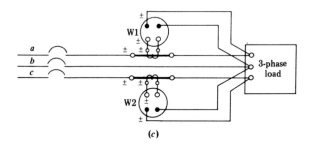

(c)

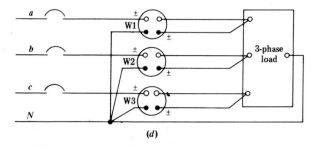

(d)

Figure 22-5 Measurement of active power in the three-phase system (balanced or unbalanced): (a) two-wattmeter method; (b) polyphase wattmeter method; (c) two wattmeters with C-Ts; (d) three-wattmeter method.

phasors. *This angle is not the power-factor angle. The power-factor angle is the angle between a branch voltage and its associated branch current.*

Wattmeter WM1 will read a value equal to

$$WM1 = E_{ab}I_a \cos\angle{\,}^{-E_{ab}}_{\quad I_a} \tag{22-20}$$

Wattmeter WM2 will read a value equal to

$$WM2 = E_{cb}I_c \cos\angle{\,}^{-E_{cb}}_{\quad I_c} \tag{22-21}$$

The total active power delivered to the three-phase load, whether balanced or unbalanced, is equal to the algebraic sum of the two wattmeter readings. Thus,

$$\boxed{P_{3\phi} = WM1 + WM2} \tag{22-22}$$

If the pointer of one wattmeter deflects in the negative direction, open the breaker or switch, interchange the two connections at the current terminals of the reversed wattmeter, and then reclose the circuit. When entering the wattmeter readings in Eq. (22-22), a minus sign must be associated with the data taken from the reversed wattmeter.

Substituting Eqs. (22-20) and (22-21) into Eq. (22-22),

$$P_{3\phi} = E_{ab}I_a \cos\angle{\,}^{E_{ab}}_{\quad I_a} + E_{cb}I_c \cos\angle{\,}^{E_{cb}}_{\quad I_c} \tag{22-23}$$

where E_{ab} = rms voltage between lines a and b
E_{cb} = rms voltage between lines c and b
I_a = rms current in line a
I_c = rms current in line c

$\angle{\,}^{E_{ab}}_{\ I_a}$ = angle measured from $\mathbf{I}_a$ to $\mathbf{E}_{ab}$

$\angle{\,}^{E_{cb}}_{\ I_c}$ = angle measured from $\mathbf{I}_c$ to $\mathbf{E}_{cb}$

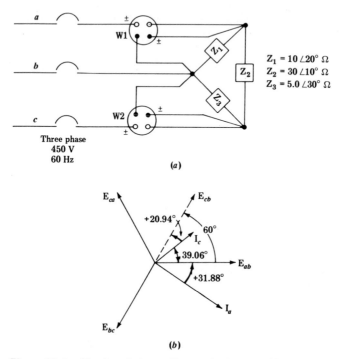

$$Z_1 = 10\,\angle 20^\circ\ \Omega$$
$$Z_2 = 30\,\angle 10^\circ\ \Omega$$
$$Z_3 = 5.0\,\angle 30^\circ\ \Omega$$

Three phase
450 V
60 Hz

(a)

(b)

Figure 22-6 Circuit and phasor diagram for Example 22-7.

Note 1: Equations (22-22) and (22-23) cannot be used with a three-phase system that includes a neutral line; such a system is called a three-phase, four-wire system, and requires three wattmeters as shown in Fig. 22-5d.

Note 2: If the wattmeters are to be corrected for instrument losses, the numerical sum of the instrument losses (taken without regard to sign) should be subtracted from the algebraic sum of the wattmeter readings.

Example 22-7 Determine the two wattmeter readings and the total three-phase power delivered to the load in Fig. 22-6a.

Solution
The standard phasor diagram for the three-phase driving voltages is shown in Fig. 22-6b.

$$\mathbf{I}_a = \frac{\mathbf{E}_{ab}}{\mathbf{Z}_1} + \frac{\mathbf{E}_{ac}}{\mathbf{Z}_2} = \frac{450\underline{/0^\circ}}{10\underline{/20^\circ}} + \frac{-(450\underline{/120^\circ})}{30\underline{/10^\circ}} = 55.84\underline{/-31.88^\circ}\ \text{A}$$

$$\mathbf{I}_c = \frac{\mathbf{E}_{ca}}{\mathbf{Z}_2} + \frac{\mathbf{E}_{cb}}{\mathbf{Z}_3} = \frac{450\underline{/120^\circ}}{30\underline{/10^\circ}} + \frac{-(450\underline{/-120^\circ})}{5\underline{/30^\circ}} = 93.78\underline{/39.06^\circ}\ \text{A}$$

$$\text{WM1} = E_{ab}I_a \cos \left< \begin{array}{l} \mathbf{E}_{ab} \\ \\ \mathbf{I}_a \end{array} \right. = (450)(55.84)\cos 31.88°$$

$$\text{WM1} = 21.3 \text{ kW}$$

$$\text{WM2} = E_{cb}I_c \cos \left< \begin{array}{l} \mathbf{E}_{cb} \\ \\ \mathbf{I}_c \end{array} \right. = (450)(93.78)\cos 20.94°$$

$$\text{WM2} = 39.4 \text{ kW}$$

$$P_{3\phi} = \text{WM1} + \text{WM2} = 21.3 + 39.4 = 60.7 \text{ kW}$$

Derivation of Equation (22-23)

Figure 22-7a shows an unbalanced load connected to a three-phase source. The total three-phase power delivered at any instant of time is

$$P_{3\phi,\text{ inst}} = e_{ak}i_a + e_{bk}i_b + e_{ck}i_c \tag{22-24}$$

The two line voltages e_{ab} and e_{cb}, expressed in terms of the phase voltages, are

$$e_{ab} = e_{ak} + e_{kb} \tag{22-25}$$

$$e_{cb} = e_{ck} + e_{kb} \tag{22-26}$$

Note: Since the load is unbalanced, k is not at the same potential as a', b', c'.

Rearranging Eqs. (22-25) and (22-26),

$$e_{ak} = e_{ab} - e_{kb} = e_{ab} + e_{bk} \tag{22-27}$$

$$e_{ck} = e_{cb} - e_{kb} = e_{cb} + e_{bk} \tag{22-28}$$

Substituting Eqs. (22-27) and (22-28) into Eq. (22-24) and expanding,

$$P_{3\phi,\text{ inst}} = (e_{ab} + e_{bk})i_a + e_{bk}i_b + (e_{cb} + e_{bk})i_c \tag{22-29}$$

$$P_{3\phi,\text{ inst}} = e_{ab}i_a + e_{cb}i_c + e_{bk}(i_a + i_b + i_c)$$

From Kirchhoff's current law applied to junction k in Fig. 22-7a,

$$i_a + i_b + i_c = 0$$

Hence,

$$P_{3\phi,\text{ inst}} = e_{ab}i_a + e_{cb}i_c \tag{22-30}$$

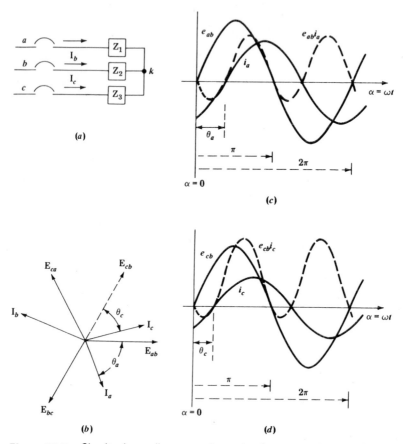

Figure 22-7 Circuit, phasor diagram, and associated current, voltage, and power waves for an unbalanced load connected to a three-phase source.

A phasor diagram showing representative line-voltage phasors and line-current phasors for the general case of an unbalanced load is shown in Fig. 22-7b, and the voltage, current, and power waves corresponding to e_{ab}, i_a, e_{cb}, and i_c are shown in Fig. 22-7c and d.

The average power delivered over one cycle may be determined by summing all the instantaneous values of power over one cycle and dividing by 2π. Thus, from Eq. (22-30),

$$P_{av} = \frac{1}{2\pi}\int_{\alpha=0}^{\alpha=2\pi} (e_{ab}i_a + e_{cb}i_c)d\alpha$$

$$P_{av} = \frac{1}{2\pi}\int_{\alpha=0}^{\alpha=2\pi} e_{ab}i_a\, d\alpha + \frac{1}{2\pi}\int_{\alpha=0}^{\alpha=2\pi} e_{cb}i_c\, d\alpha$$

Defining,

$$P_A = \frac{1}{2\pi} \int_0^{2\pi} e_{ab} i_a \, d\alpha \tag{22-31}$$

$$P_C = \frac{1}{2\pi} \int_0^{2\pi} e_{cb} i_c \, d\alpha \tag{22-32}$$

$$P_{av} = P_A + P_C \tag{22-33}$$

From Fig. 22-7c,

$$e_{ab} = E_{ab, \, max} \sin \alpha$$

$$i_a = I_{a, \, max} \sin(\alpha - \theta_a)$$

$$\theta_a = \text{phase angle between } e_{ab} \text{ and } i_a$$

Substituting into Eq. (22-31),

$$P_A = \frac{1}{2\pi} \int_0^{2\pi} (E_{ab, \, max} \sin \alpha)[I_{a, \, max} \sin(\alpha - \theta_a)] d\alpha$$

Simplifying,

$$P_A = \frac{E_{ab, \, max} I_{a, \, max}}{2\pi} \int_0^{2\pi} \sin \alpha(\sin \alpha - \theta_a) d\alpha$$

From trigonometry,

$$\sin(\alpha - \theta_a) = \sin \alpha \cos \theta_a - \cos \alpha \sin \theta_a$$

Thus

$$P_A = \frac{E_{ab, \, max} I_{a, \, max}}{2\pi} \int_0^{2\pi} (\sin^2 \alpha \cos \theta_a - \sin \alpha \cos \alpha \sin \theta_a) d\alpha$$

$$P_A = \frac{E_{ab, \, max} I_{a, \, max} \cos \theta_a}{2\pi} \int_0^{2\pi} \sin^2 \alpha \, d\alpha$$

$$+ \frac{E_{ab, \, max} I_{a, \, max} \sin \theta_a}{2\pi} \int_0^{2\pi} (-\sin \alpha \cos \alpha) d\alpha$$

Integrating,

$$P_A = \frac{E_{ab, \, max} I_{a, \, max} \cos \theta_a}{2\pi} [\tfrac{1}{2}(\alpha - \sin \alpha \cos \alpha)]_0^{2\pi}$$

$$- \frac{E_{ab, \, max} I_{a, \, max} \sin \theta_a}{2\pi} (\tfrac{1}{2} \sin^2 \alpha)_0^{2\pi}$$

Substituting the limits and evaluating,

$$P_A = \frac{E_{ab,\,max} I_{a,\,max} \cos \theta_a}{2\pi} \; [\tfrac{1}{2}(2\pi)] - \frac{E_{ab,\,max} I_{a,\,max} \sin \theta_a}{2\pi}(0)$$

$$P_A = \frac{E_{ab,\,max} I_{a,\,max} \cos \theta_a}{2} = \frac{E_{ab,\,max}}{\sqrt{2}} \frac{I_{a,\,max}}{\sqrt{2}} \cos \theta_a$$

$$P_A = E_{ab,\,rms} I_{a,\,rms} \cos \theta_a$$

Thus,

$$P_A = E_{ab} I_a \cos \sphericalangle \begin{matrix} \mathbf{E}_{ab} \\ \mathbf{I}_a \end{matrix}$$

Using Fig. 22-7d and a similar derivation,

$$P_C = E_{cb} I_c \cos \sphericalangle \begin{matrix} \mathbf{E}_{cb} \\ \mathbf{I}_c \end{matrix}$$

Thus,

$$P_{av\,3\phi} = E_{ab} I_a \cos \sphericalangle \begin{matrix} \mathbf{E}_{ab} \\ \mathbf{I}_a \end{matrix} + E_{cb} I_c \cos \sphericalangle \begin{matrix} \mathbf{E}_{cb} \\ \mathbf{I}_c \end{matrix}$$

22-6 DETERMINING PHASE ANGLES OF LINE CURRENTS IN A BALANCED THREE-PHASE SYSTEM OF UNKNOWN CONFIGURATION

If a balanced wye load and a balanced delta load are equivalent in active and reactive power, the branch power-factor angles of both are the same. Thus, if the circuit configuration (wye or delta) of a balanced three-phase system is not known, or the balanced system contains a combination of wye and delta loads, the phase angles of the system line currents may be determined by assuming the system to be an equivalent wye load whose active power and power factor are that of the actual system.

In the equivalent wye load, the magnitude and phase angle of each line current are identical to the magnitude and phase angle of the corresponding branch current. Thus, if the branch-current phasors are included in the standard phasor diagram of driving voltages, with the branch currents displaced from the respective branch voltages by the power-factor angle, the

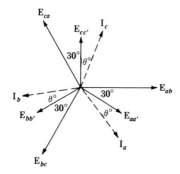

Figure 22-8 Phasor diagram for determining phase angles of line currents in a balanced three-phase system of unknown configuration by assuming an equivalent wye load.

phase angles of the line currents may be determined. This is shown in Fig. 22-8, where the branch currents are shown lagging their corresponding branch voltages by a power-factor angle of $\theta°$. The three line currents and their associated phase angles, as determined from the phasor diagram, are

$$\mathbf{I}_a = I_a \underline{/-30° - \theta°}$$

$$\mathbf{I}_b = I_b \underline{/-150° - \theta°}$$

$$\mathbf{I}_c = I_c \underline{/90° - \theta°}$$

Example 22-8 A 220-V, 50-hp, 850-rpm, three-phase induction motor operating at $\frac{1}{2}$ rated hp, rated voltage, and rated frequency has an efficiency and power factor of 87 percent and 71 percent respectively. Determine (*a*) the power input to the motor; (*b*) the line current; (*c*) the readings of each of two wattmeters placed in lines *a* and *c*, respectively.

Solution

The circuit is shown in Fig. 22-9*a*.

(*a*) The input power to the motor is

$$P_{in} = \frac{hp × 746}{eff.} = \frac{(50/2)746}{0.87} = 21{,}437 \text{ W}$$

(*b*) The line current drawn by the motor is

$$P_{3\phi,\,bal} = \sqrt{3} \, V_{line} I_{line} \, pf$$

$$21{,}437 = \sqrt{3} \, (220) \, I_{line} \, (0.71)$$

$$I_{line} = 79.2 \text{ A}$$

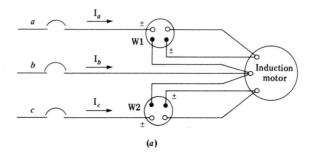

(a)

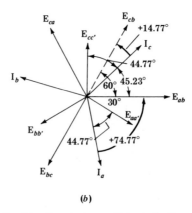

(b)

Figure 22-9   Circuit and phasor diagram for Example 22-8.

(c) The power-factor angle for a balanced three-phase load is the cosine of the angle between the branch voltage and its respective branch current. Thus,

$$\cos \angle\begin{array}{c} \mathbf{V}_{br} \\ \theta \\ \mathbf{I}_{br} \end{array} = pf$$

$$\cos \angle\begin{array}{c} \mathbf{V}_{br} \\ \theta \\ \mathbf{I}_{br} \end{array} = 0.71$$

$$\theta = 44.77°$$

The standard phasor diagram for the line and branch voltages is shown in Fig. 22-9b. Assuming an *equivalent wye load*, the phasor line currents are equal to their corresponding branch currents, and lag their respective branch voltages by 44.77°. Note that induction motors draw a lagging current. Substituting the rms line voltage, rms line current, and the appropriate phase angles obtained from the phasor diagram,

$$WM1 = E_{ab}I_a \cos \overset{\displaystyle E_{ab}}{\underset{\displaystyle I_a}{<}}$$

$$WM1 = 220(79.24)\cos 74.77° = 4580$$

$$WM2 = E_{cb}I_c \cos \overset{\displaystyle E_{cb}}{\underset{\displaystyle I_c}{<}}$$

$$WM2 = 220(79.24)\cos 14.77° = 16,857$$

Note: WM1 + WM2 = 4580 + 16,857 = 21,437 W, which checks with part *a*.

22-7 SIMPLIFIED PROCEDURE FOR CALCULATING WATTMETER READINGS IN A BALANCED THREE-PHASE SYSTEM

Figure 22-10*a* shows a three-phase source supplying a balanced three-phase lagging power-factor load, and Fig. 22-10*b* shows the phasor diagram for the wye equivalent. *Wattmeter WM1 is in line "a," wattmeter WM2 is in line "c," and the phase sequence of the driving voltage is abc.* The branch currents are shown lagging their respective branch voltages by a power-factor angle of $\theta°$. The wattmeter indications are

$$WM1 = E_{ab}I_a \cos \overset{\displaystyle E_{ab}}{\underset{\displaystyle I_a}{<}} \quad = E_{ab}I_a \cos(\theta° + 30°)$$

$$WM2 = E_{cb}I_c \cos \overset{\displaystyle E_{cb}}{\underset{\displaystyle I_c}{<}} \quad = E_{cb}I_c \cos(\theta° - 30°)$$

Simplifying

$$WM1 = V_{line}I_{line} \cos(\theta° + 30°) \qquad (22\text{-}34)$$

$$WM2 = V_{line}I_{line} \cos(\theta° - 30°) \qquad (22\text{-}35)$$

where V_{line} = rms line voltage, V
   I_{line} = rms line current, A
   θ = pf angle, degrees

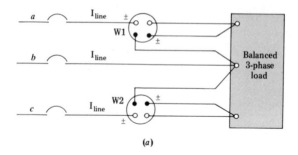

(a)

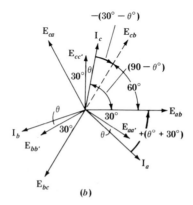

(b)

Figure 22-10 Circuit and phasor diagram for illustrating a simplified procedure for calculating wattmeter readings in a balanced three-phase system.

Note: If the power factor is lagging, θ is positive. If the power factor is leading, θ is negative. For unity power factor, $\theta = 0$ and both wattmeters will have equal deflections. For power factors below unity and above 0.5, leading or lagging, $|\theta| < 60°$, both wattmeters will read in the positive direction but one will have a greater indication than the other.

For a power factor of 0.5, leading or lagging, $|\theta| = 60°$, one wattmeter will read zero and the other will indicate the total power.

For power factors below 0.5, leading or lagging, $|\theta| > 60°$, one wattmeter will have a negative deflection and the other a positive deflection. To obtain total power in this case, the current connections or potential connections (not both) of the reversed reading instrument should be reversed. A minus sign should be recorded with the reading of the corrected instrument to indicate subtraction.

If the two-wattmeter method is to be corrected for instrument losses, the numerical sum of the instrument losses (taken without regard to sign) should be subtracted from the algebraic sum of the wattmeter readings.

Example 22-9 The line current to a lightly loaded three-phase motor, operating from a 450-V three-phase system, is 24 A. The power factor of the motor is 0.47 lagging. If the two-wattmeter method is used to measure the three-phase power supplied to the motor, what would each wattmeter read?

Solution
Since a three-phase motor is a balanced load, Eqs. (22-34) and (22-35) may be used.

$$\theta = \cos^{-1} 0.47 = 61.966°$$

$$WM1 = V_{line} I_{line} \cos(\theta° + 30°)$$

$$WM1 = (450)(24)\cos(61.966° + 30°) = -371$$

$$WM2 = V_{line} I_{line} \cos(\theta° - 30°)$$

$$WM2 = (450)(24)\cos(61.966° - 30°) = 9162$$

$$P_{3\phi} = WM1 + WM2 = -371 + 9162 = 8791 \text{ W}$$

22-8 TWO-WATTMETER METHOD FOR DETERMINING THE POWER FACTOR OF A BALANCED THREE-PHASE SYSTEM

The power-factor angle of a balanced three-phase system may be determined from the two-wattmeter readings by use of the following formula:

$$\tan \theta = \sqrt{3} \, \frac{WM2 - WM1}{WM2 + WM1} \qquad (22\text{-}36)$$

where θ is the pf angle
  WM1 is in line *a*
  WM2 is in line *c*
  The phase sequence is *abc*

Equation (22-36) is derived from Eqs. (22-34) and (22-35) as follows:

$$WM1 + WM2 = V_{line} I_{line}[\cos(\theta° + 30°) + \cos(\theta° - 30°)] \qquad (22\text{-}37)$$

$$WM1 - WM2 = V_{line} I_{line}[\cos(\theta° + 30°) - \cos(\theta° - 30°)] \qquad (22\text{-}38)$$

Expanding the trigonometric functions,

$$\cos(\theta + 30) = \cos \theta \cos 30 - \sin \theta \sin 30 \qquad (22\text{-}39)$$

$$\cos(\theta - 30) = \cos \theta \cos 30 + \sin \theta \sin 30 \qquad (22\text{-}40)$$

Substituting Eqs. (22-39) and (22-40) into Eqs. (22-37) and (22-38), and simplifying,

$$WM1 + WM2 = V_{line} I_{line}(2 \cos \theta \cos 30) \qquad (22\text{-}41)$$

$$WM1 - WM2 = V_{line} I_{line}(-2 \sin \theta \sin 30) \qquad (22\text{-}42)$$

Dividing Eq. (22-41) by Eq. (22-42) and simplifying,

$$\frac{WM1 + WM2}{WM1 - WM2} = \frac{2 \cos \theta \cos 30}{-2 \sin \theta \sin 30} = \frac{-\sqrt{3} \cos \theta}{\sin \theta} = \frac{-\sqrt{3}}{\tan \theta}$$

$$\tan \theta = \sqrt{3} \frac{WM2 - WM1}{WM2 + WM1}$$

Example 22-10 If the wattmeters in the balanced three-phase system of Fig. 22-10a read 60 kW and 40 kW for WM1 and WM2, respectively, determine (a) the system active power; (b) the system power factor; (c) the system apparent power; (d) the system reactive power.

Solution

(a) $P_{3\phi} = WM1 + WM2 = 60 + 40 = 100 \text{ kW}$

(b) $\tan \theta = \sqrt{3} \dfrac{WM2 - WM1}{WM2 + WM1} = \sqrt{3} \dfrac{40 - 60}{40 + 60} = -0.3464$

$\theta = -19.10°$

The negative angle indicates that the power factor is leading.

$pf = \cos -19.1° = 0.945 \text{ leading}$

(c) $pf = \dfrac{P_{3\phi, \text{bal}}}{S_{3\phi, \text{bal}}}$

$0.945 = \dfrac{100}{S_{3\phi, \text{bal}}}$

$S_{3\phi, \text{bal}} = 105.8 \text{ kVA}$

(d) $S = \sqrt{P^2 + Q^2}$

$105.8 = \sqrt{(100)^2 + (Q)^2}$

$Q = 34.6 \text{ leading kvar}$

2-9 **MEASUREMENT OF REACTIVE POWER IN THE THREE-PHASE THREE-WIRE SYSTEM (BALANCED OR UNBALANCED)**

The reactive power drawn by any three-phase three-wire system can be measured with two varmeters or with a polyphase varmeter. Varmeters with internal phase shifters are connected in the same manner as the wattmeters in Fig. 22-5a, b, c, and d.

As developed in Chap. 13 and expressed in Eq. (13-39), a varmeter will indicate the product of the voltage across its potential terminals, the current to its current terminals, and the sine of the angle measured *from* the current phasor *to* the voltage phasor. That is,

$$Q = V_{pc} I_{cc} \sin \angle \begin{smallmatrix} \mathbf{V}_{pc} \\ \\ \mathbf{I}_{cc} \end{smallmatrix}$$

where V_{pc} = voltage across potential coil
I_{cc} = current in current coil

Referring to Fig. 22-5a and assuming varmeters are used instead of wattmeters,

Varmeter 1 will read a value equal to

$$VAR1 = E_{ab} I_a \sin \angle \begin{smallmatrix} \mathbf{E}_{ab} \\ \\ \mathbf{I}_a \end{smallmatrix} \qquad (22\text{-}43)$$

Varmeter 2 will read a value equal to

$$VAR2 = E_{cb} I_c \sin \angle \begin{smallmatrix} \mathbf{E}_{cb} \\ \\ \mathbf{I}_c \end{smallmatrix} \qquad (22\text{-}44)$$

The total reactive power delivered to the three-phase load, whether balanced or unbalanced, is equal to the algebraic sum of the two varmeter readings. Thus,

$$Q_{3\phi} = VAR1 + VAR2 \qquad (22\text{-}45)$$

If the pointer of one varmeter deflects in the negative direction, open the breaker or switch, interchange the two connections at the current terminals of the reversed varmeter, and then reclose the circuit. When the varmeter readings are entered in Eq. (22-45), a minus sign must be associated with the reversed varmeter.

Substituting Eqs. (22-43) and (22-44) into Eq. (22-45),

$$Q_{3\phi} = E_{ab} I_a \sin\sphericalangle \begin{smallmatrix} \mathbf{E}_{ab} \\ \\ \mathbf{I}_a \end{smallmatrix} + E_{cb} I_c \sin\sphericalangle \begin{smallmatrix} \mathbf{E}_{cb} \\ \\ \mathbf{I}_c \end{smallmatrix} \qquad (22\text{-}46)$$

22-10   POWER-FACTOR IMPROVEMENT IN A BALANCED THREE-PHASE SYSTEM

The power factor of a balanced three-phase system may be improved by the addition of wye-connected or delta-connected capacitor banks as shown in Fig. 22-11a. The theory of power-factor improvement that was developed in Sec. 13-7 for single-phase systems is applicable to the three-phase system.

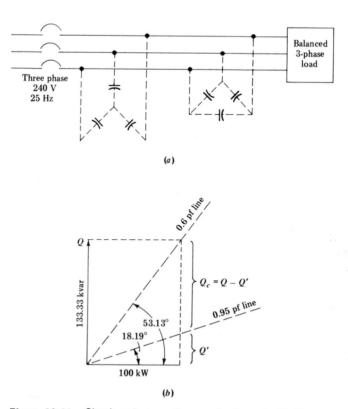

(a)

(b)

Figure 22-11   Circuit and power diagrams for Example 22-11.

Example 22-11 A 240-V, 25-Hz, three-phase system supplies 100 kW at 0.6 pf lagging to the balanced load shown in Fig. 22-11a. Determine (a) the capacitance of each leg of a wye-connected capacitor bank required to adjust the system power factor to 0.95 lagging; (b) repeat part a for a delta-connected capacitor bank.

Solution
(a) The original power-factor angle is

$$\theta = \cos^{-1} 0.6 = 53.13°$$

The desired power-factor angle is

$$\theta' = \cos^{-1} 0.95 = 18.19°$$

The reactive power of the uncorrected system may be determined from the power diagram shown in Fig. 22-11b.

$$\tan 53.13° = \frac{Q_{3\phi, bal}}{100}$$

$$Q_{3\phi, bal} = 133.33 \text{ kvar}$$

When corrected to 0.95 power factor, the kvars required from the source is

$$\tan 18.19° = \frac{Q'_{3\phi, bal}}{100}$$

$$Q'_{3\phi, bal} = 32.86 \text{ kvar}$$

Hence a three-phase capacitor bank required to change the reactive-power demand from 133.33 kvar to 32.86 kvar must have a rating of

$$Q_{C, 3\phi} = Q_{3\phi} - Q'_{3\phi} = 133.33 - 32.86 = 100.47 \text{ kvar}$$

The kvar requirement for each branch of a *wye-connected* capacitor bank is

$$Q_C = \frac{Q_{C, 3\phi}}{3} = \frac{100.47}{3} = 33.49 \text{ kvar}$$

The capacitance of each branch will be

$$Q_C = \frac{V_C^2}{X_C}$$

$$33,490 = \frac{(240/\sqrt{3})^2}{X_C}$$

$$X_C = 0.57 \ \Omega$$

$$X_C = \frac{1}{2\pi f C}$$

$$0.57 = \frac{1}{2\pi(25)C}$$

$$C = 11.12 \times 10^{-3} \text{ F}$$

(b) If a delta-connected capacitor bank is desired, the voltage across each branch will be 240 V, and the capacitance of each branch will be

$$Q_C = \frac{V_C^2}{X_C}$$

$$33,490 = \frac{(240)^2}{X_C}$$

$$X_C = 1.72 \ \Omega$$

$$X_C = \frac{1}{2\pi f C}$$

$$1.72 = \frac{1}{2\pi(25)C}$$

$$C = 3.70 \times 10^{-3} \text{ F}$$

SUMMARY OF FORMULAS

General Case, Balanced or Unbalanced

$$P_{br} = V_{br} I_{br} \cos \angle \begin{array}{c} \mathbf{V}_{br} \\ \mathbf{I}_{br} \end{array} \qquad Q_{br} = V_{br} I_{br} \sin \angle \begin{array}{c} \mathbf{V}_{br} \\ \mathbf{I}_{br} \end{array}$$

$$P_{3\phi} = \sum P_{br} \qquad Q_{3\phi} = \sum Q_{br}$$

$$P_{3\phi} = E_{ab} I_a \cos \angle \begin{array}{c} \mathbf{E}_{ab} \\ \mathbf{I}_a \end{array} + E_{cb} I_c \cos \angle \begin{array}{c} \mathbf{E}_{cb} \\ \mathbf{I}_c \end{array}$$

$$Q_{3\phi} = E_{ab} I_a \sin \angle \begin{array}{c} \mathbf{E}_{ab} \\ \mathbf{I}_a \end{array} + E_{cb} I_c \sin \angle \begin{array}{c} \mathbf{E}_{cb} \\ \mathbf{I}_c \end{array}$$

$$S_{3\phi} = \sqrt{P_{3\phi}^2 + Q_{3\phi}^2} \qquad \text{pf}_{3\phi} = \frac{P_{3\phi}}{S_{3\phi}}$$

Balanced Only

$$P_{3\phi,\,\text{bal}} = 3V_{\text{br}}I_{\text{br}}\cos\!\!\underset{\mathbf{I}_{\text{br}}}{\overset{\mathbf{V}_{\text{br}}}{\diagup}} = \sqrt{3}\,V_{\text{line}}I_{\text{line}}\cos\!\!\underset{\mathbf{I}_{\text{br}}}{\overset{\mathbf{V}_{\text{br}}}{\diagup}}$$

$$Q_{3\phi,\,\text{bal}} = 3V_{\text{br}}I_{\text{br}}\sin\!\!\underset{\mathbf{I}_{\text{br}}}{\overset{\mathbf{V}_{\text{br}}}{\diagup}} = \sqrt{3}\,V_{\text{line}}I_{\text{line}}\sin\!\!\underset{\mathbf{I}_{\text{br}}}{\overset{\mathbf{V}_{\text{br}}}{\diagup}}$$

$$S_{3\phi,\,\text{bal}} = 3V_{\text{br}}I_{\text{br}} = \sqrt{3}\,V_{\text{line}}I_{\text{line}} = \sqrt{P_{3\phi,\,\text{bal}}^2 + Q_{3\phi,\,\text{bal}}^2}$$

$$S_{3\phi,\,\text{bal}} = P_{3\phi,\,\text{bal}} + jQ_{3\phi,\,\text{bal}} = S_{3\phi,\,\text{bal}}\underline{/\theta_{\text{br}}^\circ}$$

$$\text{pf}_{\text{bal}} = \frac{P_{3\phi,\,\text{bal}}}{S_{3\phi,\,\text{bal}}} = \frac{P_{\text{br}}}{S_{\text{br}}} = \cos\!\!\underset{\mathbf{I}_{\text{br}}}{\overset{\mathbf{V}_{\text{br}}}{\diagup}}$$

$$\text{WM1} = V_{\text{line}}I_{\text{line}}\cos(\theta_{\text{br}}^\circ + 30^\circ) \qquad P_{3\phi} = \text{WM1} + \text{WM2}$$

$$\text{WM2} = V_{\text{line}}I_{\text{line}}\cos(\theta_{\text{br}}^\circ - 30^\circ) \qquad \tan\theta_{\text{br}} = \sqrt{3}\,\frac{\text{WM2} - \text{WM1}}{\text{WM2} + \text{WM1}}$$

PROBLEMS

22-1 A balanced wye-connected load draws a line current of 60 A from a 450-V, 60-Hz, three-phase system. The power factor of the load is 0.70 lagging. Determine the active power, apparent power, and reactive power drawn by the load.

22-2 A three-phase motor draws a line current of 30 A when supplied from a 450-V, three-phase, 25-Hz source. The motor efficiency and power factor are 90 percent and 75 percent, respectively. Determine (a) the active power drawn by the motor; (b) the shaft horsepower output; (c) the reactive power drawn by the motor; (d) the apparent power input to the motor.

22-3 A four-wire, 208-V, three-phase, 60-Hz system is used to supply power to a three-phase 5-hp induction motor and a single-phase 6-kW heater connected between line c and the neutral line. The operating efficiency and power factor of the motor are 81 percent and 71 percent, respectively. (a) Sketch the circuit diagram and include an ammeter in each of the four lines; (b) sketch the standardized phasor diagram and calculate the three line currents and the neutral current.

22-4 A 450-V, three-phase, three-wire, 60-Hz feeder supplies power to a 25-hp induction motor and a 30-hp induction motor. The 25-hp motor is operating at full load, is 87 percent efficient, and has a power factor of 90 percent. The 30-hp motor is operating at one-half of its rated horsepower; at this load, it is 88 percent efficient and has a power factor of 74 percent. Sketch a one-line diagram for the system and determine (a) the feeder active power; (b) the feeder reactive power; (c) the feeder apparent power; (d) the feeder power factor; (e) the feeder current.

22-5 A three-phase balanced wye-connected load totaling 10 kVA at 0.80 pf lagging and a balanced delta-connected load totaling 10 kVA at unity power factor is supplied by a three-phase, three-wire, 440-V, 25-Hz system. Sketch a one-line diagram and determine (a) the system kW; (b) the system kvars; (c) the system kVA; (d) the system pf; (e) the system line current; (f) the line current to each three-phase load.

22-6 A three-phase, three-wire, 500-V, 60-Hz source supplies a three-phase induction motor, a wye-connected capacitor bank that draws 2 kvar per phase, and a balanced three-phase heater that draws a total of 10 kW. The induction motor is operating at its rated 75-hp load and has an efficiency and power factor of 90.5 percent and 89.5 percent, respectively. Sketch a one-line diagram of the system and determine (a) the system kW; (b) the system kvars; (c) the system kVA; (d) the system pf; (e) the system line current: (f) the line current drawn by each three-phase load.

22-7 A balanced delta-connected load whose impedance is $45\underline{/70°}$ Ω per branch, a three-phase motor that draws a total of 10 kVA at 0.65 pf lagging, and a wye-connected load whose impedance is 10 Ω (resistance) per branch are supplied from a three-phase, three-wire, 208-V, 60-Hz source. Sketch the circuit and determine (a) the line current to each three-phase load; (b) the active and reactive power drawn by each three-phase load; (c) the system kW, kvars, apparent power, and pf.

22-8 A delta-connected load, consisting of $20\underline{/0°}$ Ω between lines a and b, $14\underline{/-45°}$ Ω between lines b and c, and $14\underline{/45°}$ Ω between lines c and a, are supplied by a three-phase, 208-V, 60-Hz source. Sketch the circuit showing wattmeters in lines a and c, and determine (a) the three line currents; (b) the readings of wattmeters connected in line a and line c, respectively.

22-9 The two-wattmeter method is used to measure the power drawn by a balanced delta-connected load. Each branch of the load draws 2.5 kW at a power factor of 0.80 lagging. The supply voltage is 120 V, three phase, 25 Hz. Sketch the circuit showing wattmeters in lines a and c, and determine the respective wattmeter readings.

22-10 Three parallel-connected three-phase loads are supplied by a three-phase, 240-V, 60-Hz source. The loads are three-phase, 10 kVA, 0.80 power factor lagging; three-phase, 10 kVA, unity power factor; three-phase, 10 kVA, 0.80 power factor leading. Sketch the circuit showing wattmeters in lines a and c, and determine (a) the system active power; (b) the system reactive power; (c) the system apparent power; (d) the system power factor; (e) the system line circuit; (f) the respective wattmeter readings.

22-11 A 100-kVA balanced three-phase load is operating at 0.65 pf lagging from a 450-V, 25-Hz, three-phase supply. Sketch a one-line diagram and determine (a) the active power drawn by the load; (b) the reactive power drawn by the load; (c) line current; (d) if the two-wattmeter method is used to measure the three-phase power, what the wattmeters will read; (e) the kvar rating of a delta-connected capacitor bank required to obtain a system power factor of 0.90 lagging; (f) the capacitance of each capacitor in the bank.

22-12 The following instrument readings and C-T ratios were obtained when using the two-wattmeter method for measuring power to a balanced three-phase load: $W_1 = 100$, C-T ratio = 20; $W_2 = 800$, C-T ratio = 10; $A_1 = 0.9$, C-T ratio = 20; $A_2 = 1.8$, C-T ratio = 10; $V_1 = V_2 = 450$ V, $f = 60$ Hz. Determine (a) the active power; (b) the apparent power; (c) the power factor; (d) the reactive power; (e) the kvar rating of a *delta-connected* capacitor bank required to adjust the power factor to 0.8 lagging; (f) the capacitance of each branch of a *wye*-connected capacitor bank required to adjust the system pf to 0.8 lagging.

22-13 The following instrument readings and C-T ratios were obtained when using the two-wattmeter method for measuring power to a balanced three-phase load: $W_1 = 120$, C-T ratio = 10; $W_2 = 100$, C-T ratio = 20; $A_1 = 1.16$, C-T ratio = 10; $A_2 = 0.58$; C-T ratio = 20; $V_1 = 200$, $V_2 = 200$, $f = 25$ Hz. Determine (a) the three-phase power; (b) the line current; (c) the apparent power; (d) the power factor; (e) the kvar rating of a delta-connected capacitor bank required to adjust the power factor to unity; (f) the capacitance of each branch of a delta-con-

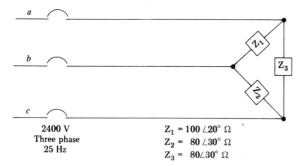

$Z_1 = 100 \angle 20° \ \Omega$
$Z_2 = 80 \angle 30° \ \Omega$
$Z_3 = 80 \angle 30° \ \Omega$

2400 V
Three phase
25 Hz

Figure 22-12 Circuit for Prob. 22-14.

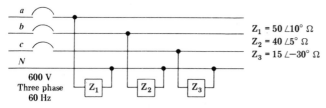

$Z_1 = 50 \angle 10° \ \Omega$
$Z_2 = 40 \angle 5° \ \Omega$
$Z_3 = 15 \angle -30° \ \Omega$

600 V
Three phase
60 Hz

Figure 22-13 Circuit for Prob. 22-15.

nected capacitor bank required to adjust the system pf to unity; (*g*) repeat part *f* for a wye-connected capacitor bank.

22-14 For the circuit in Fig. 22-12 determine (*a*) the three-line currents; (*b*) the three-phase active power; (*c*) the three-phase reactive power; (*d*) the power factor.

22-15 Determine (*a*) the active power, reactive power, and power factor of each individual load in Fig. 22-13; (*b*) the system active power, system reactive power, and system power factor; (*c*) the parameters of equivalent *series-connected* circuit elements that can be used to synthesize each impedance in Fig. 22-13; (*d*) the parameters of equivalent *parallel-connected* circuit elements that can be used to synthesize each impedance in Fig. 22-13.

CHAPTER 23

ELECTROMECHANICAL FORCES AND METER MOVEMENTS

Electromechanical devices such as meter movements, relays, automatic switches, breaker-tripping mechanisms, motors, and generators involve electromechanical forces generated by the interaction of magnetic fields.

23-1 MAGNITUDE OF THE MECHANICAL FORCE ON A CURRENT-CARRYING CONDUCTOR IN A MAGNETIC FIELD

The magnitude of the mechanical force exerted on a straight conductor that is carrying an electric current and situated within, and perpendicular to a magnetic field, as shown in Fig. 23-1a, may be determined by the following formula:

$$F_{ins} = B_{ins}\, i_{ins}\, \ell \qquad\qquad (23\text{-}1)$$

where F_{ins} = instantaneous mechanical force, newtons (N)
          B_{ins} = instantaneous flux density, teslas (T)
          i_{ins} = instantaneous current, amperes (A)
           ℓ = effective length of conductor, meters (m)

The effective length of the conductor is that portion of its length that is immersed in and perpendicular to the magnetic field of the magnet. If the

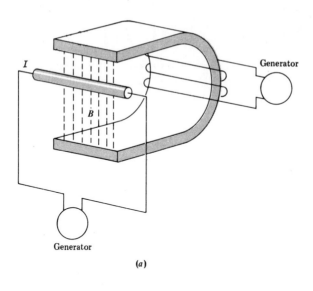

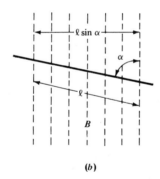

Figure 23-1 (*a*) Conductor situated within and perpendicular to a magnetic field; (*b*) determination of active length of conductor.

conductor is not perpendicular to the magnetic field, as shown in Fig. 23-1*b*, the active length of the conductor is $\ell \sin \alpha$, and Eq. (23-1) becomes

$$F_{\text{ins}} = B_{\text{ins}} i_{\text{ins}} (\ell \sin \alpha) \qquad (23\text{-}2)$$

If in Fig. 23-1*a* the current in the conductor and the flux of the magnet are sinusoidal, the mechanical force developed will not be constant but will vary with time.

Figure 23-2*a* shows the mechanical force developed when the wave representing the current in the conductor is in phase with the flux-density

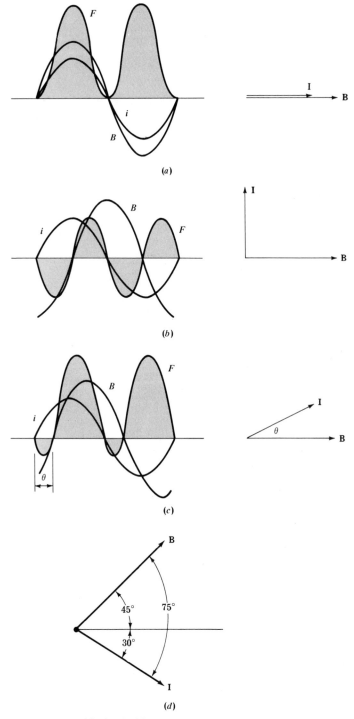

Figure 23-2 Mechanical force wave developed when current and pole flux are sinusoidal: (*a*) current in phase with flux density wave; (*b*) current 90° out of phase; (*c*) phase angle greater than zero and less than 90°; (*d*) phasor representation of current and flux-density waves in Example 23-1.

wave of the magnet. The resultant force wave, as indicated by the shaded areas, will be in pulses, all in the same direction. The force wave is the product of the flux-density wave, the current wave, and the constant ($\ell \sin \alpha$).

However, if the two generators are adjusted so that the current in the conductor is 90° out of phase with the flux-density wave, the resultant force wave will be alternating, as indicated by the shaded areas in Fig. 23-2b. Four pulses of force, equal in magnitude but alternating in direction, will occur in each cycle. Hence, the average force per cycle on the conductor will be zero.

Figure 23-2c illustrates the force wave for a phase angle other than zero or 90°. A small pulse of force in one direction is followed by a larger pulse in the opposite direction. Hence the average force in this case is neither zero nor so great as that for the in-phase condition.

The average force on the conductor in terms of the rms current and the rms flux density of the magnet is

$$F_{av\sim} = BI(\ell \sin \alpha) \cos \angle^{B}_{I} \qquad (23\text{-}3)$$

where $F_{av\sim}$ = average force, N

$\qquad\qquad B$ = rms value of flux density, T

$\qquad \ell \sin \alpha$ = effective length of conductor, m

$\qquad\qquad I$ = rms current, A

$\qquad \angle^{B}_{\theta\ I}$ = phase angle between the **B** and **I** phasors

Example 23-1 A conductor is situated in and perpendicular to a magnetic field. The current in the conductor and the flux density of the field are expressed by

$$B = 4 \sin\left(377t + \frac{\pi}{4}\right) \text{T}$$

$$i = 3 \sin\left(377t - \frac{\pi}{6}\right) \text{A}$$

If the effective length of the conductor is 0.060 m, determine the average force on the conductor.

Solution
The phasor diagram is shown in Fig. 23-2d.

$$B_{rms} = \frac{4}{\sqrt{2}} = 2.83 \text{ T} \qquad \frac{\pi}{4} = 45°$$

$$I_{rms} = \frac{3}{\sqrt{2}} = 2.12 \text{ A} \qquad \frac{\pi}{6} = 30°$$

$$F_{av\sim} = BI(\ell \sin \alpha)\cos\theta$$

$$F_{av\sim} = 2.83(2.12)(0.06)\cos 75°$$

$$F_{av\sim} = 0.093 \text{ N}$$

Example 23-2 A conductor whose resistance is 0.080 Ω is situated in the magnetic field of a horseshoe-shaped permanent magnet. The flux density in the air gap is a uniform 1.20 T. If the conductor is connected to a 1.5-V battery and the path of current through the conductor is perpendicular to the magnetic field of the magnet, what will be the mechanical force on the conductor? The active length of the conductor is 0.20 m.

Solution
The steady-state current in a DC system is the same at every instant of time, and the flux density of a permanent magnet is constant. Thus,

$$B_{ins} = 1.2 \text{ T}$$

$$i_{ins} = I = \frac{E_{bat}}{R} = \frac{1.5}{0.08} = 18.75 \text{ A}$$

$$F_{ins} = B_{ins} i_{ins}(\ell \sin \alpha)$$

$$F = 1.2 (18.75)(0.2 \sin 90°) = 4.5 \text{ N}$$

Example 23-3 A rotor conductor of a certain AC motor has an effective length of 0.5 m. It carries a current of 200 A and is situated within and perpendicular to a magnetic field of 0.10 T. The phase angle between the current in the conductor and the flux of the field is 50°. Calculate the average force exerted on the conductor.

Solution

$$F_{av\sim} = BI(\ell \sin \alpha)\cos\theta$$

$$F_{av\sim} = 0.1(200)(0.5)\cos 50°$$

$$F_{av\sim} = 6.43 \text{ N}$$

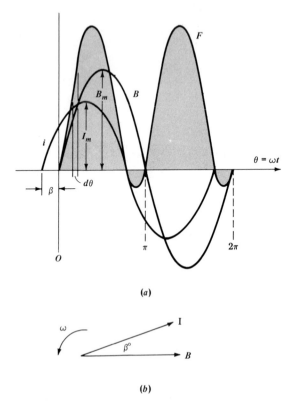

(a)

(b)

Figure 23-3 General case for sinusoidal current and flux-density waves, and the corresponding phasor diagram.

Derivation of Equation (23-3)

Figure 23-3 illustrates the general sinusoidal case for the current wave, the flux-density wave, and the resultant force wave as it applies to the circuit in Fig. 23-1a. The average value of the force wave may be determined by summing all the instantaneous $F_{\text{ins}}\, d\theta$ values over one cycle and dividing by 2π. Thus, using calculus,

$$F_{\text{av}} = \int_0^{2\pi} \frac{F_{\text{ins}}\, d\theta}{2\pi} \tag{23-4}$$

Substituting Eq. (23-2) into Eq. (23-4),

$$F_{\text{av}} = \int_0^{2\pi} \frac{B_{\text{ins}} i_{\text{ins}}(\ell \sin \alpha)\, d\theta}{2\pi}$$

$$F_{\text{av}} = \frac{\ell \sin \alpha}{2\pi} \int_0^{2\pi} B_{\text{ins}} i_{\text{ins}}\, d\theta \tag{23-5}$$

The equations of B_{ins} and i_{ins} for the given time-zero are

$$B_{\text{ins}} = B_m \sin \theta \tag{23-6}$$

$$i_{\text{ins}} = I_m \sin(\theta + \beta) \tag{23-7}$$

Substituting Eqs. (23-6) and (23-7) into Eq. (23-5),

$$F_{\text{av}} = \frac{\ell \sin \alpha}{2\pi} \int_0^{2\pi} B_m \sin \theta I_m \sin(\theta + \beta) \, d\theta$$

$$F_{\text{av}} = B_m I_m \frac{\ell \sin \alpha}{2\pi} \int_0^{2\pi} \sin \theta \sin(\theta + \beta) \, d\theta$$

Using trigonometric identities,

$$\sin(\theta + \beta) = \sin \theta \cos \beta + \cos \theta \sin \beta$$

$$\sin \theta \sin(\theta + \beta) = \sin^2 \theta \cos \beta + \sin \theta \cos \theta \sin \beta$$

$$F_{\text{av}} = B_m I_m \frac{\ell \sin \alpha}{2\pi} \int_0^{2\pi} (\sin^2 \theta \cos \beta + \sin \theta \cos \theta \sin \beta) \, d\theta$$

Integrating and evaluating the limits,

$$F_{\text{av}} = \frac{B_m I_m (\ell \sin \alpha)}{2\pi} \left[(\cos \beta) \frac{\theta - \sin \theta \cos \theta}{2} + (\sin \beta) \frac{\sin^2 \theta}{2} \right]_{\theta = 0}^{\theta = 2\pi}$$

$$F_{\text{av}} = \frac{B_m I_m (\ell \sin \alpha)}{2\pi} \left[(\cos \beta) \frac{2\pi}{2} + 0 \right]$$

Rearranging the factors,

$$F_{\text{av}} = \frac{B_m}{\sqrt{2}} \frac{I_m}{\sqrt{2}} (\ell \sin \alpha) \cos \beta$$

$$F_{\text{av}} = BI(\ell \sin \alpha) \cos \beta$$

where $\ell \sin \alpha$ = effective length of conductor, m
β = phase angle between B and i waves
B = rms flux density, T
I = rms current, A

23-2   MAGNITUDE OF THE MECHANICAL FORCE BETWEEN ADJACENT PARALLEL CONDUCTORS

The force of attraction or repulsion between two identical straight parallel conductors of round cross section in free air, as shown in Fig. 23-4, may be determined from the following equation:

$$F_{ins} = 2(10^{-7})i_1 i_2 \frac{\ell}{d}$$ (23-8)

where F_{ins} = instantaneous force, N
   i_1 = instantaneous current in conductor 1, A
   i_2 = instantaneous current in conductor 2, A
   ℓ = length of one conductor, m
   d = perpendicular distance between centers of conductors, m

Example 23-4   Two insulated round conductors are used to supply power from a 240-V 60-Hz system to a motor load. The spacing between the centers of the conductors is 0.050 m. A short circuit at the motor terminals causes the current to attain a maximum instantaneous value of 10,000 A. (*a*) Determine the maximum instantaneous force per meter of conductor length. (*b*) Will the force on the conductors be a separating force or an attracting force?

Solution
(*a*) the current in each conductor will be 10,000 or 10^4 A. Thus, for 1 m of length,

$$F_{ins} = 2(10^{-7})i_1 i_2 \frac{\ell}{d}$$

$$F_{ins} = 2(10^{-7})(10^4)(10^4)\frac{1}{0.05}$$

$$F_{ins} = 400 \text{ N/m}$$

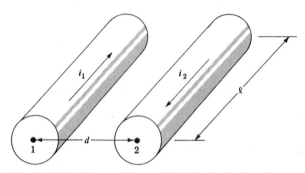

Figure 23-4   Two straight round conductors in free air.

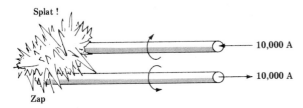

Figure 23-5 Circuit for Example 23-4.

(b) The directions of the respective fluxes in the common region are the same, as shown in Fig. 23-5. Hence a separating force is produced.

Example 23-5 Three straight circular cables are spaced 0.50 m between centers as shown in Fig. 23-6a. The instantaneous values of the currents in the conductors are 50,000 A, 70,000 A, and 65,000 A for conductors A, B, and C, respectively. Determine the magnitude and direction of the mechanical force exerted on each meter of conductor

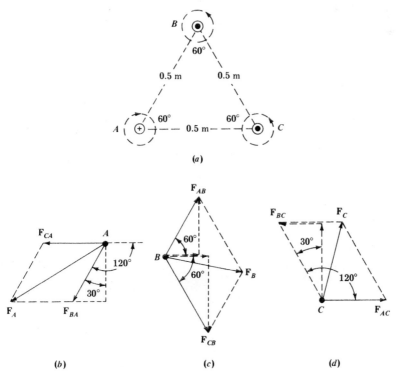

Figure 23-6 (a) Three straight circular cables spaced 0.50 m between centers; (b) force on conductor A; (c) force on conductor B; (d) force on conductor C.

length. The dot in the center of conductors B and C indicates that the direction of current is toward the reader; the cross mark on conductor A indicates that the current is in a direction away from the reader.

Solution

As determined from the relative directions of the respective fluxes around the conductors in Fig. 23-6a, a separating force will be produced between conductors A and B, a force of attraction between conductors B and C, and a separating force between conductors A and C. From Eq. (23-8), the magnitude of the instantaneous force acting on a conductor is

$$F = 2\,(10^{-7})\,i_1 i_2 \frac{\ell}{d} \quad \text{N}$$

Defining $F_{B,A}$ as the force on conductor A due to the current in B, and $F_{A,B}$ as the force on conductor B due to the current in conductor A, etc.,

$$F_{B,A} = F_{A,B} = 2(10^{-7})(5 \times 10^4)(7 \times 10^4)\frac{1}{0.5} = 1400 \text{ N/m}$$

$$F_{C,B} = F_{B,C} = 2(10^{-7})(7 \times 10^4)(6.5 \times 10^4)\frac{1}{0.5} = 1820 \text{ N/m}$$

$$F_{C,A} = F_{A,C} = 2(10^{-7})(6.5 \times 10^4)(5 \times 10^4)\frac{1}{0.5} = 1300 \text{ N/m}$$

From Fig. 23-6b,

$$\mathbf{F}_A = \mathbf{F}_{C,A} + \mathbf{F}_{B,A} = 1300\underline{/180°} + 1400\underline{/-120°}$$

$$\mathbf{F}_A = 2338.8\underline{/-148.77°} \text{ N}$$

From Fig. 23-6c,

$$\mathbf{F}_B = \mathbf{F}_{A,B} + \mathbf{F}_{C,B} = 1400\underline{/60°} + 1820\underline{/-60°}$$

$$\mathbf{F}_B = 1650.6\underline{/-12.73°} \text{ N}$$

From Fig. 23-6d,

$$\mathbf{F}_C = \mathbf{F}_{A,C} + \mathbf{F}_{B,C} = 1300\underline{/0°} + 1820\underline{/120°}$$

$$\mathbf{F}_C = 1623.7\underline{/76.10°}$$

Derivation of Equation (23-8)

Referring to Fig. 23-4, the flux density at the center of conductor 2 due to current i_1 in conductor 1 is expressed by

$$B = \frac{\mu_0 i_1}{2\pi d} \tag{23-9}$$

Equation (23-9) stems from Ampere's law, discussed in Sec. 7-3, Chap. 7. Substituting Eq. (23-9) into Eq. (23-1) and simplifying,

$$F_{\text{ins}} = \frac{\mu_0 i_1}{2\pi d} i_2 \ell = 2(10^{-7}) i_1 i_2 \frac{\ell}{d}$$

where $\mu_0 = 4\pi 10^{-7}$

23-3 **FORCE OF ATTRACTION BETWEEN A MAGNET AND A MAGNETIC MATERIAL**

Figure 23-7 shows a permanent magnet attracting a block of ferromagnetic material called the *armature*. The force of attraction between the magnet and the armature, called *tractive force*, is given by

$$\boxed{F_{\text{tr}} = \frac{B^2 A}{2\mu_0} \qquad d \approx 0} \tag{23-10}$$

where F_{tr} = total tractive force, N
   B = flux density of *air gap*, T
   A = total cross-sectional area of air gap, m^2
   μ_0 = permeability of free space $(4\pi 10^{-7})$, H/m
   d = distance between magnet face and armature (air gap), m

Equation (23-10) is exact for a zero air gap and is a very close approximation for air-gap distances approaching zero ($d \approx 0$). Thus Eq. (23-10) provides

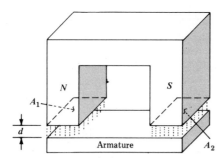

Figure 23-7 Force of attraction between a magnet and a magnetic material.

a means for calculating the *holding force of a magnet when in physical contact with the armature.*

Increasing the air gap causes fringing of the flux at the pole faces. This decreases the flux density in the air gap, which in turn reduces the tractive force.

Example 23-6 Determine the flux density required in the air gap between the magnet and the armature, as shown in Fig. 23-7, in order that it may be capable of a holding force of 6000 N. Each pole face is 0.10 m × 0.10 m.

Solution
Area of pole face is $0.1 \times 0.1 = 0.01 \text{ m}^2$

$$F_{tr} = \frac{B^2 A}{2\mu_0}$$

$$6000 = \frac{B^2(2 \times 0.01)}{2 \times 4\pi 10^{-7}}$$

$$B = 0.87 \text{ T}$$

23-4   **METERS**

The two general categories of meters used in electrical and other measurements are analog and digital. The analog types are electromechanical devices with moving pointers, springs, and moving coils or moving iron vanes. The digital types utilize electronic circuits in place of electromechanical devices, and provide a numerical readout. The accuracy and range of a well-designed digital meter is significantly better than its well-designed analog counterpart. A digital meter draws essentially no energy from the circuit being measured, and hence will not affect the measured quantity. Many digital meters, called autorangers, change scales automatically, providing the correct readout without having to change the range manually.

Analog meters have inherent limitations. They are subject to errors due to bearing friction, frequency variations, and possible loss of magnetism in the "permanent" magnets. They are also prone to user errors such as parallax error (reading from the side), interpolation error (estimating between graduations), and interpretation error (reading on the wrong scale or failing to consider a multiplying factor). However, analog instruments do have a decided advantage over digital meters in those applications where a high degree of accuracy is not as important as the ability to monitor rapidly changing variables for verification of safe or normal levels of operation, and to do this at a glance.

The three basic meter movements used in analog-type meters are the D'Arsonval, the electrodynamometer, and the iron vane.

3-5 **D'ARSONVAL MOVEMENT**

The D'Arsonval movement shown in Fig. 23-8a is used in DC instruments and consists of a movable coil situated within the magnetic field of a permanent magnet. Current in the coil sets up a magnetic field that interacts with the magnetic field of the magnet to produce a turning moment. The moving coil acts against the restraining force of two springs, which serve the additional function of carrying current to and away from the coil.

Each turn of the coil has two sides, of length ℓ, perpendicular to the magnetic field of the permanent magnet. From Eq. (23-1), the force exerted on one side of each turn is

$$F_{ins} = B_{ins} i_{ins} \ell \quad N$$

The instantaneous turning moment or torque developed by one turn-side is equal to the instantaneous force times the moment arm. That is,

$$T_{ins} = F_{ins} r = B_{ins} i_{ins} \ell r$$

where r = radius of the coil, m
$\qquad$ T = torque, $N \cdot m$
$\qquad$ ℓ = length of one turn-side, m

The total torque for N turns ($2N$ turn-sides) is

$$T_{ins} = 2N B_{ins} i_{ins} \ell r \quad N \cdot m$$

Since the flux density of the permanent magnet is constant, the instantaneous torque developed is proportional to the instantaneous current in the coil. That is,

$$T_{ins} \propto i_{ins}$$

However, because of mechanical inertia and built-in damping, the instrument pointer will not be able to follow rapid changes in developed torque caused by a rapidly fluctuating current. The *meter deflection will be proportional to the average torque, which is proportional to the average current.*

$$T_{av} \propto I_{av}$$

If a D'Arsonval meter is connected to a 60-Hz sinusoidal driver, the average current will be zero, the net torque will be zero, and the meter pointer will not deflect.

When used as a voltmeter, the coil of the D'Arsonval meter is connected in series with an internal wirewound resistor, called a *multiplier*, to limit the current. The resistance wire is made of a material such as constantan to provide it with the desired resistance-temperature characteristic. The sensitivity of the voltmeter is expressed in terms of the ratio of ohms per volt. For example, a 300-V (full-scale) voltmeter with an internal resistance of 300,000 Ω has a sensitivity of 300,000/300, or 1000 Ω/V. High-sensitivity voltmeters should be used on circuits that normally draw low values of current; otherwise the current drawn by the meter may represent a greater load than that of the circuit itself.

When used as an ammeter, a calibrated resistor of very low resistance called *a shunt* is connected in parallel with the D'Arsonval coil, as shown in Fig. 23-8b. The resistance strips of the shunt are generally made of manganin for resistance stability over a wide range of temperatures, and are terminated in large blocks of copper that act as heat sinks. Connections between shunt and meter should be made with *calibrated leads*, called *standard leads*, furnished by the manufacturer. The resistance of the calibrated leads is

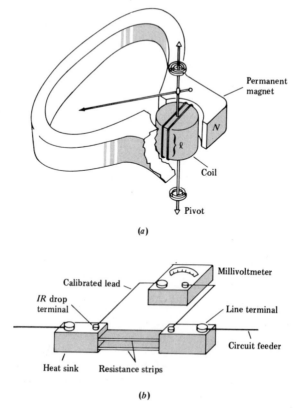

Figure 23-8   (a) D'Arsonval type of meter movement; (b) connections to an ammeter shunt.

taken into consideration during calibration of the meter. Leads with ap-preciably higher resistance will cause an appreciable error in the reading. The standard voltage ratings for ammeter shunts and their associated meters are 50 mV and 100 mV. However, a meter and its shunt must form a matching set; a 50-mV meter must be used with a 50-mV shunt, and a 100-mV meter must be used with a 100-mV shunt. The ampere rating of the shunt indicates the maximum allowable current.

The current in a shunt causes an IR drop that is measured on the milli-voltmeter. Millivoltmeters used for this purpose may have two or three scales, calibrated in amperes, to provide direct readings when used with corresponding shunts. If the millivoltmeter is not scaled in amperes or if it does not have a matching shunt, the ammeter scale factor may be determined by dividing the meter scale into the ampere rating of the shunt.

Example 23-7 A millivoltmeter with a range of 50 mV and a 100-division scale is used with a 300-A 50-mV shunt. Determine (a) the current in the shunt if the meter indicates 80 on the scale; (b) the resistance of the shunt; (c) the voltage drop across the shunt.

Solution
(a) The scale factor is

$$K_A = \frac{300}{100} = 3 \text{ A/div}$$

Hence the current is

$$I = 80 \times 3 = 240 \text{ A}$$

(b) Applying Ohm's law to the shunt rating,

$$R = \frac{V}{I} = \frac{0.050}{300} = 0.000167 \ \Omega$$

(c) $V = IR = 240(0.000167) = 0.04 \text{ V}$

3-6   ELECTRODYNAMOMETER MOVEMENT

The electrodynamometer movement shown in Fig. 23-9 consists of a moving coil, called the armature coil, that is free to move within a magnetic field set up by two stationary field coils. The interaction of the magnetic field set up by current in the armature coil, with the magnetic field set up by current in the field coils, provides the turning moment. A control spring provides the countertorque. *The instantaneous torque developed is proportional to the*

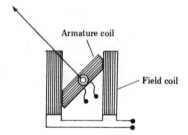

Armature coil

Field coil

Figure 23-9 Electrodynamometer type of meter movement.

product of the instantaneous current in the armature coil and the instantaneous current in the field coils. That is,

$$T_{\text{ins}} \propto i_{\text{arm}} i_{\text{field}}$$

However, because of mechanical inertia and built-in damping, the meter deflection will be proportional to the average torque.

If the current in the armature coil and that in the field coils are sinusoidal, then

$$T_{\text{av}} \propto I_{\text{arm}} I_{\text{field}} \cos\!\!\underset{\mathbf{I}_{\text{field}}}{\overset{\mathbf{I}_{\text{arm}}}{\theta}}$$

where I_{arm} and I_{field} are rms values.

If the meter is to be used as an ammeter or a voltmeter, the armature coil is connected in series with the field coils so that the same current will pass through each. When used in this manner, the instantaneous torque will be proportional to the square of the instantaneous current values, that is,

$$T_{\text{ins}} \propto i_{\text{ins}}^2$$

The deflection of the pointer will be proportional to the average of the squares of all the instantaneous current values and thus will be proportional to the rms value. To indicate rms values, the meter scale is calibrated in terms of the square root of the average of the squares (root mean square).

The electrodynamometer movement is used in wattmeters, varmeters, power-factor meters, frequency meters, synchroscopes, ammeters, voltmeters, and other applications. In wattmeter, voltmeter, and ammeter applications the driver may be AC or DC. The application of electrodynamometers, wattmeters, varmeters, and power-factor meters is discussed in Chaps. 13 and 22.

3-7 IRON-VANE MOVEMENT

The iron-vane movement shown in Fig. 23-10 is widely used for current and voltage measurements in AC circuits. The soft iron vanes are fastened to the shaft that drives the pointer and are free to move within the magnetic field set up by an inclined coil. Current in the coil induces a magnetic field in the iron vanes, causing them to turn in a direction of alignment with the magnetic

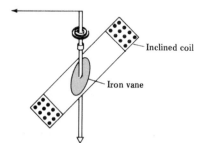

Inclined coil

Iron vane

Figure 23-10   Iron-vane type of meter movement.

axis of the coil. A control spring provides a countertorque. The magnetic field induced in the iron vanes, and the magnetic field of the coil, are both produced by the same current. Hence the turning moment at any instant of time is proportional to the square of the instantaneous coil current. That is,

$$T_{ins} \propto i_{ins}^2$$

However, because of mechanical inertia and built-in damping, the deflection of the pointer will be proportional to the average of the squares of all the instantaneous current values, and thus will be proportional to the rms value. To indicate rms values, the meter scale is calibrated in terms of the square root of the average of the squares (root mean square).

Iron-vane instruments are seldom used in DC systems because hysteresis effects in the iron vanes cause significant errors in the meter readings.

3-8 SELECTION AND APPLICATION OF BASIC COMMERCIAL INSTRUMENTS

The selection of a meter, or the method of measurement, requires a fair approximation of the magnitude and characteristics of the quantity to be measured. This determination is generally based on the rated values of the equipment under test and the available source of energy. Except for transient conditions or circuit faults, the nameplate data of the apparatus under test provide the best basis for meter selection. From the nameplate, the rated values of voltage, current, speed, frequency, power, etc., may be determined.

When data must be continuously recorded for hours or days, a recording instrument should be used.

To read an analog-type meter, the observer's eye should be positioned directly over the pointer to avoid errors due to parallax. Meters equipped with a mirror on the scale should be read with the eye positioned so that the pointer coincides with its reflection. Meters that have zero adjustments should be set to zero before they are energized. If the zero cannot be adjusted, its "zero reading" should be subtracted from the measured deflection. When selecting meters, the range chosen should be one that will result in a sizable deflection.

Connecting an electromechanical type of ammeter, voltmeter, or other similar instrument into a circuit will unfortunately affect to some extent the accuracy of the measurement. Although in most cases this inaccuracy is negligible, there are occasions, especially in high-impedance circuits, when the meter itself may overload the circuit and cause false indications. To avoid *loading errors*, high-impedance circuits should be metered with a high-resistance voltmeter such as a digital voltmeter (DVM); the very high internal impedance of a DVM will prevent loading the circuit. Another factor that should be considered is the effect of temperature on the measured values. Most conductors increase in resistance with increased temperature, but organic insulation such as paper, silk, and cotton decreases in resistance with increased temperature. Whenever two or more meters are used to measure related quantities, they should be read simultaneously, provided that one meter does not have a loading effect on the other. This is particularly important if the quantities to be measured are fluctuating.

Iron-vane and electrodynamometer voltmeters and ammeters are rms-responding meters with scales calibrated in rms values and should be used in AC systems. D'Arsonval voltmeters and ammeters are average-responding meters with scales calibrated in average values and should be used in DC systems.

A volt-ohmmeter or VOM, is a DC meter of the electromechanical type. It consists of a D'Arsonval movement with multipliers, shunts, diodes, and selector switches arranged to provide an economical means for measuring several different ranges of amperes, volts, and ohms. An internal battery and a zero-adjusting rheostat are used in connection with resistance measurement. When used as an ammeter, a shunt is switched across the meter coil; when used as a voltmeter, a multiplier is switched in series with the meter coil; when used on AC, diodes are connected between the meter coil and the circuit to provide a DC input to the meter coil.

When measuring sinusoidal current or voltage, the VOM will respond to the average value of the rectified sine wave. The current is read on the associated AC scale, which is calibrated in rms values based on a sinusoidal input. However, *if the alternating current is nonsinusoidal, the reading is meaningless.*

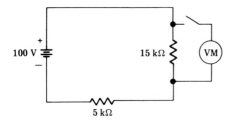

Figure 23-11   Circuit for Example 23-8.

Example 23-8 (a) Assuming the voltmeter switch in Fig. 23-11 is open, determine the voltage drop across the 15-kΩ resistor; (b) with the switch closed, determine the voltmeter reading if the resistance of the voltmeter is 10 kΩ; (c) repeat part b assuming a digital voltmeter (DVM) whose input resistance is 1 MΩ.

Solution
(a) Using the voltage-divider equation,

$$V_{15\,k\Omega} = 100\,\frac{15,000}{15,000 + 5000} = 75\text{ V}$$

(b) With the switch closed, the resistance of the voltmeter is in parallel with the 15-kΩ resistor. Thus,

$$R_{eq,\,P} = \frac{(15,000)(10,000)}{15,000 + 10,000} = 6000\ \Omega$$

The voltage drop across the parallel combination of voltmeter and 15-Ω resistor is

$$V_{15\,k\Omega} = 100\,\frac{6000}{6000 + 5000} = 54.6\text{ V}$$

(c) $R_{eq,\,P} = \dfrac{(15 \times 10^3)(10^6)}{15 \times 10^3 + 10^6} = 14{,}778\ \Omega$

$$V_{15\,k\Omega} = 100\,\frac{14,778}{14,778 + 5000} = 74.7\text{ V}$$

The 74.7-V reading of the DVM closely approaches the actual 75 V that appears across the 15-kΩ resistor when no voltmeter is connected.

High-impedance circuits such as that shown in Fig. 23-11 require a high-impedance voltmeter if realistic measurements are to be expected.

SUMMARY OF FORMULAS

$$F_{ins} = B_{ins} i_{ins}(\ell \sin \alpha)$$

$$F_{av} = BI \, (\ell \sin \alpha) \cos \overset{\textstyle B}{\underset{\textstyle I}{\diagdown}} \theta$$

$$F_{ins} = 2(10^{-7}) i_1 i_2 \frac{\ell}{d}$$

$$F_{tr} = \frac{B^2 A}{2\mu_0} \qquad d \approx 0$$

$$\mu_0 = 4\pi 10^{-7}$$

PROBLEMS

23-1 A copper conductor is immersed in a uniform magnetic field of 1.80 T. The resistance of the conductor is 0.040 Ω, and its active length is 0.25 m. Determine the DC driving voltage required to produce a mechanical force of 6.5 N on the conductor.

23-2 A conductor 0.50 m long is situated in and perpendicular to a sinusoidally varying magnetic field. The current in the conductor and the flux density of the field are expressed by $i = 30 \sin 377t$ A; $B = 1.2 \sin 377t$ T. Determine (a) the average force on the conductor; (b) the instantaneous force at $t = 0.0014$ s.

23-3 A rotor conductor of an AC motor is 0.20 m long, carries an rms current of 15 A, and is situated within and perpendicular to a magnetic field of 1.30 T (rms). The phase angle between the current in the conductor and the flux of the field is 12°. Calculate the average force acting on the conductor.

23-4 Consider two identical AC motors. The rotor conductors of motor A carry a sinusoidal current of 30 A (max) and are situated in a sinusoidal magnetic field of 2.0 T (max); the phase angle between the current and flux waves is 5°. The rotor conductors of motor B carry a sinusoidal current of 100 A (max) and are in a sinusoidal magnetic field of 2 T (max); the phase angle between the current and field is 75°. Which machine, A or B, will develop the greater torque?

23-5 Each of two parallel round conductors supplies 5000 A DC to a load. The spacing between the centers of the conductors is 3 cm. Determine the force per meter of conductor length.

23-6 Assume a coil of 100 turns and 2-Ω resistance is wound around the horseshoe-shaped magnet in Fig. 23-7. The magnet and armature are constructed of sheet steel. The dimensions of each pole face are 12 cm by 15 cm, and the mean length of the magnetic circuit including the armature is 120 cm. Determine (a) the air-gap flux density required to obtain a total tractive force of 4000 N; (b) the magnetic field intensity required to obtain part a; (c) the current required in the coil; (d) the driving voltage required to obtain the current in part c.

23-7 (a) Referring to Fig. 23-7, determine the flux density in the air gap between the magnet and the armature in order that it may be able to provide a holding force of 3600 N. Assume the dimensions of each pole face are 20 cm by 25 cm.

23-8 A 50-mV meter with a 0 to 600 scale is used with a 40-A 50-mV shunt. Sketch the circuit and determine (a) the scale factor; (b) the current through the shunt if the millivoltmeter indicates 255 on its scale; (c) the resistance of the shunt.

23-9 A 750-A 50-mV shunt is used with a 50-mV meter that has a 0 to 200 scale. Determine (*a*) the scale factor; (*b*) the current if the meter indicates 160 on its scale; (*c*) the shunt resistance.

23-10 A 240-V DC voltage source is connected to a 40-kΩ resistor load. The source resistance is 1500 Ω. Sketch the circuit and determine (*a*) the voltage drop across the load; (*b*) assuming a voltmeter is connected across the load, the voltmeter reading if its internal resistance is 200 kΩ.

23-11 A certain 90-V source with an internal resistance of 100 kΩ has a voltmeter connected across its output terminals. Determine the voltmeter reading if the resistance of the voltmeter is (*a*) 10 kΩ; (*b*) 100 kΩ; (*c*) 2 MΩ.

CHAPTER 24

INTRODUCTION TO CIRCUITS WITH MULTIFREQUENCY DRIVERS

Circuits containing two or more sinusoidal drivers of different frequencies will have a *nonsinusoidal current wave*. The equivalent rms value of this wave, and the average power drawn by the load, may be determined by using techniques developed in this chapter.

CURRENT, VOLTAGE, AND POWER IN MULTIFREQUENCY CIRCUITS

Figure 24-1a shows two sinusoidal generators of different frequencies connected in series with a load containing resistance and inductance. The time-domain equations for the sinusoidal driving voltages are given as

$$e_{10\sim} = 30 \sin 2\pi 10t \tag{24-1}$$

$$e_{20\sim} = 50 \sin 2\pi 20t \tag{24-2}$$

The resultant voltage across the load is

$$e_T = e_{10\sim} + e_{20\sim} \tag{24-3}$$

Because of the inductance of the load, the impedance will have different values at different frequencies, and current calculations will have to be done by superposition as outlined in Sec. 15-4, Chap. 15.

Using superposition, the circuit in Fig. 24-1a is divided into two components, each with its own sinusoidal driver, as shown in Fig. 24-1b and c.

For Fig. 24-1b	*For Fig. 24-1c*
$\mathbf{Z}_{10\sim} = R + jX_{L,\,10\sim}$	$\mathbf{Z}_{20\sim} = R + jX_{L,\,20\sim}$
$\mathbf{Z}_{10\sim} = 4 + j2\pi(10)(0.028)$	$\mathbf{Z}_{20\sim} = 4 + j2\pi(20)(0.028)$
$\mathbf{Z}_{10\sim} = 4 + j1.76$	$\mathbf{Z}_{20\sim} = 4 + j3.52$
$\mathbf{Z}_{10\sim} = 4.37\underline{/23.75^\circ}\ \Omega$	$\mathbf{Z}_{20\sim} = 5.33\underline{/41.34^\circ}\ \Omega$
$\mathbf{I}_{10\sim} = \dfrac{\mathbf{E}_{10\sim}}{\mathbf{Z}_{10\sim}}$	$\mathbf{I}_{20\sim} = \dfrac{\mathbf{E}_{20\sim}}{\mathbf{Z}_{20\sim}}$
$\mathbf{E}_{10\sim} = \dfrac{30\underline{/0^\circ}}{\sqrt{2}} = 21.21\ \text{V}$	$\mathbf{E}_{20\sim} = \dfrac{50\underline{/0^\circ}}{\sqrt{2}} = 35.36\ \text{V}$
$\mathbf{I}_{10\sim} = \dfrac{21.21\underline{/0^\circ}}{4.37\underline{/23.75^\circ}}$	$\mathbf{I}_{20\sim} = \dfrac{35.36\underline{/0^\circ}}{5.33\underline{/41.34^\circ}}$
$\mathbf{I}_{10\sim} = 4.85\underline{/-23.75^\circ}\ \text{A}$	$\mathbf{I}_{20\sim} = 6.63\underline{/-41.34^\circ}\ \text{A}$

The phasor components $(\mathbf{I}_{10\sim}, \mathbf{I}_{20\sim})$ *of the circuit current rotate at different angular velocities. Hence phasor addition cannot be used to obtain the resultant rms current.*

The resultant instantaneous value can be determined by expressing the component currents in the time domain, and adding. Thus,

$$i_{10\sim} = 4.85\sqrt{2}\,\sin{(2\pi10t - 23.75^\circ)} \tag{24-4}$$

$$i_{20\sim} = 6.63\sqrt{2}\,\sin{(2\pi20t - 41.34^\circ)} \tag{24-5}$$

From the superposition theorem,

$$i_T = i_{10\sim} + i_{20\sim} \tag{24-6}$$

Hence,

$$i_T = 6.86\,\sin{(2\pi10t - 23.75^\circ)} + 9.38\,\sin{(2\pi20t - 41.34^\circ)} \tag{24-7}$$

A plot of the current expressed by Eq. (24-7) is shown in Fig. 24-1d. The *nonsinusoidal current was caused by the addition of sinusoidal drivers of different frequencies.*

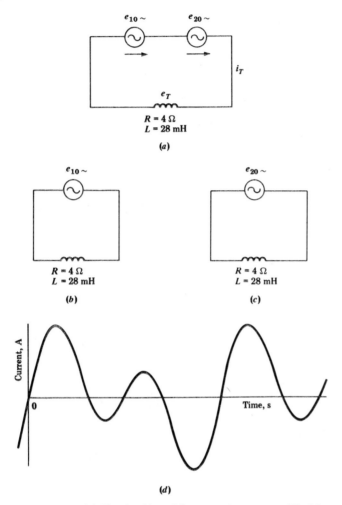

Figure 24-1 (a) Circuit with multifrequency generators; (b), (c) component circuits; (d) plot of circuit current versus time.

The power drawn by the load, at any instant of time, is equal to the product of the instantaneous voltage across the load times the instantaneous current to the load. Thus, for Fig. 24-1a,

$$P_{ins} = e_T i_T \tag{24-8}$$

Substituting Eqs. (24-3) and (24-6) into Eq. (24-8) and expanding,

$$P_{ins} = (e_{10\sim} + e_{20\sim})(i_{10\sim} + i_{20\sim}) \tag{24-9}$$

$$P_{ins} = e_{10\sim} i_{10\sim} + e_{20\sim} i_{20\sim} + e_{10\sim} i_{20\sim} + e_{20\sim} i_{10\sim} \tag{24-10}$$

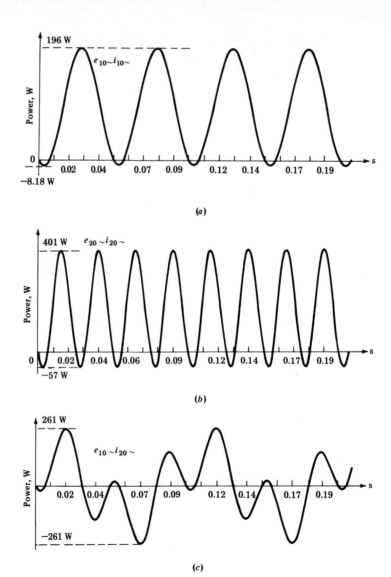

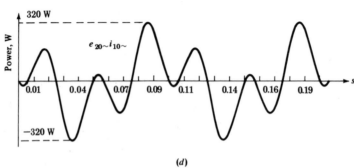

Figure 24-2 Component power waves expressed in Eq. (24-10).

Sketches of the four component power waves in Eq. (24-10) are shown in Fig. 24-2.

In Fig. 24-2c and d, the voltage and current waves have different frequencies, and the average power calculated over one or more periods of the lower-frequency wave is zero.

In Fig. 24-2a and b, the current and voltage waves have the same frequency, and the average power calculated over one or more periods is not zero. A comparison of Fig. 24-2a, b, c, and d indicates that *only current and voltage waves of the same frequency provide a nonzero value for the average power.* The average value for each of the power waves shown in Fig. 24-2a and b may be calculated by using the relationships developed in Chap. 13. Thus, for the circuit in Fig. 24-1a,

$$P_{10\sim} = E_{10\sim} I_{10\sim} \cos \left\langle \begin{matrix} \mathbf{E}_{10\sim} \\ \mathbf{I}_{10\sim} \end{matrix} \right. \qquad P_{20\sim} = E_{20\sim} I_{20\sim} \cos \left\langle \begin{matrix} \mathbf{E}_{20\sim} \\ \mathbf{I}_{20\sim} \end{matrix} \right.$$

$$P_{10\sim} = (21.21)(4.85) \cos 23.75° \qquad P_{20\sim} = (35.36)(6.63) \cos 41.34°$$

$$P_{10\sim} = 94 \text{ W} \qquad P_{20\sim} = 176 \text{ W}$$

The total power supplied to the load is

$$P_T = P_{10\sim} + P_{20\sim}$$

$$P_T = 94 + 176 = 270 \text{ W}$$

The power may also be calculated by using the I^2R relationship. Thus,

$$P_{10\sim} = I_{10\sim}^2 R = (4.85)^2 4 = 94 \text{ W}$$

$$P_{20\sim} = I_{20\sim}^2 R = (6.63)^2 4 = 176 \text{ W}$$

$$P_T = P_{10\sim} + P_{20\sim} = 270 \text{ W}$$

It is also possible to calculate the power by using the E^2/R relationship. *However, the voltage used in this expression must be the voltage drop across the resistance component alone, not the voltage drop across the entire impedance.*

To summarize, in a series circuit with multifrequency drivers, the current components, power components, and total power may be determined by means of the following formulas:

$$I_{g\sim} = \frac{E_{g\sim}}{Z_{g\sim}} \tag{24-11}$$

$$Z_{g\sim} = R + j\left(2\pi f_{g\sim} L - \frac{1}{2\pi f_{g\sim} C}\right) \tag{24-12}$$

$$P_{g\sim} = E_{g\sim} I_{g\sim} \cos \angle\left(E_{g\sim}, I_{g\sim}\right) \tag{24-13}$$

$$P_{g\sim} = I_{g\sim}^2 R \tag{24-14}$$

$$P_T = \sum P_{1\sim} + P_{2\sim} + \cdots + P_{g\sim} + \cdots + P_{n\sim} \tag{24-15}$$

where $g\sim = g$ Hz
$I_{g\sim}$ = rms current at g Hz
$E_{g\sim}$ = rms driving voltage at g Hz
$f_{g\sim}$ = frequency of g Hz
$Z_{g\sim}$ = complex impedance at g Hz
R = resistance
$P_{g\sim}$ = active power delivered to load by the g-Hz generator
P_T = total active power delivered to the load by all generators

24-2 **ROOT SUM SQUARE (RSS) OF MULTIFREQUENCY CURRENTS AND VOLTAGES**

For the general case of a set of multifrequency (MF) drivers in series with a resistor or in series with an impedance containing a resistor, the total active power drawn by the circuit is

$$P_{\text{MF}} = I_{f1}^2 R + I_{f2}^2 R + \cdots + I_{fg}^2 R + \cdots + I_{fn}^2 R \tag{24-16}$$

Defining I_{eq} as the rms value of an equivalent single-frequency current that would cause the same power dissipation in R as do the multifrequency components in Eq. (24-16), then

$$P_{\text{MF}} = I_{\text{eq}}^2 R \tag{24-17}$$

Equating Eqs. (24-16) and (24-17), and solving for I_{eq},

$$I_{eq}^2 R = (I_{f1}^2 + I_{f2}^2 + \cdots + I_{fg}^2 + \cdots + I_{fn}^2)R$$

$$I_{eq} = \sqrt{I_{f1}^2 + I_{f2}^2 + \cdots + I_{fg}^2 + \cdots + I_{fn}^2} \qquad (24\text{-}18)$$

Since I_{eq} is the square root of the sum of the squares of the component currents, it is defined as the *root-sum-square current or rss current.* Thus Eqs. (24-18) and (24-17) may be written as

$$\boxed{I_{rss} = \sqrt{I_{f1}^2 + I_{f2}^2 + \cdots + I_{fg}^2 + \cdots + I_{fn}^2}} \qquad (24\text{-}19)$$

$$\boxed{P_{MF} = I_{rss}^2 R} \qquad (24\text{-}20)$$

where I_{f1}, I_{f2}, etc., are rms values

In a similar manner, the total active power drawn by the impedance may be calculated from the E^2/R relationship, *provided that the multifrequency voltage drops across the resistance component are known.* Thus,

$$P_{MF} = \frac{E_{f1}^2}{R} + \frac{E_{f2}^2}{R} + \cdots + \frac{E_{fg}^2}{R} + \cdots + \frac{E_{fn}^2}{R} \qquad (24\text{-}21)$$

Defining E_{eq} as the rms value of an *equivalent single-frequency voltage drop across the resistance component* that would cause the same power dissipation in R as does the multifrequency components in Eq. (24-21), then

$$P_{MF} = \frac{E_{eq}^2}{R} \qquad (24\text{-}22)$$

Equating Eqs. (24-22) and (24-21) and solving for E_{eq},

$$\frac{E_{eq}^2}{R} = \frac{E_{f1}^2}{R} + \frac{E_{f2}^2}{R} + \cdots + \frac{E_{fg}^2}{R} + \cdots + \frac{E_{fn}^2}{R}$$

$$E_{eq} = \sqrt{E_{f1}^2 + E_{f2}^2 + \cdots + E_{fg}^2 + \cdots + E_{fn}^2} \qquad (24\text{-}23)$$

Since E_{eq} is the square root of the sum of the squares of the component voltages across the resistor, it is defined as the *root-sum-square voltage, or rss voltage.* Thus, Eqs. (24-23) and (24-22) may be written as

$$E_{\text{rss}, R} = \sqrt{E_{f1}^2 + E_{f2}^2 + \cdots + E_{fg}^2 + \cdots + E_{fn}^2} \qquad (24\text{-}24)$$

$$P_{\text{MF}} = \frac{E_{\text{rss}, R}^2}{R} \qquad (24\text{-}25)$$

where. E_{f1}, E_{f2}, etc., are rms values

$E_{\text{rss}, R}$ is the root-sum-square voltage drop across the resistance component of the impedance

Similarly, the rss voltage drops across L, C, and Z, respectively, in a circuit containing these elements are

$$E_{\text{rss}, L} = \sqrt{E_{L1}^2 + E_{L2}^2 + E_{Lg}^2 + \cdots + E_{Ln}^2}$$

$$E_{\text{rss}, C} = \sqrt{E_{C1}^2 + E_{C2}^2 + E_{Cg}^2 + \cdots + E_{Cn}^2}$$

$$E_{\text{rss}, Z} = \sqrt{E_{Z1}^2 + E_{Z2}^2 + E_{Zg}^2 + \cdots + E_{Zn}^2}$$

where $E_{Lg} = I_{fg}(2\pi f_g L)$

$$E_{Cg} = I_{fg} \frac{1}{2\pi f_g C}$$

$$E_{Zg} = I_{fg}(Z_{fg})$$

Note 1: The root-sum-square values for multifrequency current and multifrequency voltage are the respective ammeter and voltmeter indications on rms-responding meters.
Note 2: In a multifrequency circuit, if two or more series-connected generators have the same frequency, it is necessary to calculate the resultant voltage at that frequency before attempting to calculate the rss values. Failure to do so will result in erroneous answers.
Note 3: The reactive power in multifrequency circuits cannot be determined by using root-sum-square values of current or voltage. The inductive reactance and capacitive reactance are frequency-dependent, having different values at different frequencies. Thus reactive power contributions would have to be determined separately for each frequency, if such information is desired.

24-3 **APPARENT POWER AND POWER FACTOR IN CIRCUITS WITH MULTIFREQUENCY DRIVERS**

The apparent power delivered to a multifrequency circuit is defined as the product of the rss current to the circuit and the rss voltage across the circuit.

$$S_{MF} = I_{rss} E_{rss}$$ (24-26)

The definition of power factor for a multifrequency system is the same as for a single-frequency system. *However, the power factor angle of a multifrequency system is nonexistent.*

$$pf_{MF} = \frac{P_{MF}}{S_{MF}}$$ (24-27)

Example 24-1 For the circuit shown in Fig. 24-3, determine (*a*) the voltmeter reading; (*b*) the ammeter reading; (*c*) the active power dissipated in the coil.

Solution

(*a*) $E_{rss} = \sqrt{(120)^2 + (40)^2 + (300)^2} = 326$ V

(*b*) $\mathbf{Z}_{60\sim} = 10 + j[2\pi60(0.05)] = 10 + j18.85 = 21.34\underline{/62.05°}\ \Omega$

$\mathbf{Z}_{20\sim} = 10 + j[2\pi(20)(0.05)] = 10 + j6.28 = 11.81\underline{/32.14°}\ \Omega$

$\mathbf{Z}_{0\sim} = 10 + j[2\pi(0)(0.05)] = 10 + j0 = 10\underline{/0°}\ \Omega$

$\mathbf{I}_{60\sim} = \dfrac{120\underline{/0°}}{21.34\underline{/62.05°}} = 5.62\underline{/-62.05°}$ A

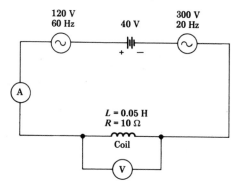

Figure 24-3 Circuit for Example 24-1.

$$I_{20\sim} = \frac{300/\underline{0°}}{11.81/\underline{32.14°}} = 25.40/\underline{-32.4°} \text{ A}$$

$$I_{0\sim} = \frac{40/\underline{0°}}{10/\underline{0°}} = 4.0/\underline{0°} \text{ A}$$

$$I_{rss} = \sqrt{(5.62)^2 + (25.40)^2 + (4)^2} = 26.32 \text{ A}$$

(c) $P_{MF} = I_{rss}^2 R = (26.32)^2(10) = 6927.4 \text{ W}$

Example 24-2   The current and voltage to a load are expressed by

$$e_T = 20 + 30\sin(377t) + 50\sin(1130t + 20°)$$

$$i_T = 15\sin(377t) + 14\sin(1130t - 36°)$$

Determine (a) the frequency of each component of the driving voltage; (b) the rss current; (c) the rss voltage; (d) the active power drawn by the load; (e) the apparent power; (f) the power factor; (g) the impedance offered by the load to each frequency.

Solution
(a) The three component frequencies are

1.   0 Hz (DC)

2. $2\pi f = 377$

$$f = 60 \text{ Hz}$$

3. $2\pi f = 1130$

$$f = 180 \text{ Hz}$$

(b) $I_{rss} = \sqrt{\left(\dfrac{15}{\sqrt{2}}\right)^2 + \left(\dfrac{14}{\sqrt{2}}\right)^2} = 14.5 \text{ A}$

(c) $V_{rss} = \sqrt{(20)^2 + \left(\dfrac{30}{\sqrt{2}}\right)^2 + \left(\dfrac{50}{\sqrt{2}}\right)^2} = 45.8 \text{ V}$

(d) $P_{MF} = P_{0\sim} + P_{60\sim} + P_{180\sim}$

$$P_{0\sim} = V_{0\sim}I_{0\sim} = 20(0) = 0$$

$$P_{60\sim} = V_{60\sim}I_{60\sim}\cos \overbrace{}^{V_{60\sim}}_{I_{60\sim}}$$

$$P_{60\sim} = \frac{30}{\sqrt{2}} \frac{15}{\sqrt{2}} \cos 0° = 225 \text{ W}$$

$$P_{180\sim} = V_{180\sim} I_{180\sim} \cos$$

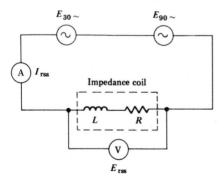

$$P_{180\sim} = \frac{50}{\sqrt{2}} \frac{14}{\sqrt{2}} \cos 56° = 196 \text{ W}$$

$$P_{MF} = 0 + 225 + 196 = 421 \text{ W}$$

(e) $S_{MF} = V_{rss} I_{rss} = (45.83)(14.51) = 665 \text{ VA}$

(f) $\text{pf}_{MF} = \dfrac{P_{MF}}{S_{MF}} = \dfrac{421}{665} = 0.63$

(g) $Z_{0\sim} = \dfrac{V_{0\sim}}{I_{0\sim}} = \dfrac{20}{0} \; \infty\,\Omega$

$$Z_{60\sim} = \frac{V_{60\sim}}{I_{60\sim}} = \frac{30/\sqrt{2}}{15/\sqrt{2}} = 2 \; \Omega$$

$$Z_{180\sim} = \frac{V_{180\sim}}{I_{180\sim}} = \frac{50/\sqrt{2}}{14/\sqrt{2}} = 3.57 \; \Omega$$

Example 24-3 The voltage and current to a coil are expressed by

$$e_T = 141.42 \sin 2\pi 30t + 141.42 \sin 2\pi 90t$$

$$i_T = 9.80 \sin(2\pi 30t - 30°) + 5.66 \sin(2\pi 90t - 60°)$$

Determine (a) I_{rss}; (b) E_{rss} across the coil; (c) $Z_{30\sim}$, (d) $Z_{90\sim}$; (e) the resistance and inductance of the coil; (f) the active power drawn by the coil; (g) the apparent power; (h) the power factor.

Figure 24-4 Circuit for Example 24-3.

Solution
The circuit is shown in Fig. 24-4.

(a) $I_{30\sim} = \dfrac{9.80}{\sqrt{2}} = 6.93$ A

$I_{90\sim} = \dfrac{5.66}{\sqrt{2}} = 4.00$ A

$I_{rss} = \sqrt{(6.93)^2 + (4.00)^2} = 8.00$ A

(b) $E_{30\sim} = \dfrac{141.42}{\sqrt{2}} = 100$ V

$E_{90\sim} = \dfrac{141.42}{\sqrt{2}} = 100$ V

$E_{rss, Z} = \sqrt{(100)^2 + (100)^2} = 141.42$ V

(c) $\mathbf{Z}_{30\sim} = \dfrac{\mathbf{E}_{30\sim}}{\mathbf{I}_{30\sim}} = \dfrac{100\underline{/0^\circ}}{6.93\underline{/-30^\circ}}$

$\mathbf{Z}_{30\sim} = 14.44\underline{/30^\circ} = (12.5 + j7.22)\ \Omega$

(d) $\mathbf{Z}_{90\sim} = \dfrac{\mathbf{E}_{90\sim}}{\mathbf{I}_{90\sim}} = \dfrac{100\underline{/0^\circ}}{4\underline{/-60^\circ}}$

$\mathbf{Z}_{90\sim} = 25\underline{/60^\circ} = (12.5 + j21.65)\ \Omega$

(e) The resistance and inductance of the coil may be determined from the impedance values calculated in parts c or d. Using the calculations in part c,

$\mathbf{Z}_{30\sim} = (12.5 + j7.22)\ \Omega$

thus,

$R = 12.5\ \Omega$

$X_{L\,30\sim} = 7.22\ \Omega$

$2\pi(30)\,L = 7.22$

$L = 0.0383$ H   or 38.3 mH

Using the calculations in part d,

$\mathbf{Z}_{90\sim} = (12.5 + j21.65)\ \Omega$

$R = 12.5\ \Omega$

$X_{L90\sim} = 21.65\ \Omega$

$2\pi(90)\ L = 21.65$

$L = 0.0383\ \text{H}$ or $38.3\ \text{mH}$

(f) The active power drawn by the coil may be calculated by five different methods:

Method 1:

$P_{30\sim} = E_{30\sim} I_{30\sim} \cos\!\!\begin{array}{l} \nearrow \mathbf{E}_{30\sim} \\ \searrow \mathbf{I}_{30\sim} \end{array}$

$P_{30\sim} = (100)(6.93)\cos 30°$

$P_{30\sim} = 600\ \text{W}$

$\quad P_{\text{MF}} = 600 + 200 = 800\ \text{W}$

$P_{90\sim} = E_{90\sim} I_{90\sim} \cos\!\!\begin{array}{l} \nearrow \mathbf{E}_{90\sim} \\ \searrow \mathbf{I}_{90\sim} \end{array}$

$P_{90\sim} = (100)(4.00)\cos 60°$

$P_{90\sim} = 200\ \text{W}$

Method 2:

$P_{30\sim} = I_{30\sim}^2\, R$

$P_{30\sim} = (6.93)^2(12.5)$

$P_{30\sim} = 600\ \text{W}$

$\quad P_{\text{MF}} = 600 + 200 = 800\ \text{W}$

$P_{90\sim} = I_{90\sim}^2\, R$

$P_{90\sim} = (4.00)^2(12.5)$

$P_{90\sim} = 200\ \text{W}$

Method 3:

$P_{\text{MF}} = I_{\text{rss}, R}^2\, R$

$P_{\text{MF}} = (8)^2(12.5) = 800\ \text{W}$

Method 4:

$P_{\text{MF}} = \dfrac{E_{\text{rss}, R}^2}{R}$

where $E_{rss, R}$ is the root-sum-square of the multifrequency voltage drop across the resistance component.

$$E_{30\sim, R} = I_{30\sim} R = 6.93(12.5) = 86.63 \text{ V}$$

$$E_{90\sim, R} = I_{90\sim} R = 4.00(12.5) = 50.00 \text{ V}$$

$$E_{rss, R} = \sqrt{(86.63)^2 + (50.00)^2} = 100 \text{ V}$$

thus,

$$P_{MF} = \frac{(100)^2}{12.5} = 800 \text{ W}$$

Method 5:

$$P_{MF} = E_{rss, R} I_{rss, R}$$

$$P_{MF} = (100)(8) = 800 \text{ W}$$

(g) $S_{MF} = E_{rss} I_{rss}$

$$S_{MF} = (141.42)(8.00) = 1131 \text{ V A}$$

(h) $\text{pf}_{MF} = \dfrac{P_{MF}}{S_{MF}} = \dfrac{800}{1131} = 0.71$

Example 24-4 A 10-Ω resistor is connected in series with the following series-connected generators:

$$e_1 = 50 \sin(377t + 40°)$$

$$e_2 = 40 \sin(377t + 20°)$$

$$e_3 = 80 \sin 150t$$

Determine (*a*) the frequency of the individual generators; (*b*) the rss voltage across the resistor; (*c*) the rss current to the resistor; (*d*) the active power delivered to the resistor.

Solution

(a) $f_1 = \dfrac{377}{2\pi} = 60 \text{ Hz}$

$f_2 = \dfrac{377}{2\pi} = 60 \text{ Hz}$

$f_3 = \dfrac{150}{2\pi} = 23.87 \text{ Hz}$

(b) Since voltages e_1 and e_2 have the same frequency, it is necessary to calculate the resultant voltage at that frequency before calculating the rss value. Thus,

$$\mathbf{E}_1 + \mathbf{E}_2 = \frac{50\underline{/40°}}{\sqrt{2}} + \frac{40\underline{/20°}}{\sqrt{2}}$$

$$\mathbf{E}_1 + \mathbf{E}_2 = (27.08 + j22.73) + (26.58 + j9.67)$$

$$\mathbf{E}_1 + \mathbf{E}_2 = (53.66 + j32.40) = 62.7\underline{/31.12°} \text{ V}$$

The rms value of the 23.87 hertz wave is $E_3 = 80/\sqrt{2} = 56.6$ V.

$$E_{\text{rss}, R} = \sqrt{E_{60\sim}^2 + E_{23.87\sim}^2}$$

$$E_{\text{rss}, R} = \sqrt{(62.7)^2 + (56.6)^2} = 84.5 \text{ V}$$

(c) $\quad I_{60\sim} = \dfrac{V_{60\sim}}{Z_{60\sim}} = \dfrac{62.68}{10} = 6.268$ A

$$I_{23.87\sim} = \frac{56.57}{10} = 5.657 \text{ A}$$

$$I_{\text{rss}} = \sqrt{(6.268)^2 + (5.657)^2} = 8.44 \text{ A}$$

(d) $P_{\text{MF}} = I_{\text{rss}, R}^2 = (8.44)^2(10) = 712.3$ W

Example 24-5 A series circuit containing a 295-μF capacitor and a coil whose resistance and inductance are 3 Ω and 4.42 mH, respectively, are supplied by the following series-connected generators: 35 V at 60 Hz, 10 V at 180 Hz, and 8 V at 300 Hz. Determine (a) the rss driving voltage; (b) the rss current; (c) the reading of an rms voltmeter connected across the coil; (d) the rss voltage across the capacitor; (e) the system active power; (f) the system apparent power; (g) the system power factor; (h) the time-domain current.

Solution

(a) $E_{\text{rss}} = \sqrt{(35)^2 + (10)^2 + (8)^2} = 37.3$ V

(b) *At 60 Hz*

$$X_L = 2\pi(60)(4.42 \times 10^{-3}) = 1.67 \ \Omega$$

$$X_C = \frac{1}{2\pi 60(295 \times 10^{-6})} = 8.99 \ \Omega$$

$$\mathbf{Z}_{60} = R + j(X_L - X_C) = 3 + j(1.67 - 8.99)$$

$$\mathbf{Z}_{60} = 7.91\underline{/-67.72^\circ}\ \Omega$$

$$\mathbf{I}_{60} = \frac{\mathbf{E}_{60}}{\mathbf{Z}_{60}} = \frac{35\underline{/0^\circ}}{7.91\underline{/-67.72^\circ}} = 4.42\underline{/67.72^\circ}\ A$$

$$I_{60} = 4.42\ A$$

At 180 Hz:

$$X_L = 2\pi(180)(4.42 \times 10^{-3}) = 5\ \Omega$$

$$X_C = \frac{1}{2\pi(180)(295 \times 10^{-6})} = 3\ \Omega$$

$$\mathbf{Z}_{180} = 3 + j(5 - 3) = (3 + j2)$$

$$\mathbf{Z}_{180} = 3.61\underline{/33.69^\circ}\ \Omega$$

$$\mathbf{I}_{180} = \frac{\mathbf{E}_{180}}{\mathbf{Z}_{180}} = \frac{10\underline{/0^\circ}}{3.61\underline{/33.69^\circ}} = 2.77\underline{/-33.69^\circ}\ A$$

$$I_{180} = 2.77\ A$$

At 300 Hz:

$$X_L = 2\pi(300)(4.42 \times 10^{-3}) = 8.33\ \Omega$$

$$X_C = \frac{1}{2\pi(300)(295 \times 10^{-6})} = 1.80\ \Omega$$

$$\mathbf{Z}_{300} = R + j(X_L - X_C) = 3 + j(8.33 - 1.80)$$

$$\mathbf{Z}_{300} = 7.19\underline{/65.33^\circ}\ \Omega$$

$$\mathbf{I}_{300} = \frac{\mathbf{E}_{300}}{\mathbf{Z}_{300}} = \frac{8\underline{/0^\circ}}{7.19\underline{/65.33^\circ}} = 1.11\underline{/-65.33^\circ}\ A$$

$$I_{300} = 1.11\ A$$

$$I_{rss} = \sqrt{I_{60}^2 + I_{180}^2 + I_{300}^2}$$

$$I_{rss} = \sqrt{(4.42)^2 + (2.77)^2 + (1.11)^2} = 5.33\ A$$

(c) $\mathbf{V}_{\text{coil }60} = \mathbf{I}_{60}\mathbf{Z}_{\text{coil }60} = \mathbf{I}_{60}(R_{\text{coil}} + jX_{L\text{coil}})_{60}$

$\mathbf{V}_{\text{coil }60} = 4.42\underline{/67.72°}\,(3 + j1.67) = (4.42\underline{/67.72°})(3.43\underline{/29.10°})$

$\mathbf{V}_{\text{coil }60} = 15.16\underline{/96.82°}$ V

Similarly,

$\mathbf{V}_{\text{coil }180} = (2.77\underline{/-33.69°})(3 + j5) = (2.77\underline{/-33.69°})(5.83\underline{/59.04°})$

$\mathbf{V}_{\text{coil }180} = 16.15\underline{/25.35°}$ V

$\mathbf{V}_{\text{coil }300} = (1.11\underline{/-65.33°})(3 + j8.33) = (1.11\underline{/-65.33°})(8.85\underline{/70.19°})$

$\mathbf{V}_{\text{coil }300} = 9.82\underline{/4.86°}$ V

$V_{\text{rss coil}} = \sqrt{(15.16)^2 + (16.15)^2 + (9.82)^2}$

$V_{\text{rss coil}} = 24.2$ V

(d)

$V_{C60} = I_{C60}X_{C60}$	$V_{C180} = I_{C180}X_{C180}$	$V_{C300} = I_{C300}X_{C300}$
$V_{C60} = 4.42(8.99)$	$V_{C180} = 2.77(3)$	$V_{C300} = (1.11)(1.80)$
$V_{C60} = 39.74$ V	$V_{C180} = 8.31$ V	$V_{C300} = 2.00$ V

$V_{\text{rss},C} = \sqrt{(39.74)^2 + (8.31)^2 + (2.00)^2} = 40.65$ V

(e) $P_{\text{MF}} = RI_{\text{rss},R}^2 = 3(5.33)^2 = 85.23$ W

(f) $S_{\text{MF}} = E_{\text{rss}}I_{\text{rss}} = (37.27)(5.33) = 198.65$ V A

(g) $\text{pf}_{\text{MF}} = \dfrac{P_{\text{MF}}}{S_{\text{MF}}} = \dfrac{85.23}{198.65} = 0.43$

(h) $i = i_{60} + i_{180} + i_{300}$

$i = 4.42\sqrt{2}\sin(2\pi 60t + 67.72°) + 2.77\sqrt{2}\sin(2\pi 180t - 33.69°)$

$\qquad + 1.11\sqrt{2}\sin(2\pi 300t - 65.33°)$

$i = 6.25\sin(377t + 67.72°) + 3.92\sin(1131t - 33.69°)$

$\qquad + 1.57\sin(1885t - 65.33°)$

SUMMARY OF FORMULAS

Multifrequency Sinusoidal Drivers (Fig. 24-5)

1. E and I are rms
2. P is average power in watts
3. g denotes the frequency in hertz
4. Instruments are rms-responding

$$\mathbf{I}_g = \frac{\mathbf{E}_g}{\mathbf{Z}_g} \qquad \mathbf{Z}_g = R + j\left(2\pi f_g L - \frac{1}{2\pi f_g C}\right)$$

$$\mathbf{E}_{Rg} = \mathbf{I}_g R \qquad \mathbf{E}_{Lg} = \mathbf{I}_g(jX_{Lg}) \qquad \mathbf{E}_{Cg} = \mathbf{I}_g(-jX_{Cg})$$

$$P_g = E_g I_g \cos\!\!\measuredangle{}_{\mathbf{I}_g}^{\mathbf{E}_g}$$

$$P_g = I_{R,g}^2 R = \frac{E_{R,g}^2}{R} = E_{R,g} I_{R,g}$$

$$I_{\text{rss}} = \sqrt{I_0^2 + I_1^2 + I_2^2 + \cdots + I_g^2}$$

$$E_{\text{rss},R} = \sqrt{E_{R1}^2 + E_{R2}^2 + E_{R3}^2 + \cdots + E_{R,g}^2}$$

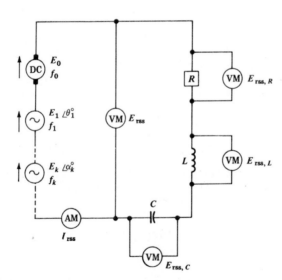

Figure 24-5 Circuit for interpreting formulas in summary.

$E_{\text{rss}, L}, E_{\text{rss}, C}, E_{\text{rss}, z}$ similar to preceding formula

$$P_{\text{MF}} = P_0 + P_1 + P_2 + \cdots + P_k$$

$$P_{\text{MF}} = E_{\text{rss}, R} I_{\text{rss}, R} = R I_{\text{rss}, R}^2 = \frac{E_{\text{rss}, R}^2}{R}$$

$$S_{\text{MF}} = E_{\text{rss}} I_{\text{rss}} \qquad \text{pf}_{\text{MF}} = \frac{P_{\text{MF}}}{S_{\text{MF}}}$$

PROBLEMS

24-1 The voltage impressed across a 10-Ω resistor is $e = (6 \sin 188.50t + 4 \sin 18,850t)$V. Sketch the circuit and determine (a) the frequency of each voltage component; (b) E_{rss}; (c) I_{rss}; (d) the average power dissipated in the resistor.

24-2 Two sinusoidal generators, an rms-reading ammeter, and a coil, are connected in series. The inductance and resistance of the coil are 0.010 H and 6.0 Ω, respectively. The generator voltages are expressed by $e_1 = 10 \sin 377t$ and $e_2 = 25 \sin 754t$, respectively. Sketch the circuit and determine (a) the ammeter reading; (b) the average heat power dissipated in the resistor.

24-3 A 100/0°-V 120-Hz generator and an 80/0°-V 60-Hz generator are connected in series with a 60-V battery and a coil. The resistance and inductance of the coil are 3.0 Ω and 2.65 mH, respectively. Sketch the circuit and determine (a) the impedance at each frequency; (b) I_{rss}.

24-4 Three sinusoidal generators and a battery are connected in series with a coil whose resistance and inductance are 8.0 Ω and 26.53 mH, respectively. The frequency and rms voltage of the respective generators are 20 Hz, 15 V; 60 Hz, 30 V; and 80 Hz, 50 V. The battery voltage is 6 V. Sketch the circuit and determine (a) the rss voltage; (b) the rss current; (c) the heat power supplied to the circuit; (d) the apparent power supplied to the circuit; (e) the power factor of the system.

24-5 A coil whose resistance and inductance are 50 Ω and 88 mH, respectively, is connected in series with three sinusoidal generators. The generator voltages are $e_1 = 400 \sin 377t$, $e_2 = 100 \sin 754t$, $e_3 = 50 \sin 1131t$. Sketch the circuit and determine (a) the frequency of each generator; (b) the rss voltage across the coil; (c) the rss current; (d) the average heat power expended in the coil.

24-6 A 30-Ω resistor is connected in series with a 500-μF capacitor, a 120-V battery, an ammeter, and three sinusoidal generators. The generator voltages are 100 sin 30t, 50 sin 80t, and 70 sin 100t V. Sketch the circuit and include a voltmeter across the capacitor-resistor combination. Determine (a) the voltmeter reading; (b) the ammeter reading; (c) the average power delivered to the resistor.

24-7 A load consisting of an 8.842-μF capacitor and a coil whose inductance and resistance are 88.4 mH and 5 Ω, respectively, is connected in series with three sinusoidal generators. The generator voltages are 100 sin 377t, 100 sin (377t + 50°) and 100 sin 1131t V. Sketch the circuit and determine (a) the frequency of each generator; (b) the rss voltage across the load; (c) the rss current; (d) the average power expended in the load.

CHAPTER 25

CIRCUITS WITH NONSINUSOIDAL PERIODIC SOURCES

Problems involving nonsinusoidal drivers such as that shown in Fig. 25-1a may be solved by converting the nonsinusoidal driver into an equivalent series circuit of sinusoidal drivers of different frequencies that produce the same current and deliver the same power to the load as did the original nonsinusoidal driver. An equivalent series circuit of *sinusoidal* drivers whose algebraic sum produces the same nonsinusoidal output voltage is shown in Fig. 25-1b, where

$$e_{gen} = e_{f1} + e_{f2} + e_{f3} + \cdots + e_{fn}$$

Once the sinusoidal components that make up the nonsinusoidal wave are known, the respective current and power components may be determined using the techniques developed in Chap. 24.

However, if only the effective value of the periodic wave is required and the equation of the nonsinusoidal wave is known, the effective value may be determined by substitution into one of the following equations that were derived in Sec. 9-4 of Chap. 9:

$$E_{rms} = \sqrt{\frac{1}{2\pi} \int_0^{2\pi} [f(\alpha)]^2 \, d\alpha} \qquad e = f(\alpha)$$

$$E_{rms} = \sqrt{\frac{1}{T} \int_0^T [f(t)]^2 \, dt} \qquad e = f(t)$$

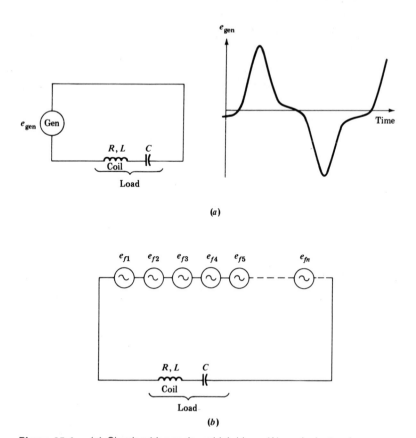

(a)

(b)

Figure 25-1 (*a*) Circuit with nonsinusoidal driver; (*b*) equivalent series circuit of sinusoidal drivers.

25-1   FOURIER-SERIES EXPANSION OF NONSINUSOIDAL PERIODIC WAVES

Nonsinusoidal periodic waves encountered in the real world of electric circuits, vibration, sound, etc., that can be defined over a 2π interval can be expanded into a series of sine and cosine waves of different frequencies, and a constant term called the 0-Hz or DC term. The series is called a Fourier series and is named after Fourier, a French mathematician and physicist.

The Fourier series for the general case of a nonsinusoidal periodic driving voltage of frequency f_1 is

$$
\begin{aligned}
e_{\text{gen}} = {} & A_0 + A_1 \cos 2\pi f_1 t + A_2 \cos [2\pi(2f_1)t] + \cdots \\
& + A_k \cos [2\pi(kf_1)t] + \cdots + A_n \cos [2\pi(nf_1)t] \\
& + B_1 \sin 2\pi f_1 t + B_2 \sin [2\pi(2f_1)t] + \cdots \\
& + B_k \sin [2\pi(kf_1)t] + \cdots + B_n \sin [2\pi(nf_1)t]
\end{aligned}
\tag{25-1}
$$

where e_{gen} = nonsinusoidal voltage wave

f_1 = frequency of e_{gen}

A_0 = DC term (0-Hz term)

k = a whole-number multiplier $(1, 2, 3, \ldots, n)$

A_k = amplitude of the cosine term whose frequency is k times the frequency of the nonsinusoidal wave e_{gen}

B_k = amplitude of the sine term whose frequency is k times the frequency of the nonsinusoidal wave e_{gen}

Note: The frequencies of the sine and cosine terms of the Fourier series are whole-number multiples of the frequency of the original nonsinusoidal wave.

The frequency of the nonsinusoidal wave is called the fundamental frequency, and all multiples of the fundamental frequency are called harmonic frequencies or harmonics. Thus,

f_1 = first harmonic (fundamental)

$2f_1$ = second harmonic

kf_1 = kth harmonic

The infinite number of terms in the Fourier series provides for all possible sine and cosine terms that may be represented in any given nonsinusoidal driver. However, the actual number of terms that constitute a particular nonsinusoidal driver depends on the shape of that particular wave.

Equation (25-1) may be simplified by defining

$$2\pi f_1 t = \omega t = \alpha \text{ rad}$$

then,

$$2\pi(2f_1)t = 2\omega t = 2\alpha$$

$$2\pi(kf_1)t = k\omega t = k\alpha$$

Substituting into Eq. (25-1),

$$\begin{aligned}
e_{gen} = f(\alpha) = {}& A_0 + A_1 \cos \alpha + A_2 \cos 2\alpha + \cdots + A_k \cos k\alpha + \cdots \\
& + A_n \cos n\alpha + B_1 \sin \alpha + B_2 \sin 2\alpha + \cdots + B_k \sin k\alpha + \cdots \\
& + B_n \sin n\alpha
\end{aligned} \tag{25-2}$$

Determination of the constant term A_0 and the coefficients $A_1, A_2, \ldots, A_n$, $B_1, B_2, \ldots, B_n$ may be accomplished by a number of different methods: If the equation representing the nonsinusoidal wave is known, the coefficients may be determined by integration using integral tables or by computer. If the equation representing the nonsinusoidal wave is not known, the coefficients may be determined by tedious longhand calculations and tabulations or by computer. If the actual nonsinusoidal voltage is available, the coefficients may be determined with a wave analyzer.

25-2   DETERMINATION OF COEFFICIENTS WHEN THE EQUATION OF THE NONSINUSOIDAL WAVE IS KNOWN

Using calculus, the constant term may be determined by evaluating one of the following integrals:

$$A_0 = \frac{1}{2\pi} \int_0^{2\pi} f(\alpha)\, d\alpha$$

or   (25-3)

$$A_0 = \frac{1}{2\pi} \int_{-\pi}^{+\pi} f(\alpha)\, d\alpha$$

Although any 2π interval (for example, $-\pi/2$ to $3\pi/2$) may be used, the 2π intervals in equation set (25-3) will be used in this chapter.

The integral $\int_0^{2\pi} f(\alpha)\, d\alpha$ is the net area between zero and 2π enclosed by the curve and the horizontal axis. Hence, if the shape of the curve is rectangular, triangular, or some other known configuration, the constant term may be calculated from the geometry of the wave by using the following equation:

$$A_0 = \frac{1}{2\pi} (\text{net area})_0^{2\pi}$$ (25-4)

If the net area is zero, that is, if the area above the horizontal axis is equal to the area below the horizontal axis, the constant term is zero ($A_0 = 0$).

The coefficient of any cosine term may be determined by evaluating one of the following integrals:

$$A_k = \frac{1}{\pi} \int_0^{2\pi} f(\alpha)\cos(k\alpha)\, d\alpha$$ (25-5)

or

$$A_k = \frac{1}{\pi} \int_{-\pi}^{\pi} f(\alpha)\cos(k\alpha)\, d\alpha$$ (25-6)

The coefficient of any sine term may be determined by evaluating one of the following integrals:

$$B_k = \frac{1}{\pi} \int_0^{2\pi} f(\alpha)\sin(k\alpha)d\alpha \tag{25-7}$$

or

$$B_k = \frac{1}{\pi} \int_{-\pi}^{\pi} f(\alpha)\sin(k\alpha)d\alpha \tag{25-8}$$

Example 25-1 For the nonsinusoidal wave in Fig. 25-2, determine (a) coefficients A_0, A_3, and B_4; (b) the frequency of the fourth harmonic if the period of the non-sinusoidal wave is 0.04 s

Solution
(a) The constant term can be easily determined from the net area of the wave for one cycle. Thus,

$$A_0 = \frac{1}{2\pi}(\text{net area})_0^{2\pi} = \frac{1}{2\pi}\left(4 \times \frac{4\pi}{3}\right) = 2.67$$

$$A_3 = \frac{1}{\pi}\int_0^{4\pi/3} 4\cos(3\alpha)d\alpha = \frac{4}{\pi}\left(\frac{\sin 3\alpha}{3}\right)_0^{4\pi/3}$$

$$A_3 = \frac{4}{3\pi}(\sin 4\pi) = 0$$

$$B_4 = \frac{1}{\pi}\int_0^{4\pi/3} 4\sin(4\alpha)d\alpha = \frac{4}{\pi}\left(-\frac{\cos 4\alpha}{4}\right)_0^{4\pi/3}$$

$$B_4 = -\frac{1}{\pi}\left(\cos\frac{16\pi}{3} - \cos 0\right) = -\frac{1}{\pi}(-0.5 - 1) = 0.48$$

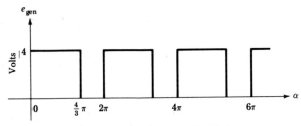

Figure 25-2   Periodic wave for Example 25-1.

(b) The frequency of the nonsinusoidal wave is

$$f = \frac{1}{T} = \frac{1}{0.04} = 25 \text{ Hz}$$

The fourth harmonic is four times the frequency of the nonsinusoidal wave. Hence,

$$f_4 = 4(25) = 100 \text{ Hz}$$

Example 25-2 Determine the Fourier-series coefficients A_0 and B_2 for the periodic voltage wave in Fig. 25-3.

Solution

$$A_0 = \frac{1}{2\pi} (\text{net area})_0^{2\pi} = \frac{1}{2\pi} (\text{area of large triangle} - \text{area of small triangle})$$

$$A_0 = \frac{1}{2\pi} \left[\frac{1}{2} \left(\frac{7\pi}{4} \times 140 \right) - \frac{1}{2} \left(\frac{\pi}{4} \times 20 \right) \right] = 60$$

If the DC term is to be calculated by the use of calculus, the equation for the straight-line function between $\alpha = 0$ and $\alpha = 2\pi$ must be determined. The general equation for a straight line is

$$y = mx + b \tag{25-9}$$

where y = vertical variable
         x = horizontal variable
         m = slope of the line
         b = y intercept (point of intersection of line with vertical axis)

For the straight line in Fig. 25-3, v corresponds to y and α corresponds to x. Hence Eq. (25-9) may be written as

$$v = m\alpha + b$$

The slope of the line is

$$m = \frac{140 + 20}{2\pi} = \frac{80}{\pi}$$

The vertical axis intercept is

$$b = -20$$

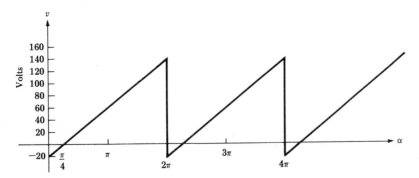

Figure 25-3   Periodic wave for Example 25-2.

Thus, the equation for the function in Fig. 25-3, between 0 and 2π, is

$$v = \frac{80}{\pi}\alpha - 20 \qquad\qquad (25\text{-}10)$$

Substituting Eq. (25-10) into Eq. (25-3) and integrating,

$$A_0 = \frac{1}{2\pi}\int_0^{2\pi}\left(\frac{80}{\pi}\alpha - 20\right)d\alpha$$

$$A_0 = \frac{1}{2\pi}\left(\int_0^{2\pi}\frac{80\alpha}{\pi}\,d\alpha - \int_0^{2\pi}20\,d\alpha\right)$$

$$A_0 = \frac{1}{2\pi}\left(\frac{80}{\pi}\frac{\alpha^2}{2} - 20\alpha\right)_0^{2\pi}$$

$$A_0 = \frac{1}{2\pi}\left[\frac{80(2\pi)^2}{2\pi} - 20(2\pi)\right] = 60$$

$$B_2 = \frac{1}{\pi}\int_0^{2\pi}f(\alpha)\sin 2\alpha\,d\alpha = \frac{1}{\pi}\int_0^{2\pi}\left(\frac{80}{\pi}\alpha - 20\right)\sin 2\alpha\,d\alpha$$

$$B_2 = \frac{20}{\pi}\left(\int_0^{2\pi}\frac{4\alpha}{\pi}\sin 2\alpha\,d\alpha - \int_0^{2\pi}\sin 2\alpha\,d\alpha\right)$$

Using integral tables,

$$B_2 = \frac{20}{\pi}\left\{\frac{4}{\pi}\left[\left(\frac{1}{2^2}\sin 2\alpha\right)_0^{2\pi} - \left(\frac{1}{2}\alpha\cos 2\alpha\right)_0^{2\pi}\right] - \left(-\frac{1}{2}\cos 2\alpha\right)_0^{2\pi}\right\}$$

$$B_2 = \frac{20}{\pi}\left\{\frac{4}{\pi}\left[0 - \frac{1}{2}(2\pi)\right] + 0\right\} = -25.46$$

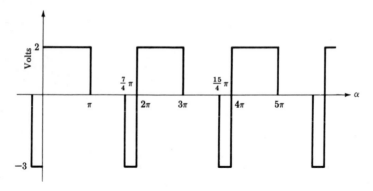

Figure 25-4 Periodic wave for Example 25-3.

Example 25-3 (a) Determine the Fourier coefficients A_0, A_k, and B_k for the periodic voltage wave in Fig. 25-4; (b) evaluate the first three harmonics.

Solution

(a) $\quad A_0 = \dfrac{1}{2\pi} \text{(net area)}_0^{2\pi} = \dfrac{1}{2\pi}\left(2\pi - 3\dfrac{\pi}{4}\right) = 0.63$

$$A_k = \frac{1}{\pi}\int_0^{2\pi} f(\alpha)\cos(k\alpha)\,d\alpha \qquad\qquad (25\text{-}11)$$

where $f(\alpha) = 2$ for $\alpha = 0$ to $\alpha = \pi$
$\qquad\quad f(\alpha) = 0$ for $\alpha = \pi$ to $\alpha = \frac{7}{4}\pi$
$\qquad\quad f(\alpha) = -3$ for $\alpha = \frac{7}{4}\pi$ to $\alpha = 2\pi$

Expanding Eq. (25-11),

$$A_k = \frac{1}{\pi}\int_0^{\pi} 2\cos(k\alpha)\,d\alpha + \frac{1}{\pi}\int_{\pi}^{7\pi/4} 0\cos(k\alpha)\,d\alpha + \frac{1}{\pi}\int_{7\pi/4}^{2\pi}(-3)\cos(k\alpha)\,d\alpha$$

Integrating,

$$A_k = \frac{1}{\pi}\left[\frac{2\sin k\alpha}{k}\Big|_0^{\pi} + 0 + \frac{(-3)\sin k\alpha}{k}\Big|_{7\pi/4}^{2\pi}\right]$$

$$A_k = \frac{1}{\pi}\left[\left(\frac{2\sin k\pi}{k} - 0\right) - \frac{3}{k}\left(\sin 2\pi k - \sin\frac{7\pi k}{4}\right)\right]$$

Recognizing that $\sin k\pi = 0$ and $\sin 2\pi k = 0$ for all values of k ($k = 1, 2, 3, \ldots, n$),

$$A_k = \frac{3}{k\pi}\left(\sin\frac{7k\pi}{4}\right) \tag{25-12}$$

Similarly,

$$B_k = \frac{1}{\pi}\int_0^{2\pi} f(\alpha)\sin(k\alpha)\,d\alpha$$

$$B_k = \frac{1}{\pi}\int_0^{\pi} 2\sin(k\alpha)\,d\alpha + \frac{1}{\pi}\int_{\pi}^{7\pi/4} 0\,\sin(k\alpha)\,d\alpha + \frac{1}{\pi}\int_{7\pi/4}^{2\pi} -3\sin(k\alpha)\,d\alpha$$

Integrating,

$$B_k = \frac{1}{\pi}\left[\frac{2(-\cos k\alpha)}{k}\Big|_0^{\pi} + 0 - 3\frac{-\cos k\alpha}{k}\Big|_{7\pi/4}^{2\pi}\right]$$

$$B_k = \frac{1}{k\pi}\left[-2(\cos k\pi - \cos 0) + 3\left(\cos 2k\pi - \cos\frac{7k\pi}{4}\right)\right]$$

$\cos 2k\pi = 1$ for all values of k ($k = 1, 2, 3, \ldots$)

$\cos 0 = 1$

Hence,

$$B_k = \frac{1}{k\pi}\left[-2(\cos k\pi - 1) + 3\left(1 - \cos\frac{7k\pi}{4}\right)\right] \tag{25-13}$$

(b) The first three terms (sine and cosine) may be obtained by substituting 1, 2, and 3, respectively, for k in Eqs. (25-12) and (25-13).

$$A_1 = \frac{3}{\pi}\sin\frac{7\pi}{\pi} = -0.68$$

$$A_2 = \frac{3}{2\pi}\sin\frac{14\pi}{4} = -0.48$$

$$A_3 = \frac{3}{3\pi}\sin\frac{21\pi}{4} = -0.23$$

$$B_1 = \frac{1}{\pi}\left(-2\cos\pi + 2 + 3 - 3\cos\frac{7\pi}{4}\right) = 1.55$$

$$B_2 = \frac{1}{2\pi}\left(-2\cos 2\pi + 2 + 3 - 3\cos\frac{14\pi}{4}\right) = 0.48$$

$$B_3 = \frac{1}{3\pi}\left(-2\cos 3\pi + 2 + 3 - 3\cos\frac{21\pi}{4}\right) = 0.97$$

Thus the DC term and the first three harmonics of the driving voltage in Fig. 25-4 are

$$v = 0.63 - 0.68\cos\alpha + 1.55\sin\alpha - 0.48\cos 2\alpha + 0.48\sin 2\alpha$$

$$-0.23\cos 3\alpha + 0.97\sin 3\alpha$$

Derivation of Equation (25-3)

The equation for the DC term may be obtained by multiplying both sides of Eq. (25-2) by $d\alpha$ and integrating between 0 and 2π. Thus,

$$\int_0^{2\pi} f(\alpha)\,d\alpha = \int_0^{2\pi} A_0\,d\alpha + \int_0^{2\pi} A_1\cos(\alpha)\,d\alpha + \int_0^{2\pi} A_2\cos(2\alpha)\,d\alpha + \cdots$$

$$+ \int_0^{2\pi} A_k\cos(k\alpha)\,d\alpha + \cdots + \int_0^{2\pi} A_n\cos(n\alpha)\,d\alpha$$

$$+ \int_0^{2\pi} B_1\sin(\alpha)\,d\alpha + \int_0^{2\pi} B_2\sin(2\alpha)\,d\alpha + \cdots$$

$$+ \int_0^{2\pi} B_k\sin(k\alpha)\,d\alpha + \cdots$$

The integration of all sine terms and cosine terms between zero and 2π is zero. Hence,

$$\int_0^{2\pi} f(\alpha)\,d\alpha = \int_0^{2\pi} A_0\,d\alpha$$

$$\int_0^{2\pi} f(\alpha)\,d\alpha = A_0\alpha\Big|_0^{2\pi} = 2\pi A_0$$

$$A_0 = \frac{1}{2\pi}\int_0^{2\pi} f(\alpha)\,d\alpha$$

Derivation of Equation (25-5)

Equation (25-5) may be obtained by multiplying both sides of Eq. (25-2) by $\cos(k\alpha)d\alpha$ and integrating between 0 and 2π. Thus,

$$\int_0^{2\pi} f(\alpha)\cos(k\alpha)\,d\alpha = \int_0^{2\pi} A_0\,\cos(k\alpha)\,d\alpha + \int_0^{2\pi} A_1\,\cos(\alpha)(k\alpha)\,d\alpha$$

$$+ \int_0^{2\pi} A_2\,\cos(2\alpha)\cos(k\alpha)\,d\alpha + \cdots + \int_0^{2\pi} A_k\,\cos(k\alpha)\cos(k\alpha)\,d\alpha + \cdots$$

$$+ \int_0^{2\pi} B_1\,\sin(\alpha)\cos(k\alpha)\,d\alpha + \int_0^{2\pi} B_2\,\sin(2\alpha)\cos(k\alpha)\,d\alpha + \cdots$$

$$+ \int_0^{2\pi} B_k\,\sin(k\alpha)\cos(k\alpha)\,d\alpha + \cdots \tag{25-14}$$

Except for the A_k term, all terms on the right-hand side of the equal sign integrate to zero. Hence Eq. (25-14) reduces to

$$\int_0^{2\pi} f(\alpha)\cos(k\alpha)\,d\alpha = \int_0^{2\pi} A_k\,\cos^2(k\alpha)\,d\alpha$$

$$\int_0^{2\pi} f(\alpha)\cos(k\alpha)\,d\alpha = A_k\pi$$

$$A_k = \frac{1}{\pi}\int_0^{2\pi} f(\alpha)\cos(k\alpha)\,d\alpha$$

Derivation of Equation (25-7)

Equation (25-7) may be obtained by multiplying both sides of Eq. (25-2) by $\sin(k\alpha)d\alpha$ and integrating between 0 and 2π. Thus,

$$\int_0^{2\pi} f(\alpha)\sin(k\alpha)\,d\alpha = \int_0^{2\pi} A_0\,\sin(k\alpha)\,d\alpha + \int_0^{2\pi} A_1\,\cos\alpha\,\sin(k\alpha)\,d\alpha$$

$$+ \int_0^{2\pi} A_2\,\cos 2\alpha\,\sin(k\alpha)\,d\alpha + \cdots + \int_0^{2\pi} A_k\,\cos k\alpha\,\sin(k\alpha)\,d\alpha + \cdots$$

$$+ \int_0^{2\pi} B_1\,\sin\alpha\,\sin(k\alpha)\,d\alpha + \int_0^{2\pi} B_2\,\sin 2\alpha\,\sin(k\alpha)\,d\alpha + \cdots$$

$$+ \int_0^{2\pi} B_k\,\sin k\alpha\,\sin(k\alpha)\,d\alpha + \cdots \tag{25-15}$$

Except for the B_k term, all terms on the right-hand side of the above equation integrate to zero. Hence Eq. (25-15) reduces to

$$\int_0^{2\pi} f(\alpha)\sin(k\alpha)\,d\alpha = \int_0^{2\pi} B_k \sin^2(k\alpha)\,d\alpha$$

$$\int_0^{2\pi} f(\alpha)\sin(k\alpha)\,d\alpha = B_k\pi$$

$$B_k = \frac{1}{\pi}\int_0^{2\pi} f(\alpha)\sin(k\alpha)\,d\alpha$$

25-3 EFFECT OF SYMMETRY ON THE NUMBER OF TERMS IN THE FOURIER SERIES

A nonsinusoidal wave that is symmetrical about the vertical axis, symmetrical about the horizontal axis, or symmetrical about both axes will have significantly fewer terms in its Fourier series. Simple symmetry checks of the nonsinusoidal wave can determine the *absence* of even-numbered terms, cosine terms, or sine terms, from its equivalent Fourier series, and thus avoid unnecessary calculations.

Half-Wave Symmetry

If the bottom half of a nonsinusoidal wave is the mirror image of the top half but if displaced from it by π rad as shown in Fig. 25-5a, the wave is said to have *half-wave symmetry*. A wave possessing half-wave symmetry will have no DC term and no even harmonics. That is, if $f(\alpha) = -f(\alpha + \pi)$,

$$A_0 = 0 \qquad A_2, A_4, A_6, \ldots, A_n = 0$$

$$B_2, B_4, B_6, \ldots, B_n = 0$$

Half-wave symmetry is unaffected by the location of the vertical axis.

Even Symmetry

If the symmetry of the nonsinusoidal wave about the vertical axis is similar to that shown in Fig. 25-5b, the wave is said to have *even symmetry*. A wave possessing even symmetry will have no sine terms. That is, if $f(\alpha) = f(-\alpha)$.
$$B_1, B_2, B_3, \ldots, B_n = 0$$

Odd Symmetry

If the symmetry of the nonsinusoidal wave about the vertical axis is similar to that shown in Fig. 25-5c, the wave is said to have *odd symmetry*. A wave possessing odd symmetry will have no cosine terms. That is, if $f(\alpha) = -f(-\alpha)$,
$$A_1, A_2, A_3, \ldots, A_n = 0$$

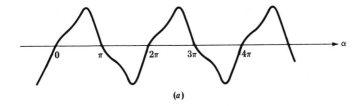

(a)

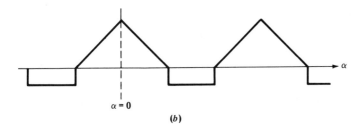

$\alpha = 0$

(b)

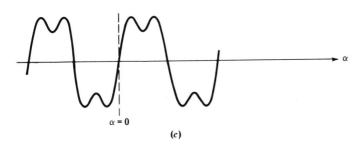

$\alpha = 0$

(c)

Figure 25-5 Symmetrical waves: (a) half-way symmetry; (b) even symmetry; (c) odd symmetry.

The three sketches in Fig. 25-6 illustrate how shifting the vertical axis can reduce the number of terms in a Fourier series.

In Fig. 25-6a the wave has only half-wave symmetry. Hence *only odd harmonics (sine and cosine) are present; there are no even harmonics and no constant term.*

In Fig. 25-6b the wave has both half-wave symmetry and even symmetry. Hence *only odd cosine terms are present; there are no sine terms, no even harmonics, and no constant term.*

In Fig. 25-6c the wave has both half-wave symmetry and odd symmetry. Hence *only odd sine terms are present; there are no cosine terms, no even harmonics, and no constant term.*

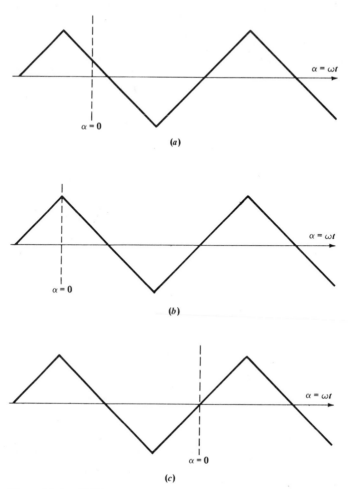

Figure 25-6   Shifting the vertical axis to reduce the number of terms in a Fourier series.

25-4   FREQUENCY SPECTRUM OF A NONSINUSOIDAL WAVE

The frequency spectrum of a nonsinusoidal periodic wave is a plot of harmonic amplitude versus harmonic frequency. The plot is in the form of a bar graph or line graph as shown in Fig. 25-7. No frequencies exist between the plotted lines. Waves with smooth curves such as those shown in Fig. 25-7a have a rapidly converging series. That is, only the first few harmonics are dominant; the amplitude of the higher-order harmonics decreases rapidly with increasing frequency. On the other hand, periodic waves with sharp bends such as the sawtooth wave shown in Fig. 25-7b have a slowly converging series; such waves have many dominant harmonics.

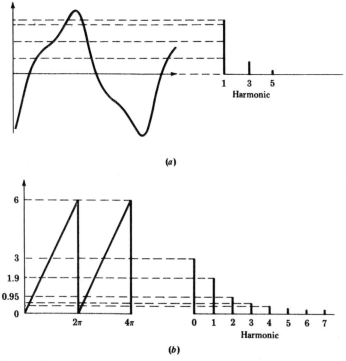

(a)

(b)

Figure 25-7   Frequency spectrum of periodic waves with (a) smooth curves; (b) sharp bends.

The amplitude and phase angle of any harmonic may be obtained by combining the respective sine and cosine components for that frequency. For example, the addition of the two components of the k harmonic is

$$A_k \cos k\omega t + B_k \sin k\omega t = C_k \sin(k\omega t + \phi_k^\circ)$$

where C_k = amplitude of harmonic k
ϕ_k = phase angle of harmonic k

Converting the cosine term to an equivalent sine term,

$$A_k \sin(k\omega t + 90°) + B_k \sin k\omega t = C_k \sin(k\omega t + \phi_k^\circ) \qquad (25\text{-}16)$$

Expressing Eq. (25-16) in phasor form,

$$A_k\underline{/90°} + B_k\underline{/0°} = C_k\underline{/\phi_k^\circ}$$

$$jA_k + B_k = C_k\underline{/\phi_k^\circ} \qquad (25\text{-}17)$$

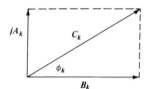

Figure 25-8   Phasor diagram for determining the magnitude of a harmonic from its sine and cosine components.

The phasor diagram corresponding to Eq. (25-17) is shown in Fig. 25-8. From the geometry of the phasor diagram,

$$C_k = \sqrt{A_k^2 + B_k^2} \qquad\qquad (25\text{-}18)$$

$$\phi_k = \tan^{-1} \frac{A_k}{B_k} \qquad\qquad (25\text{-}19)$$

Note: The calculated values for the A_k and B_k components are dependent on the location of the $\alpha = 0$ axis and will be different for different $\alpha = 0$ locations (see Fig. 25-6). However, the actual amplitudes of the harmonics (C_k), as determined from Eq. (25-18), are independent of the $\alpha = 0$ location. This must be so, because the amplitude and frequency of the harmonics present in a given nonsinusoidal wave cannot change with the method of calculation. Positioning the vertical axis to simplify the calculations cannot change the frequency or amplitude of the harmonics.

Example 25-4 (*a*) Write the Fourier-series equation represented by the following Fourier coefficients of a certain 20-Hz nonsinusoidal periodic voltage wave; (*b*) plot the frequency spectrum; (*c*) determine the rss voltage.

(*a*) $A_0 = 8.00$

$$A_1 = -6.63 \qquad B_1 = 10.00$$

$$A_2 = 0.26 \qquad B_2 = -4.99$$

$$A_3 = -0.97 \qquad B_3 = 3.33$$

$$A_4 = 0.26 \qquad B_4 = -2.49$$

$$A_5 = -0.52 \qquad B_5 = 1.99$$

$$A_6 = 0.26 \qquad B_6 = -1.65$$

Solution

(a) Using Eqs. (25-18) and (25-19), the calculated values for C_k are ϕ_k are:

k	f, Hz	C_k, V	ϕ_k, degrees
0	0	8.00	
1	20	12.00	− 33.54
2	40	5.00	+ 177.02
3	60	3.47	− 16.24
4	80	2.50	+ 174.04
5	100	2.06	− 14.64
6	120	1.67	+ 171.05

Using the calculated data, the Fourier-series representation of the nonsinusoidal voltage wave is

$$e = 8 + 12 \sin(2\pi 20t - 33.54°) + 5.0 \sin(2\pi 40t + 177.02°) + 3.47 \sin(2\pi 60t - 16.24°)$$

$$+ 2.5 \sin(2\pi 80t + 174.04°) + 2.06 \sin(2\pi 100t - 14.64°)$$

$$+ 1.67 \sin(2\pi 120t + 171.05°)$$

(b) The frequency spectrum is plotted in Fig. 25-9.

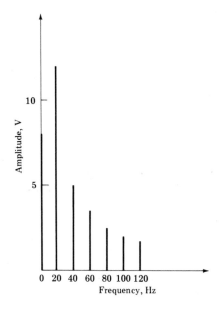

0 20 40 60 80 100 120
Frequency, Hz

Figure 25-9 Frequency spectrum for Example 25-4.

$(c)\ E_{rss} = \sqrt{(8)^2 + \left(\dfrac{12}{\sqrt{2}}\right)^2 + \left(\dfrac{5}{\sqrt{2}}\right)^2 + \left(\dfrac{3.47}{\sqrt{2}}\right)^2 + \left(\dfrac{2.5}{\sqrt{2}}\right)^2 + \left(\dfrac{2.06}{\sqrt{2}}\right)^2 + \left(\dfrac{1.67}{\sqrt{2}}\right)^2}$

$E_{rss} = 12.69\ V$

SUMMARY OF FORMULAS

Fourier-Series Coefficients

$$A_0 = \frac{1}{2\pi}\ (\text{net area})_0^{2\pi} = \frac{1}{2\pi}\int_0^{2\pi} f(\alpha)\ d\alpha$$

$$A_k = \frac{1}{\pi}\int_0^{2\pi} f(\alpha)\cos(k\alpha)\ d\alpha$$

$$B_k = \frac{1}{\pi}\int_0^{2\pi} f(\alpha)\sin(k\alpha)\ d\alpha$$

$$C_k = \sqrt{A_k^2 + B_k^2} \qquad \phi_k = \tan^{-1}\frac{A_k}{B_k}$$

$$C_k = B_k + jA_k$$

$$\int \alpha \sin(k\alpha)\ d\alpha = \frac{1}{k^2}\sin k\alpha - \frac{\alpha}{k}\cos k\alpha$$

$$\int \alpha \cos(k\alpha)\ d\alpha = \frac{1}{k^2}\cos k\alpha + \frac{\alpha}{k}\sin k\alpha$$

$$\int \sin(k\alpha)\ d\alpha = -\frac{\cos k\alpha}{k}$$

$$\int \cos(k\alpha)\ d\alpha = \frac{\sin k\alpha}{k}$$

PROBLEMS

25-1 One cycle of a nonsinusoidal periodic voltage wave has a value of 1 V from 0 to $\pi/2$, 0 V from $\pi/2$ to π, and -0.5 V from π to 2π. Sketch the wave and determine (a) the DC term; (b) the coefficient of the third harmonic.

25-2 For the periodic voltage wave in Fig. 25-10, determine the constant term and the coefficient of the fifth harmonic.

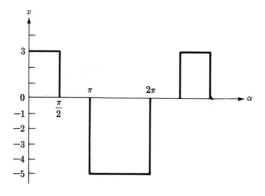

Figure 25-10 Circuit for Prob. 25-2.

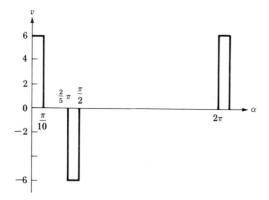

Figure 25-11 Circuit for Prob. 25-3.

25-3 In Fig. 25-11, determine the constant term and the coefficient of the first harmonic.

25-4 (a) Determine A_0, A_3, B_3, and C_3 for the 300-Hz periodic voltage wave in Fig. 25-12; (b) write the sinusoidal equation for the third harmonic as a function of time; (c) determine the rms value of the third harmonic.

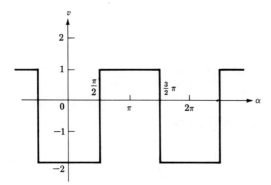

Figure 25-12 Circuit for Prob. 25-4.

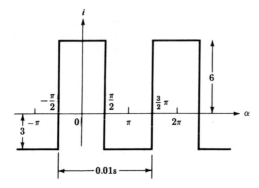

Figure 25-13 Circuit for Prob. 25-5.

25-5 For the periodic wave in Fig. 25-13, determine (*a*) the DC term; (*b*) the amplitude of the fifth harmonic; (*c*) the frequency of the fifth harmonic.

25-6 The following Fourier coefficients represent the components of a certain 20-Hz periodic current wave in a 6-Ω resistor:

$A_0 = -8.70$
$A_1 = -4.64$ $B_1 = -6.06$
$A_2 = 4.14$ $B_2 = 0.47$
$A_3 = -0.59$ $B_3 = -0.24$
$A_4 = 0.57$ $B_4 = -0.38$
$A_5 = 0.22$ $B_5 = 0.26$

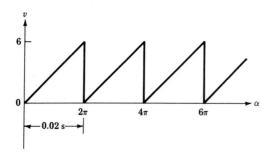

Figure 25-14 Circuit for Prob. 25-7.

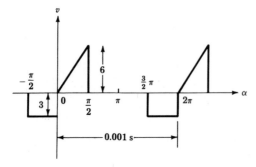

Figure 25-15 Circuit for Prob. 25-8.

(*a*) Plot the frequency spectrum; (*b*) determine the rss current in the resistor; (*c*) calculate the average power drawn by the resistor.

25-7 Determine (*a*) the DC term and the coefficients of the first and second harmonics of the periodic wave shown in Fig. 25-14; (*b*) write the time-domain equation represented by the coefficients in part *a*.

25-8 (*a*) Calculate the DC term and the coefficient of the fifth harmonic for the periodic voltage wave in Fig. 25-15; (*b*) write the time-domain equation represented by the coefficients in part *a*.

CHAPTER 26

TRANSIENTS IN SOURCE-FREE CIRCUITS

The circuit analysis presented in the preceding chapters dealt only with the steady-state behavior of electric circuits that are driven by a DC or sinusoidal source. Chapters 26 and 27 deal with the *complete behavior of the circuit, for all time, including the behavior during the first five time constants after the switch is operated.*

The behavior of a circuit during the first five time constants is the response of the circuit to energy stored or being stored in capacitors, inductors, or both. This behavior, which is of short duration, is called the *source-free* or *natural response.*

26-1 SOURCE-FREE RESPONSE OF AN RL CIRCUIT

Figure 26-1a shows a coil of L H and R Ω in series with a battery, and a resistor of R_1 Ω. Assuming the circuit is at steady state, and the switch is open, the current in the coil is

$$i_{coil} = \frac{E_{bat}}{R + R_1} \tag{26-1}$$

When the switch is closed, the coil is shorted, and although it no longer receives current from the battery, the inductance of the circuit delays the change in current. Hence the current through the coil the instant after the

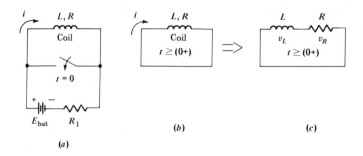

(a)

(b)

(c)

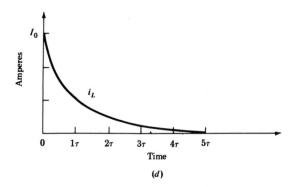

(d)

Figure 26-1   Source-free response of an *RL* circuit.

switch is closed will be equal to the current through the coil the instant before
the switch was closed. Expressed symbolically,

$$\left. i_L \right|_{t=(0+)} = \left. i_L \right|_{t=(0-)} = \frac{E_{\text{bat}}}{R + R_1} = I_0 \tag{26-2}$$

where $t = (0-)$ is the infinitesimal time before the switch was closed
        $t = (0+)$ is the infinitesimal time after the switch is closed
        $I_0 = $ current at $t = (0+)$

*The instantaneous values of current and voltage at $t = (0+)$ are called the
initial conditions (IC) of the circuit.*

The *RL* circuit formed by the coil and switch at $t = (0+)$ is shown in
Fig. 26-1*b*, and the equivalent series circuit model is shown in Fig. 26-1*c*.
Figure 26-1*d* shows the behavior of the coil current after the switch is closed.
The current decays from its initial value of I_0 A to essentially 0 A in five time

constants. The time constant of an *RL* circuit is derived in Sec. 8-5 and is equal to

$$\tau = \frac{L}{R} \quad \text{s}$$

where R = resistance, Ω
$\quad\quad$ L = inductance, H

Applying Kirchhoff's voltage law to Fig. 26-1*c* and recognizing that there is no driver in the loop,

$$0 = v_L + v_R \tag{26-3}$$

Expressing Eq. (26-3) in terms of current,

$$0 = L\frac{di}{dt} + iR \tag{26-4}$$

The solution of differential equations such as Eq. (26-4) or integro-differential equations that will be discussed later, whose parameters (R, L, C) are constants, can have only exponential terms, sinusoidal terms, or products of an exponential term and a sinusoidal term. These are the only mathematical expressions whose derivative and integral have the same form as the function itself, and are therefore the only possible solutions. The exponential and the sinusoid are related by Euler's formula, given in Sec. 26-3.
The solution of Eq. (26-4) is the exponential function

$$i = A\varepsilon^{pt} \tag{26-5}$$

where A = a constant that depends on IC
$\quad\quad$ p = a constant that depends on circuit parameters
$\quad\quad$ ε = 2.718 (base for natural logarithms)
$\quad\quad$ t = elapsed time since switch was closed, s

The exponential function represented by Eq. (26-5) describes the current curve in Fig. 26-1*d*.
The constant p may be determined by substituting Eq. (26-5) into Eq. (26-4), performing the indicated operations, and solving for p. Thus,

$$0 = L\frac{d}{dt}(A\varepsilon^{pt}) + R(A\varepsilon^{pt})$$

$$0 = LAp\varepsilon^{pt} + RA\varepsilon^{pt}$$

$$\boxed{0 = Lp + R} \tag{26-6}$$

Equation (26-6) is called the *characteristic equation* because, as will be seen later, its root indicates the characteristic behavior of the response.

Solving for the root of the characteristic equation,

$$p = -\frac{R}{L} \tag{26-7}$$

Substituting Eq. (26-7) into Eq. (26-5),

$$\boxed{i = A\varepsilon^{-(R/L)t}} \tag{26-8}$$

Equation (26-8) is the *general solution of differential equation* (26-4).

The coefficient A is determined by substituting the initial conditions $i|_{t=0+} = I_0$ into Eq. (26-8) and evaluating. Thus,

$$I_0 = A\varepsilon^{-(R/L)(0+)}$$

$$I_0 = A$$

Hence Eq. (26-8) becomes

$$\boxed{i = I_0\varepsilon^{-(R/L)t}} \tag{26-9}$$

$$\text{where} \quad I_0 = \frac{E_{bat}}{R + R_1}$$

Equation (26-9) is the *specific solution of Eq. (26-4) for the initial conditions specified in the problem.*

To check the validity of the specific solution, substitute Eq. (26-9) into Eq. (26-4) and perform the indicated operations:

$$0 = L\frac{d}{dt}(I_0\varepsilon^{-(R/L)t}) + I_0\varepsilon^{-(R/L)t}R$$

$$0 = L\left[I_0\left(-\frac{R}{L}\right)\varepsilon^{-(R/L)t}\right] + I_0R\varepsilon^{-(R/L)t}$$

$$0 = -I_0R\varepsilon^{-(R/L)t} + I_0R\varepsilon^{-(R/L)t}$$

$$0 = 0$$

Thus Eq. (26-9) is indeed the specific solution.

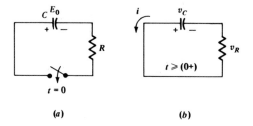

(a)   (b)

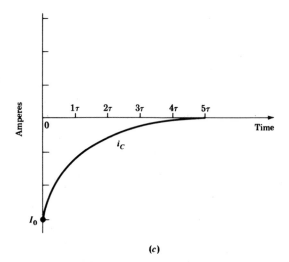

(c)

Figure 26-2   Source-free response of an *RC* circuit.

6-2   SOURCE-FREE RESPONSE OF AN RC CIRCUIT

Figure 26-2a shows a charged ideal capacitor in series with a switch and a resistor. The voltage across the capacitor is E_0 V.

The capacitance of a capacitor delays a change in the voltage across it. Hence the voltage across the capacitor at the instant the switch is closed will be equal to the voltage across it before the switch was closed. That is,

$$v_C \Big|_{t=(0+)} = v_C \Big|_{t=(0-)} = E_0$$

The circuit after the switch is closed is shown in Fig. 26-2b. Applying Kirchhoff's voltage law to the loop, and recognizing that there is no driver,

$$0 = v_C + v_R \tag{26-10}$$

At $t = (0+)$, Eq. (26-10) becomes

$$0 = E_0 + I_0 R$$

where $E_0 = $ voltage across the capacitor at $t = (0+)$
   $I_0 = $ capacitor current at $t = (0+)$

Solving for the current,

$$I_0 = -\frac{E_0}{R}$$

The negative sign indicates that the capacitor is discharging.

Figure 26-2c shows the behavior of the capacitor current after the switch is closed; the current decays from its I_0 A to essentially 0 A in five time constants. The time constant of an RC circuit is derived in Sec. 6-6 and is equal to

$$\tau = RC \qquad \text{s}$$

where $R = $ resistance, Ω
   $C = $ capacitance, F

The current at any instant of time after the switch is closed may be determined by expressing Eq. (26-10) in terms of the circuit current and then solving for the current. Thus,

$$0 = v_C + v_R$$

$$0 = \frac{1}{C} \int i \, dt + iR \tag{26-11}$$

Taking the derivative of Eq. (26-11) with respect to time to obtain a pure differential equation,

$$0 = \frac{1}{C} i + R \frac{di}{dt} \tag{26-12}$$

The solution of Eq. (26-12) is the exponential function

$$i = A\varepsilon^{pt} \tag{26-13}$$

Substituting the exponential function into Eq. (26-12) and performing the indicated operations,

$$0 = \frac{1}{C}(A\varepsilon^{pt}) + R\frac{d}{dt}(A\varepsilon^{pt})$$

$$0 = \frac{A}{C}\varepsilon^{pt} + ARp\varepsilon^{pt}$$

$$0 = \frac{1}{C} + Rp \qquad (26\text{-}14)$$

Equation (26-14) is the *characteristic equation* for the circuit in Fig. 26-2b. Solving for the root,

$$p = -\frac{1}{RC}$$

Substituting the root of Eq. (26-14) into the exponential function represented by Eq. (26-13),

$$i = A\varepsilon^{-(1/RC)t} \qquad (26\text{-}15)$$

Equation (26-15) is the *general solution* for the source-free response of the RC circuit shown in Fig. 26-2b.

Coefficient A may be determined from the initial condition of the problem. Substituting the IC into Eq. (26-15),

$$I_0 = A\varepsilon^{-(1/RC)(0+)}$$

Evaluating,

$$A = I_0$$

Hence Eq. (26-15) becomes

$$i = I_0\varepsilon^{(-1/RC)t} \qquad (26\text{-}16)$$

where

$$I_0 = -\frac{E_0}{R}$$

Equation (26-16) is the *specific solution of Eq. (26-12) for the initial conditions specified in the problem.*

26-3 SOURCE-FREE RESPONSE OF AN LC CIRCUIT

Figure 26-3a shows a charged ideal capacitor in series with an ideal inductor ($R_{coil} = 0$) and a switch. The voltage across the capacitor is E_0 V. When the switch is closed, the capacitor is connected across the ideal inductor, as shown in Fig. 26-3b.

The capacitance of the capacitor delays a change in voltage across it, and the inductance of the inductor delays a change in the current through it. Thus,

$$v_C\big|_{t=(0+)} = v_C\big|_{t=(0-)} = E_0 \qquad \text{and} \qquad i_L\big|_{t=(0+)} = i_L\big|_{t=(0-)} = I_0 = 0$$

Applying Kirchhoff's voltage law to the circuit in Fig. 26-3b and recognizing that there is no driver,

$$0 = v_L + v_C \tag{26-17}$$

Expressing Eq. (26-17) in terms of the circuit current,

$$0 = L\frac{di}{dt} + \frac{1}{C}\int i\,dt \tag{26-18}$$

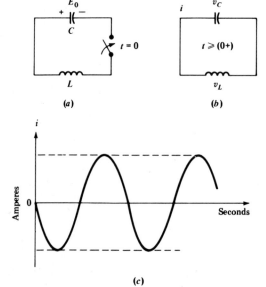

(a)

(b)

(c)

Taking the derivative of Eq. (26-18) with respect to time to obtain a pure differential equation,

$$0 = L\frac{d^2i}{dt^2} + \frac{1}{C}i \qquad (26\text{-}19)$$

The solution of Eq. (26-19) is the exponential function

$$i = A\varepsilon^{pt} \qquad (26.20)$$

Substituting Eq. (26-20) into Eq. (26-19) and performing the indicated operations,

$$0 = L\frac{d^2}{dt^2}(A\varepsilon^{pt}) + \frac{1}{C}(A\varepsilon^{pt})$$

$$0 = LAp^2\varepsilon^{pt} + \frac{A}{C}\varepsilon^{pt}$$

$$\boxed{0 = Lp^2 + \frac{1}{C}} \qquad (26\text{-}21)$$

Equation (26-21) is the *characteristic equation* for the circuit shown in Fig. 26-3b. Solving for the roots of the characteristic equation,

$$p^2 = -\frac{1}{LC}$$

$$p = \pm\sqrt{\frac{-1}{LC}}$$

$$p = +j\sqrt{\frac{1}{LC}} \quad , \quad -j\sqrt{\frac{1}{LC}}$$

Defining,

$$\sqrt{\frac{1}{LC}} = \omega$$

$$p = +j\omega \quad , \quad -j\omega \qquad (26\text{-}22)$$

Substituting the roots of the characteristic equation into Eq. (26-20) results in

$$i = A_1 \varepsilon^{j\omega t} + A_2 \varepsilon^{-j\omega t} \tag{26-23}$$

From Euler's formula,

$$\varepsilon^{j\omega t} = \cos \omega t + j \sin \omega t \tag{26-24}$$

$$\varepsilon^{-j\omega t} = \cos \omega t - j \sin \omega t \tag{26-25}$$

Substituting Eqs. (26-24) and (26-25) into Eq. (26-23),

$$i = A_1(\cos \omega t + j \sin \omega t) + A_2(\cos \omega t - j \sin \omega t)$$

Rearranging the terms,

$$i = (A_1 + A_2)\cos \omega t + j(A_1 - A_2)\sin \omega t \tag{26-26}$$

The current i to the left of the equal sign in Eq. (26-26) is real. Hence the quantities $(A_1 + A_2)$ and $j(A_1 - A_2)$ must be real numbers. To satisfy this requirement, A_1 and A_2 must be complex conjugates. For example, if

$$A_1 = 2 + j3 \qquad \text{and} \qquad A_2 = 2 - j3$$

then

$$A_1 + A_2 = (2 + j3) + (2 - j3) = 4$$

$$j(A_1 - A_2) = j[(2 + j3) - (2 - j3)] = j(j6) = -6$$

Both 4 and -6 are real numbers. Thus, defining,

$$B_1 = A_1 + A_2 \qquad \text{and} \qquad B_2 = j(A_1 - A_2)$$

where B_1 and B_2 are real numbers, Eq. (26-26) becomes

$$\boxed{i = B_1 \cos \omega t + B_2 \sin \omega t} \tag{26-27}$$

Equation 26-27 is the *general solution* for the source-free response of the *LC* circuit shown in Fig. 26-3b.

Coefficients B_1 and B_2 may be determined from the initial conditions of the problem. Substituting the IC into Eq. (26-27) and evaluating,

$$0 = B_1 \cos[\omega(0+)] + B_2 \sin[\omega(0+)]$$

$$0 = B_1 + 0$$

$$B_1 = 0 \tag{26-28}$$

To determine B_2, a second set of IC is required. This is obtained by examining Kirchhoff's voltage equation at $t = (0+)$. Thus,

$$0 = v_L \Big|_{t=(0+)} + v_C \Big|_{t=(0+)}$$

$$0 = L \frac{di}{dt} \Big|_{t=(0+)} + E_0$$

$$\frac{di}{dt} \Big|_{t=(0+)} = - \frac{E_0}{L} \tag{26-29}$$

Taking the derivative of Eq. (26-27) to obtain di/dt,

$$\frac{di}{dt} = -B_1 \omega \sin \omega t + B_2 \omega \cos \omega t$$

Substituting the initial conditions expressed in Eq. (26-29) and knowing that $B_1 = 0$,

$$\frac{-E_0}{L} = B_2 \omega \cos[\omega(0+)]$$

$$\frac{-E_0}{L} = B_2 \omega$$

$$B_2 = \frac{-E_0}{\omega L}$$

Hence Eq. (26-27) becomes

$$\boxed{i = \frac{-E_0}{\omega L} \sin \omega t} \tag{26-30}$$

Equation (26-30) is the *specific solution of Eq. (26-19) for the initial conditions specified in the problem.* The minus sign indicates the initial discharge current is opposite in direction to the current that originally charged the capacitor.

Figure 26-3c shows the oscillatory response of the circuit. The relationship between the frequency in hertz and the radian frequency is

$$\omega = 2\pi f$$

26-4   **OBTAINING THE CHARACTERISTIC EQUATION BY INSPECTION**

Comparing differential equations (26-4), (26-12), and (26-19) with their corresponding characteristic equations (26-6), (26-14), and (26-21) indicates that the differential equations can be transformed into their corresponding characteristic equations by making the following substitutions:

p^0 or 1 for i

p^1 or p for $\dfrac{di}{dt}$

p^2 for $\dfrac{d^2 i}{dt^2}$ (26-31)

p^3 for $\dfrac{d^3 i}{dt^3}$

p^n for $\dfrac{d^n i}{dt^n}$

This technique, called the *p*-operator method, can be used only with those circuits whose R, L, and C parameters are constants.

Example 26-1 Convert the following differential equation into its corresponding characteristic equation:

$$40\frac{d^2 i}{dt^2} + 6\frac{di}{dt} + 20i = 0$$

Solution
Using the *p*-operators tabulated in equation set (26-31),

$$40p^2 + 6p + 20 = 0$$

Depending on the characteristic equation, its roots may be real, imaginary, or complex. Thus the roots of a characteristic equation that represents the source-free response of an electric circuit may have one or more of the following terms:

$$p = -a$$

$$p = -a, -b$$

$$p = -a, -a \tag{26-32}$$

$$p = (-a + j\omega), (-a - j\omega)$$

$$p = +j\omega, -j\omega$$

As indicated in equation set (26-32), all real roots and all real parts of complex roots are *negative*. A negative root or negative real part for p indicates the presence of the decaying exponential ε^{-at}, which in turn implies the exponential decay of current in the circuit. In a "driverless" circuit containing RL, RC, or RLC, the current decays as the stored energy in the inductor and/or capacitor is converted to heat and dissipated in the resistance of the circuit.

Example 26-2 A 240-V DC generator supplies current to a parallel circuit consisting of a resistor and a coil as shown in Fig. 26-4a. The system is at steady state. Determine the current in the coil, the voltage *induced* in the coil, and the voltage across the coil, 1 s after the breaker is tripped.

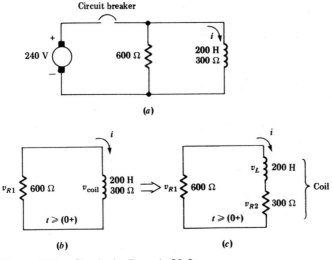

Figure 26-4   Circuits for Example 26-2.

Solution
The new circuit formed after the breaker is tripped is shown in Fig. 26-4b, and its equivalent circuit is shown in Fig. 26-4c. The current in the inductor cannot change instantaneously. Hence,

$$i_L\big|_{t=(0+)} = i_L\big|_{t=(0-)} = \frac{240}{300} = 0.8 \text{ A}$$

Applying Kirchhoff's voltage law to the circuit in Fig. 26-4c, and recognizing that there is no driver,

$$0 = v_{R1} + v_{R2} + v_L$$

$$0 = iR_1 + iR_2 + L\frac{di}{dt}$$

$$0 = i(600 + 300) + 200\frac{di}{dt}$$

Converting to *p*-operator form, the characteristic equation is

$$\boxed{0 = 900 + 200\,p}$$

$$p = -4.5$$

Hence the general solution is

$$\boxed{i = A\varepsilon^{-4.5t}}$$

To obtain the specific solution, substitute the initial conditions into the general solution:

$$i\big|_{t=(0+)} = 0.8$$

$$0.8 = A\varepsilon^{-4.5(0+)}$$

$$0.8 = A$$

Substituting the value of *A* into the general solution results in the specific solution:

$$\boxed{i = 0.8\varepsilon^{-4.5t}}$$

thus, at $t = 1$ s,

$$i| \atop {t=1} = 0.8\varepsilon^{-4.5(1)} = 0.0089 \text{ A}$$

To obtain the voltage *induced* in the coil, apply Kirchhoff's voltage law to Fig. 26-4c. *Recognizing that there is no driver*,

$$v_{R1} + v_{R2} + v_L = 0$$

$$iR_1 + iR_2 + v_L = 0$$

at $t = 1$ s,

$$0.0089(600 + 300) + v_L| \atop {t=1} = 0$$

$$v_L| \atop {t=1} = -8.01 \text{ V}$$

The *negative voltage drop* indicates that the induced voltage is acting as a voltage rise or driver which tends to keep the current from decreasing.

To obtain the voltage across the coil, apply Kirchhoff's voltage law to Fig. 26-4b.

$$v_{\text{coil}} + v_{R1} = 0$$

$$v_{\text{coil}} + iR_1 = 0$$

at $t = 1$ s,

$$v_{\text{coil}}| \atop {t=1} + 0.0089(600) = 0$$

$$v_{\text{coil}}| \atop {t=1} = -5.34 \text{ V}$$

The difference between v_L and v_{coil} is the *IR* drop caused by coil resistance. The minus sign indicates a negative voltage drop which tends to keep the current from decreasing.

Example 26-3　A 1000-μF capacitor charged to 120 V is connected in series with a 2000-Ω resistor and a switch as shown in Fig. 26-5a. Determine the current to the resistor 2 s after the switch is closed.

Solution
The circuit formed after the switch is closed is shown in Fig. 26-5b. The voltage across a capacitor cannot change instantaneously. Hence,

$$v_C| \atop {t=(0+)} = v_C| \atop {t=(0-)} = 120 \text{ V}$$

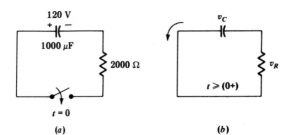

v_C

Figure 26-5 Circuits for Example 26-3.

(a)   *(b)*

Applying Kirchhoff's voltage law to the circuit in Fig. 26-5b,

$$v_R + v_C = 0$$

$$iR + \frac{1}{C}\int i\, dt = 0$$

Taking the derivative,

$$R\frac{di}{dt} + \frac{i}{C} = 0$$

$$2000\frac{di}{dt} + \frac{i}{1000 \times 10^{-6}} = 0$$

The characteristic equation is

$$2000p + 1000 = 0$$

$$p = -0.5$$

Thus the general solution is

$$i = A\varepsilon^{-0.5t}$$

Substituting the initial conditions into Kirchhoff's voltage equation,

$$v_C + v_R = 0$$

$$120 + i|_{t=(0+)}(2000) = 0$$

$$i|_{t=(0+)} = -0.06 \text{ A}$$

The negative sign indicates that the capacitor is discharging. Substituting the initial conditions into the general solution,

$$-0.06 = A\varepsilon^{-0.5(0+)}$$

$$A = -0.06$$

Thus the specific solution is

$$i = -0.06\varepsilon^{-0.5t}$$

The current at $t = 2$ s is

$$i\Big|_{t=2} = -0.06\varepsilon^{-0.5(2)} = -0.0221 \text{ A}$$

Example 26-4   A 100-μF capacitor charged to 24 V is connected in series with a 200-μF uncharged capacitor, a 1000-Ω resistor, and a switch, as shown in Fig. 26-6a. Determine (a) the current 0.1 s after the switch is closed; (b) the voltage drop across each of the circuit elements at $t = 0.1$ s.

Solution
(a) The circuit formed after the switch is closed is shown in Fig. 26-6b. The initial conditions are

$$v_{C1}\Big|_{t=(0+)} = v_{C1}\Big|_{t=(0-)} = 24 \text{ V} \qquad v_{C2}\Big|_{t=(0+)} = v_{C2}\Big|_{t=(0-)} = 0$$

Applying Kirchhoff's voltage law to the circuit in Fig. 26-6b,

$$v_{C2} + v_{C1} + v_R = 0$$

$$\frac{10^6}{200}\int i \, dt + \frac{10^6}{100}\int i \, dt + iR = 0$$

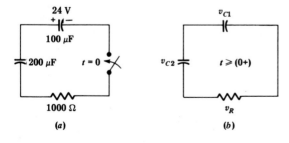

Figure 26-6   Circuits for Example 26-4.

Differentiating with respect to time,

$$\frac{10^6}{200}i + \frac{10^6}{100}i + 1000\frac{di}{dt} = 0$$

$$15 \times 10^3 i + 10^3 \frac{di}{dt} = 0$$

$$15i + \frac{di}{dt} = 0$$

The characteristic equation is

$$15 + p = 0$$

$$p = -15$$

Thus, the general solution is

$$i = A\varepsilon^{-15t}$$

Applying the initial conditions to Kirchhoff's voltage equation,

$$v_{C2} + v_{C1} + v_R = 0$$

$$0 + 24 + i|_{t=(0+)}(1000) = 0$$

$$i|_{t=(0+)} = -0.024 \text{ A}$$

Substituting the initial conditions into the general solution,

$$-0.024 = A\varepsilon^{-1.5(0+)}$$

$$A = -0.024$$

Hence the specific solution is

$$i = -0.024\varepsilon^{-15t}$$

The current at $t = 0.1$ s is

$$i|_{t=(0.1)} = -0.024\varepsilon^{-15(0.1)} = -0.00536 \text{ A}$$

(b) The voltage drop across the resistor at $t = 0.1$ s is

$$v_R\big|_{t=(0.1)} = iR = -0.00536(1000) = -5.36 \text{ V}$$

The voltage drop across the 200-μF capacitor at $t = 0.1$ s is

$$v_{C2}\big|_{t=(0.1)} = \frac{10^6}{200} \int_0^{0.1} i \, dt = 0.5 \times 10^4 \int_0^{0.1} (-0.024\varepsilon^{-15t}) dt$$

$$v_{C2}\big|_{t=(0.1)} = -(0.5)(0.024)10^4 \left(\frac{\varepsilon^{-15t}}{-15} \right)_0^{0.1}$$

$$v_{C2}\big|_{t=(0.1)} = 8(\varepsilon^{-1.5} - \varepsilon^0) = -6.215 \text{ V}$$

To obtain the voltage drop across v_{C1} at $t = 0.1$ s, substitute the calculated values of v_R and v_{C1} into Kirchhoff's voltage equation and solve for v_{C1}. Thus,

$$v_{C1} + v_{C2} + v_R = 0$$

$$v_{C1}\big|_{t=(0.1)} + (-6.215) + (-5.36) = 0$$

$$v_{C1}\big|_{t=(0.1)} = 11.575 \text{ V}$$

26-5 SOURCE-FREE RESPONSE OF AN RLC CIRCUIT

Figure 26-7a shows a charged capacitor in series with a coil and a switch. When the switch is closed, the capacitor is connected across the coil as shown in Fig. 26-7b. The equivalent series circuit model is shown in Fig. 26-7c. The initial conditions as determined by the capacitor and the inductor are

$$v_C\big|_{t=(0+)} = v_C\big|_{t=(0-)} = E_0$$

$$i_L\big|_{t=(0+)} = i_L\big|_{t=(0-)} = 0$$

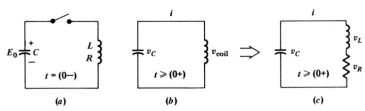

(a) (b) (c)

Figure 26-7 Source-free RLC circuit.

Applying Kirchhoff's voltage law to the circuit in Fig. 26-7c,

$$v_L + v_R + v_C = 0$$

$$L\frac{di}{dt} + iR + \frac{1}{C}\int i\,dt = 0$$

Taking the derivative with respect to time,

$$L\frac{d^2i}{dt^2} + R\frac{di}{dt} + \frac{i}{C} = 0$$

The characteristic equation is

$$Lp^2 + Rp + \frac{1}{C} = 0$$

or

$$p^2 + \frac{R}{L}p + \frac{1}{LC} = 0$$

the roots are

$$p = \frac{-R/L \pm \sqrt{(R/L)^2 - 4/LC}}{2}$$

$$p = -\frac{R}{2L} \pm \sqrt{\left(\frac{R}{2L}\right)^2 - \frac{1}{LC}}$$

Defining,

$$\sigma = \frac{R}{2L}$$

$$\omega = \sqrt{r}$$

$$r = \left[\left(\frac{R}{2L}\right)^2 - \frac{1}{LC}\right]$$

$$p = -\sigma \pm \sqrt{r}$$

Depending on the respective values of resistance, inductance, and capacitance, the radicand (expression under the square-root sign) may be positive, zero, or negative.

Case 1—Radicand Is Positive

If the radicand is positive, the two roots of the characteristic equation are different but real.

$$p = (-\sigma + \sqrt{r}), (-\sigma - \sqrt{r})$$

$$p = -a, -b$$

Substituting the roots into the expected form for the general response results in

$$i = A_1 \varepsilon^{-at} + A_2 \varepsilon^{-bt}$$

Case 2—Radicand Is Zero

If the radicand is zero, the two roots are real and equal.

$$p = (-\sigma + \sqrt{0}), (-\sigma - \sqrt{0})$$

$$p = -\sigma, -\sigma$$

However, in this case, substituting the two roots into the exponential response equation does not provide the two terms necessary for the correct solution; it results in a single term as shown in the following substitution:

$$i = A_1 \varepsilon^{-\sigma t} + A_2 \varepsilon^{-\sigma t} \tag{26-33}$$

The two terms in Eq. (26-33) are in effect a single term that may be expressed as

$$i = (A_1 + A_2)\varepsilon^{-\sigma t} \tag{26-34}$$

which, in effect, is the single term

$$i = A_3 \varepsilon^{-\sigma t} \tag{26-35}$$

Equations (26-33), (26-34), and (26-35) are incorrect. *The general solution for the source-free response of a second-order differential equation must have two separate terms that cannot be combined.* The correct form for the general solution is

$$i = A_1 \varepsilon^{-\sigma t} + A_2 t \varepsilon^{-\sigma t}$$

Case 3—Radicand Is Negative

If the radicand is negative, the two roots of the characteristic equation are complex conjugates.

$$p = (-\sigma + \sqrt{-r}), (-\sigma - \sqrt{-r})$$

or

$$p = (-\sigma + j\sqrt{r}), (-\sigma - j\sqrt{r})$$

Substituting the roots into the exponential response equation results in

$$i = A_1\varepsilon^{(-\sigma + j\sqrt{r})t} + A_2\varepsilon^{(-\sigma - j\sqrt{r})t}$$

Simplifying,

$$i = \varepsilon^{-\sigma t}(A_1\varepsilon^{j\sqrt{r}t} + A_2\varepsilon^{-j\sqrt{r}t}) \tag{26-36}$$

The sum of the exponential terms in Eq. (26-36) is converted to a sum of trigonometric terms by using Euler's formula, as outlined in Sec. 26-3. Thus the general solution may be expressed as

$$i = \varepsilon^{-\sigma t}(B_1 \cos \omega t + B_2 \sin \omega t) \tag{26-37}$$

A graph of Eq. (26-37), with $B_1 = 2$, $B_2 = 3$, $\omega = 1$, and $\sigma = 0.0796$ is the damped sinusoid shown in Fig. 26-8. The damped frequency of oscillation may be determined from

$$\omega = 2\pi f$$

where

$$\omega = \sqrt{r}$$

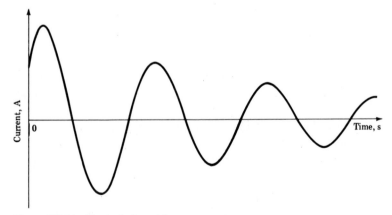

Figure 26-8   Damped sinusoid.

Table 26-1 Source-free response

Row	Roots	General solution
1	$p = -a$	$i = A\varepsilon^{-at}$
2	$p = -a, -b$	$i = A_1\varepsilon^{-at} + A_2\varepsilon^{-bt}$
3	$p = -a, -a$	$i = A_1\varepsilon^{-at} + A_2t\varepsilon^{-at}$
4	$p = j\omega, -j\omega$	$i = B_1 \cos \omega t + B_2 \sin \omega t$ $i = C \sin(\omega t + \phi)$
5	$p = (-\sigma + j\omega), (-\sigma - j\omega)$	$i = \varepsilon^{-\sigma t}(B_1 \cos \omega t + B_2 \sin \omega t)$ $i = C\varepsilon^{-\sigma t} \sin(\omega t + \phi)$

Note: For rows 4 and 5,

$$C = \sqrt{B_1^2 + B_2^2} \qquad \phi = \tan^{-1}\frac{B_1}{B_2}$$

A summary of the relationship between the roots of the characteristic equations and the respective general solution of the source-free response is given in Table 26-1.

Example 26-5 A 14.28-mF capacitor, charged to 50 V, is connected in series with a switch and a coil whose inductance and resistance are 5 H and 45 Ω, respectively. Determine the current 0.5 s after the switch is closed.

Solution
The circuit before the switch is closed is shown in Fig. 26-9a, and the equivalent series circuit for $t \geq (0+)$ is shown in Fig. 26-9b. The initial conditions are as follows:

Because of the inductance,

$$i\Big|_{t=(0+)} = i\Big|_{t=(0-)} = 0$$

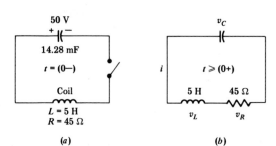

(a)   (b)

Figure 26-9 Circuits for Example 26-5.

Because of the capacitance,

$$v_C\Big|_{t=(0+)} = v_C\Big|_{t=(0-)} = 50 \text{ V}$$

Applying Kirchhoff's voltage law to Fig. 26-9b and recognizing that there is no driver,

$$v_L + v_R + v_C = 0$$

$$L\frac{di}{dt} + iR + \frac{1}{C}\int i\, dt = 0$$

$$5\frac{di}{dt} + 45i + \frac{1}{0.01428}\int i\, dt = 0$$

Differentiating with respect to t,

$$5\frac{d^2i}{dt^2} + 45\frac{di}{dt} + 70i = 0$$

Substituting the p-operator to obtain the characteristic equation, and then solving for p,

$$5p^2 + 45p + 70 = 0$$

$$p^2 + 9p + 14 = 0$$

$$p = -7, -2$$

Thus the general response is

$$i = A_1\varepsilon^{-7t} + A_2\varepsilon^{-2t} \tag{26-38}$$

Substituting the initial condition $i\Big|_{t=(0+)} = 0$,

$$0 = A_1\varepsilon^{-7(0+)} + A_2\varepsilon^{-2(0+)}$$

$$0 = A_1 + A_2$$

$$A_1 = -A_2 \tag{26-39}$$

A second set of initial conditions is obtained by examining Kirchhoff's voltage equation at $t = (0+)$:

$$v_L\Big|_{t=(0+)} + v_R\Big|_{t=(0+)} + v_C\Big|_{t=(0+)} = 0$$

$$L\frac{di}{dt}\Big|_{t=(0+)} + i\Big|_{t=(0+)}R + v_C\Big|_{t=(0+)} = 0$$

$$5 \left. \frac{di}{dt} \right|_{t=(0+)} + (0+)(45) + 50 = 0$$

$$\left. \frac{di}{dt} \right|_{t=(0+)} = -10$$

Taking the derivative of Eq. (26-38) and substituting the initial conditions,

$$\frac{di}{dt} = -7A_1 \varepsilon^{-7t} - 2A_2 \varepsilon^{-2t}$$

$$-10 = -7A_1 \varepsilon^{-7(0+)} - 2A_2 \varepsilon^{-2(0+)}$$

$$-10 = -7A_1 - 2A_2 \qquad\qquad (26\text{-}40)$$

Substituting Eq. (26-39) into Eq. (26-40) and solving for A_1 and A_2,

$$A_1 = +2$$

$$A_2 = -2$$

Hence the specific solution is

$$i = +2\varepsilon^{-7t} - 2\varepsilon^{-2t}$$

The current at 0.5 s after the switch is closed is

$$\left. i \right|_{t=(0.5)} = +2\varepsilon^{-7(0.5)} - 2\varepsilon^{-2(0.5)}$$

$$\left. i \right|_{t=0.5} = +0.0604 - 0.7358 = -0.6754 \text{ A}$$

Example 26-6 In Fig. 26-10*a*, a 0.04-F capacitor charged to 20 V is connected in series with a switch and a coil. The resistance and inductance of the coil are 10 Ω and 1.0 H, respectively. Determine the current 1 s after the switch is closed.

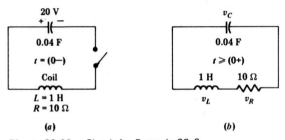

Figure 26-10 Circuit for Example 26-6.

Solution

The equivalent series circuit for $t \geq (0+)$ is shown in Fig. 26-10b. The initial conditions are as follows:

Because of the inductance,

$$i\Big|_{t=(0+)} = \quad i\Big|_{t=(0-)} = 0$$

Because of the capacitance,

$$v_C\Big|_{t=(0+)} = v_C\Big|_{t=(0-)} = 20 \text{ V}$$

Applying Kirchhoff's law to Fig. 26-10b,

$$v_L + v_R + v_C = 0$$

$$1\frac{di}{dt} + 10i + \frac{1}{0.04}\int i \, dt = 0$$

Differentiating,

$$\frac{d^2i}{dt^2} + 10\frac{di}{dt} + 25i = 0$$

The characteristic equation is

$$p^2 + 10p + 25 = 0$$

The roots are

$$p = -5, -5$$

Since the roots are equal, the general solution for the source-free response, as indicated in Table 26-1, is

$$i = A_1\varepsilon^{-5t} + A_2 t\varepsilon^{-5t} \tag{26-41}$$

Substituting the initial condition $i\Big|_{t=(0+)} = 0$,

$$0 = A_1\varepsilon^{-5(0+)} + A_2(0+)\varepsilon^{-5(0+)}$$

$$0 = A_1$$

With $A_1 = 0$, Eq. (26-41) becomes

$$i = A_2 t\varepsilon^{-5t} \tag{26-42}$$

The second set of initial conditions is obtained from Kirchhoff's voltage equation at $t = (0+)$,

$$v_L\Big|_{t=(0+)} + v_R\Big|_{t=(0+)} + v_C\Big|_{t=(0+)} = 0$$

$$1\frac{di}{dt}\Big|_{t=(0+)} + (0+)10 + 20 = 0$$

$$\frac{di}{dt}\Big|_{t=(0+)} = -20$$

Taking the derivative of Eq. (26-42) and substituting the second set of initial conditions,

$$\frac{di}{dt} = A_2[(t)(-5\varepsilon^{-5t}) + \varepsilon^{-5t}]$$

At $t = (0+)$,

$$-20 = A_2[(0+)(-5\varepsilon^{-5(0+)}) + \varepsilon^{-5(0+)}]$$

$$A_2 = -20$$

Hence the specific solution is

$$i = -20t\varepsilon^{-5t}$$

The current 1 s after the switch is closed is

$$i\Big|_{t=1} = -20(1)(\varepsilon^{-5(1)}) = -0.135 \text{ A.}$$

Example 26-7 For the circuit shown in Fig. 26-11a, determine the current 0.1 s after the switch is closed.

Solution
The equivalent series circuit is shown in Fig. 26-11b.

Because of the inductance,

$$i\Big|_{t=(0+)} = i\Big|_{t=(0-)} = 0$$

Because of the capacitance,

$$v_C\Big|_{t=(0+)} = v_C\Big|_{t=(0-)} = 100 \text{ V}$$

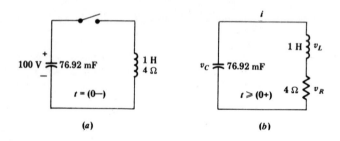

(a)   (b)

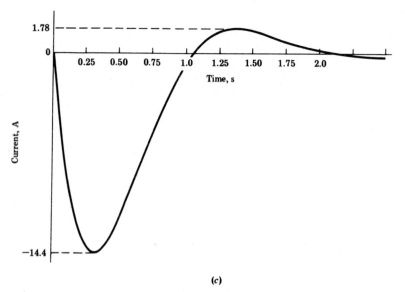

(c)

Figure 26-11 Circuits and current wave for Example 26-7.

Applying Kirchhoff's law to Fig. 26-11b,

$$v_L + v_R + v_C = 0$$

$$1\frac{di}{dt} + 4i + \frac{1000}{76.92}\int i\,dt = 0$$

$$\frac{d^2i}{dt^2} + 4\frac{di}{dt} + 13i = 0$$

The characteristic equation and its roots are

$$p^2 + 4p + 13 = 0$$

$$p = (-2 + j3), (-2 - j3)$$

The general solution corresponding to these roots, as determined from Table 26-1, is

$$i = \varepsilon^{-2t}(B_1 \cos 3t + B_2 \sin 3t) \tag{26-43}$$

Substituting the initial condition $i\big|_{t=(0+)} = 0$,

$$0 = \varepsilon^{-2(0+)}[B_1 \cos 3(0+) + B_2 \sin 3(0+)]$$

$$0 = B_1$$

Thus Eq. (26-43) reduces to

$$i = B_2 \varepsilon^{-2t} \sin 3t \tag{26-44}$$

The second set of initial conditions is obtained from Kirchhoff's voltage equation at $t = (0+)$:

$$v_L\big|_{t=(0+)} + v_R\big|_{t=(0+)} + v_C\big|_{t=(0+)} = 0$$

$$1 \frac{di}{dt}\bigg|_{t=(0+)} + 4(0+) + 100 = 0$$

$$\frac{di}{dt}\bigg|_{t=(0+)} = -100$$

Taking the derivative of Eq. (26-44) and substituting the second set of initial conditions,

$$\frac{di}{dt} = B_2[\varepsilon^{-2t}(3 \cos 3t) + (\sin 3t)(-2\varepsilon^{-2t})]$$

At $t = (0+)$,

$$-100 = B_2\{\varepsilon^{-2(0+)}[3 \cos 3(0+)] + [\sin 3(0+)](-2\varepsilon^{-2(0+)})\}$$

$$-100 = B_2(3 + 0)$$

$$B_2 = -33.33$$

Hence the specific solution is

$$i = -33.33\varepsilon^{-2t} \sin 3t \tag{26-45}$$

The current 0.1 s after the switch is closed is

$$i|_{t=(0.1)} = -33.33\varepsilon^{-2(0.1)} \sin[3(0.1)]^R$$

$$i|_{t=(0.1)} = -8.06 \text{ A}$$

A plot of Eq. (26-45) is shown in Fig. 26-11c. The frequency of oscillation is determined from ω.

$$\omega = 2\pi f$$

$$3 = 2\pi f$$

$$f = 0.48 \text{ Hz}$$

26-6   PLOTTING THE SYSTEM RESPONSE

A plot of the system response may be approximated by sketching the component parts of the system equation, adding and/or multiplying as indicated. To simplify the plotting procedure, determine the time constants of the exponentials and the periods of the sinusoidal waves, and then mark off the time axis in time constants $1\tau, 2\tau, \ldots, 5\tau$ and in periods $1T, 2T, \ldots$, etc. To demonstrate the procedure, Eq. (26-46) will be sketched from $t = 0$ to $t = 12$ s.

$$i = 6\varepsilon^{-0.5t} \sin 4.186t \tag{26-46}$$

Rewriting Eq. (26-46) to show the three components,

$$i = (6)(1\varepsilon^{-0.5t})(1 \sin 4.186t) \tag{26-47}$$

The time constant of the exponential component is

$$\tau = \frac{1}{0.5} = 2 \text{ s}$$

The period of the sinusoidal component is

$$\omega = 2\pi f = 4.186$$

$$f = 0.6667 \text{ Hz}$$

$$T = \frac{1}{f} = \frac{1}{0.6667} = 1.5 \text{ s}$$

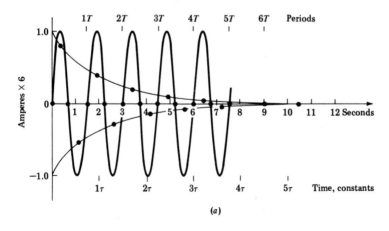

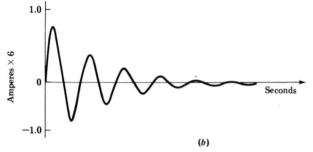

Figure 26-12 Plotting the system response: (*a*) component curves; (*b*) composite curve.

Scaling of the graph paper is shown in Fig. 26-12*a*. The current axis is scaled from 0 to 1 A. A scale factor of 6 accounts for the constant term in Eq. (26-47).

The time scale is marked off at intervals corresponding to the period of the sine wave, and at intervals corresponding to the time constant of the exponential.

The sinusoidal component is plotted by using its period as a guide, and the exponential component is plotted by using the time constant as a guide. The exponential term falls from its value of 1.0 A at $t = 0$ to 0.368 A at $t = 1$ time constant, and approximates its steady-state value of 0 A in five time constants. The plotting procedure may be aided by sketching the mirror image of the exponential component, as shown in Fig. 26-12*a*. To draw the curve,

1. Circle all points on the time scale that are intersected by the sine wave.
2. Circle all points on the exponential curve and its mirror image that correspond to the peak positive values and peak negative values, respectively, of the sine wave.
3. Draw a smooth curve joining all the circled points as indicated in Fig. 26-12*b*.

Example 26-8 Sketch, to approximate scale, the curve representing the current equation

$$i = \varepsilon^{-0.5t}(6 \cos 11.06t - 1.63 \sin 11.06t).$$

Solution

To simplify the plotting procedure, convert the cosine term to a sine term, express the sinusoidal functions in complex numbers, combine, and then reconvert to the time domain.

$$\cos 11.06t \Rightarrow \sin(11.06t + 90°)$$

$$[6 \sin(11.06t + 90°) - 1.63 \sin 11.06t] \Rightarrow 6/\underline{90°} - 1.63/\underline{0°}$$

$$(j6 - 1.63) \Rightarrow 6.22/\underline{105.2°}$$

$$6.22/\underline{105.2°} \Rightarrow 6.22 \sin(11.06t + 105.2°)$$

Thus,

$$i = \varepsilon^{-0.5t}[6.22 \sin(11.06t + 105.2°)]$$

Rearranging terms for plotting,

$$i = (6.22)(1\varepsilon^{-0.5t})[1 \sin(11.06t + 105.2°)]$$

The time constant is

$$\tau = \frac{1}{0.5} = 2 \text{ s}$$

$$5\tau = 10 \text{ s}$$

The frequency and period of the sinusoidal component are

$$2\pi f = 11.06$$

$$f = 1.76 \text{ Hz}$$

$$T = \frac{1}{f} = \frac{1}{1.76} = 0.57 \text{ s}$$

Sketch a sine wave (sine-wave templates are available), locate the time-zero line as shown in Fig. 26-13, and mark the time axis in seconds to correspond to the period of

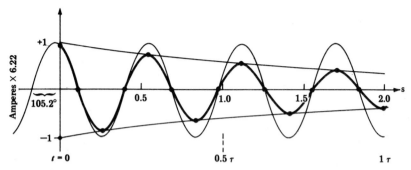

Figure 26-13   Plotting the response curve for Example 26-8.

the wave. Then mark the time axis in terms of the time constant, and sketch the exponential and its mirror image. In this example, a more accurate sketch requires the evaluation of the exponential at several additional points. Thus,

t	$\varepsilon^{-0.5t}$
0.5	0.78
1.0	0.61
1.5	0.47
2.0	0.37

Circle all points on the time scale that are intersected by the sine wave, and all points on the exponential curve and its mirror image that correspond to the peak positive values and peak negative values, respectively, of the sine wave. The smooth curve joining the circled points represents the graph of the current wave.

PROBLEMS

26-1   In Fig. 26-14, the battery voltage is 24 V, the resistance R_1 is 2.0 Ω, and the parameters of the coil are 10 H and 3.0 Ω. The switch is closed, and the circuit current is at steady state. At $t = 0$, the switch is opened. Determine (*a*) the coil current at $t = (0-)$; (*b*) the coil current at $t = (0+)$; (*c*) the current in the 2.0-Ω resistor at $t = (0-)$; (*d*) the current in the 2.0-Ω resistor at $t = (0+)$; (*e*) the current in coil at $t = 1.5$ s; (*f*) sketch the current versus time curve.

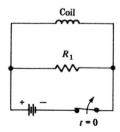

Figure 26-14   Circuit for Probs. 26-1, 26-2, and 26-3.

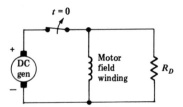

Figure 26-15 Circuit for Probs. 26-4 and 26-5.

26-2 Repat Prob. 26-1, assuming $R_1 = 1200\ \Omega$, $E_{bat} = 360$ V, and the coil parameters are 200 H and 90 Ω.

26-3 Repeat Prob. 26-1, assuming $R_1 = 50\ \Omega$, $E_{bat} = 100$ V, and the coil parameters are 100 H and 200 Ω.

26-4 In Fig. 26-15, a field discharge resistor is connected in parallel with the motor field winding to discharge the energy in the magnetic field when the switch is opened. This allows for a gradual discharge of the energy and thus avoids damage to the switch and to the coil when the switch is opened. The energy stored in the magnetic field is dissipated as heat energy in R_D and in the resistance of the field windings. Assuming R_D is 1000 Ω, the generator is operating at 120 V, the parameters of the field winding are 100 H and 94 Ω, and the circuit current is at steady state, determine (a) the current in the discharge resistor at $t = (0-)$; (b) the current in the discharge resistor at $t = (0+)$; (c) the field current 0.5 s after the switch is opened; (d) the voltage across the field windings at $t = (0+)$; (e) the elapsed time required for the voltage across the field winding to decay to 40 V; (f) sketch the current versus time curve, and the field-winding voltage versus time curve on a common time axis.

26-5 Repeat Prob. 26-4, assuming R_D is 10 Ω, the generator is operating at 24 V, and the parameters of the field winding are 2.0 H and 4.0 Ω.

26-6 In Fig. 26-16, the discharge element is a diode. Assuming the diode has negligible resistance in the forward direction, all the energy stored in the field will be dissipated as heat in the resistance of the winding when the switch is opened. The parameters of the winding are 200 H and 350 Ω, and the battery voltage is 250 V. Determine (a) the current in the winding at the instant the switch is opened; (b) the current 0.20 s after the switch is opened; (c) the voltage across the winding at $t = (0-)$; (d) the voltage across the winding at $t = (0+)$.

26-7 Repeat Prob. 26-6, assuming the parameters of the winding are 100 H and 200 Ω, and the battery voltage is 100 V.

26-8 In Fig. 26-17, the two paralleled field windings are protected by a common discharge resistor. The parameters of winding 1 are 300 H and 200 Ω, the parameters of winding 2 are 100 H and 200 Ω, R_D is 600 Ω, and the generator voltage is 240 V. Determine (a) the steady-state current in each winding and in the resistor before the switch is opened; (b) the voltage across each winding at $t = (0-)$; (c) the voltage across each winding at $t = (0+)$; (d) the value of R_D that will limit the voltage across the windings to a maximum of 240 V when the switch is opened.

26-9 Repeat Prob. 26-8, assuming the parameters of winding 1 are changed to 300 H and 400 Ω and those of winding 2 are changed to 100 H and 250 Ω.

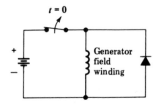

Figure 26-16 Circuit for Probs. 26-6 and 26-7.

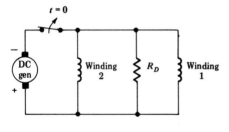

Figure 26-17   Circuit for Probs. 26-8 and 26-9

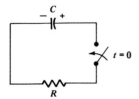

Figure 26-18   Circuit for Prob. 26-10.

26-10   The circuit parameters in Fig. 26-18 are 1000 Ω and 100 μF, and the capacitor is charged to 24 V. Determine (a) the voltage across the capacitor at $t = (0+)$; (b) the circuit current at $t = (0+)$; (c) the current 0.02 s after the switch is closed; (d) the voltage across the capacitor at $t = 0.02$ s; (e) sketch the current versus time curve and the v_c versus time curve on a common time axis.

26-11   The circuit in Fig. 26-19 was at steady state before the switch was opened. The circuit parameters are $R_1 = 1.0$ Ω, $R_2 = 2.0$ Ω, and $C = 0.167$ F. The battery voltage is 24 V. Determine (a) the voltage across the capacitor at $t = (0-)$; (b) the voltage across the capacitor at $t = (0+)$; (c) the current in the 2.0-Ω resistor at $t = (0+)$; (d) the current in the 2-Ω resistor 1 s after the switch is opened.

26-12   The circuit parameters in Fig. 26-20 are $C = 2.4$ F and $R = 5.0$ Ω. The battery voltage is 100 V. Assuming the circuit is at steady state, determine (a) the current in the resistor, and (b) the energy stored in the capacitor 10 s after the switch is opened.

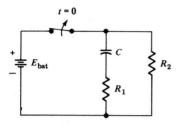

Figure 26-19   Circuit for Prob. 26-11.

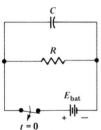

Figure 26-20   Circuit for Prob. 26-12.

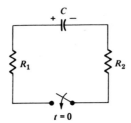

Figure 26-21 Circuit for Prob. 26-13.

26-13 The circuit parameters in Fig. 26-21 are $R_1 = 3.0 \, \Omega$, $R_2 = 7.0 \, \Omega$, and $C = 0.20$ F. The capacitor is charged to 100 V. Determine (a) the current at $t = (0+)$; (b) the current after an elapsed time of 6 s; (c) the voltage across the 3.0-Ω resistor at $t = 6$ s; (d) the voltage across the capacitor at $t = 6$ s; (e) sketch the curves of current and capacitor voltage as functions of time on a common time axis.

26-14 In Fig. 26-22, the coil parameters are 1.0 H and 6.0 Ω. The capacitor is rated at 0.20 F and is initially charged to 24 V. Determine (a) the current at $t = (0+)$; (b) the current 1 s after the switch is closed; (c) sketch the current versus time curve.

26-15 Assume the coil parameters in Fig. 26-22 are 2.0 H and 12 Ω. The capacitor is rated at 0.0625 F and is initially charged to 100 V. Determine (a) the current at $t = (0+)$; (b) the current 0.5 s after the switch is closed; (c) the current 10 s after the switch is closed.

26-16 Assume the coil parameters in Fig. 26-22 are 4 H and 40 Ω. The capacitor is rated at 0.010 F and is charged to 600 V. Determine (a) the current at $t = (0+)$; (b) the current 0.010 s after the switch is closed; (c) the final value of current (steady state); (d) the charge in coulombs at steady state.

26-17 For the circuit shown in Fig. 26-23, the parameters of coil 1 are 1.50 H and 8.0 Ω, and the parameters of coil 2 are 0.50 H and 4.0 Ω. The capacitor is rated at 1/18 F and has an initial charge of 100 V. Determine (a) the current at $t = (0+)$; (b) the current 0.2 s after the switch is closed; (c) the voltage across coil 2 at $t = 0.2$ s; (d) sketch the current versus time characteristic for five time constants.

26-18 A 2.55-μF capacitor charged to 60 V is connected in series with a switch and an "ideal" inductance of 200 mH. Sketch the circuit and determine (a) the current 0.40 s after the switch

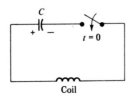

Figure 26-22 Circuit for Probs. 26-14, 26-15, 26-16, 26-19, 26-20, and 26-21.

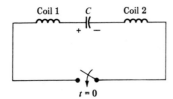

Figure 26-23 Circuit for Prob. 26-17.

is closed; (b) the voltage across the capacitor at $t = 0.40$ s; (c) the frequency of oscillation; (d) sketch the current versus time characteristic.

26-19 Assume the coil parameters in Fig. 26-22 are 2.0 H and 16 Ω. The capacitor is rated at 0.0122 F and is charged to 480 V. Determine (a) the current at $t = (0+)$; (b) the current at $t = 0.5$ s; (c) the time constant; (d) the frequency in hertz; (e) the voltage across the coil at $t = 0.5$ s; (f) the voltage across the capacitor at $t = 0.5$ s; (g) sketch the current versus time characteristic.

26-20 Assume the coil parameters in Fig. 26-22 are 1.0 H and 12 Ω. The capacitor is rated at 0.010 F and is charged to 60 V. Determine (a) the time constant of the circuit and the frequency in hertz; (b) the current 0.2 s after the switch is closed; (c) the voltage across the coil at $t = 0.2$ s; (d) the voltage across the capacitor at $t = 0.2$ s; (e) sketch the current versus time characteristic.

26-21 Assume the coil parameters in Fig. 26-22 are 2.0 H and 1.0 Ω. The capacitor is rated at 1.0 F and is charged to 100 V. Determine (a) the time constant of the circuit and the frequency of oscillation; (b) the current at $t = 1$ time constant; (c) sketch the current versus time characteristic.

26-22 Sketch to approximate scale the curve of $i = 6\varepsilon^{-0.5t} \sin 3.142t$ for $t = 0$ to $t = 10$ s.

26-23 Sketch to approximate scale the curve representing $i = 10\varepsilon^{-2.5t} \sin (12.56t + 30°)$ over a time interval equal to five time constants.

CHAPTER 27

TRANSIENTS IN DRIVEN SYSTEMS

Chapter 26 discussed transients in source-free circuits and concentrated only on circuit behavior during the *discharge* of energy-storage elements. This chapter is concerned with the response of circuits to the more common types of drivers, and includes both transient and steady-state components.

27-1 RESPONSE OF AN ELECTRIC CIRCUIT TO A DRIVING VOLTAGE

The response of a circuit to a driving voltage is defined as the behavior of the circuit after the switch is closed, from $t = (0+)$ to $t \to (\infty)$. It includes a transient component (source-free response) and a driven component (forced response). *The sum of the components is the circuit response.*

The source-free component accounts for the time delay inherent in RC, RL, and RLC circuits and is a transient of relatively short duration. The time delay and frequency (if oscillating) are strictly a function of the circuit parameters R, L, and C; they are independent of the type and magnitude of the driver, and independent of the initial conditions.

The behavior of the driven component is similar in form to that of the driver. If the driver is a battery or constant-voltage generator, the driven component will be a constant or zero; if the driver is sinusoidal, the driven component will be sinusoidal and have the same frequency; if the driver is an exponential function, the driven component will be an exponential and have the same exponent; if the driver is a linearly changing voltage, called a *ramp*,

Table 27-1 Common driving voltages and the associated driven response

Driving voltage	Driven response
$V_T = B$ (constant)	$i_D = K$ †
$V_T = Bt$ (ramp)	$i_D = K_0 + K_1 t$
$V_T = Bt^2$ (parabola)	$i_D = K_0 + K_1 t + K_2 t^2$
$V_T = Bt^3$	$i_D = K_0 + K_1 t + K_2 t^2 + K_3 t^3$
$V_T = B\varepsilon^{-gt}$	$i_D = K_0 \varepsilon^{-gt}$ ‡
$V_T = V_{max} \sin(\omega t + \phi)$	$i_D = K_1 \cos \omega t + K_2 \sin \omega t$
	$i_D = I_{max} \sin(\omega t + \theta)$ †

† If the driver is a constant DC voltage or a sinusoidal voltage, the driven component will be more easily obtained by using phasors and Ohm's law, as outlined in previous chapters.
‡ If a term in the driven response has the same exponent as a term in the source-free response, $i_D = K_0 t\varepsilon^{-gt}$.

the driven component will also be a ramp; etc. The driver "commands" and the driven component "obeys."

A summary of common driving voltages and the associated driven response is given in Table 27-1.

The driven component may be obtained by selecting the appropriate driven-response equation from Table 27-1, substituting it into Kirchhoff's equation for the circuit at $t \geq (0+)$, performing the indicated operations, and then solving for the respective K values.

27-2 GENERAL PROCEDURE FOR OBTAINING THE COMPLETE RESPONSE

1. Sketch the circuit corresponding to $t = (0-)$.
2. Sketch the circuit corresponding to $t \geq (0+)$.
3. Determine the initial conditions.
4. Write Kirchhoff's equation for the circuit corresponding to $t \geq (0+)$.
5. Select the appropriate driven-response equation from Table 27-1 and solve for the driven component.
6. To obtain the source-free response, substitute zero for the driver in the equation obtained in step 4, differentiate if necessary to remove all integrals, and then obtain the characteristic equation.
7. Solve for the roots of the characteristic equation.
8. Using Table 26-1 as a guide, write the source-free component in *general form. Do not substitute the initial conditions at this time.*

9. Add the driven component to the source-free component.
10. Substitute the initial conditions into the equation obtained in step 9 and solve for the specific response.

Example 27-1 A 250-V DC generator is used to supply a 40-A motor load, as shown in Fig. 27-1a. The circuit breaker, with negligible inertia, has an instantaneous trip setting that operates at a current of 2000 A. If an accidental short circuit occurs at the breaker terminals, how long will it take for the breaker to trip? The resistance of the short is 0.001 Ω. Assume the resistance and inductance of the generator windings are negligible compared with those of the short circuit and the connecting cables.

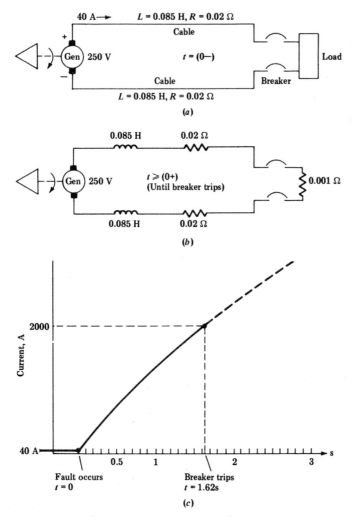

(a)

(b)

(c)

Figure 27-1 Circuits and plot for Example 27-1.

Solution
The equivalent circuit corresponding to $t \geq (0+)$ is shown in Fig. 27-1b. The *initial* conditions are

$$i\Big|_{t=(0+)} = i\Big|_{t=(0-)} = 40 \text{ A}$$

Applying Kirchhoff's voltage law to Fig. 27-1b,

$$L_T \frac{di}{dt} + iR_T = 250$$

where $L_T = 0.085 + 0.085 = 0.170$ H
$R_T = 0.02 + 0.001 + 0.02 = 0.041 \ \Omega$

$$0.170 \frac{di}{dt} + 0.041i = 250 \qquad (27\text{-}1)$$

Obtaining the driven component: Since the driver is a battery, the driven component can be determined from Ohm's law and the circuit impedance.

$$\mathbf{Z} = R + jX_L$$

$$\mathbf{Z} = 0.041 + j(2\pi)(0)(0.170)$$

$$\mathbf{Z} = 0.041 \ \Omega$$

Hence

$$I_D = \frac{V}{Z} = \frac{250}{0.041} = 6098 \text{ A}$$

$$i_D = 6098 \text{ A}$$

A steady-state short-circuit current of 6098 A would be attained if the circuit breaker did not trip!

Obtaining the source-free component: Substitute zero for the driver in Eq. (27-1) and then obtain the characteristic equation and the corresponding source-free response:

$$0.170 \frac{di}{dt} + 0.041i = 0$$

$$0.170p + 0.041 = 0$$

$p = -0.2412$

$i_{SF} = A\varepsilon^{-0.2412t}$

The complete general response is

$i = i_{SF} + i_D$

$i = A\varepsilon^{-0.2412t} + 6098$ $\qquad$ (27-2)

Substituting the initial conditions to obtain the specific response,

$40 = A\varepsilon^{-0.2412(0+)} + 6098$

$A = -6058$

Thus the complete response to the specific initial conditions is

$i = -6058\varepsilon^{-0.2412t} + 6098$ $\qquad$ (27-3)

To determine the elapsed time before the breaker trips, substitute 2000 A in Eq. (27-3) and solve for t.

$2000 = -6058\varepsilon^{-0.2412t} + 6098$

$\varepsilon^{-0.2412t} = 0.676$

$\ln(\varepsilon^{-0.2412t}) = \ln(0.676)$

$-0.2412t = -0.3916$

$t = 1.62 \text{ s}$

A plot of Eq. (27-3) is shown in Fig. 27-1c; the broken section represents the continued rise in current that would occur if the breaker did not trip.

Example 27-2 The primary circuit of a gasoline-engine ignition system is shown in Fig. 27-2a. The contacts, called points, are closed and opened by a rotating cam. Assume the cam closes for a period of 0.003 s (called dwell), after which the contacts open. Determine (a) the coil current after 0.003 s of dwell; (b) the coil current 0.001 s after the contacts open.

Solution
(a) Figure 27-2b represents the circuit during the dwell period ($0.003 \geq t \geq 0+$). The capacitor is shorted by the points and is not in the circuit. Because of inductance,

$i_L \big|_{t=(0+)} = i_L \big|_{t=(0-)} = 0 \text{ A}$

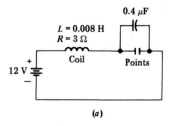

(a)

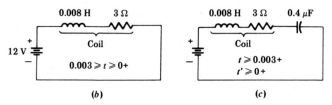

(b)   (c)

Figure 27-2   Circuits for Example 27-2.

Kirchhoff's equation for the circuit in Fig. 27-2b is

$$v_L + v_R = E_{\text{bat}}$$

$$0.008 \frac{di}{dt} + 3i = 12$$

Driven response:

$$i_D = \frac{E_{\text{bat}}}{R} = \frac{12}{3} = 4 \text{ A}$$

Source-free response:

$$0.008 \frac{di}{dt} + 3i = 0$$

$$0.008P + 3 = 0$$

$$P = -375$$

$$i_{\text{SF}} = A\varepsilon^{-375t}$$

$$i = i_D + i_{\text{SF}}$$

$$i = 4 + A\varepsilon^{-375t}$$

Substituting the initial conditions,

$0 = 4 + A\varepsilon^0$

$A = -4$

$i = 4 - 4\varepsilon^{-375t}$

$i\big|_{t=0.003} = 4 - 4\varepsilon^{-375(0.003)} = 2.70 \text{ A}$

(b) When the points open, the capacitor is inserted into the circuit as shown in Fig. 27-2c. In analyzing this circuit, time is measured from the instant the points separate. Thus time-zero for the circuit in Fig. 27-2c represents 0.003 s of real time. To avoid confusion, all variables involved in part b will be primed variables (t', v', i'). Because of the effect of inductance,

$i'_L\big|_{t'=(0+)} = i'_L\big|_{t'=(0-)} = i_L\big|_{t=(0.003)} = 2.70 \text{ A}$

Because of the effect of capacitance,

$v'_C\big|_{t'=(0+)} = v'_C\big|_{t'=(0-)} = 0$

Kirchhoff's equation for the circuit in Fig. 27-2c is

$v'_L + v'_R + v'_C = E_{\text{bat}}$

$0.008 \dfrac{di'}{dt'} + 3i' + \dfrac{1}{0.4 \times 10^{-6}} \displaystyle\int i'\, dt' = 12$

Driven response: With the capacitor in the circuit, the driven response is zero.

$i'_D = 0$

Source-free response:

$0.008 \dfrac{di'}{dt'} + 3i' + 2.5 \times 10^6 \displaystyle\int i'\, dt' = 0$

Differentiating, converting to operator form, and solving for the roots,

$0.008 \dfrac{d^2 i'}{dt'^2} + 3 \dfrac{di'}{dt'} + 2.5 \times 10^6 i' = 0$

$0.008 P^2 + 3P + 2.5 \times 10^6 = 0$

$P = (-187.5 + j17{,}676.7), (-187.5 - j17{,}676.7)$

$i' = \varepsilon^{-187.5t'}(B_1 \cos 17{,}676.7t' + B_2 \sin 17{,}676.7t')$

Substituting the first set of initial conditions,

$$2.70 = \varepsilon^0(B_1 \cos 0 + B_2 \sin 0)$$

$$B_1 = 2.7$$

To obtain the second set of initial conditions, apply Kirchoff's law to Fig. 27-2c. At $t' = 0+$,

$$v'_L + v'_R + v'_C = E_{bat}$$

At $t' = 0+$,

$$0.008 \left.\frac{di'}{dt'}\right|_{t'=(0+)} + (2.7)(3) + 0 = 12$$

$$\left.\frac{di'}{dt'}\right|_{t'=(0+)} = 487.5$$

Taking the derivative of the current equation to obtain di'/dt',

$$\frac{di'}{dt'} = \varepsilon^{-187.5t'}(-17{,}676.7B_1 \sin 17{,}676.7t' + 17{,}676.7B_2 \cos 17{,}676.7t')$$

$$+ (B_1 \cos 17{,}676.7t' + B_2 \sin 17{,}676.7t')(-187.5\varepsilon^{-187.5t'})$$

Substituting the second set of initial conditions, noting that $B_1 = 2.7$, and evaluating,

$$487.5 = \varepsilon^{0+}(0 + 17{,}676.7B_2) + (2.7 + 0)(-187.5\varepsilon^{0+})$$

$$B_2 = 0.056$$

Thus,

$$i' = \varepsilon^{-187.5t'}(2.7 \cos 17{,}676.7t' + 0.056 \sin 17{,}676.7t')$$

$$\left. i' \right|_{t'=0.001} = \varepsilon^{-187.5(0.001)}[2.7 \cos(17{,}676.7 \times 0.001)^R + 0.056 \sin(17{,}676.7 \times 0.001)^R]$$

$$\left. i' \right|_{t'=0.001} = 0.82 \text{ A}$$

Example 27-3   Figure 27-3a shows a turbine-driven generator, operating at 280 V, supplying power to an electromagnet whose resistance and inductance are 13 Ω and 1.5 H, respectively. If the turbine is tripped, the generator slows down and its voltage decays exponentially in the following manner:

$$e_{gen} = 280\varepsilon^{-2t}$$

Determine the current to the magnet 1 s after the turbine is tripped.

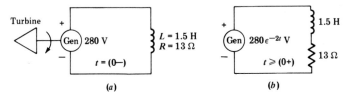

Figure 27-3 Circuits for Example 27-3.

Solution

The equivalent circuit corresponding to $t \geq (0+)$ is shown in Fig. 27-3b. The initial condition, as determined by the circuit inductance, is

$$\left. i \right|_{t=(0+)} = \left. i \right|_{t=(0-)} = \frac{280}{13} = 21.54 \text{ A}$$

Applying Kirchhoff's voltage law to the circuit in Fig. 27-3b,

$$v_L + v_R = e_{\text{gen}}$$

$$1.5 \frac{di}{dt} + 13i = 280\varepsilon^{-2t} \qquad (27\text{-}4)$$

Since the driver is an exponential, the driven component will be an exponential and will have the same exponent as the driver. Thus the general equation for the driven component, as obtained from Table 27-1, is

$$i_D = K\varepsilon^{-2t} \qquad (27\text{-}5)$$

Substituting Eq. (27-5) into Eq. (27-4), performing the indicated operations, and solving for K,

$$1.5 \frac{d}{dt} (K\varepsilon^{-2t}) + 13(K\varepsilon^{-2t}) = 280\varepsilon^{-2t}$$

$$1.5(-2K\varepsilon^{-2t}) + 13K\varepsilon^{-2t} = 280\varepsilon^{-2t}$$

$$-3K + 13K = 280$$

$$K = 28$$

Thus,

$$i_D = 28\varepsilon^{-2t} \qquad (27\text{-}6)$$

To obtain the source-free component, substitute zero for the driver in Eq. (27-4), and solve,

$$1.5 \frac{di}{dt} + 13i = 0$$

$$1.5p + 13 = 0$$

$$p = \frac{-13}{1.5} = -8.67$$

$$i_{SF} = A\varepsilon^{-8.67t}$$

Adding the two components,

$$i = i_D + i_{SF}$$

$$i = 28\varepsilon^{-2t} + A\varepsilon^{-8.67t}$$

Substituting the initial conditions,

$$i\Big|_{t=(0+)} = 21.54 \text{ A}$$

$$21.54 = 28\varepsilon^{-2(0+)} + A\varepsilon^{-8.67(0+)}$$

$$A = -6.46$$

Hence the specific response for the system is

$$i = 28\varepsilon^{-2t} - 6.46\varepsilon^{-8.67t}$$

One second after the turbine is tripped the current will be

$$i\Big|_{t=1} = 28\varepsilon^{-2(1)} - 6.46\varepsilon^{-8.67(1)}$$

$$i\Big|_{t=1} = 28(0.135) - 6.46(0.00017) = 3.78 \text{ A}$$

Example 27-4 The build-up of "inrush current" to a transformer or motor depends on the value of the driving voltage at the instant the switch is closed. Figure 27-4a illustrates the equivalent circuit of a loaded transformer connected to a sinusoidal driving voltage. It is assumed that the voltage wave is at its peak value of 2400 V when the switch is closed. (a) Determine the equation for the current as a function of time; (b) repeat part a assuming the switch is closed at the instant the voltage wave passes through 0 V.

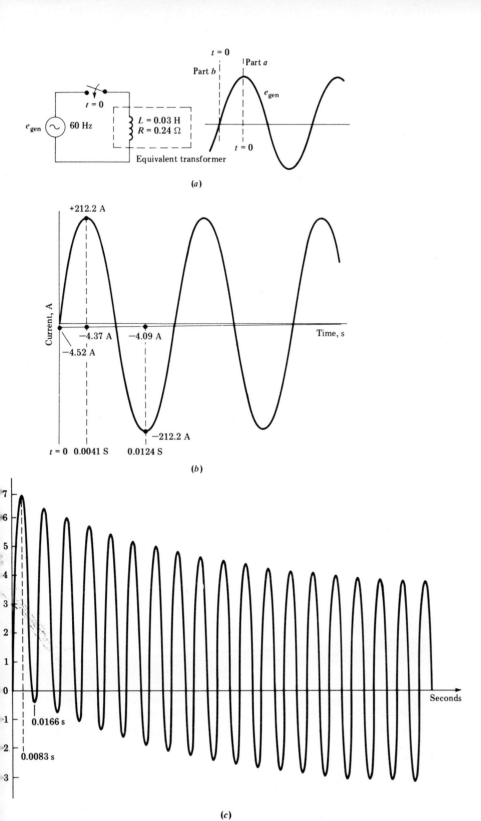

Figure 27-4   Circuits and plot for Example 27-4.

Solution

(a) The initial conditions are dictated by the circuit inductance:

$$i|_{t=(0+)} = i|_{t=(0-)} = 0$$

Applying Kirchhoff's voltage law to the circuit,

$$v_L + v_R = e_{gen}$$

$$0.03 \frac{di}{dt} + 0.24i = 2400 \sin(2\pi 60t + 90°) \qquad (27\text{-}7)$$

The driven response associated with the sinusoidal driver is obtained by using phasors and Ohm's law.

$$X_L = 2\pi f L = 2\pi(60)(0.03) = 11.31 \ \Omega$$

$$\mathbf{Z} = R + jX_L = 0.24 + j11.31 = 11.313 \underline{/88.78°} \ \Omega$$

$$V_{rms} = \frac{2400}{\sqrt{2}} = 1697.06 \text{ V}$$

$$\mathbf{I} = \frac{\mathbf{V}}{\mathbf{Z}} = \frac{1697.06 \underline{/90°}}{11.313 \underline{/88.78°}}$$

$$\mathbf{I} = 150.01 \underline{/1.22°} \text{ A}$$

$$i_D = 150.01\sqrt{2} \sin(377t + 1.22°)$$

$$i_D = 212.2 \sin(377t + 1.22°)$$

To obtain the source-free response, substitute zero for the driver in Eq. (27-7) and solve for i_{SF}.

$$0.03 \frac{di}{dt} + 0.24i = 0$$

$$0.03p + 0.24 = 0$$

$$p = -8$$

$$i_{SF} = A\varepsilon^{-8t}$$

Adding the driven and source-free components,

$$i = i_D + i_{SF}$$

$$i = 212.2 \sin(337t + 1.22°) + A\varepsilon^{-8t}$$

Substituting the initial conditions to obtain the specific response,

$$0 = 212.2 \sin[377(0) + 1.22°] + A\varepsilon^{-8(0)}$$

$$0 = 4.52 + A$$

$$A = -4.52$$

$$i = 212.2 \sin(377t + 1.22°) - 4.52\varepsilon^{-8t} \tag{27-8}$$

Figure 27-4b shows a representative but nonscale plot of the two components of Eq. 27-8. Note that the two components are subtractive at $t = 0.004$J s but additive at $t = 0.0124$ s. However, after five time constants, the transient component decays to insignificance and the steady-state sinusoidal current has a peak value of 212.2 A.
(b) The phasor current obtained by closing the switch at the instant that the wave of driving voltage passes through zero is

$$\mathbf{I} = \frac{\mathbf{V}}{\mathbf{Z}} = \frac{1697.06\underline{/0°}}{11.313\underline{/88.78°}}$$

$$\mathbf{I} = 150.01\underline{/-88.78°} \text{ A}$$

$$i_D = 212.2 \sin(337t - 88.78°)$$

The source-free response is the same as that determined in part a.

$$i_{SF} = A\varepsilon^{-8t}$$

$$i = i_D + i_{SF}$$

$$i = 212.2 \sin(377t - 88.78°) + A\varepsilon^{-8t}$$

Substituting initial conditions,

$$0 = 212.2 \sin[377(0) - 88.78°] + A\varepsilon^{-8(0)}$$

$$A = 212.15$$

Thus,

$$i = 212.2 \sin(377t - 88.78°) + 212.15\varepsilon^{-8t} \tag{27-9}$$

A properly scaled curve of Eq. (27-9) is shown in Fig. 27-4c. A peak inrush current of 410.71 A occurs 0.0083 s after the switch is closed. However, the rapid decay of the transient term causes the current to decrease to approximately its steady-state peak of 212.2 A in five time constants ($5 \times \frac{1}{8} = 0.625$ s).

A comparison of the current waves in Fig. 27-4b and c indicates the significant effect that the phase angle of the driving voltage has on the inrush current to loads such as transformers and motors. A somewhat similar behavior occurs when energizing capacitor banks from sinusoidal sources.

PROBLEMS

27-1 Assuming the coil parameters in Fig. 27-5 are 10 H and 2 Ω, resistor $R = 6 \Omega$ and $E_{bat} = 12$ V, determine (a) the current at $t = (0+)$; (b) the current at $t = 0.1$ s; (c) the voltage drop across the coil at $t = 0.1$ s; (d) sketch the current versus time curve.

27-2 The coil parameters in Fig. 27-5 are 150 H and 200 Ω. Resistor $R = 100 \Omega$ and $E_{bat} = 240$ V. Determine (a) the current at $t = (0+)$; (b) the current at $t = 0.5$ s; (c) the voltage across the 100-Ω resistor at $t = 0.5$ s; (d) the voltage across the coil at $t = 0.5$ s; (e) the energy stored in the magnetic field of the coil at steady state; (f) sketch the current versus time curve.

27-3 In Fig. 27-6, the coil parameters are 6 H and 10 Ω, $R = 14 \Omega$, and $E_{bat} = 24$ V. Determine (a) the current at $t = (0-)$; (b) the current at $t = (0+)$; (c) the current 0.1 s after the switch is opened.

27-4 The coil parameters for the circuit shown in Fig. 27-7 are 8 H and 4 Ω. Resistor $R = 12$ Ω and $E_{bat} = 24$ V. Determine (a) the current in the coil at $t = (0+)$; (b) the current in the

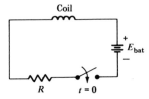

$R \qquad t = 0$	**Figure 27-5** Circuit for Probs. 27-1 and 27-2.

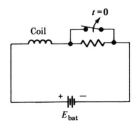

Figure 27-6 Circuit for Prob. 27-3.

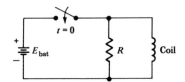

Figure 27-7 Circuit for Prob. 27-4.

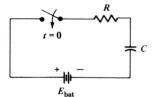

Figure 27-8   Circuit for Prob. 27-5.

resistor at $t = (0+)$; (c) the current in the coil and the current in resistor R 0.3 s after the switch is closed; (d) the current in the coil and the current in resistor R at steady state. When the system reaches steady state, the switch is opened. Determine (e) the current in the coil and the current in resistor R at the instant the switch is opened; (f) the current in the coil 0.3 s after the switch is opened; (g) the energy in the magnetic field of the coil at $t = 0.3$ s.

27-5   In Fig. 27-8, $R = 600\ \Omega$, $C = 400\ \mu F$, and $E_{bat} = 12$ V. Determine (a) the current at $t = (0+)$; (b) the current 0.1 s after the switch is closed; (c) the voltage across the capacitor at $t = 0.1$ s; (d) the current at steady state.

27-6   For the circuit in Fig. 27-9 determine the current to the capacitor (a) at $t = (0-)$; (b) at $t = (0+)$; (c) 21 s after the switch is opened.

27-7   In Fig. 27-10, a charged capacitor is connected in series with a battery and a resistor. Determine (a) the current at $t = (0+)$; (b) the current 0.5 s after the switch is closed; (c) the voltage across the capacitor at steady state; (d) the accumulated charge in the capacitor at steady state.

27-8   For the circuit shown in Fig. 27-11, determine (a) the current in the coil and the voltage across the capacitor at $t = (0-)$; (b) the current in the coil and the voltage across the capacitor at $t = (0+)$; (c) the capacitor current and coil current at $t = 1$ s.

27-9   In Fig. 27-12, the coil parameters are 5 H and 80 Ω, $R = 20\ \Omega$, $C = 184\ \mu F$; and $E_{bat} = 470$ V. (a) Determine the current 0.01 s after the switch is closed; (b) sketch the current versus time characteristic.

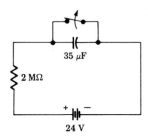

Figure 27-9   Circuit for Prob. 27-6.

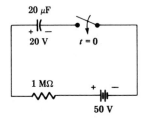

Figure 27-10   Circuit for Prob. 27-7.

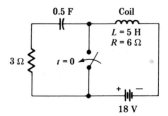

Figure 27-11 Circuit for Prob. 27-8.

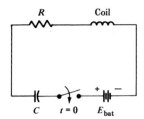

Figure 27-12 Circuit for Probs. 27-9 and 27-10.

27-10 In Fig. 27-12 the coil parameters are 2 H and 0.20 Ω, $R = 0.80\ \Omega$, $C = 200\ \mu F$, and $E_{bat} = 50$ V. Determine (a) the current at $t = 0.5$ s; (b) the voltage induced in the coil at $t = 0.5$ s; (c) the voltage across the capacitor at $t = 0.5$ s.

27-11 Figure 27-13 represents the primary circuit of a gasoline-engine ignition system. The contacts called "points" are closed and opened by a rotating cam. Assume the cam closes the points for a "dwell" period of 0.004 s, after which the contacts open. Determine (a) the coil current 0.004 s after the contacts close; (b) the current 0.001 s after the contacts open; (c) the frequency of oscillation of the current wave in part b.

27-12 The circuit in Fig. 27-14 is at its steady-state condition before the switch is opened. Determine (a) the coil current and the voltage across the capacitor at $t = (0-)$; (b) the coil current and the voltage across the capacitor at $t = (0+)$; (c) the current 0.4 s after the switch is opened; (d) the voltage *generated* in the coil, and the voltage drop across the capacitor at $t = 0.4$ s; (e) the energy stored in the magnetic field of the coil, and in the capacitor at $t = 0.4$ s; (f) the capacitor charge in coulombs at $t = 0.4$ s; (g) the voltage drop across the coil at $t = 0.4$ s.

27-13 A turbine-driven DC generator driven at 30 V feeds power to an electromagnet whose inductance and resistance are 6.0 H and 3.0 Ω, respectively. The resistance and inductance of the generator are negligible. Sketch the circuit and (a) determine the circuit current. (b) Tripping the turbine causes an exponential decay in turbine speed and voltage. Assuming the decay of

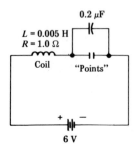

Figure 27-13 Circuit for Prob. 27-11.

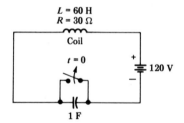

$L = 60$ H
$R = 30\ \Omega$

Coil

$t = 0$

120 V

1 F

Figure 27-14   Circuit for Prob. 27-12.

generator voltage may be expressed by $e_{gen} = 30\varepsilon^{-0.5t}$, determine the current 5 s after the turbine tripped.

27-14   A series circuit consisting of a 250-μF capacitor and a 1000-Ω resistor is connected to a generator whose voltage is represented by $e_{gen} = 60\varepsilon^{-7t}$ V. The capacitor is initially discharged, and the generator voltage is applied at $t = 0$. Sketch the circuit, and determine the current at $t = 0.2$ s.

27-15   A 60-Hz 2400-V (rms) generator is connected in series with a switch and a load whose resistance and inductance are 0.24 Ω and 0.030 H, respectively. Assume the impedance of the generator is negligible. Sketch the circuit. If the switch is closed at the instant the voltage wave passes through 0 V rising in the positive direction, determine (a) the current 0.10 s after the switch is closed; (b) the steady-state rms current.

27-16   Repeat Prob. 27-15 assuming that the switch is closed at the instant the voltage wave attains its peak positive value.

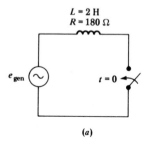

$L = 2$ H
$R = 180\ \Omega$

e_{gen}

$t = 0$

(a)

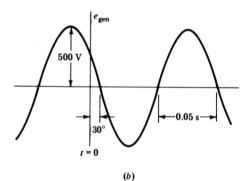

e_{gen}

500 V

$30°$

$t = 0$

0.05 s

(b)

Figure 27-15   Circuit for Prob. 27-17.

27-17 For the conditions indicated in Fig. 27-15, determine the current 0.01 s after the switch closes.

27-18 A 450-V 60-Hz generator supplies a series circuit consisting of a switch, a 40-Ω capacitive reactance, and a coil whose inductive reactance and resistance are 50 Ω and 2.0 Ω, respectively. The switch is closed 20° after the voltage wave passes through zero rising in the positive direction. (*a*) Sketch the circuit; (*b*) sketch the voltage wave and indicate time-zero; (*c*) determine the steady-state current; (*d*) determine the current 0.01 s after the switch is closed; (*e*) calculate the value of capacitance that should be paralleled with the original capacitor or connected in series with the original capacitor so that resonance will occur at the driving frequency; (*f*) determine the steady-state current for the conditions in part *e*.

APPENDIXES

Appendix 1 Allowable ampacities of representative insulated single copper conductors in free air

Based on ambient temperature of 30°C, 86°F †

SIZE	TEMPERATURE RATING OF CONDUCTOR AND TYPE LETTER							
	60°C (140°F)	75°C (167°F)	85°C (185°F)	90°C (194°F)	110°C (230°F)	125°C (257°F)	200°C (392°F)	
AWG or kcmil	T	RH	V	TA RHH THHN	AVL	AIA	A	Bare and covered conductors
AWG								
14	20	20	30	30‡	40	40	45	30
12	25	25	40	40‡	50	50	55	40
10	40	40	55	55‡	65	70	75	55
8	55	65	70	70	85	90	100	70
6	80	95	100	100	120	125	135	100
4	105	125	135	135	160	170	180	130
3	120	145	155	155	180	195	210	150
2	140	170	180	180	210	225	240	175
1	165	195	210	210	245	265	280	205
0	195	230	245	245	285	305	325	235
00	225	265	285	285	330	355	370	275
000	260	310	330	330	385	410	430	320
0000	300	360	385	385	445	475	510	370
kcmil								
250	340	405	425	425	495	530	—	410
300	375	445	480	480	555	590	—	460
350	420	505	530	530	610	655	—	510
400	455	545	575	575	665	710	—	555
500	515	620	660	660	765	815	—	630
600	575	690	740	740	855	910	—	710
700	630	755	815	815	940	1005	—	780
750	655	785	845	845	980	1045	—	810
800	680	815	880	880	1020	1085	—	845
900	730	870	940	940	—	—	—	905
1000	780	935	1000	1000	1165	1240	—	965
1250	890	1065	1130	1130				
1500	980	1175	1260	1260	1450	—	—	1215
1750	1070	1280	1370	1370				
2000	1155	1385	1470	1470	1715	—	—	1405

Extracted by permission from the 1978 National Electrical Code (Table 310-17), copyright 1977, National Fire Protection Association, Boston, Mass.

† For ambient temperatures over 30°C, see correction factors in Appendix 3.
‡ The ampacities for types RHH and THHN conductors for sizes 14, 12, and 10 shall be as designated for 75°C conductors.

Appendix 2 Allowable ampacities of representative insulated copper conductors in raceways or cables

Not more than three conductors or direct burial (based on ambient temperature of 30°C, 86°F)†

SIZE	TEMPERATURE RATING OF CONDUCTOR AND TYPE LETTER						
	60°C (140°F)	75°C (167°F)	85°C (185°F)	90°C (194°F)	110°C (230°F)	125°C (257°F)	200°C (392°F)
AWG or kcmil	T	RH	V	TA RHH THHN	AVL	AIA	A
AWG							
14	15	15	25	25‡	30	30	30
12	20	20	30	30‡	35	40	40
10	30	30	40	40‡	45	50	55
8	40	45	50	50	60	65	70
6	55	65	70	70	80	85	95
4	70	85	90	90	105	115	120
3	80	100	105	105	120	130	145
2	95	115	120	120	135	145	165
1	110	130	140	140	160	170	190
0	125	150	155	155	190	200	225
00	145	175	185	185	215	230	250
000	165	200	210	210	245	265	285
0000	195	230	235	235	275	310	340
kcmil							
250	215	255	270	270	315	335	
300	240	285	300	300	345	380	
350	260	310	325	325	390	420	
400	280	335	360	360	420	450	
500	320	380	405	405	470	500	
600	355	420	455	455	525	545	
700	385	460	490	490	560	600	
750	400	475	500	500	580	620	
800	410	490	515	515	600	640	
900	435	520	555	555			
1000	455	545	585	585	680	730	
1250	495	590	645	645			
1500	520	625	700	700	785		
1750	545	650	735	735			
2000	560	665	775	775	840		

Extracted by permission from the 1978 National Electrical Code (Table 310-16), copyright 1977, National Fire Protection Association, Boston, Mass.

† For ambient temperatures over 30°C, see correction factors in Appendix 3.
‡ The ampacities for types RHH and THHN conductors for sizes 14, 12, and 10 shall be the same as designated for 75°C conductors.

Appendix 3   Correction factors for ambient temperatures over 30°C, 86°F

For use with Appendix 1 and Appendix 2

°C	°F	60°C (140°F)	75°C (167°F)	85°C (185°F)	90°C (194°F)	110°C (230°F)	125°C (257°F)	200°C (392°F)
40	104	0.82	0.88	0.90	0.90	0.94	0.95	
45	113	0.71	0.82	0.85	0.85	0.90	0.92	
50	122	0.58	0.75	0.80	0.80	0.87	0.89	
55	131	0.41	0.67	0.74	0.74	0.83	0.86	
60	140	—	0.58	0.67	0.67	0.79	0.83	0.91
70	158	—	0.35	0.52	0.52	0.71	0.76	0.87
75	167	—	—	0.43	0.43	0.66	0.72	0.86
80	176	—	—	0.30	0.30	0.61	0.69	0.84
90	194	—	—	—	—	0.50	0.61	0.80
100	212	—	—	—	—	—	0.51	0.77
120	248	—	—	—	—	—	—	0.69
140	284	—	—	—	—	—	—	0.59

Extracted by permission from the 1978 National Electrical Code (Article 310), copyright 1977, National Fire Protection Association, Boston, Mass.

Appendix 4 Multiplying factors for converting DC resistance to 60-Hz AC resistance

Size	MULTIPLYING FACTOR			
	FOR NONMETALLIC SHEATHED CABLES IN AIR OR NONMETALLIC CONDUIT		FOR METALLIC SHEATHED CABLES OR ALL CABLES IN METALLIC RACEWAYS	
	Copper	*Aluminum*	*Copper*	*Aluminum*
AWG				
3	1.000	1.000	1.00	1.00
2	1.000	1.000	1.01	1.00
1	1.000	1.000	1.01	1.00
0	1.001	1.000	1.02	1.00
00	1.001	1.001	1.03	1.00
000	1.002	1.001	1.04	1.01
0000	1.004	1.002	1.05	1.01
kcmil				
250	1.005	1.002	1.06	1.02
300	1.006	1.003	1.07	1.02
350	1.009	1.004	1.08	1.03
400	1.011	1.005	1.10	1.04
500	1.018	1.007	1.13	1.06
600	1.025	1.010	1.16	1.08
700	1.034	1.013	1.19	1.11
750	1.039	1.015	1.21	1.12
800	1.044	1.017	1.22	1.14
1000	1.067	1.026	1.30	1.19
1250	1.102	1.040	1.41	1.27
1500	1.142	1.058	1.53	1.36
1750	1.185	1.079	1.67	1.46
2000	1.233	1.100	1.82	1.56

Extracted by permission from the 1978 National Electrical Code. (Table 9), copyright 1977, National Fire Protection Association, Boston, Mass.

ANSWERS TO PROBLEMS

Chapter 1

1. 6.594×10^2, 1.73×10^1, 1.00750×10^2, 4.5×10^{-4}, 8×10^{-4}, 1.06×10^{-1}
2. 10 kV, 500 V, 10 MV, 6 μV, 125.8 μV, 1 nV
3. 46×10^4, 25×10^2, 66×10^2, 37×10^2, 15×10^{-3}, 1.6, 42×10^{-5}, 81×10^{-4}
4. (a) 1646.8 (b) 34.08
5. (a) 1.9×10^5, (b) 13.64, (c) 0.324, (d) 6.7×10^2

Chapter 2

1. 50.8 Ω
2. 74.2 Ω
3. 140 Ω
4. 0.020 Ω
5. (a) 3.82×10^6 cmil, (b) 543 $\mu\Omega$
6. 212 $\mu\Omega$
7. 1.06 mΩ
8. 24.4 Ω
9. 509.8 mΩ
10. 2×10^{-5} S, 2.15 mS, 55.6 S, 2500 S, 1.92 nS
11. 244 Ω, 0.308 Ω, 5.5386 mΩ, 1.0 mΩ, 2.50 Ω

Chapter 3

1. 5 parallel branches each containing 10 series-connected batteries.
2. 12 parallel branches each containing 3 series-connected batteries.
3. 49.4 V
4. 4.29 A
5. (a) 50 mA, (b) 17 mA
6. 39.0 kΩ
7. 1500 V
8. 26.3 μA
9. (a) 20 A, (b) 5Ω, (c) 16 kW · h
10. (a) 500 W, (b) 4.17 A, (c) 28.75 Ω
11. 80 A for 0.5 s

Chapter 4

1. 51 Ω
2. 300 Ω
3. (a) 5.4 mA, (b) 0.108 V, 1.08 V, 10.8 V, 108 V, (c) 0.648 W
4. (a) 2.4 A, (b) 14.4 V, (c) 34.6 W, (d) 691 W · h, (e) 0
5. 3000 Ω
6. 62.5 V
7. 3.69 V
8. 2 Ω
9. 1.18 Ω
10. (a) 1.44 A, (b) 15.4 W
11. (a) 44.4 Ω, (b) 4.95 Ω, (c) 1000 Ω, (d) 0.0099 Ω
12. 10.8 A
13. 66.7 A
14. (a) 18.2 A, (b) 7.08 V, (c) 112.9 V, (d) 128.9 V, (e) 308.5 V
15. (a) 48.9 V, (b) 50 V, (c) 2.25 percent
16. (a) 20 V, (b) 220 V, (c) 10.0 percent
17. 8 Ω

Chapter 5

1. (a) 9.98 Ω; (b) 0.30 A, (c) v_{16} = 4.8 V, v_{14} = 4.2 V, v_{21} = v_{19} = 3 V, (d) 0.14 A
2. 4.25 A
3. (a) 391.7 Ω, (b) 0.613 A, (c) 0.357 A, (d) 147.1 W, (e) 5.30 kW · h
4. (a) 245 Ω, (b) 2.45 A, (c) 86.3 mA, (d) 14.9 W, (e) 232.5 V
5. 40.3 Ω
6. (a) 1882 Ω, (b) 6.375 mA, (c) 4.25 mA, (d) 3.19 V, (e) 3.5 V
7. 9000 Ω

Chapter 6

1. 354 pF
2. 9.39 nF

3. 1.73 μF
4. 220 μF
5. 63.2 μF
6. (a) 795 pF, (b) $v_1 = 596$ V, $v_2 = 0.143$ V, $v_3 = 0.427$ V, $v_4 = 3.18$ V,
   $v_5 = 0.315$ V, $v_6 = 0.255$ V
7. (a) 8 V, (b) 240 μC, (c) 8 V, (d) 500 μA
8. (a) 7.82 V, (b) 782 μC, (c) 0.85 A
9. (a) 30.000 s, (b) 6 μA, 0 V, 60 V, (c) 2.21 μA, 37.9 V, 22.1, V, (d) 0.114 C, (e) 5.4 J
10. (a) 20 s, (b) 368 μA, 12.6 V, 7.36 V, (c) 0 A, 20 V, 0 V, (d) 0 W, (e) 0.20 J
11. (a) 15 A, (b) 0, (c) 240 V, (d) 0.04 s, (e) 5.52 A, (f) 88.32 V, (g) 151.7 V,
    (h) 837.3 W, (i) 72 J, (j) 186.5 V
12. (a) 0.024 A, 0 V, (b) 0 A, (c) 8.8 mA, (d) 8.8 V, (e) 0 V, (f) 0.043 J, (g) 3.6 mC,
    (h) 133.5 mW, (i) 77.4 mW
13. (a) 0.04 F, (b) 4 C
14. (a) 0.20 A, (b) 32.83 V

Chapter 7

1. (a) 24.8 mT, (b) 19.84 mT, (c) 4.133×10^{-4} T
2. (a) 0 T, (b) 8.53 mT, (c) 2 mT
3. (a) 2 Ω, (b) 0.4 Wb
4. 2.5 mH/m
5. (a) 7073.4 A $\cdot$ t/Wb, (b) 0.085 Wb, (c) 0.34 T
6. 0.112 Wb
7. (a) 1.2 A, (b) 72 V, (c) 2.8×10^{-3} H/m
8. 0.345 A
9. 7.23 A
10. 4.13 A

Chapter 8

1. (a) 250 A $\cdot$ t/Wb, (b) 3.2 V
2. 9.38 H
3. (a) 0.08 Wb, (b) 160 H
4. 118 H
5. 0.773 H
6. 20 H
7. 0.96 H
8. (a) 0.63 s, (b) 0.5 s, (c) 6 A, (d) 3.79 A, (e) 104.8 V, (f) 96 V
9. (a) 0 A, (b) 1 A, (c) 6 V, (d) 0.632 A, (e) 1.26 V, (f) 20 Ω
10. (a) 0 A, (b) 0.8 A, (c) 0.2 s, (d) 0.51 A, 5.1 V, (e) 1.92 J
11. (a) 0.5 s, (b) 0.17 s, (c) 0 A, (d) 4 A, (e) 48 J, (f) 76.7 W, (g) 230 W, (h) 134 W,
    (i) 24 H
12. (a) 2 s, (b) 0 A, (c) 12 A, (d) 720 J, (e) 172.6 W, (f) 167.5 J, (g) 8.18 s
13. (a) 0.5 s, (b) 0 A, (c) 6 A, (d) 3.79 A, (e) 12 V, (f) 36 V, (g) 72 J, (h) 17.7 V,
    (i) 67.0 W, (j) 86.2 W

Chapter 9

1. (a) $e = 300 \sin 157.08t$, (b) 0 V
2. (a) 32 A, (b) 59.68 Hz, (c) 0.017 s, (d) -15.64 A
3. (a) 100 V, (b) 70.71 V, (c) 0 V, (d) 24.99 Hz, (e) 0.04 s, (f) 0.136 V
4. (a) 29.92 Hz, (b) 29.92 Hz, (c) 100 V, (d) 14.14 A, (e) 20 A, (f) 90°
5. (a) 21.63 W, (b) 43.26 W, (c) 30.64 W, (d) 1.471 A, 1.324 A
6. (a) 0.12 A, (b) 28.8 W
7. (a) 0.12 A, (b) 28.8 W
8. (a) 144 Ω, (b) 0.83 A, (c) 1.18 sin 377t
9. (a) 22.73 A, (b) 32.15 A, (c) 20 kW, (d) 10 kW, (e) 0.04 s, (f) 19.36 Ω
10. (a) 22.73 A, (b) 22.73 A, (c) 10,000 W, (d) 10,000 W, (e) 19.36 Ω
11. (a) $30 \sin(125.66t + 30°)$, $10 \sin(125.66t - 90°)$, (b) -10 V, (c) 7.07 A, (d) 0 A, (e) 6.37 A
12. (a) $140 \sin(251.3t + 60°)$, $80 \sin(251.3t - 150°)$, (b) 121.24 V, -40 A, (c) 98.99 V, (d) 0 V, (e) 89.13 V
13. (b) 662.9 V, (c) $+12.5°$ from $\mathbf{E}_2$, (e) $339.5 \sin 377t$, $339.5 \sin(377t + 25°)$, $662.9 \sin(377t + 12.5°)$
14. 0 V
15. 0 V
16. $222.11\underline{/38.2°}$ V
17. $439.7\underline{/165.36°}$ V
18. $25.98\underline{/150°}$
19. $414.7\underline{/37.9°}$
20. (a) 25 Hz, (b) $10 \sin(157t - 30°)$, $5 \sin(157t + 60°)$ (c) 7.07 V, (e) $11.18\underline{/-3.44°}$ V, (f) 37 mW
21. (a) 17.68 V, (b) 95.5 Hz, (c) 0.01 s, (e) $93.11\underline{/35.6°}$
22. (c) 15.88 V, (d) 1.12 A, (e) 20 Hz, (f) 1.59 A
23. (c) $100\underline{/90°}$, $100 \sin(125.7t + 90°)$, (d) 70.71 V, (e) 1.77 A, (f) 125 W, (g) 1.82 A
24. (b) 56.99 V, (c) 32.48 W
25. (a) 223.61 V, (b) 22.36 A, (c) $141.42 \sin(377t + 30°)$, $282.84 \sin(377t - 60°)$, (d) sketch, (e) 25.26 A, (f) 166.5 J
26. (c) 145.46 V, 9.9°, (d) 10.29 A, 1058 W, (e) 63.5 J

Chapter 10

1. (a) 76.5 mA, (b) 9.18 W, (c) 169.7 sin 157t, 0.108 sin 157t
2. (a) 159 Hz, (b) 344 mA, (c) 10.81 W, (d) 18.22 W, (e) 1.02 kW·h
3. 64 mΩ
4. (a) 1885 Ω, (b) 127 mA
5. (a) 452.4 Ω, (b) 0.46 A, (c) 254 mJ
6. (a) 1200 Ω, (b) 58.9 mA, (c) $0.0833 \sin(400t - 90°)$, (d) 0.014 J, (e) 0 W, (f) 400 A
7. (a) 120 A, (b) 169.7 A, (c) 5.3 mH, (d) sketch (e) $339.4 \sin(377t)$, $169.7 \sin(377t - 90°)$, (f) 76.32 J
8. (a) 105.8 Ω, (b) 0.334 A, (c) sketch, (d) $0.047 \sin(10584t - 90°)$, (e) ∞ A
9. (a) 2.65 Ω, (b) 45.2 A

10. (a) 15.92 Ω, (b) 13.1 A, (c) 17.3 J, (d) ∞ Ω, 0 A
11. (a) 5.31 Ω, (b) 45.3 A, (c) 339.4 V, (d) 170 mC, (e) 28.8 J
12. (a) 246 mΩ, (b) 13.22 A
13. (a) 4.45 Ω, (b) 40.43 A, (c) 165.5 μC, (d) 21.1 mJ
14. (a) 11.79 V, (b) 16.67 sin(30t − 90°), (c) 0.25 C, (d) 2.78 J, (e) 0 A

Chapter 11

1. (a) 8.68 + j49.24, (b) 28.19 + j10.26, (c) − 120 + j0, (d) 0 − j70,
   (e) 106.07 + j106.07
2. (a) 7.62/66.80°, (b) 10/− 53.13°, (c) 4/90°, (d) 5.83/−59.04°, (e) 6.32/−108.43°
3. (a) 3/−70°, (b) 5/40°, (c) 2 − j5, (d) −5 + j6, (e) 4 + j0
4. (a) 1.41/−45°, (b) 16.64/−57.26°, (c) 5.24/3.06°
5. (a) 36.05/33.70°, (b) 2.52/−63.78°
6. (a) 56.23/125.65°, (b) 2.76/72.40°

Chapter 12

1. (a) 7.21/33.69°, (b) 1.96/− 33.69° A, (c) 25 Hz
2. (a) 7/0° Ω, (b) 50/73.13° V
3. 4.33 Ω, 6.6 mH
4. (a) 10.44/16.70° Ω, (b) 7.22/−33.7° V, (c) 37.69/72.48° V
5. (a) 8.06/60.26° Ω, (b) 111.52/37.48° V, (c) 13.84/− 22.78° A
6. 17.15 A
7. (a) 2.24/− 26.57° Ω, (b) 107.3 A, (c) 322 V, (d) 0 A, (e) 240 V
8. 45.28 sin(125.66t − 38.66°)
9. (a) 5.66/−45°, (b) 35.4 A, (c) sketch. (d) 212.1 V, (e) 158 V, (f) 198.9 mJ.
   (g) 1.33 mC
10. (a) I_L = 10.61 A, I_R = 12 A, I_C = 15.83 A, (b) 13.1/23.5°, (c) 9.17/−23.51°,
    (d) 23.51°
11. (a) 0.417/53.13° S, (b) 2.4/− 53.13°, (c) 100/53.13° A,
    (d) v = 339 sin 188.5t, i = 141 sin(188.5t + 53.13°)
12. (a) 0.283/28° S, (b) 3.53/− 28° Ω, (c) 113.3/28° A,
    (d) 160.2 sin(314.2t + 28°), (e) ∞
13. (a) 4.77 Hz, (b) 120 Ω, (c) 0.59/− 50° A, (d) 0.83 sin(30t − 50°), (e) 9.8 J,
    (f) 1396 W
14. (a) 0.391/− 50.19° S, (b) 48/90° A, 60/0° A, 120/−90° A (c) 93.7/−50.19° A,
    (d) 5.3 mH, 530.5 μF, (e) 76.32 J
15. 41.45 A, 14.14 A, 54.47 A
16. (a) 4.99/69.34° A, (b) 6.57/−21.8° A, (c) 8.17/15.83° A
17. 1.79/−14.51°
18. 203.7/45° A
19. (a) 0.477/−55° S, (b) 190.8/−55°, (c) 113 sin(314.2t − 53.13°)
20. (a) 4.99/69.3° A, (b) 6.57/−21.8° A, (c) 8.17/15.8° A, (d) 1.2 A
21. (a) 90/60° Ω, (b) 45 Ω, 207 mH, (c) 180 Ω, 276 mH
22. (a) 5.77 Ω, 26.5 mH, (b) 4.33 Ω, 6.63 mH

23. $A_1 = 10$ A, $A_2 = 0$ A, $A_3 = 5$ A, $A_4 = 5$ A
24. 48.65 A
25. $I_1 = 32.5$ A, $I_6 = 7.23$ A,
26. (a) $0.845\underline{/-55.4°}$ A,   (b) $6.76\underline{/4.58°}$ V
27. $9.7\underline{/-17.8°}$ Ω
28. (a) 120 V,   (b) 3.88 A,   (c) $i = 28.3 \sin(377t + 30°)$
29. (a) $2.24\underline{/-26.6°}$,   (b) 448 mS,   (c) $107.3\underline{/56.6°}$ A,   (d) $151.8\underline{/11.6°}$ V
30. $A_1 = 150.4$ A, $A_2 = 58.2$ A, $A_3 = 38.7$ A, $A_4 = 29.4$ A, $V_1 = 208.6$ V, $V_2 = 73.3$ V
31. (a) $I_A = 33.2\underline{/3.69°}$ A, $I_B = 13.42\underline{/123.4°}$ A,   (b) $29.0\underline{/27.4°}$ A
32. (a) $2.83\underline{/0°}$ Ω,   (b) $159\underline{/0°}$ A,   (c) $132.5\underline{/36.9°}$ A,   (d) $477\underline{/-19.4°}$ V,   (e) $159\underline{/90°}$ V,
    (f) sketch,   (g) $i = 135.1 \sin(157t - 56.3°)$,   (h) 112.5 A
33. (a) $V_1 = 50$ V, $V_2 = 9.78$ V, $V_3 = 56.4$ V, $I = 1.63$ A,
    (b) $V_1 = 50$ V, $V_2 = 0$ V, $V_3 = 50$ V, $I = 0$ A

Chapter 13

1. (a) $2.94\underline{/11.3°}$ Ω,   (b) $I_R = 40$ A, $I_L = 20$ A, $I_C = 12$ A,   (c) $40.8\underline{/-11.3°}$ (d) 4800 W,
   (e) 960 var,   (f) 4895 VA,   (g) 0.98
2. (a) $589.3\underline{/32°}$,   (b) $0.20\underline{/-32°}$,   (c) 20.4 W,   (d) 24 VA,   (e) 12.7 var,   (f) 0.85
3. (a) 14 kW,   (b) 14.28 kvar
4. (a) 34 kW,   (b) 32 kvar,   (c) 46.7 kVA,   (d) 0.73
5. (a) 299.6 W,   (b) 130.9 var,   (c) 327 VA,   (d) 0.92
6. (a) 400 A   (b) 82.9 A,   (c) $314\underline{/67.2°}$ A,   (d) 149.1 kW,   (e) 63.3 kvar,
   (f) 162 kVA,   (g) 0.92
7. (a) 150 kW,   (b) 60 kvar,   (c) 161.6 kVA,   (d) 0.93,   (e) 367.2 A,
   (f) 50 kW · h,   (g) 20 kvar · h.
8. (a) 125 kVA,   (b) 75 kvar,   (c) 1940 μF
9. (a) 17.45 kW,   (b) 23.58 kVA,   (c) 15.86 kvar,   (d) 7.41 kvar,   (e) 30.9 A,
   (f) 341.2 μF
10. (a) 13.82 kW,   (b) 6.85 kvar,   (c) 15.42 kVA,   (d) 0.90,   (e) 6.85 kvar
11. (a) 6.39 kW,   (b) 4.23 kvar,   (c) 7.66 kVA,   (d) 9.02 kvar,   (e) 415 μF
12. (a) 195 kW,   (b) 156.1 kvar,   (c) 249.8 kVA,   (d) 0.78,   (e) 61.7 kvar   (f) 808 μF
13. (a) 17.55 kW, 15.0 kvar,   (b) 24.56 kW, 17.2 kvar,   (c) 42.13 kW,   (d) 32.22 kvar,
    (e) 53.04 kVA,   (f) 11.82 kvar,   (g) 155 μF
14. (a) 5 kVA,   (b) 4.33 kW,   (c) 2.5 kvar,   (d) 0.87
15. (a) 4400 VA,   (b) 4135 W,   (c) 1505 var,   (d) 20 A
16. (a) 3000 VA,   (b) 0 W,   (c) 3000 var, 0.0
17. (a) 1000 VA,   (b) 766 W,   (c) 642.8 var,   (d) 0.77,   (e) 7.66, 412.7 μF
18. (a) 3000 VA,   (b) 2298 W,   (c) 1928 var,   (d) 0.77,   (e) $3.33\underline{/40°}$ Ω
19. (a) 7812 W, 2846 var,   (b) 0.94
20. (a) 6.71 A,   (b) 6.67, 85 mH
21. (a) 7.2 kVA.   (b) 58.72 kVA,   (c) 489 A.   (d) 0.95
22. (a) $192\underline{/20°}$,   (b) 180.4 Ω, 418 mH
23. 4.33 Ω, 1061 μF

24. (*a*) $11.18/-63.4°$ Ω, (*b*) 4.5 kW
25. (*a*) 3285 VA, (*b*) 2797 W, (*c*) 0.85, (*d*) 1722.8 var, (*e*) 803.5 var (*f*) 25.3 μF
26. (*a*) 384.3 VA, (*b*) 1.47 W, (*c*) 223.5 W, (*d*) 0.58, (*e*) 312.6 var, (*f*) 145 var,
    (*g*) 31 μF

Chapter 14

1. 120 V, 2 Ω
2. 20 A, 2.5 Ω
3. (*a*) 30 A, 300 V, (*b*) 30 A, 300 V
4. (*a*) $15/-31.1°$ A, $150/-11.1°$ V, (*b*) $15/-31.1°$ A, $150/-11.1°$ V
5. (*a*) 60 A, (*b*) 300 V, (*c*) 13.59 kW, 6.34 kvar, 15 kVA, (*d*) 5.44 Ω, 6.74 mH
6. (*a*) $100 = 8\mathbf{I}_1 + 6\mathbf{I}_2$, $50 = 6\mathbf{I}_1 + 10\mathbf{I}_2$, (*b*) $A_1 = 15.9$ A, $A_2 = 4.6$ A, $A_3 = 11.36$ A,
    (*c*) 68.16 V
7. (*a*) $100/0° = (1.02 + j3)\mathbf{I}_1 + (1 + j3)\mathbf{I}_2$
    $\quad 100/30° = (1 + j3)\mathbf{I}_1 + (1.02 + j3)\mathbf{I}_2$
    (*b*) $A_1 = 1309$ A, $A_2 = 1280$ A, $A_3 = 30.6$ A, (*c*) $96.7/15.5°$ V
8. (*a*) $\quad 40 + j60 = (8 + j8)\mathbf{I}_1 + (-5 - j6)\mathbf{I}_2$
    $\quad 80 - j60 = (-5 - j6)\mathbf{I}_1 + (6 + j5)\mathbf{I}_2$
    (*b*) $A_1 = 30.1$ A, $A_2 = 35.1$ A, $A_3 = 7.8$ A, (*c*) $60.7/165.6°$ V,
    (*d*) 302 W, 362 var, 472 VA
9. 0.52 A
10. (*a*) $1000/90°$ V, $10/90°$ Ω, (*b*) $29.7/-162°$ A
11. (*a*) $10 + j0 = (10 - j2)\mathbf{I}_1 + (0 + j0)\mathbf{I}_2 + (4 - j2)\mathbf{I}_3$
    $\quad 5.2 + j3 = (0 + j0)\mathbf{I}_1 + (6 - j6)\mathbf{I}_2 + (-3 + j0)\mathbf{I}_3$
    $\quad 0 + j0 = (4 - j2)\mathbf{I}_1 + (-3 + j0)\mathbf{I}_2 + (9 + j2)\mathbf{I}_3$
    (*b*) $A_1 = 1.03$ A, $A_2 = 0.75$ A, $A_3 = 0.87$ A, $A_4 = 0.32$ A, $A_5 = 0.67$ A
12. $A_1 = 0.91$ A, $A_2 = 0$ A, $A_3 = 0$ A, $A_4 = 0.91$ A, $A_5 = 0.91$ A
13. 39 A
14. 4.1 A, 205 W
15. 8.8 A
16. 12.6 A
17. 3.5 A, 17.6 A
18. 0 A, 0 V
19. $A_1 = 11.4$ A, $A_2 = 2.34$ A
20. $A_1 = 0$ A, $A_2 = 3.9$ A
21. (*a*) 1.33 A, 9 Ω, (*b*) 10.69 V
22. 689.7 V
23. 96.5 V, 30.6 A
24. (*a*) $8/10°$ A, $5/20°$ Ω,
    (*b*) $8/10° = (0.5069 - j0.2408)\mathbf{V}_1 - (0.069 - j0.1724)\mathbf{V}_2$
    $\quad - 100/0° = -(0.0690 - j0.1724)\mathbf{V}_1 + (0.094 - j0.2724)\mathbf{V}_2$, (*c*) 326.85 V, 60.69 A
25. 3.87 V, 0.95 V, 3.01 V
26. $\mathbf{V}_1 = 39.00/97.36°$, $\mathbf{V}_2 = 10.00/-90°$, $\mathbf{V}_3 = 38.98/82.64°$
27. (*a*) 39.1 V, 6.5 A

Chapter 15

1. 60 V, 8 Ω
2. 19.2 V, 2.31 Ω
3. $15.8\underline{/18.4°}$ V, $7.28\underline{/2.01°}$ Ω
4. (a) 50 V, 12 Ω, (b) 12 Ω, (c) 51.9 W
5. (a) 50 V, 12 Ω, (b) 12 Ω, (c) 51.9 W
6. (a) $21.2\underline{/-34.9°}$ V, $2.75\underline{/-56°}$ Ω, (b) 2.1 A, 22.7 V, (c) 26.5 W, 39.7 var, 47.7 VA, (d) 0.55
7. (a) $150\underline{/0°}$ V, $5.66\underline{/45°}$ Ω, (b) 10.8 A, 92.3 V, (c) 350 W, 933 var, (d) 0.35
8. (a) 150 V, 4 Ω, (b) 21.4 A, 64.3 V, (c) 4.87 J
9. (a) $23.4\underline{/-19.8°}$ V, $5\underline{/11.6°}$, (b) $A_1 = 1.11$ A, $A_2 = 1.55$ A, 15.54 V
10. $3\underline{/0°}$ A, $20\underline{/30°}$ Ω
11. $20\underline{/40°}$ A, $5\underline{/20°}$ Ω
12. 8.66 Ω, 6.67 V
13. 40 V, 8 Ω
14. 4.19 V, 3.49 Ω
15. 4.68 V, 4.05 Ω
16. 30 V, 2.4 Ω
17. 40.9 V, 1.09 Ω
18. $6.8\underline{/5.87°}$ A
19. $A_1 = 10$ A, $A_2 = 3.75$ A, $A_3 = 2.5$ A
20. $A_1 = 0$ A, $A_2 = 4.81$ A, $A_3 = 129.8$ A
21. 59.73 A
22. $i_1 = 6 - 8.91 \sin(6.28t + 155.3°)$
    $i_2 = 6 + 28.3 \sin(6.28t + 65.3°)$
    $i_3 = 29.64 \sin(6.28t + 82.78°)$
23. $i = 0.058 + 0.0042 \sin(45,238t - 52.8°)$
24. (a) 4.5 A, (b) 65.5 W, 20.6 var, 68.7 VA
25. (a) 4.5 A, (b) 72 W, 0 var, 72 VA
26. 14.76 V, 0.369 Ω
27. $52.8\underline{/2.08°}$ V, $1.27\underline{/55.7°}$ Ω
28. (a) 6.7 A, (b) $17.9\underline{/41°}$, 4.5 A
29. (a) $2.67\underline{/25°}$ Ω, (b) $36\underline{/40°}$ Ω
30. (a) 30.3 A, (b) 8.15 V, (c) 16.6 W
31. $2.59\underline{/3.72°}$ Ω

Chapter 16

1. (a) 33.6 kHz, (b) 40 Ω, (c) 1.2 A, (d) 12.6 kV
2. (a) 450 Hz, (b) 11.1 kΩ, (c) 0.0013
3. (a) 338 nF, (b) 0.6 A, (c) 1.08 mJ, (d) 4.7
4. (a) $2.24\underline{/-26.6°}$ Ω, (b) $4.47\underline{/56.6°}$ A, (c) sketch,
    (d) $e = 141.4 \sin(377t + 30°)$
    $i = 63.2 \sin(377t + 56.6°)$
    (e) $50\underline{/30°}$ A, (f) $173.7\underline{/-60°}$ V, (g) $200\underline{/90°}$ V

5. (a) $5.97\underline{/-39.6°}\ \Omega$, (b) $20.1\underline{/39.6°}$ A, (c) $65\underline{/61.4°}$ V, (d) 40.9 Hz

6. 50.27

7. (a) 31.53 kHz, (b) 1268, (c) 24.9, (d) 24 A, (e) 152 kV, (f) 0 A

8. (a) 105.5 μF, (b) 20 kA, (c) 150.8 kV, (d) 2.4 MJ, (e) 377, (f) 0.53, (g) 200.27 Hz, 199.74 Hz, (h) 0 A

9. (a) 80,000, (b) 44.72 kHz, (c) 0.56, (d) 2 mH, (e) 6.3 nF

10. (a) 1000 Hz, (b) 730 Hz, 1550 Hz, (c) 820 Hz, 1.22

11. (a) $A_1 = 12$ A, $A_2 = 30$ A, $A_3 = 30$ A, $A_4 = 12$ A, $A_5 = 0$ A
(b) $A_1 = 12$ A, $A_2 = 30$ A, $A_3 = 30$ A, $A_4 = 12$ A, $A_5 = 32.3$ A
(c) $A_1 = 12$ A, $A_2 = 30$ A, $A_3 = 30$ A, $A_4 = 12$ A, $A_5 = 32.3$ A

12. (a) 480 A, (b) 31.83 Hz, (c) 339.4 V

13. (a) 159.2 Hz, (b) 4 MΩ, (c) 25 μA, (d) 100 A

14. 668 Hz

15. 8.23 mH

16. (a) $6.70\underline{/-13.7°}\ \Omega$, (b) $73.1\underline{/13.7°}$ A, (c) $33.8\underline{/30.2°}$ A, (d) no

17. (a) 3578 Hz, (b) 9.83, (c) 6854, (d) 8.62, (e) 415, (f) 3376 Hz, 3791 Hz

18. (a) 259 pF, (b) 4.28 V, 11.5 mA

19. (a) 26.6 pF, (b) 28 mA

20. (a) 3.33, (b) 1.08 μF, (c) 3.33, (d) 300 Hz, (e) 255 mA

21. (a) 4518 Hz, (b) 2.81, (c) 2.77, (d) 1631 Hz, (e) 2.41 V (f) 7.63 mA

Chapter 17

1. 99.5 nF

2. 995 Ω

3. 149 mH

4. 628 Ω

5. 79.6 nF

6. 27.2 kΩ

7. 25.5 mH

8. 811 pF

9. (a) 750,644 Hz, (b) 329 Hz, (c) 750,834 Hz, 750, 495 Hz, (d) 60 V, 39.3 V, 44.5 V, 2.7 mV

10. (a) 24,558 Hz, (b) 1707 Hz, (c) 25,412 Hz, 23,705 Hz, (d) 0.44 V, 30 V, 30 V, 30 V

11. (a) 109 pF, (b) 0.27 V

12. $L_1 = 1.207$ mH, $L_2 = 284\ \mu$H

13. $L_1 = 693\ \mu$H, $C_2 = 746\ \mu$F

Chapter 18

1. 1.6 H

2. (a) 24 A, (b) 3.7 H, (c) 33.4 H

3. (a) 16.7 A, (b) 7.5 H, 46.9 H

4. (a) 1.5 H, (b) 0.83 H

5. (a) 0.95 H, (b) 4 H, 0.25 H

6. (a) 0.94 Wb, (b) 146.4 H, (c) 118.1 V, (d) 120 V
7. (a) 0.075 Wb, (b) 5 H, (c) 225 V, (d) 150 V
8. (a) 0.1 Wb, (b) 0.08 Wb, (c) 667 mH, (d) 167 mH, (e) 267 mH
9. 4.5 mWb
10. 131,868 V
11. (a) 23.3 mWb, (b) 0.0233 sin 157t
12. (a) $22.8/-60.8°$ A, (b) $19.1/58.7°$ A, (c) $2.1/-88.2°$, (d) 66.7 A, (e) 0 A
13. (a) $5.99/32.7°$ A, (b) $0.92/-86.6°$ A
14. (a) $5.68/43.1°$ Ω, (b) $704.3/-43.1°$ A
15. (a) $17/21.8°$ Ω, (b) $135/-21.8°$ A
16. (a) 2271 V, (b) 1429 A, 0 A
17. (a) 300 V, (b) 10 A, (c) 30 A, (d) 30 A, 10 A, 20 A, 10 A
18. (a) 12 V, (b) 2000 A, (c) 100 A, (d) 1900 A
19. (a) 80 A, (b) 20 kVA, (c) 250 V, (d) 12.5 kVA, (e) 7.5 kVA
20. (a) 89, (b) 250 V, (c) 27.8 A, (d) 1.39 kVA, (e) 11.1 kVA, (f) 12.5 kVA
21. (a) 150 V, (b) 200 A, (c) 50 A
22. 35.4
23. 37.5
24. 35.4
25. (a) $2.22/30°$ Ω, (b) 15.9 A, (c) 5.31 A, (d) 106 V, (e) 487.8 W
26. (a) 60 A, (b) 40 V, (c) 180 A, (d) 0.22 Ω

Chapter 19

5. $10/53.1°$ Ω
6. $8.06/7.1°$ Ω
7. (a) 3.36 H, (b) 3.36 H, (c) 12.64 H, (d) 12.64 H
8. $10 + j5$
9. (a) 5 Ω, (b) 13.3 mH
   (c) $43.3 + j25 = (6 + j4.5)I_1 + (0 + j5)I_2$
   $10.0 + j17.32 = (0 + j5)I_1 + (5 + j15)I_2$
   (d) 1.12 A, (e) $6.03/-60.4°$ V
10. (a) $A_1 = 5.12$ A, $A_2 = 1.09$ A, (b) $9.37/23.4°$ V, (c) 10.21 VA, 5.94 W, 8.31 var
11. (a) $A_1 = 15$ A, $A_2 = 10$ A, (b) 50 V, (c) 500 VA, 500 W, 0 var
12. (a) $100 + j0 = (6 + j9)I_1 + (-5 - j3)I_2$
    $-100 + j0 = (-5 - j3)I_1 + (12 + j1)I_2$
    $A_1 = 7.56$ A, $A_2 = 9.29$ A, $A_3 = 6.32$ A
    (b) $G_1 = 173.7$ W, 735.8 var, $G_2 = 594.8$ W, -213.8 var   (c) same as in part a
13. (a) 18.9 A, (b) 0.55, (c) 328.6
14. (a) 16.7 A
15. (a) 21.7 A, (b) 42.3 V, (c) 2 Ω, 10.6 mH
16. 61.7 A
17. $A_1 = 8.41$ A, $A_2 = 13.98$ A, $A_3 = 10.5$ A
18. (a) 6.13 A, (b) 30.7 V, (c) 0.188 kvar
19. (a) 8.71 A, (b) 550.7 W, 674.8 var

20. (a) 12.9 A, (b) 27.7 V, (c) 408 mJ

21. (a) 1.6 A, (b) 32 VA, 30.3 W, 10.4 var

22. (a) 4.44 A, (b) 88.8 W

23. $Z_A = 0.99\ \Omega$, $Z_B = j2.99\ \Omega$, $Z_C = j2.01\ \Omega$

24. $Z_A = 4 + j6.83\ \Omega$, $Z_B = 2 + j4.83\ \Omega$, $Z_C = -j2.83\ \Omega$

25. (a) $Z_A = 5 - j1\ \Omega$, $Z_B = 5 + j5\ \Omega$, $Z_C = j3\ \Omega$, (b) 39.4/66.8° V, (6.09 + j7.54) Ω, (c) (6.09 − j7.54) Ω, (d) 63.9 W, 79.1 var

26. (a) $Z_A = 6 + j14\ \Omega$, $Z_B = 6 + j14\ \Omega$, $Z_C = -j7\ \Omega$, (b) 20/−73.7° V, (9.03 + j13.6) Ω, (c) 1.02 A

27. (a) $Z_A = 1 + j2\ \Omega$, $Z_B = 3 + j1\ \Omega$, $Z_C = j1\ \Omega$, (b) 42.4/−16.9° V, 4.62/27.7° Ω, (c) 7.55/26.4° A, (d) 0.907 J

Chapter 20

1. $Z_{11} = 4000\ \Omega$, $Z_{12} = 3000\ \Omega$, $Z_{21} = 3000\ \Omega$, $Z_{22} = 13{,}000\ \Omega$

2. $Z_{11} = 5 - j6\ \Omega$, $Z_{12} = 2 - j6\ \Omega$, $Z_{21} = 0 - j6\ \Omega$, $Z_{22} = 4 - j3\ \Omega$

3. (a) $Z_{11} = 12 + j17\ \Omega$, $Z_{12} = -9.08 - j9.73\ \Omega$, $Z_{21} = -9.08 - j9.73\ \Omega$, $Z_{22} = 18 + j8.6\ \Omega$, (b) 0.59/−16.4° A

4. $y_{11} = 450\ \mu S$, $y_{12} = -200\ \mu S$, $y_{21} = -200\ \mu S$, $y_{22} = 700\ \mu S$

5. (a) 227.3 mA, 5.17 W, (b) 473 mA

6. (a) −1.33 V, (b) 666 μA, (c) 66.7

7. (a) 260 mV, (b) 260

8. (a) 1125 Ω, (b) 11.63 kΩ, (c) −66.7, (d) 37.5

9. (a) 37.54 Ω, (b) 1.65 MΩ, (c) 260, (d) −0.98

Chapter 21

1. (a) sketch, (b) $A_A = 127.2$ A, $A_B = 118.9$ A, $A_C = 80$ A

2. (a) sketch, (b) $A_A = 62.3$ A, $A_B = 90.1$ A, $A_C = 68.1$ A

3. $A_A = 49.2$ A, $A_B = 48.3$ A, $A_C = 12.4$ A

4. $A_A = 189.6$ A, $A_B = 87.7$ A, $A_C = 220.7$ A, $A_N = 52$ A

5. 31.4 A

6. $A_A = 52.3$ A, $A_{38} = 11.8$ A, $A_Y = 6.5$ A, $A_\Delta = 39$ A

7. (a) sketch, (b) $A_A = 108.5$ A, $A_B = 166.3$ A, $A_C = 108.5$ A, $A_N = 60$ A, (c) $A_A = 108.5$ A, $A_B = 122.6$ A, $A_C = 65.5$ A, $A_N = 60$ A

8. $A_1 = 26$ A, $A_2 = 45$ A, $A_3 = 26$ A, $A_4 = 62$ A

9. $A_1 = 93.5$ A, $A_2 = 129.5$ A, $A_3 = 199.4$ A, $A_4 = 250.5$ A

10. $A_1 = 68.1$ A, $A_2 = 0$ A, $A_3 = 170.7$ A, $A_4 = 191$ A

11. $A_1 = 289$ A, $A_2 = 182.6$ A, $A_3 = 68.7$ A, $A_4 = 138.5$ A

12. (a) $A_1 = 63.5$ A, $A_2 = 57.1$ A, $A_3 = 24.7$ A, (b) $A_1 = 37.7$ A, $A_2 = 51.9$ A, $A_3 = 50$ A

Chapter 22

1. (a) 32.74 kW, 46.77 kVA, 33.4 kvar

2. (a) 17.54 kW, (b) 21.16 hp, 15.47 kvar, (d) 23.38 kVA

3. (a) sketch, (b) line A = 18 A, line B = 18 A, line C = 63.5 A, line N = 50 A

4. (a) 34.15 kW, (b) 21.94 kvar, (c) 40.59 kVA, (d) 0.84, (e) 52.1 A
5. (a) 18 kW, (b) 6 kvar, (c) 18.97 kVA, (d) 0.95, (e) 24.9 A, (f) 13.12 A, 13.12 A
6. (a) 71.8 kW, (b) 24.8 kvar, (c) 75.98 kVA, (d) 0.95, (e) 87.7 A,
   (f)I_M = 79.8 A, I_C = 6.9 A, I_H = 11.6 A
7. (a) I_Y = 12 A, I_Δ = 8 A, I_M = 27.8 A, (b) P_Y = 4.32 kW, P_Δ = 986 W, P_M = 6.5 kW,
   Q_Y = 0, Q_Δ = 2.7 kvar, Q_M = 7.6 kvar, (c) 11.8 kW, 10.31 kvar, 15.67 kVA, 0.75
8. (a) I_a = 15.8 A, I_b = 15.8 A, I_c = 28.7 A, (b) W_a = 1363 W, W_c = 5170 W
9. W_a = 2126 W, W_b = 5374 W
10. (a) 26 kW, (b) 0 kvar, (c) 26 kVA, (d) 1.0, (e) 62.6 A,
    (f) W_1 = 13 kW, W_2 = 13 kW
11. (a) 65 kW, (b) 76 kvar, (c) 128.3 A, (d) W_1 = 10.56 kW, W_2 = 54.44 kW,
    (e) 44.51 kvar, (f) 466.5 μF
12. (a) 10 kW, (b) 14.03 kVA, (c) 0.71, (d) 9.84 kvar, (e) 2.34 kvar, (f) 30.7 μF
13. (a) 3.2 kW, (b) 11.6 A, (c) 4.02 kVA, (d) 0.80, (e) 2.43 kvar, (f) 128.9 μF,
    (g) 382 μF
14. (a) I_a = 44.4 A, I_b = 49 A, I_c = 52 A, (b) 178.83 kW, (c) 91.71 kvar, (d) 0.89
15. (a) S_1 = 2400 VA, S_2 = 2999 VA, S_3 = 7999 VA
    P_1 = 2364 W, P_2 = 2988 W, P_3 = 6927 W
    Q_1 = 417 var, Q_2 = 261 var, Q_3 = −3999 var
    pf$_1$ = 0.985, pf$_2$ = 0.996, pf$_3$ = 0.866
    (b) 12.28 kW, −3.32 kvar, 0.965
    (c) R_1 = 49.24, L_1 = 23 mH, R_2 = 39.85, L_2 = 9.25 mH,
    R_3 = 12.99, C_3 = 353.7 μF
    (d) R_1 = 50.77 Ω, L_1 = 0.764 H
    R_2 = 40.15 Ω, L_2 = 1.22 H
    R_3 = 17.32 Ω, C_3 = 88.42 μF

Chapter 23

1. 0.578 V
2. (a) 9 N, (b) 4.56 N
3. 3.81 N
4. A
5. 1.667 N
6. (a) 528 mT, (b) 75 A · t/m, (c) 0.90 A, (d) 1.8 V
7. 0.3 T
8. (a) 0.067 A/div, (b) 17 A, (c) 1.25 mΩ
9. (a) 3.75 A/div, (b) 600 A, (c) 66.7 $\mu\Omega$
10. (a) 231.3 V, (b) 229.7 V
11. (a) 8.18 V, (b) 45 V, (c) 85.7 V

Chapter 24

1. (a) 30 Hz, 3 kHz, (b) 5.1 V, (c) 0.51 A, (d) 2.6 W
2. (a) 2.09 A, (b) 26.21 W
3. (a) Z_{120} = 3 + j2 Ω, Z_{60} = 3 + j1 Ω, Z_0 = 3 + j0 Ω, (b) 42.6 A
4. (a) 60.5 V, (b) 4.4 A, (c) 154.9 W, (d) 266.2 VA, (e) 0.58
5. (a)f_1 = 60 Hz, f_2 = 120 Hz, f_3 = 180 Hz, (b) 293.7 V, (c) 4.8 A, 1152 W

6. (*a*) 152 V, (*b*) 1.9 A, (*c*) 109.4 W
7. (*a*) $f_1 = 60$ Hz, $f_2 = 60$ Hz, $f_3 = 180$ Hz, (*b*) 146.4 W, (*c*) 14.2 A, (*d*) 1 kW

Chapter 25

1. (*a*) 0, (*b*) 0.237
2. −1.75, 0.85
3. 0, 0.703
4. (*a*) $A_0 = -0.5$, $A_3 = 0.637$, $B_3 = 0$, $C_3 = 0.637$,
   (*b*) 0.637 sin (5655*t* + 90°), (*c*) 0.45 V
5. (*a*) 1.5, (*b*) 1.15, (*c*) 500 Hz
6. (*a*) plot, (*b*) 10.68, (*c*) 684.3 W
7. (*a*) $A_0 = 3.0$, $C_1 = 1.93$, $C_2 = 0.955$,
   (*b*) $v = 3.0 + 1.93 \sin(314.2t + 171°) + 0.955 \sin(628.3t + 180°)$
8. (*a*) $A_0 = 0$, $C_5 = 0.292$, (*b*) $v = 0.292 \sin(6283t + 29.1°)$

Chapter 26

1. (*a*) 8 A, (*b*) 8 A, (*c*) 12 A, (*d*) 8 A, (*e*) 3.8 A
2. (*a*) 4 A, (*b*) 4 A, (*c*) 0.3 A, (*d*) 4 A, (*e*) 251 μA
3. (*a*) 0.5 A, (*b*) 0.5 A, (*c*) 2 A, (*d*) 0.5 A, (*e*) 12 mA
4. (*a*) 0.12 A, (*b*) 1.28 A, (*c*) 5.4 mA, (*d*) 1280 V, (*e*) 317 ms
5. (*a*) 2.4 A, (*b*) 6 A, (*c*) 181 mA, (*d*) 60 V, (*e*) 58 ms
6. (*a*) 714 mA, (*b*) 503 mA, (*c*) 250 V, (*d*) 0 V
7. (*a*) 0.5 A, (*b*) 0.34 A, (*c*) 100 V, (*d*) 0 V
8. (*a*) $A_1 = 1.2$ A, $A_2 = 1.2$ A, $A_D = 0.4$ A, (*b*) 240 V, (*c*) 1440 V, (*d*) 100 Ω
9. (*a*) $A_1 = 0.6$ A, $A_2 = 0.96$ A, $A_D = 0.4$ A, (*b*) 240 V, (*c*) 936 V, (*d*) 153.9 Ω
10. (*a*) 24 V, (*b*) −24 mA, (*c*) −19.6 mA, (*d*) 19.6 V
11. (*a*) 24 V, (*b*) 24 V, (*c*) −8 A, (*d*) −1.08 A
12. (*a*) −8.69 A, (*b*) 2265 J
13. (*a*) −10 A, (*b*) −0.5 A, (*c*) −1.5 V, (*d*) 5 V
14. (*a*) 0, (*b*) −2.17 A
15. (*a*) 0, (*b*) −5.81 A, (*c*) 0 A
16. (*a*) 0, (*b*) −1.43 A, (*c*) 0 A, (*d*) 0
17. (*a*) 0, (*b*) −5.49 A, (*c*) −27.4 V
18. (*a*) −153 mA, (*b*) 41.9 V, (*c*) 222.8 Hz
19. (*a*) 0, (*b*) −3.89 A, (*c*) 0.25 s, (*d*) 0.8 Hz, (*e*) 22.6 V, (*f*) −22.6 V
20. (*a*) 167 ms, 1.27 Hz, (*b*) −2.26 A, (*c*) −13.08 V, (*d*) 13.08 V
21. (*a*) 4 s, 105.3 mHz, (*b*) −13.24 A
22. sketch
23. sketch

Chapter 27

1. (*a*) 0, (*b*) 115 mA, (*c*) 11.3 V
2. (*a*) 0, (*b*) 506 mA, (*c*) 50.6 V, (*d*) 189.4 V, (*e*) 48 J

3. (a) 2.4 A, (b) 2.4 A, (c) 1.94 A
4. (a) 0, (b) 2 A, (c) $i_R = 2$ A, $i_{coil} = 4.66$ A, (d) $i_R = 2$ A, $i_{coil} = 6$ A, (e) $i_R = 6$ A, $i_{coil} = 6$ A, (f) 3.29 A, (g) 43.3 J
5. (a) 20 mA, (b) 13 mA, (c) 4.2 V, (d) 0
6. (a) 0, (b) 12 μA, (c) 8.9 μA
7. (a) 70 μA, (b) 68.3 μA, (c) 50 V, (d) 1 mC
8. (a) 0 A, 18 V, (b) 0 A, 18 V, (c) -3.08 A, 2.1 A
9. (a) 836 mA, (b) sketch
10. (a) -58 mA, (b) 43.6 V, (c) 6.47 V
11. (a) 3.3 A, (b) 2.51 A, (c) 5093 Hz
12. (a) 4 A, 0 V, (b) 4 A, 0 V, (c) 4 A, (d) -9.8 V, (e) $W_\phi = 477.6$ J, $W_C = 50.8$ J, (f) 10.1 C, (g) 110 V,
13. (a) 10 A, (b) 2.87 A
14. -1.42 mA
15. (a) -165.2 A, (b) 212.2 A
16. (a) 3.78 A, (b) 212.2 A
17. 260 mA
18. (a) sketch, (b) sketch, (c) 44.1 A, (d) -17.85 A, (e) 261 μF, (f) 225 A

INDEX

UNIT CONVERSION RATIOS

$$\frac{1 \text{ m}}{\text{mm} \times 10^3} = 1$$

$$\frac{1 \text{ kWh}}{\text{Btu} \times 3413} = 1$$

$$\frac{1 \text{ m}}{\text{in} \times 39.37} = 1$$

$$\frac{1 \text{ J}}{\text{ft} \cdot \text{lb} \times 0.7736} = 1$$

$$\frac{1 \text{ m}}{\text{yd} \times 1.094} = 1$$

$$\frac{1 \text{ J}}{\text{Btu} \times 9.486 \times 10^{-4}} = 1$$

$$\frac{1 \text{ m}}{\text{ft} \times 3.281} = 1$$

$$\frac{1 \text{ J}}{\text{Wh} \times 2.778 \times 10^{-4}} = 1$$

$$\frac{1 \text{ m}}{\text{mi} \times 62140} = 1$$

$$\frac{1 \text{ Wb}}{\text{maxwells} \times 10^8} = 1$$

$$\frac{1 \text{ m}^2}{\text{mm}^2 \times 10^6} = 1$$

$$\frac{1 \text{ Wb}}{\text{lines} \times 10^8} = 1$$

$$\frac{1 \text{ m}^2}{\text{cm}^2 \times 10^4} = 1$$

$$\frac{1 \text{ T}}{\text{G} \times 10^4} = 1$$

$$\frac{1 \text{ m}^2}{\text{in}^2 \times 1550} = 1$$

$$\frac{1 \text{ A} \cdot \text{t}}{\text{Gb} \times 1.257} = 1$$